# Biofilms in Bioremediation

Current Research and Emerging Technologies

Edited by

Gavin Lear

School of Biological Sciences
The University of Auckland
Auckland
New Zealand

Caister Academic Press

Caister Academic Press
Norfolk, UK

www.caister.com

British Library Cataloguing-in-Publication Data
A catalogue record for this book is available from the British Library

ISBN: 978-1-910190-29-6 (paperback)
ISBN: 978-1-910190-30-2 (ebook)



Front cover image: A dead sea bird following the Rena oil spill, New Zealand. Photo by Rambo Estrada, 2011.

Back cover image: Scanning electron micrograph image of a mixed freshwater microbial biofilm community. Designed by Andrew Dopheide, 2011.

# Contents

# Contributors

**Wan Azlina Ahmad**
Department of Chemistry
Faculty of Science
Universiti Teknologi Malaysia
Johor
Malaysia

azlina@kimia.fs.utm.my

**Wan Haslinda Wan Ahmad**
Department of Chemistry
Faculty of Science and Mathematics
Universiti Pendidikan Sultan Idris
Perak
Malaysia

wan.haslinda@fsmt.upsi.edu.my

**Pieter Albers**
Division of Soil and Water Management
Faculty of Bioscience Engineering
KU Leuven
Leuven
Belgium

pieter.albers@ees.kuleuven.be

**Dina M. Al-Mailem**
Department of Biological Sciences
Faculty of Science
Kuwait University
Safat
Kuwait

dina.almailem@ku.edu.kw

**Audrey S. Commault**
University of Technology Sydney
Plant Function Biology and Climate Change Cluster
Sydney, NSW
Australia

audrey.commault@gmail.com

**Marc A. Demeter**
Department of Biological Sciences
University of Calgary
Calgary, AB
Canada

mademete@ucalgary.ca

**Laura Ferreira**
Department of Chemical Engineering
University of Vigo
Vigo
Spain

ferreira@uvigo.es

**Michael Harbottle**
Cardiff School of Engineering
Cardiff University
Cardiff
UK

harbottlem@cardiff.ac.uk

**Joe J. Harrison**
Department of Biological Sciences
University of Calgary
Calgary, AB
Canada

jjharris@ucalgary.ca

**Benjamin Horemans**
Division of Soil and Water Management
Faculty of Bioscience Engineering
KU Leuven
Leuven
Belgium

benjamin.horemans@ees.kuleuven.be

**Koki Kurashina**
Division of Biosphere Science
Graduate School of Environmental Science
Hokkaido University
Sapporo
Japan

twpgww@docomo.ne.jp

**Gavin Lear**
School of Biological Sciences
The University of Auckland
Auckland
New Zealand

g.lear@auckland.ac.nz

**Joseph A. Lemire**
Department of Biological Sciences
University of Calgary
Calgary, AB
Canada

jalemire@ucalgary.ca

**Sindhu Mathew**
Department of Biotechnology
Lund University
Lund
Sweden

sindhumathew@hotmail.com

**Kyoko Miwa**
Division of Biosphere Science
Graduate School of Environmental Science
Hokkaido University
Sapporo
Japan

miwakyoko@ees.hokudai.ac.jp

**Kazuhiro Mori**
Department of Research
Interdisciplinary Graduate School of Medicine and Engineering
University of Yamanashi
Yamanashi
Japan

mori@yamanashi.ac.jp

**Masaaki Morikawa**
Division of Biosphere Science
Graduate School of Environmental Science
Hokkaido University
Sapporo
Japan

morikawa@ees.hokudai.ac.jp

**Marta Pazos**
Department of Chemical Engineering
University of Vigo
Vigo
Spain

mcurras@uvigo.es

**Samir S. Radwan**
Department of Biological Sciences
Faculty of Science
Kuwait University
Safat
Kuwait

samir.radwan@ku.edu.kw

**Gunaratna K. Rajarao**
School of Biotechnology
Royal Institute of Technology
Stockholm
Sweden

gkr@kth.se

**Emilio Rosales**
Department of Chemical Engineering
University of Vigo
Vigo
Spain

emiliorv@uvigo.es

**Maria Ángeles Sanromán**
Department of Chemical Engineering
University of Vigo
Vigo
Spain

sanroman@uvigo.es

**Youngwoo (Young) Seo**
Department of Civil Engineering;
Department of Chemical and Environmental Engineering
University of Toledo
Toledo, OH
USA

Youngwoo.Seo@utoledo.edu

**Dirk Springael**
Division of Soil and Water Management
Faculty of Bioscience Engineering
KU Leuven
Leuven
Belgium

dirk.springael@ees.kuleuven.be

**Kazuya Suzuki**
Division of Biosphere Science
Graduate School of Environmental Science
Hokkaido University
Sapporo
Japan

kazuya_suzuki@nissui.co.jp

**Tadashi Toyama**
Department of Research
Interdisciplinary Graduate School of Medicine and Engineering
University of Yamanashi
Yamanashi
Japan

ttohyama@yamanashi.ac.jp

**Raymond J. Turner**
Department of Biological Sciences
University of Calgary
Calgary, AB
Canada

turnerr@ucalgary.ca

**Diana L. Vullo**
Área Química
Instituto de Ciencias
Universidad Nacional de General Sarmiento-CONICET
Buenos Aires
Argentina

dvullo@ungs.edu.ar

**Kenji Washio**
Division of Biosphere Science
Graduate School of Environmental Science
Hokkaido University
Sapporo
Japan

washi@ees.hokudai.ac.jp

**Richard J. Weld**
Lincoln Agritech Ltd
Lincoln University
Christchurch
New Zealand

Richard.Weld@lincolnagritech.co.nz

**Fumiko Yamaga**
Division of Biosphere Science
Graduate School of Environmental Science
Hokkaido University
Sapporo
Japan

fumikoy218@gmail.com

**Zainul Akmar Zakaria**
Institute of Bioproduct Development
and Department of Bioprocess Engineering
Faculty of Chemical Engineering
Universiti Teknologi Malaysia
Johor
Malaysia

zainul@ibd.utm.my

# Current Books of Interest

# Preface

The continued release of contaminants into the environment, predominantly by industrial and agricultural activities, means that there is an ongoing need to remediate sites exposed to elevated concentrations of a wide range of pollutants. A number of chemical and physical processes are currently used to encourage the degradation of, or altered mobility of pollutants in field, laboratory and bioreactor settings. However, the microbial bioremediation of troublesome contaminants is increasingly seen as being both cost-effective and reliable, and a number of approaches are in widespread commercial use. Microbial bioremediation largely capitalizes on the metabolic activities of biofilm-dwelling microorganisms which are widely accepted as being responsible for the majority of pollutant degradation in natural environments. Common methods for the *in situ* biological treatment of contaminated sites include the use of permeable reactive biobarriers in which the natural movement of water is used to direct subsurface contaminants though engineered 'walls' of enhanced biofilm microbial activity. Alternatively, contaminated media can be extracted and manipulated within biofilm bioreactors. In this way, a greater level of experimental control is provided and the fate of pollutants, as well as their degradation products may be more easily monitored. Where the biomass and activity of potential degraders in the natural community is too low, new microbial strains can also be introduced. Regardless of the approach used, it is evident that biofilm-dwelling microbial communities rather than planktonic organisms dominate the bioremediation of most pollutants. Thus, an increased understanding of the complex structure of microbial biofilms and the communication and cooperation among individual microbial cells will inevitably aid their successful use in bioremediation applications. Fortuitously, recent advances in both molecular and microscopy-based methods have revolutionized our understanding of the microbial biofilm 'mode-of-life'. The three-dimensional structure of biofilms and their coating by extracellular polymeric substances have been interrogated using a broad range of microscopy-based techniques to reveal how this mode of community organization provides microbial cells with substantial additional protection from toxic substances and mechanical stresses. Gradients in the availability of nutrients, toxicants and gases across biofilm structures have also been observed and characterized. The complex spatial organization of biofilms provides a variety of microniches to support an increased diversity of microbial life and their associated metabolic potential. Evidence of this is provided by the outputs of DNA sequencing studies, while advanced isotope labelling methodologies have permitted the accurate identification of even low-abundance taxa involved in key metabolic processes. As advances in biofilm research continue, the scientific community is finding ever more applications and ways to manipulate the degradation of pollutants by biofilms present in soils, natural waters, and on the surfaces of other organisms frequenting polluted environments, including plants and even fish!

In the first part of this book we provide an up-to-date review of the latest scientific research that has contributed to our understanding of the vital importance of microbial biofilms for the biological remediation of contaminated environments. In part two, the results of a variety of key case studies are presented to highlight the broad range of treatment approaches and applications at our disposal. Finally, as the application of biofilms in bioremediation continues to increase, we seek to predict the future trends and likely growth areas in biofilm-related research.

Gavin Lear

# Part I

# Introduction

# Engineering Successful Bioremediation

1

Michael Harbottle

## Abstract

Bioremediation has become a well-established tool in the armoury of engineers wishing to address the problems of contaminated land, water or waste. Successful delivery of bioremediation requires a combination of expertise from such diverse fields as civil engineering, soil science and environmental microbiology; whilst the contribution from engineering may appear to be straightforward, a detailed understanding of often complex ground conditions (physical, chemical and biological) is needed to avoid failure. In this chapter the common techniques for the implementation of bioremediation are described, alongside an introduction to the effects of real materials and environments on the behaviour of contaminants and microorganisms, particularly those of natural heterogeneity in geological materials, and the potentially detrimental effects these may have on achieving a successful outcome within a reasonable period of time. The chapter concludes with a brief summary of potential enhancements for the practical implementation of bioremediation currently at the research stage.

## Bioremediation – a geoenvironmental tool

Risks arising from contamination and pollution of the geo- and hydro-environment are often conceptualized using the notion of a pollutant linkage. This comprises a source of contamination, a receptor that is affected by the contamination and a pathway between the two that allows contamination to reach the receptor. If one or more of these components are missing, then land or water may not be deemed contaminated in law (although this will depend on the local legal system of the reader). Bioremediation is a versatile option for remediation of contamination that, depending on its nature, offers the possibility either of permanent breakdown or alteration of the target contaminant (source depletion or elimination), or immobilization or sequestration of non-degradable contaminants such as metals (breaking the pathway). Bioremediation has been defined as 'the use of living organisms to reduce or eliminate environmental hazards resulting from accumulations of toxic chemicals or other hazardous wastes' (Gibson and Sayler, 1992).

Most commonly, bioremediation occurs through biodegradation brought about by the actions of microorganisms (e.g. bacteria) on organic chemicals. It can be applied to soils, waters and wastes (solid or liquid), and implemented *in situ* or *ex situ*. Such biological methods for tackling environmental problems have been used for many years, through dealing with waste water and sewage to treating oil spills (Atlas, 1991). Bioremediation has been applied since the 1980s, but initially had something of a negative image due to early failures in implementation. It is now, however, a relatively well-established technology, with application on high-profile projects (e.g. Hellings *et al.*, 2011). This is especially true of *ex situ* remediation due to the greater degree of control that can be exerted on the process. *In situ* remediation, in contrast, is still sometimes seen as developmental. Application of bioremediation has increased greatly in recent years through a combination of factors, including:

- A drive to reduce waste going to landfill (for example in the EU) has increased the cost of excavation and disposal, or 'dig and dump', previously by far the most popular method of tackling contaminated land problems, and has made process-based techniques including bioremediation more competitive.

- Greater corporate and social awareness of the need for sustainable practices; demand for a reduction in energy/material use and waste production may favour processes such as bioremediation.

It is, however, in competition with a wide range of other technologies for removal, containment or treatment of contaminated land or groundwater. These include the original civil engineering methods (e.g. excavation and disposal, cover systems, barrier walls), physical and chemical methods (e.g. soil washing, chemical oxidation) and thermal methods (e.g. thermal desorption, incineration). The choice generally depends on cost, feasibility (i.e. is implementation possible at the site in question?) and applicability (i.e. is the technique appropriate for the contaminant and the soil/groundwater regime present at the site?). There is a wide range of biological remediation techniques available, discussed later in this chapter, which can be applied to varied situations and thus such techniques are options in many pollution scenarios. In addition, they tend to be less costly than other techniques (Summersgill, 2006). Advantages and disadvantages of the various techniques are discussed more fully later in this chapter. In practice, bioremediation is often only part of the solution as contaminants are often mixed (e.g. metals and organics together) or site conditions are varied. For example, the main site for London Olympic Games in 2012 was subjected to a mix of soil washing, chemical stabilization and bioremediation amongst others (Hellings *et al.*, 2011).

There exists a great deal of literature on bioremediation at the laboratory and field scale, for a wide range of contaminants. This chapter limits itself simply to the implementation of bioremediation in the geoenvironment from the point of view of an engineer, and so deliberately avoids discussion of particular contaminants, organisms or the contribution of molecular biology.

## Bioremediation at the small scale – mechanisms and behaviour important for engineering

### Contaminant behaviour in the ground

The ground is a complex, heterogeneous environment, consisting of solid, liquid and gaseous phases as well as the biosphere. It is impossible to generalize as to the nature of the geoenvironment at a given site, and a huge range of factors and interactions impact upon the suitability or otherwise of bioremediation. However, several factors are of particular interest and in this section a number of the most important are discussed.

The form a contaminant takes governs its fate and transport within the contaminated matrix, and so is fundamental to the success or otherwise of remediation. Inorganic contamination, including metals, will be present either dissolved in the liquid phase, attached or closely associated with mineral surfaces or charged organic matter, or in the form of a precipitate, often located on mineral surfaces. Organic contamination may again be present dissolved in groundwater, sorbed to surfaces or associated with organic matter, depending on its hydrophobicity and polarity or charge. In addition, they may be present as an entirely separate phase, particularly as a non-aqueous phase liquid (NAPL), either light or dense depending on whether it floats on or sinks through groundwater. NAPLs may move through the ground via mechanisms analogous to the flow of groundwater itself, albeit often considerably more slowly.

For remediation of any type where the source is to be removed, the contaminant should be available for transport to whatever means of remediation are employed. In this case, that means the organisms that are expected to deal with the contaminants but in other situations may mean transport to boreholes, wells or treatment zones, which may be some distance away. Dissolved or volatile contaminants may be transported in the liquid phase or the gaseous phase respectively, where the primary mechanisms of transportation are similar, consisting of advection (transport along with bulk flow of a fluid), diffusion (transport along concentration gradients in particular) and dispersion (spreading occurring due to variations in flow in porous media).

Most contamination interacts with the solid phase (be that soil grains, sediment or rock) to a degree, particularly with any clay or organic fractions present. These materials are particularly important not just because they tend to have a large surface area in a given volume, but because they are relatively chemically active due to the presence of charged surfaces or moieties, or because organic materials may offer hydrophobic regions into which organic contamination may partition preferentially. Sorption may also occur on other surfaces, including colloidal particles suspended in solution and microbial cells. In these cases, the surfaces are on transportable objects and so colloidal or microbial transport in advective flow, or microbial motility, may contribute to contaminant transport (Sen and Khilar, 2006).

Contamination will attenuate and disperse naturally via both transport mechanisms as well as natural sinks. These may include sorption to the solid phase, but this is not a true sink as sequestration is likely to be only temporary. Other mechanisms may be dependent on the contaminant type, such as radioactive decay, volatilization, photodegradation and biodegradation. However, the rate at which these occur is often insufficient to remove contamination to an acceptable level within a feasible timescale, and thus remedial activities are required. The persistence of contamination depends on the ability of the contaminants to become mobile or available to degrading mechanisms, the ability of the soil/groundwater/gas system to transport any mobile contamination, and the rate of any degradative processes.

## Microbial interactions with contaminated soil and groundwater environments

It is well known that microorganisms are widely distributed and highly diverse in the geo-environment. Their survival, activity and motility are governed by soil and groundwater chemistry (pH, redox potential etc.; e.g. Baker *et al.*, 2009), physical structure of the porous medium (pore structure, pore and pore throat size, particle size; e.g. Rebata-Landa and Santamarina, 2006) and location (particularly depth, which governs accessibility to important requirements such as oxygen and organic matter).

Microorganisms and chemical solutes in particulate media are subject to similar transport processes, including advection via groundwater flow and sorption to surfaces. The majority of microbial cells present, in their natural state, are likely to be sessile – attached to solid surfaces and sequestered within relatively inaccessible pores or structures that protect against predation by other soil organisms. This inaccessibility governs and restricts their activity, as their access to chemical species vital for growth and operation is therefore limited. For example, the ability of aerobic organisms to access sufficient oxygen can be severely restricted not only because of decreasing oxygen concentrations with depth but also because of limiting diffusion rates of oxygen into these small pores.

The inability of soil organisms to access sufficient nutrients, electron acceptors and other vital chemical species is likely to be mirrored by an inability to access the contaminant itself. For these reasons, biodegradation in soil is often hindered both by a lack of microbial activity and a lack of *bioavailability*. The various forms of bioremediation technology discussed later in this chapter are therefore largely focused on removing these barriers to growth and accessibility through enhanced mass transfer.

Microbial interactions with contaminants are also affected by the nature of the medium in which they are contained (Stroud *et al.*, 2007). Bioavailability is a complex concept that has differing definitions in different contexts. In the context of bioremediation, the fraction of contaminants that is bioavailable to microorganisms has been defined as 'that which is freely available to cross an organism's cellular membrane from the medium the organism inhabits at a given time' (Semple *et al.*, 2004). Similarly, the *bioaccessible* fraction is that which might at some point become available to the organism. Where one tries to improve contaminant bioavailability through engineering, the purpose is essentially to remove the barriers preventing bioaccessible contamination becoming bioavailable. It is a mass transfer problem, aimed at enhancing contact between contaminant and degrading organisms, but in such a way that does not negatively impact on the biodegradative process, for example through delivering contamination in excessive quantities all at once.

The majority of sessile bacteria, and often fungi, on soil surfaces will be encapsulated within biofilm structures that protect against oligotrophic environments and predation. Extracellular polymeric substances exuded by these organisms typically make up a substantial proportion (greater than 90%) of the biomass in soils, and the increase in volume can substantially alter hydraulic flow within the pore space (Baveye *et al.*, 1998). Even relatively small volumes can block pore throats and small pores, leading to flow diversion and preferential flow (discussed below). In the presence of NAPLs, biofilms have been shown to develop at the interface between groundwater and NAPL (discussed in Alexander, 1999), degrading both dissolved and non-aqueous phase contamination. Similarly, biofilms can develop at soil gas–groundwater boundaries where organisms take advantage of nutrients and chemical species from both phases.

## Heterogeneity in porous media

As a natural material, the ground is a highly heterogeneous environment on multiple levels from the microscopic to the macroscopic. Physical heterogeneity due to the arrangement of particles leads to variation in hydrogeological, biological and chemical parameters, all of which impinge on successful implementation of remediation practices in the field. Observed behaviour in realistic systems does not often match what might be expected based on an understanding of phenomena in homogeneous media, and means that attempting to mimic full-scale implementation at the smaller laboratory scale may provide at best limited understanding and at worst misleading information as to performance at the field scale (Davis *et al.*, 2003). Ultimately, attempts to enhance the process of bioremediation in such media may fail unless the heterogeneity is addressed (Song and Seagren, 2008).

Physical heterogeneity arises from the multi-scale nature of soil structure. At the micro-scale, the pore space will contain a range of pore sizes, with variable connectivity to the remaining pore space: some may be highly connected to other pores whilst others will be dead-ends or isolated from much of the rest of the pore volume. At the meso-scale, the nature of pore channels will be variable: some channels will be wide, continuous and readily accept fluid movement whilst others will be narrow or discontinuous and experience only restricted fluid movement. Flow is therefore largely limited to specific regions of the ground where larger channels exist. Aggregation of particles in surface soils, for example, leads to two broad phases of porosity, with small, less accessible, intra-aggregate pores and larger, more easily drained inter-aggregate pores which contribute to the majority of moisture movement. At the macro-scale, soil strata vary in hydraulic conductivity (the ability to conduct water flow for a given gradient in water pressure), and so strata of higher hydraulic conductivity (typically coarse-grained soils such as sands and gravels) may be subject to higher water flows than strata consisting of finer-grained materials such as silts and clays. Fig. 1.1 illustrates how physical heterogeneity may arise at multiple scales, and its effect on hydrogeological flow.

The tendency for moisture to travel a particular path of least resistance to flow, be it at the large scale through particular strata or at the small scale through individual pore channels, is known as preferential flow. This phenomenon has great importance in a number of fields involving subsurface flow such as *in situ* remediation (including bioremediation) (Brusseau, 1994), as flow is channelled via these

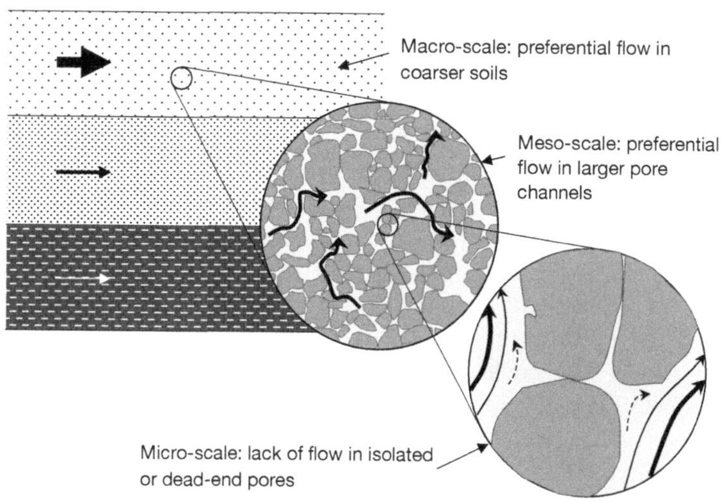


**Figure 1.1** Heterogeneity at multiple scales in the subsurface.

paths, leaving large portions of the subsurface unaffected. Rapid advective movement of solutes (required for remediation, amongst other things) therefore occurs along these preferential flow paths, whilst solute transport in the remainder of the soil body occurs primarily via diffusion, which is usually a much slower process.

The nature of the soil solid material, its attendant liquid and gas phases, and its physical location are all highly influential on the behaviour of microorganisms and contamination; all these parameters are also subject to a degree of heterogeneity. Individual soil strata may be very different, with clay soils offering high sorptive capacity for contamination and high moisture impermeability, limiting advective transport, whilst sand or silt soils are more permeable but have a lower sorptive capacity meaning contamination is more mobile. In mixed soil gradings (e.g. clayey sands), contaminants tend to be associated with finer fractions due to sorption (Scherr *et al.*, 2007). Soil strata are variable in physical structure too – they are not always horizontally bedded, and may contain features such as lenses (isolated zones of another soil type) or bands (thin layers within the bulk matrix). Although often only a tiny part of the overall stratum, they may have a substantial impact upon the behaviour of that stratum, for example bands of sandy soil within a clay stratum offer a significant preferential flow path to groundwater and can therefore substantially increase the hydraulic conductivity of these strata.

The physical location of a soil stratum, in particular its depth, will determine the nature of pore contents, which play a significant role in the behaviour of contaminants, other solutes, and microorganisms. Below the water table, pores will largely be water-filled, limiting the activities of purely aerobic organisms (including fungi and some bacteria). Above the water table, in the vadose zone, the situation is more complex. Immediately above the water table lies the capillary fringe, where capillary action in smaller pores causes water to rise to a height governed by the pore size (and therefore soil type). Nearer to the surface, the moisture regime will be governed by climatic conditions and rainfall, and so will be considerably more variable with time. Similarly, soil gas contents vary with depth; oxygen levels decrease with depth whilst gases such as carbon dioxide, carbon monoxide and possibly methane may become more prevalent.

The ability of soils to support microbial communities is again dependent not only on individual soil properties but also the physical location of that soil. It is well known that microbial numbers and diversity are greatest near the surface, with its greater concentration of organic matter and access to oxygen. The vadose zone is particularly complex, with a range of microenvironments present ranging from fully saturated to fully gas-filled pores, and so offers a wide range of niches for different species (Chau *et al.*, 2011). Soil grading can play a significant role, particularly where fine-grained soils are present, which may limit suitable pore space for organisms to thrive at depth (Rebata-Landa and Santamarina, 2006). However, such soils tend to be higher in chemical activity and may contain a greater degree of organic matter so tend to act as repositories for basic nutrients and have concomitantly higher microbial activity (Scherr *et al.*, 2007).

Contaminant heterogeneity arises due to the nature of the chemical, the nature of the ground and the process which caused contamination in the first place. Association of contaminants with particle surfaces occurs most substantially in fine-grained soils due to greater surface charge on clay as well as a larger total surface area (Scherr *et al.*, 2007). Organic contaminants with low water solubility will partition (effectively, dissolve) preferentially into organic matter. Physical variability in the location of these soil fractions will therefore impart a degree of heterogeneity to the distribution of contamination. Transport of contaminants via preferential flow paths means that contamination is often located close to these paths, at least initially. NAPL contamination is often highly heterogeneous, stemming from a specific contamination incident and because of slow and limited transport is often physically close to this location.

Contamination events often occur at a specific location or locations, for example the leak of an underground storage tank. So-called 'hotspots' of contamination result, meaning that highly contaminated regions are spatially limited and in close proximity to uncontaminated zones. A plume of contamination will then arise over time stemming from this source. The distribution of contamination on a given site is therefore governed by the distribution of contaminating processes or installations on or under that site. Site investigation of such sites often takes place in a targeted manner, using information on the distribution of these processes.

Over time, various processes occur to both increase and decrease heterogeneous contaminant distribution, which are collectively known as the phenomenon of *ageing*. Physically heterogeneous distributions of contamination, such as high concentrations located close to preferential flow paths, may be dissipated as the slow process of diffusion transports contamination into zones away from the main regions of fluid flow. Biodegradation may readily occur where contaminants are accessible and at a tolerable concentration, but excessively high or low concentrations hinder biological activity – zones of high contaminant concentration may therefore persist whilst degradation occurs in surrounding regions of lower toxicity. Coherent bodies of NAPL may break down into isolated ganglia – small reservoirs in pores away from main flow paths.

## Implementation of bioremediation in practice

The following, taken from Alexander (1999), illustrates what is needed for bioremediation:

- Microorganisms with the necessary catabolic capabilities and with capacity to reduce the contaminant to levels that meet site requirements at reasonable rates.
- Must not generate significant toxic products.
- Materials inhibitory to the degrading species have to be diluted or otherwise dealt with.
- Target contaminant(s) must be available to the microorganisms.
- Conditions in the ground, excavated material, water or waste must be conducive to microbial growth or activity.
- Cost of implementation must be less than or equal to that of other techniques where similar results may be obtained.

### Design of the process

Because of the potential variability in conditions (contaminants, chemistry/biology, site conditions, etc.), it is not possible for one technology to remediate all materials. A technology may be defined by (i) whether it eliminates or contains the contamination; (ii) where it is applied, for example in the ground or externally, and onsite or offsite. Certain techniques, or variations thereof, may be applied to surface soils, aquifer material, sediments, groundwater, soil gas, or a combination of these.

Earlier, the concept of a pollutant linkage was defined. Bioremediation is usually considered as source control, i.e. removing the contamination itself by separating contaminants and environmental matrix (soil, groundwater, etc.) – this is defined as *remediation*, and the majority of technologies available can be described in this way. There are examples of biological techniques used for *containment* of contamination, where no depletion occurs but the pathway between contaminant source and receptor is broken by isolating the contamination from its environment.

The term '*in situ*' refers to the soil/water remaining in the ground whereas '*ex situ*' treatments or processes require excavation or extraction. They have different advantages:

*In situ*

- Requires relatively low effort and little surface disturbance.
- May require long time periods.

*Ex situ*

- *Ex situ* methods are usually easier to perform – they allow greater control over the medium to be treated.
- Can often be performed relatively quickly.
- May involve more effort (e.g. due to considerable earthworks).

The terms onsite and offsite are different – it is possible to have *ex situ* onsite treatment, e.g. if soil is excavated and treated on the surface without leaving the site. Offsite treatment is possible at remediation 'hubs' that accept contaminated material for clean-up. For soils, these tend to be associated with landfill sites, where the remediated material is often used as daily cover.

Fig. 1.2 illustrates the more commonly used bioremediation technologies available, although many variations on these exist. They are broadly divided into *in situ* and *ex situ* techniques. Details of these are provided later.

Even the worst contamination incidents will be naturally remediated given enough time. Biodegradation occurs naturally, where conditions allow; however, the process is usually too slow to allow

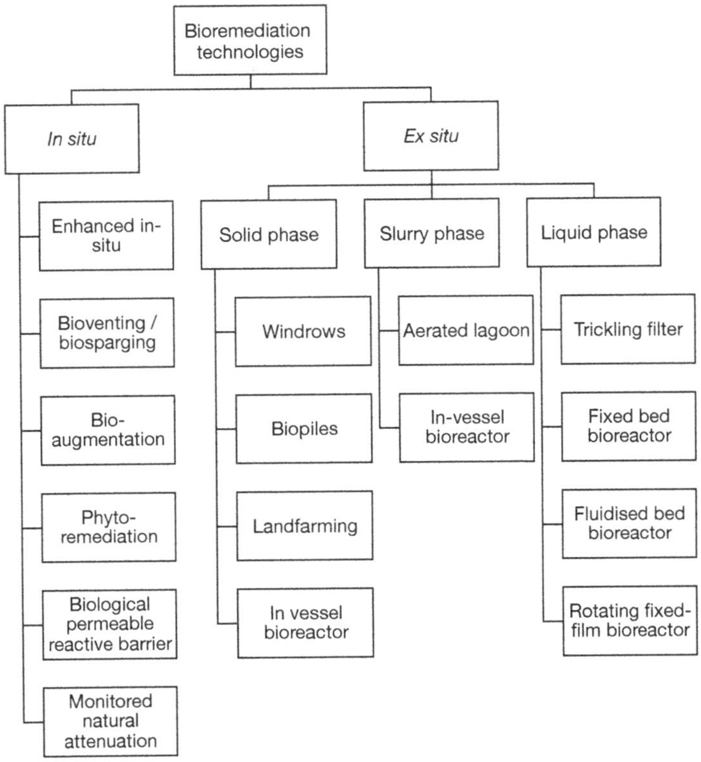


**Figure 1.2** Typical categories of bioremediation for different media.

development or to prevent harm to receptors, and so the process of bioremediation involves identifying any limiting factors and correcting them such that remediation occurs at an appropriate rate. The processes of identifying and correcting these issues are well established and discussed in considerable detail elsewhere (Cookson, 1995; Wong *et al.*, 1997), so are only briefly presented here.

Project design will be similar to that of any remediation project – it is essential to characterize the environment fully to ensure it can be manipulated correctly. A typical procedure may be:

- site investigation
  - to identify contaminants and their extent and distribution
  - to identify soil physical, chemical and biological properties that impact upon chemical and microbial behaviour and transport
  - to determine whether bioremediation is a feasible option and ...
  - ... to identify what is limiting natural biodegradation at the site;
- design calculations, e.g.
  - fate and transport of contaminants
  - biodegradation kinetics;
- further studies, e.g.
  - laboratory testing
  - column degradation tests;
- selection and design of suitable technology.

Standard geotechnical methods are useful to characterize the physical environment, including groundwater transport and soil classification. Chemical characterization can determine parameters such as:

- types and concentrations of contaminants, their availability and likely transport or fate;
- sorption and desorption characteristics, and whether leaching is likely to occur;
- soil properties such as pH, redox potential, temperature, moisture and oxygen levels;
- available nutrients, electron acceptors, electron donors, etc.

Biological properties may be ascertained from

an engineering point of view in terms of whether there is a suitable degradative capability amongst the community present. When combined with an understanding of site attributes, the nature of contamination and generation of metabolites, it will be possible to determine the relevant degradation pathways, and so understand any rate-limiting factors. This in turn will narrow the list of potential technical solutions depending on the form of the limited material. For example, a common critical factor is a lack of oxygen, which may suggest the use of techniques which either deliver or otherwise improve access to it. In the case of insufficient degradative capability, cultures may be prepared for addition to the site, a process known as bioaugmentation (Tyagi *et al.*, 2011; Vogel, 1996). Alternatively, the process of cometabolism allows contamination to be targeted that otherwise would not stimulate the degradative capability of the existing community; addition of specific molecules (sometimes contaminants such as toluene) may stimulate microorganisms to generate degradative enzymes which target both the added molecule as well as the contaminant (Frascari *et al.*, 2015).

Laboratory-scale and a limited degree of field-scale testing can be employed at a simple level to determine biodegradation kinetics, and may be utilized to develop simple models for use at the field scale. These tests may involve simple experiments *in vitro*, static microcosms (tanks containing a representative sample of the site conditions) or column degradation studies. The latter may be used to understand dynamic systems where transport (e.g. of a contaminant) is occurring. The testing programme should give a prediction of the time required for risks to be acceptably reduced in both unamended and amended conditions. However, laboratory and even field tests may not give entirely accurate predictions. Problems of heterogeneity will be far more prevalent at field-scale than laboratory scale, for example. It is therefore easier to design remediation processes that allow better control of soil conditions, as this means better control of heterogeneity.

Alongside the requirement for the remedial technology to be able to deliver amendments in a specific form (e.g. gaseous, dissolved), it must also be appropriate for the particular site conditions. For example, a number of techniques require flow of groundwater to effect transport of either amendment or contaminant, and therefore can only effectively be applied in relatively permeable soils or sediments. Remediation of fine-grained soils is often somewhat challenging for this reason. Alternatively, the depth of contamination may limit the applicability of *ex situ* remediation methods due to the difficulty of excavation.

### *In situ* techniques

*In situ* remediation simply means treating the contaminated ground where it lies, without excavation. This is often achieved by introducing materials into the subsurface. The material depends on what is limiting natural bioremediation, be it certain nutrients or a lack of degrading organisms. It has the advantage of minimizing disturbance and use of land, but it may be difficult to control what happens in the subsurface, and as such some remediation projects may last for several years. There follows a brief description of the basic techniques used currently to bring about *in situ* remediation. Here we focus on engineering aspects – other studies, such as by Pandey *et al.* (2009), provide a greater focus on microbiological issues.

#### Enhanced *in situ* bioremediation (EISB)

Also known as biostimulation, this is a relatively simple method whereby amendments that are soluble or suspendable in groundwater can be introduced to a contaminated zone either by applying directly to the ground (percolation) or by inducing an artificial flow of water, and introducing amendments this way (Fig. 1.3) (Hohener *et al.*, 1998). In the latter case, a range of borehole locations and depths may be required to ensure sufficient groundwater flow through the zone of contamination. Inducing fluid flow may raise the issue of enhanced contaminant mobility, and therefore potential transport of contaminants to new receptors, which must be taken into account in design. Uniform rates of treatment are unlikely due to heterogeneity in ground conditions and possible preferential flow.

Although nominally simple, there are potential problems with this method. For example, introduction of nutrients at a specific point can induce bioclogging, where growth of microorganisms is encouraged around the point of introduction such that they hinder or prevent further introduction (Taylor and Jaffe, 1991). There can be a lack of control due to soil heterogeneity, for example

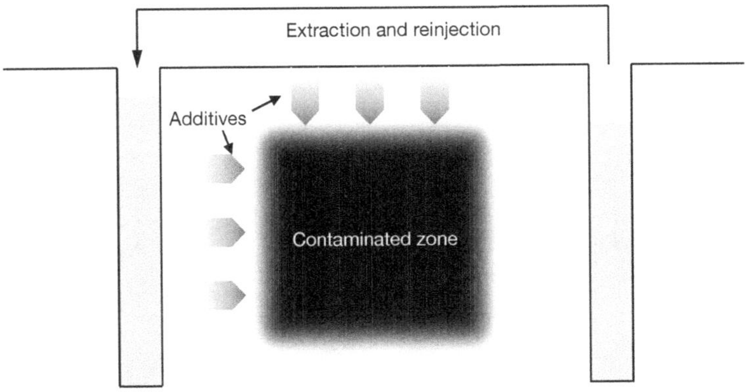


**Figure 1.3** Schematic of typical enhanced *in situ* bioremediation scheme including enhanced groundwater flow through pumped extraction and reinjection upstream.

with nutrients being transported via preferential flow paths and not entering many of the soil pores (Davis *et al.*, 2003). The technique is limited to coarser-grained soils, as flow rates in clays and silts may be insufficient.

Bioaugmentation is the addition of microorganisms, usually species (or a combination of species) that are known to deal with the contaminants at hand, to the ground when the existing community may be incapable of degrading particular contaminants at a sufficient rate (Tyagi *et al.*, 2011; Vogel, 1996). A number of commercial products containing specific consortia of microorganisms exist. However well the organisms may degrade the contaminant, a particular problem is ensuring that they are applicable to the soil conditions. Survivability and competition with indigenous soil microorganisms can be major problems, as well as ensuring delivery to the correct area; the technique is again limited to coarse-grained soils through which transport of the microorganisms can occur, and delivery methods may mimic those of EISB, discussed above.

### Bioventing/biosparging

Air or oxygen pumped through contaminated material can have several effects – it can encourage volatilization of volatile contaminants and it can also encourage microbial growth (Fig. 1.4). Bioventing and biosparging are techniques that induce gas flow through a contaminated zone to stimulate microbial activity, with the former operated above the water table and the latter below. Rather than direct injection of air, vacuum extraction may often be used to induce gas flow in the ground from boreholes drilled as vents to a central extraction well. This ensures extracted gases, which may contain volatile contaminants, are collected and treated appropriately. An impermeable plastic membrane or other cap is used to maximize the zone of influence, ensuring gas enters the ground at a reasonable distance from the extraction well rather than adjacent to it.

This technique is limited to coarse-grained soils and may only have a limited area of influence. As such, it may be necessary to have access to the whole contaminated area to ensure it is all treated adequately.

### Permeable reactive barriers (PRBs)

PRBs are used where contaminants are transported by groundwater. They are reactive zones through which contaminated water is channelled, in which contaminants react to form less toxic products. Often these involve chemical treatment but biological versions exist where the barrier is essentially a bioreactor, with specific conditions maintained within it. The physical arrangement of the barrier depends on the specific site conditions. Two options are shown in Fig. 1.5. A continuous PRB may be impractical as it would require a large reactive zone volume. Alternatively, a 'funnel and gate' system channels contaminated groundwater towards a more limited reactive barrier through judicious location of impermeable barriers. In this case, the barrier itself may be contained within an in-ground treatment vessel (Phillips *et al.*, 2010). In recent years, biofilm barriers have been considered,

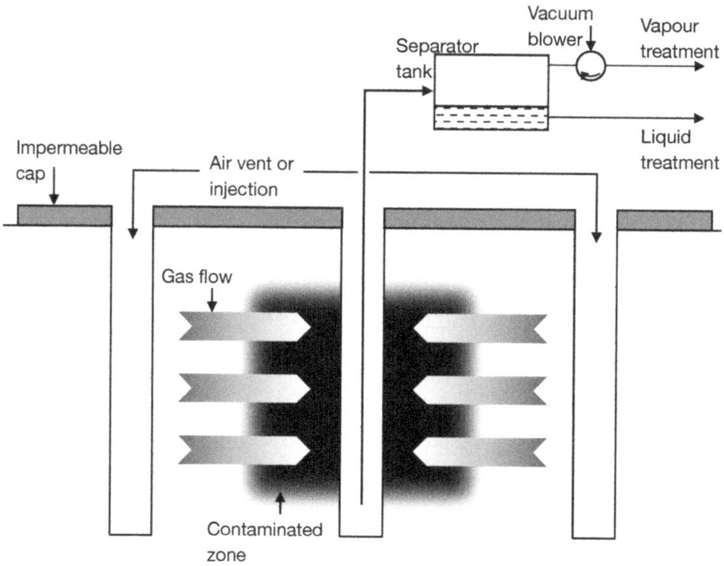


**Figure 1.4** Simple schematic for bioventing or biosparging (vacuum mode of operation).

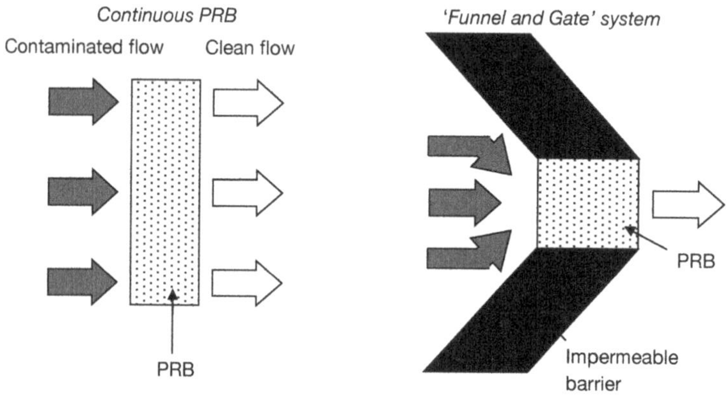


**Figure 1.5** Possible PRB configurations.

both for impermeable walls used for containment or flow control but also in the reactive zone (Komlos *et al.*, 2004). This latter application is discussed in depth in Chapter 11.

## Phytoremediation

Phytoremediation, the use of higher plants to instigate contaminant removal, can be applied to many contaminant types, organic and inorganic. For example, organics can be degraded efficiently by high numbers of microorganisms in the root system (rhizosphere), a process called phytodegradation. Aspects of degradation via biofilm communities associated with plants are discussed further in Chapters 13 and 14. Phytostabilization involves changes in soil chemistry (e.g. pH, redox) due to the presence of the plant, perhaps making contaminants less soluble and thus less likely to move. Plants themselves can take up contaminants, particularly metals, and store them in their above- or below-ground structures in the phytoextraction process (Page *et al.*, 2014).

A range of plant taxa have been used in these various forms of remediation, with different structural forms and hydrogeological impacts affecting contaminant transport and degradation. Grasses, for example, have a dense root system over shallow depths whilst tree species have a more widely dispersed root system, albeit one which operates at greater depths. The latter can also induce significant groundwater movement and thus affect contaminant transport. Constructed wetlands employing aquatic plants have been used to treat contaminated water. For example, Heathrow Airport (London, UK) has installed reedbed systems for the treatment of run-off containing pollutants such as aircraft de-icing fluid (Murphy *et al.*, 2015).

### Monitored natural attenuation (MNA)

Contaminants may be attenuated naturally by physical, chemical or biological processes including biological degradation, sequestration (e.g. sorption) and dispersion (e.g. volatilization). MNA simply involves monitoring these processes to ensure no risks develop as this attenuation goes on (Rivett and Thornton, 2008). It is a passive technique and in some cases may be the only feasible option if other technologies are projected to take many years. The only infrastructure required is construction of boreholes for monitoring purposes.

## *Ex situ* techniques

*Ex situ* bioremediation involves excavation/extraction of contaminated material (solid or liquid) from its location *in situ* prior to treatment. Once extracted, contaminated material may be treated on or off site. Typically, the techniques are relatively simple and involve the addition of the requisite nutrients, oxygen etc., as with *in situ* treatments. Similarly, processes such as bioaugmentation can be employed – the delivery mechanism may be different but the process is the same. Extensive site works and large land areas may be needed, particularly for processing solid phase material, but they allow greater control and interactivity with the material and so can be easier to operate successfully. A brief description of common methods of *ex situ* remediation follows.

### *Ex situ* remediation of soils

The process of *ex situ* remediation of soils does not vary significantly between the various technologies, differing only in the physical structure of the excavated soil. These processes are described for contaminated soils and groundwater, but may be applicable to a range of contaminated waste materials also; indeed, such methods are commonly used to compost the organic fractions of municipal solid waste. Biopiles are simply large constructed mounds of contaminated soil (Fig. 1.6A) (Pelaez *et al.*, 2013). They may be static piles or involve an element of composting, where additional organic matter is added. Composting can generate significant temperatures, encouraging biodegradation or volatilisation of contaminants. Additional amendments may be added where necessary to enhance degradation rate or extent. Usually a constructed base with suitable bunding and drainage is required to ensure any contaminated leachate is collected, and air may be drawn through the soil by vacuum pumps (allowing treatment of gases produced) in order to stimulate aerobic microorganisms.

Typical operations may require 3–6 months for a single biopile. They are commonly used for PAHs and petroleum contaminants, and not for 'difficult' pollutants (e.g. PCBs, dioxins). Operations are suitable for many soils though excessive fine-grained

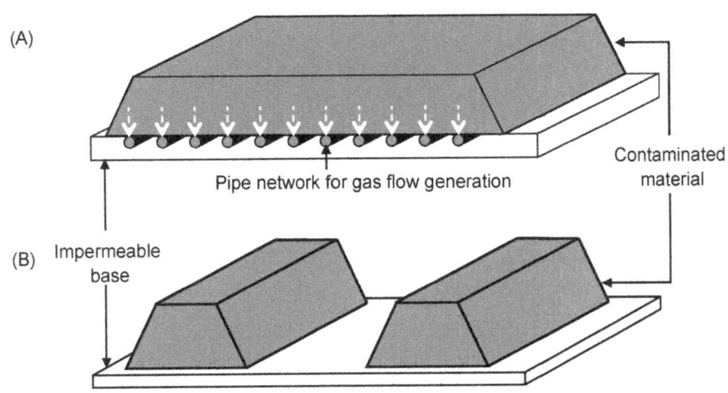


**Figure 1.6** Simplified schematics for biopile (A) and windrow (B) installations.

materials may cause problems in air flow and contaminant retention.

Windrows are similar to biopiles. Instead of a pile, soil is arranged in long rows, usually with a prepared base but without any aeration (Fig. 1.6B). Instead of aeration, the windrows are mixed, or turned, regularly using specialized plant (a windrow turner).

Contaminated soil can simply be tilled using standard agricultural equipment to ensure good mixing of soil as well as introducing oxygen and nutrients – a process known as landfarming (Lin *et al.*, 2011). It is excavated and spread over a prepared bed or liner to prevent escape of contamination. In this case a large area may be required to prepare the bed. Otherwise, conditions and amendments do not differ significantly to other *ex situ* methods.

### Bioreactors and in-vessel treatment

Bioreactors vary in form from simple lagoons to engineered closed vessels. They may treat soils, slurries and particularly water (be that contaminated surface or groundwater or process wastewater). Lagoons of various types are used for treatment of a range of contaminated liquids, ranging from the removal of inorganic contamination from mine drainage in treatment ponds with varying pH and redox conditions (Whitehead and Neal, 2005) to constructed wetlands for treatment of agricultural run-off (Gregoire *et al.*, 2009).

Greater control may be achieved when materials are treated within contained vessels. Reactors are usually more efficient than other *ex situ* processes but are limited in volume (e.g. to tens of cubic metres). When dealing with large volumes of soils and slurries in particular, treatment may be carried out in batches – this extends the treatment time considerably. Groundwater and slurry reactors may be designed to operate under continuous flow conditions which as well as batch conditions. At its simplest, the bioreactor is simply a vessel to hold the contaminated medium whilst biodegradation occurs under controlled conditions, following addition of the requisite amendments. In many cases, the material is stirred or mixed, mechanically or through the introduction of aerating bubbles, to facilitate increased contact between contaminants and degrading organisms. The microorganisms themselves may be suspended in liquid or immobilized on a solid surface in some way. Soils may be treated with relatively little pre-treatment through the process of in-vessel composting (Antizar-Ladislao *et al.*, 2007). The process simply optimizes the environment and availability of all required nutrients and other factors, which is easier to achieve in this closed system. Alternatively, soils may be slurried prior to introduction into the vessel (Robles-Gonzalez *et al.*, 2008). Slurries are introduced with the necessary amendments and are thoroughly mixed for the treatment period. Only smaller particle sizes can be treated as coarse-grained materials settle readily, so some pre-treatment of the soil may be required.

There is a range of options for the treatment of contaminated liquid wastes and groundwater (Fig. 1.7). The aim is to maximize contact between a degrading organism and contaminant and so in immobilized systems (sometimes called fixed-film reactors) a solid phase is usually introduced with a large surface area, colonized with these organisms. Although contaminant/degrader interaction may not be maximized, such systems have the advantage that the organisms are retained when material exits the reactor. Such materials may be particulate in form, such as in trickling filters, fixed bed reactors or fluidized bed reactors. The former involves liquid applied to the surface of a bed of particulate media on which a biofilm has developed and flowing across this film under gravity whilst aerobic degradation of the contamination occurs. The filter remains unsaturated, maximizing access to oxygen, but to achieve this relatively low fluid flow volumes are necessary and forced aeration may be required. A fixed bed bioreactor is similar although in this case a continuous flow of liquid passes through the pore space of the bed. Over time in both of these systems, biofilm sloughing off the solid surfaces and accumulating may block the reactor. A fluidized bed reactor gets around this by having particulates suspended in the fluid flow. Reactor materials range from rock, sand and gravels seen in trickling filters to purpose-made plastic objects in various forms (e.g. hollow cylindrical, spherical) and granular activated carbon (which has a large surface area, which confers the advantage of a large sorption capacity for many contaminants as well as an ability to support a large microbial community). In all three cases, however, biofilm material exiting the reactor following sloughing means that further treatment of the effluent may be required through

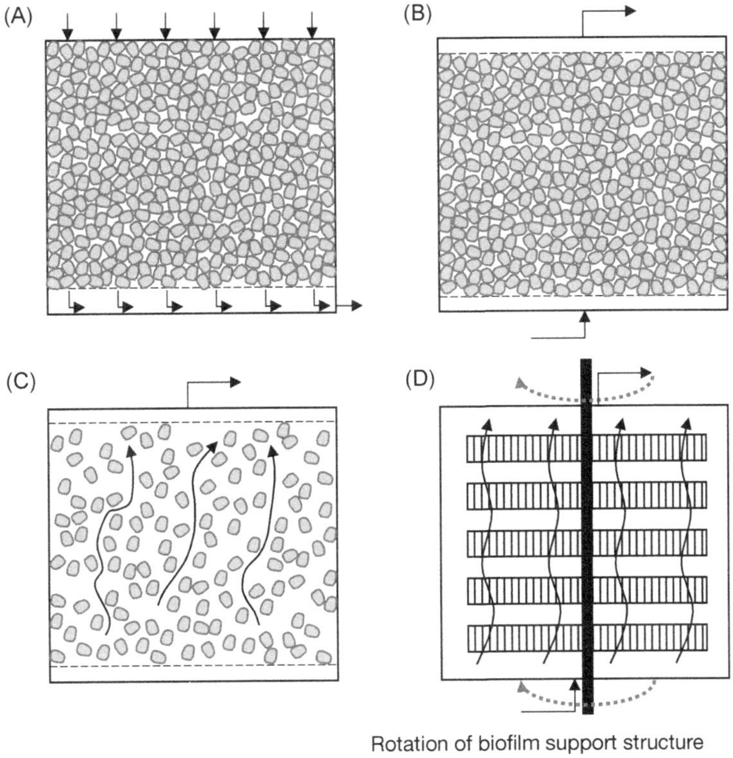


**Figure 1.7** Simplified bioreactor schmematic diagrams. (A) Trickling filter. (B) Fixed-bed reactor. (C) Fluidized bed reactor. (D) Rotating fixed-film reactor. Solid arrows represent typical fluid flow directions.

clarification and settlement. Immobilized biofilms may also be found in rotating fixed film bioreactors, where they are attached to some form of rotating frame which not only maximizes degrader/contaminant contact but also stirs the fluid.

## Combining bioremediation with other technologies

A significant proportion of contaminated sites have mixed contamination such as combined metallic and organic pollutants, and/or have complex geological or hydrogeological environments. These are likely to require more than one remediation technology as different contaminants/environments may need to be treated in different ways. It is therefore important to understand the limitations of bioremediation and where other technologies can be combined with it to achieve the best solution. There is a wide range of potential combinations, of two or more technologies – some examples of where bioremediation may benefit other technologies or vice versa are presented in Table 1.1.

## Waste bioremediation and bioprocessing

Many anthropogenic wastes are treated using bioremediation approaches. The techniques used are often identical or very similar to those discussed above for contaminated soil and groundwater, and in the context of this chapter there is little further to discuss here. To illustrate this, however, examples of the applicability of these various technologies to a varied selection of waste types are presented in Table 1.2.

## Practice and pitfalls

A short summary of some advantages and disadvantages of biological remediation methods is presented in Table 1.3. Where bioremediation is applied to an appropriate problem with proper planning and forethought it is very successful in achieving remedial targets and so has become a popular choice for certain categories of contamination problem.

**Table 1.1** Examples of mutually beneficial combination of bioremediation with other techniques

| Combined technologies | Benefits |
|---|---|
| Soil washing with slurry bioreactor (e.g. Zhang *et al.*, 2000) | Soil washing separates large grains from small grains, with contamination tending to be concentrated in the latter. The latter can then be treated in a slurry bioreactor |
| Pump and treat with enhanced *in situ* bioremediation (USEPA, 1996) | Pump and treat removes soluble or suspended contamination by extracting groundwater and treating it. The inducement of flow in EISB essentially does this, and so combined *in situ* soil treatment with *ex situ* groundwater treatment results |
| Soil vapour extraction (SVE) with bioventing (Soares *et al.*, 2012) | SVE is essentially the same as bioventing, albeit with the purpose of extracting volatile contaminants for treatment *ex situ*. By aerating the soil, it induces biodegradation |
| Chemical treatment | Pretreatment with a chemical process can render contamination with limited biodegradability suitable for bioremediation. For example, Piskonen and Itavaara (2004) suggest that Fenton oxidation can enhance subsequent PAH degradation |
| Excavation and disposal in landfill with *ex situ* bioremediation in windrows | A site contaminated with PCBs, PAHs and petroleum hydrocarbons was treated partially with excavation and disposal at a landfill to deal with PCB-contaminated material, whilst the remainder was treated using windrows (Harbottle *et al.*, 2008). PCBs would not have been easy to biodegrade in this situation |

**Table 1.2** Examples of the bioremediation of wastes

| Waste type | Remediation |
|---|---|
| Biodegradable municipal waste | Organic municipal waste may be either source-segregated or mechanically segregated (the latter in mechanical – biological treatment plants). Both may then be composted, typically in windrows. Compost from the former is used widely, whilst that from the latter, known as compost-like organics (CLO) in the UK, is more problematic as it may contain metallic contamination and so has more limited markets (Page *et al.*, 2014) |
| Sewage/wastewater | Bioreactors such as trickling filters or lagoons are commonly used to treat such material |
| Mine drainage | Mine drainage from closed or abandoned mines is problematic in former deep mining areas, producing large volumes of metal-contaminated water often with low pH. Biological control of pH and redox conditions is used to reduce metal contamination in solution. Because of the requirement for continuous treatment, passive treatment-train systems have been employed, such as at the Wheal Jane mine in Cornwall, UK (Whitehead and Neal, 2005) |
| Mineral residues from mining | Mineral residues are varied, but bioremediation processes can be employed to alleviate extremes of salinity, pH, etc., as well as amending the physical form of such deposits. An example is the treatment of bauxite residues from aluminium mining, commonly known as red muds, which occupy vast lagoons but are contaminated and unstable (Santini *et al.*, 2015) |
| Construction waste | Waste streams from construction processes contain wood products, much of which is treated in some form, for example by applications of wood preservatives [e.g. creosote or CCA (copper–chromium–arsenic)]. Both the biodegradability and potential toxicity of the waste cause issues with its disposal, so biodegradation and bioremediation are considered potentially useful. Composting methods have been proposed (McMahon *et al.*, 2009) |
| Metal cutting fluids | Organic wastes such as cutting fluids used to cool machines used in metal fabrication and working can be treated in typical bioreactors – the main issue is ensuring a suitable community is present as these may comprise a variety of contaminants (van der Gast *et al.*, 2004) |

At field scale, implementation of remediation *in situ* can be difficult to control in complex geological environments due to factors arising from the natural heterogeneity of the site (discussed earlier in this chapter), even where the 'average' conditions may seem suitable (Sutton *et al.*, 2013). Sturman *et al.* (1995) offer a thorough discussion of the practical effects of scaling up bioremediation, and quote a reduction in the biodegradation rate of 4–10 times at field scale compared to laboratory investigations owing to the variation in conditions from the ideal state usually measured in the laboratory. Important issues include heterogeneity of microbial distribution and activity, heterogeneity

**Table 1.3** Advantages and disadvantages of biological remediation methods

| Advantages | Disadvantages |
|---|---|
| Applicable to wide range of pollutants | Considerable disturbance (*ex situ*) |
| 'Remote' control (*in situ*) | Sometimes toxic by-products |
| Little site disturbance (*in situ*) | Difficulty in attaining homogenous control (*in situ*) |
| Positive image (considered 'sustainable') | Some compounds are recalcitrant – may take a long time to degrade (if at all) |
| Source removal – requires no further removal, concentration, treatment or disposal | Can be subject to external constraints (e.g. weather) |
| Usually produces harmless products from harmful chemicals | Typically longer timescales (e.g. than excavation and disposal) |
| Relatively cheap | May be hindered by contaminant mixtures |

of contaminant distribution and physical heterogeneity limiting delivery of the chosen remedial technology. The first two may lead to a lack of bioavailability – biodegradation cannot occur or is limited in extent or rate as the contaminant is separated from the organism, for example by irregular distribution of suitable organisms and contaminant; a lack of contaminant solubility, strong sorption or association with organic matter; or physical sequestration away from microbial habitats (Ortega-Calvo *et al.*, 2013). Aleer *et al.* (2014) explored the spatial heterogeneity of microbial species on a hydrocarbon-contaminated site, and determined that species diversity and potential for hydrocarbon removal were correlated, indicating that the variability in bacterial diversity noted would have implications for remediation. Contaminants that are largely unavailable, particularly heavier oil fractions or recalcitrant species such as polychlorinated biphenyls, may persist for some time, even if degrading organisms are present. Although contamination is often initially present within more permeable zones within the ground (as these are the routes by which they travel), where contamination has been present for some time, mass transfer via diffusion into less permeable strata or zones may have occurred (Mackay and Cherry, 1989). In flow-based remediation methods such as enhanced *in situ* bioremediation, the available material in coarser fractions or in the vicinity of preferential flow paths may be degraded more readily first, as amendments will be able to reach these zones relatively easily, whilst the sequestered fraction may persist until it is able to diffuse to such zones where treatment may occur. For this reason, *in situ* methods have been known to last for long periods, sometimes decades. Methods do exist to attempt to overcome a lack of bioavailability, however, such as use of surfactants to enhance solubility (Ortega-Calvo *et al.*, 2013), although their use is somewhat limited to date. If the available concentration is below a certain level, degradation may not be instigated or continued, whilst high levels of contamination may be sufficiently toxic to limit microbial activity.

The presence of mixed contaminants is known to be problematic, as organisms do not often have wide-spectrum capabilities for resisting different classes of pollutant and so whilst they may be able to tolerate or degrade one type, the presence of another will not be tolerable. For example, Thavamani *et al.* (2012) show the effect of mixed metallic and organic (PAH) contamination on the diversity and activity of a soil microbial community, determining that mixed contaminants repressed both of these factors as measured by enzyme activity.

Even when biodegradation occurs, problems may arise. For example, intermediate degradation products may be just as much, or more, of a problem as the original contaminant, especially if these products are recalcitrant and are not rapidly consumed (Ganey and Boyd, 2005; Kelce *et al.*, 1995). Remediation of certain types of contaminant may even be reversible – for example the oxidation or reduction of metallic species. Sethunathan *et al.* (2005), for example, report the ability of microorganisms to reverse the reduction of chromium when environmental conditions changed, leading to the metal becoming toxic and available once more.

Underground fluid flow and contaminant transport induced as part of remediation *in situ* (including bioremediation) can be difficult to control. Flow outside the treatment zone, for example when flow inadvertently travels downstream of an extraction well, may lead to enhanced dispersion of contamination (Keely, 1989). The effect of heterogeneity of fluid flow, arising because of variability in the structure of the particulate medium, has been demonstrated to cause heterogeneity in *in situ* remediation where fluid flows are generated. This effect was illustrated by Englert *et al.* (2009) in a metal bioremediation project, arising due to variability in the physical distribution of chemical amendments. In a case study in a highly geologically complex conglomerate deposit, Verce *et al.* (2015) demonstrate the difficulty of developing a conceptual model of such sites, with only a limited degree of information on the contaminated strata and the convoluted flow paths therein. However, they illustrate that such sites can be successfully tackled given enough time. The act of remediation can, however, alter fluid flow to the detriment of the project as the introduction of amendments that encourage microbial activity will influence the behaviour of organisms at all locations the flow passes, and those closest to the source have the first opportunity to access it. Therefore, clogging or fouling of injection wells through growth of organisms and production of extracellular materials arises as organisms in the vicinity of the well take advantage of the windfall (Baveye *et al.*, 1998; Taylor and Jaffe, 1991). Not only does this reduce the availability of amendments to the target organisms in the majority of the contaminated zone, it also alters hydrogeological flow patterns and can significantly reduce the rate of flow to the treatment zone.

Remediation of contaminated material *ex situ*, whether in a bioreactor or on the ground surface, generally allows greater control of conditions within the soil mass. Thus, issues around heterogeneity and bioavailability can be controlled to an extent and *ex situ* remediation is usually quicker and more reliable, hence it is more popular. However, new issues arise if material is excavated. Additional pathways may be generated when contamination is brought to the surface, for example through the release of volatile contamination into the air (Tomei and Daugulis, 2013). Treatment methods that take place in the open are also exposed to climatic changes – temperature and moisture levels are important in terms of microbial activity and other factors. Weather/climate can therefore affect the rapidity of remediation and in extreme conditions (e.g. flooding) can cause significant practical as well as microbiological difficulties. This is the case with *ex situ* techniques on prepared beds, where floodwater or excessive rainfall can mean escape of contamination (for example, the *ex situ* bioremediation case study reported by Harbottle *et al.* (2008) was subjected to flooding which held up remediation delivery by some weeks). Although bioremediation can take place at low temperature (Van Stempvoort and Biggar, 2008), unexpected cold weather will hinder processes that are otherwise unprepared. Bioremediation may therefore be deliberately scheduled for spring or summer where possible.

## Where is bioremediation engineering headed?

Has the implementation of bioremediation been improved as far as it can reasonably go? From the engineer's point of view there may be some truth in this, particularly with *ex situ* bioremediation. As long as time and care are taken in site investigation and design, even complex *in situ* challenges can be tackled successfully and the basic technologies to implement remediation described above have not significantly changed in form for many years. However, issues still remain which can and are being tackled in order to improve the rapidity or comprehensiveness of bioremediation. Some improvements are used in a limited fashion whilst many are at the research stage (and may remain so) but most involve relatively small changes to the operation of existing remediation methods rather than entirely new technologies. Several types of potential improvement are described.

Control of fluid flow or amendment delivery is key in *in situ* remediation, but as we have seen this is not always a simple task. The phenomena that arise when an electric field is applied to saturated or partially saturated ground can move soluble or suspended materials, including contaminants and bacteria, and can do so in all soil types, fine and coarse grained. This gives an advantage over hydraulic head as a delivery mechanism and so these so-called *electrokinetic* phenomena have been applied in various ways to enhance bioremediation (Gill *et al.*, 2014; Wick *et al.*, 2007), without harming or possibly even stimulating microorganisms (Lear *et al.*, 2004, 2007). The simplest application is delivery of electron acceptors or nutrients to a subsurface treatment zone (e.g. Elektorowicz and Boeva, 1996). Alternatively, movement of both contaminants and bacteria may contribute to enhanced bioavailability (Harbottle *et al.*, 2009). Recently, one of the first field applications of this technique at a site in Denmark demonstrated enhanced dechlorination of tetrachloroethene through delivery of lactate as an electron donor (Cox *et al.*, 2015).

The availability of electron acceptors (such as oxygen) or donors are often key limitations to successful bioremediation, hence the need for technologies such as biosparging. An alternative is to use an electrical circuit to supply or remove electrons, so that an electrode replaces the need to deliver chemical amendments. These microbial fuel cells are proposed for electricity generation at low levels, but can also be used in bioremediation (Lovley, 2006; Morris and Jin, 2012).

Soil additives such as organic matter are commonly used in bioremediation, but the application of novel materials, often recycled or some form of waste product, continue to be explored. In particular, biochar is considered as a soil amendment to improve aeration and fertility as well as a possible carbon sink. It also can demonstrate a high sorptive capacity for a range of chemical species, metal and organic, and support a considerable microbial community, due in part to a large surface area. Its ability to sorb metals whilst supporting soil microorganisms has led to biochar being identified as a potentially useful soil amendment in mixed contaminated soils (Sneath *et al.*, 2013).

Microbial treatment of heavy metal and radionuclide contamination may either take the form of contaminant mobilization (for ease of removal) or immobilization (for sequestration). The former has been explored at the research stage for contaminated land (e.g. White *et al.*, 1998) but application is limited. It is, however, well known as an industrial tool for leaching of valuable minerals (Rawlings and Johnson, 2007). Application in bioremediation may be limited to *ex situ* work where any fugitive leachate may more easily be captured. Biological immobilization occurs through the processes of uptake or mineralization. Accumulation of metals in bacteria, fungi and higher plants and animals is quite well understood and application occurs particularly through phytoremediation. By encouraging microorganisms to change the pH or the redox conditions in the soil, for example, contaminants can be made more or less soluble and therefore more or less available (e.g. Lee and Saunders, 2003). Biomineralization, the formation of minerals through microbial means which sequesters contaminants of interest, has been applied to radionuclide contamination (Fujita *et al.*, 2010) and there has been considerable exploration in the laboratory of various mechanisms for heavy metal pollutants (Barkay and Schaefer, 2001). The process could be applied *in situ* or *ex situ* although may have more applicability in the former situation where such contaminants may otherwise be difficult to treat – delivery mechanisms will require subsurface fluid flow and will be based upon those used for EISB. It is also known as *in situ* biocontainment – because the process is one of containment rather than source removal. Concerns remain over the durability of such techniques given that the contaminant-holding mineral could break down at some point in the future.

## Concluding remarks

The wide range of pollution or waste treatment scenarios that exist warrant the variety of different bioremediation technologies to tackle different media, contaminants and settings. From an engineering point of view, the implementation of these methods is relatively straightforward, but the complexity of the materials to which they are applied means that success is not guaranteed unless this is taken into account. The problems of heterogeneity at a range of scales dominate the challenges faced, particularly with *in situ* remediation. The dominance of the established methods described in this chapter is unlikely to significantly change although enhancements to bring about improved bioavailability and overcome heterogeneity *in situ* may make *in situ* remediation a more reliable, cost-effective and therefore acceptable choice in an increasing number of situations.

## References


Aleer, S., Adetutu, E.M., Weber, J., Ball, A.S., and Juhasz, A.L. (2014). Potential impact of soil microbial heterogeneity on the persistence of hydrocarbons in contaminated subsurface soils. J. Environ. Manage. *136*, 27–36.

Alexander, M. (1999). Biodegradation and Biomediation, 2nd edn (Academic Press, San Diego, CA, USA).

Antizar-Ladislao, B., Beck, A.J., Spanova, K., Lopez-Real, J., and Russell, N.J. (2007). The influence of different temperature programmes on the bioremediation of polycyclic aromatic hydrocarbons (PAHs) in a coal-tar contaminated soil by in-vessel composting. J. Hazard. Mater. *144*, 340–347.

Atlas, R.M. (1991). Microbial hydrocarbon degradation – bioremediation of oil-spills. J. Chem. Technol. Biotechnol. *52*, 149–156.

Baker, K.L., Langenheder, S., Nicol, G.W., Ricketts, D., Killham, K., Campbell, C.D., and Prosser, J.I. (2009). Environmental and spatial characterisation of bacterial community composition in soil to inform sampling strategies. Soil Biol. Biochem. *41*, 2292–2298.

Barkay, T., and Schaefer, J. (2001). Metal and radionuclide bioremediation: issues, considerations and potentials. Curr. Opin. Microbiol. *4*, 318–323.

Baveye, P., Vandevivere, P., Hoyle, B.L., DeLeo, P.C., and de Lozada, D.S. (1998). Environmental impact and mechanisms of the biological clogging of saturated soils and aquifer materials. Crit. Rev. Env. Sci. Tech. *28*, 123–191.

Brusseau, M.L. (1994). Transport of reactive contaminants in heterogeneous porous-media. Rev. Geophys. *32*, 285–313.

Chau, J.F., Bagtzoglou, A.C., and Willig, M.R. (2011). The effect of soil texture on richness and diversity of bacterial communities. Environ. Forensics *12*, 333–341.

Cookson, J.T. (1995). Bioremediation Engineering: Design and Application (McGraw-Hill, New York, NY, USA).

Cox, E., Wang, J., Durant, N., Reynolds, D., and Gent, D. (2015). Electrokinetically enhanced remediation – an innovative solution for source area remediation. In 13th International UFZ-Deltares Conference on Sustainable Use and Management of Soil, Sediment and Water Resources (Copenhagen, Denmark).

Davis, C., Cort, T., Dai, D.P., Illangasekare, T.H., and Munakata-Marr, J. (2003). Effects of heterogeneity and experimental scale on the biodegradation of diesel. Biodegradation *14*, 373–384.

Elektorowicz, M., and Boeva, V. (1996). Electrokinetic supply of nutrients in soil bioremediation. Environ. Technol. *17*, 1339–1349.

Englert, A., Hubbard, S.S., Williams, K.H., Li, L., and Steefel, C.I. (2009). Feedbacks between hydrological heterogeneity and bioremediation induced biogeochemical transformations. Environ. Sci. Technol. *43*, 5197–5204.

Frascari, D., Zanaroli, G., and Danko, A.S. (2015). In situ aerobic cometabolism of chlorinated solvents: A review. J. Hazard. Mater. *283*, 382–399.

Fujita, Y., Taylor, J.L. Wendt, L.M., Reed, D.W., and Smith, R.W. (2010). Evaluting the potential of native ureolytic microbes to remediate a Sr-90 contaminated environment. Environ. Sci. Technol. *44*, 7652–7658.

Ganey, P.E., and Boyd, S.A. (2005). An approach to evaluation of the effect of bioremediation on biological activity of environmental contaminants: Dechlorination of polychlorinated biphenyls. Environ. Health Persp. *113*, 180–185.

Gibson, D.T., and Sayler, G.S. (1992). Scientific Foundations of Bioremediation – Current Status and Future Needs (The American Academy of Microbiology).

Gill, R.T., Harbottle, M.J., Smith, J.W.N., and Thornton, S.F. (2014). Electrokinetic-enhanced bioremediation of organic contaminants: A review of processes and environmental applications. Chemosphere *107*, 31–42.

Gregoire, C., Elsaesser, D., Huguenot, D., Lange, J., Lebeau, T., Merli, A., Mose, R., Passeport, E., Payraudeau, S., Schutz, T., *et al.* (2009). Mitigation of agricultural nonpoint-source pesticide pollution in artificial wetland ecosystems. Environ. Chem. Lett. *7*, 205–231.

Harbottle, M.J., Al-Tabbaa, A., and Evans, C.W. (2008). Sustainability of land remediation. Part 1: overall analysis. P. I. Civil Eng-Geotec. *161*, 75–92.

Harbottle, M.J., Lear, G., Sills, G.C., and Thompson, I.P. (2009). Enhanced biodegradation of pentachlorophenol in unsaturated soil using reversed field electrokinetics. J. Environ. Manage. *90*, 1893–1900.

Hellings, J., Lass, M., Apted, J., and Mead, I. (2011). Delivering London 2012: geotechnical enabling works. P. I. Civil Eng-Civ. En. *164*, 5–10.

Hohener, P., Hunkeler, D., Hess, A., Bregnard, T., and Zeyer, J. (1998). Methodology for the evaluation of engineered *in situ* bioremediation: lessons from a case study. J. Microbiol. Methods *32*, 179–192.

Keely, J.F. (1989). Performance Evaluations of Pump-and-Treat Remediations (United States Environmental Protection Agency Office of Research and Development/Office of Solid Waste & Emergency Response, Washington, DC, USA).

Kelce, W.R., Stone, C.R., Laws, S.C., Gray, L.E., Kemppainen, J.A., and Wilson, E.M. (1995). Persistent DDT metabolite p,p'-DDE is a potent androgen receptor antagonist. Nature *375*, 581–585.

Komlos, J., Cunningham, A.B., Camper, A.K., and Sharp, R.R. (2004). Biofilm barriers to contain and degrade dissolved trichloroethylene. Environ. Prog. *23*, 69–77.

Lear, G., Harbottle, M.J., van der Gast, C.J., Jackman, S.A., Knowles, C.J., Sills, G., and Thompson, I.P. (2004). The effect of electrokinetics on soil microbial communities. Soil Biol. Biochem. *36*, 1751–1760.

Lear, G., Harbottle, M.J., Sills, G., Knowles, C.J., Semple, K.T., and Thompson, I.P. (2007). Impact of electrokinetic remediation on microbial communities within PCP contaminated soil. Environ. Pollut. *146*, 139–146.

Lee, M.K., and Saunders, J.A. (2003). Effects of pH on metals precipitation and sorption: field bioremediation and geochemical modeling approaches. Vadose Zone J. *2*, 177–185.

Lin, T.C., Pan, P.T., Young, C.C., Chang, J.S., Chang, T.C., and Cheng, S.S. (2011). Evaluation of the optimal strategy for *ex situ* bioremediation of diesel oil-contaminated soil. Environ. Sci. Pollut. Res. *18*, 1487–1496.

Lovley, D.R. (2006). Bug juice: harvesting electricity with microorganisms. Nat. Rev. Microbiol. *4*, 497–508.

Mackay, D.M., and Cherry, J.A. (1989). Groundwater contamination – pump-and-treat remediation. Environ. Sci. Technol. *23*, 630–636.

McMahon, V., Garg, A., Aldred, D., Hobbs, G., Smith, R., and Tothill, I.E. (2009). Evaluation of the potential of applying composting/bioremediation techniques to wastes generated within the construction industry. Waste Manage. *29*, 186–196.

Morris, J.M., and Jin, S. (2012). Enhanced biodegradation of hydrocarbon-contaminated sediments using microbial fuel cells. J. Hazard. Mater. *213*, 474–477.

Murphy, C., Wallace, S., Knight, R., Cooper, D., and Sellers, T. (2015). Treatment performance of an aerated constructed wetland treating glycol from de-icing operations at a UK airport. Ecol. Eng. *80*, 117–124.

Ortega-Calvo, J.J., Tejeda-Agredano, M.C., Jimenez-Sanchez, C., Congiu, E., Sungthong, R., Niqui-Arroyo, J.L., and Cantos, M. (2013). Is it possible to increase bioavailability but not environmental risk of PAHs in bioremediation? J. Hazard. Mater. *261*, 733–745.

Page, K., Harbottle, M.J., Cleall, P.J., and Hutchings, T.R. (2014). Heavy metal leaching and environmental risk from the use of compost-like output as an energy crop growth substrate. Sci. Total Environ. *487*, 260–271.

Pandey, J., Chauhan, A., and Jain, R.K. (2009). Integrative approaches for assessing the ecological sustainability of *in situ* bioremediation. FEMS Microbiol. Rev. 33, 324–375.

Pelaez, A.I., Lores, I., Sotres, A., Mendez-Garcia, C., Fernandez-Velarde, C., Santos, J.A., Gallego, J.L.R., and Sanchez, J. (2013). Design and field-scale implementation of an 'on site' bioremediation treatment in PAH-polluted soil. Environ. Pollut. *181*, 190–199.

Phillips, D.H., Van Nooten, T., Bastiaens, L., Russell, M.I., Dickson, K., Plant, S., Ahad, J.M.E., Newton, T., Elliot, T., and Kalin, R.M. (2010). Ten year performance evaluation of a field-scale zero-valent iron permeable reactive barrier installed to remediate trichloroethene contaminated groundwater. Environ. Sci. Technol. *44*, 3861–3869.

Piskonen, R., and Itavaara, M. (2004). Evaluation of chemical pretreatment of contaminated soil for improved PAH bioremediation. Appl. Microbiol. Biotechnol. *65*, 627–634.

Rawlings, D.E., and Johnson, D.B. (2007). The microbiology of biomining: development and optimization of mineral-oxidizing microbial consortia. Microbiol-Sgm *153*, 315–324.

Rebata-Landa, V., and Santamarina, J.C. (2006). Mechanical limits to microbial activity in deep sediments. Geochem. Geophy. Geosy. *7*, Q11006.

Rivett, M.O., and Thornton, S.F. (2008). Monitored natural attenuation of organic contaminants in groundwater: principles and application. P. I. Civil Eng.-Wat. M. *161*, 381–392.

Robles-Gonzalez, I.V., Fava, F., and Poggi-Varaldo, H.M. (2008). A review on slurry bioreactors for bioremediation of soils and sediments. Microb. Cell Fact. *7*, 5.

Santini, T.C., Kerr, J.L., and Warren, L.A. (2015). Microbially-driven strategies for bioremediation of bauxite residue. J. Hazard. Mater. *293*, 131–157.

Scherr, K., Aichberger, H., Braun, R., and Loibner, A.P. (2007). Influence of soil fractions on microbial degradation behavior of mineral hydrocarbons. Eur. J. Soil Biol. *43*, 341–350.

Semple, K.T., Doick, K.J., Jones, K.C., Burauel, P., Craven, A., and Harms, H. (2004). Peer reviewed: defining bioavailability and bioaccessibility of contaminated soil and sediment is complicated. Environ. Sci. Technol. *38*, 228A–231A.

Sen, T.K., and Khilar, K.C. (2006). Review on subsurface colloids and colloid-associated contaminant transport in saturated porous media. Adv. Colloid Interfac. *119*, 71–96.

Sethunathan, N., Megharaj, M., Smith, L., Kamaludeen, S.P.B., Avudainayagam, S., and Naidu, R. (2005). Microbial role in the failure of natural attenuation of chromium(VI) in long-term tannery waste contaminated soil. Agr. Ecosyst. Environ. *105*, 657–661.

Sneath, H.E., Hutchings, T.R., and de Leij, F.A.A.M. (2013). Assessment of biochar and iron filing amendments for the remediation of a metal, arsenic and phenanthrene co-contaminated spoil. Environ. Pollut. *178*, 361–366.

Soares, A.A., Pinho, M.T., Albergaria, J.T., Domingues, V., Alvim-Ferraz, M.D.M., De Marco, P., and Delerue-Matos, C. (2012). Sequential application of soil vapor extraction and bioremediation processes for the remediation of ethylbenzene-contaminated soils. Water Air Soil Pollut. *223*, 2601–2609.

Song, X., and Seagren, E.A. (2008). In situ bioremediation in heterogeneous porous media: Dispersion-limited scenario. Environ. Sci. Technol. *42*, 6131–6140.

Stroud, J.L., Paton, G.I., and Semple, K.T. (2007). Microbe-aliphatic hydrocarbon interactions in soil: implications for biodegradation and bioremediation. J. Appl. Microbiol. *102*, 1239–1253.

Sturman, P.J., Stewart, P.S., Cunningham, A.B., Bouwer, E.J., and Wolfram, J.H. (1995). Engineering scale-up of *in situ* bioremediation processes – a review. J. Contam. Hydrol. *19*, 171–203.

Summersgill, M. (2006). Remediation Technology Costs in the UK & Europe; Drivers and Changes from 2001 to 2005. In 5th International Congress on Environmental Geotechnics, H.R. Thomas, ed. (Thomas Telford, Cardiff, UK), pp. 310–317.

Sutton, N.B., van Gaans, P., Langenhoff, A.A.M., Maphosa, F., Smidt, H., Grotenhuis, T., and Rijnaarts, H.H.M. (2013). Biodegradation of aged diesel in diverse soil matrixes: impact of environmental conditions and bioavailability on microbial remediation capacity. Biodegradation *24*, 487–498.

Taylor, S.W., and Jaffe, P.R. (1991). Enhanced Insitu Biodegradation and Aquifer Permeability Reduction. J. Environ. Eng-Asce *117*, 25–46.

Thavamani, P., Malik, S., Beer, M., Megharaj, M., and Naidu, R. (2012). Microbial activity and diversity in long-term mixed contaminated soils with respect to polyaromatic hydrocarbons and heavy metals. J. Environ. Manage. *99*, 10–17.

Tomei, M.C., and Daugulis, A.J. (2013). Ex situ bioremediation of contaminated soils: an overview of conventional and innovative technologies. Crit. Rev. Env. Sci. Tech. *43*, 2107–2139.

Tyagi, M., da Fonseca, M.M.R., and de Carvalho, C.C.C.R. (2011). Bioaugmentation and biostimulation strategies to improve the effectiveness of bioremediation processes. Biodegradation *22*, 231–241.

USEPA (1996). Pump-and-Treat Ground-Water Remediation. A Guide for Decision Makers and Practitioners (United States Environmental Protection Agency Office of Research and Development, Washington, DC, USA).

van der Gast, C.J., Whiteley, A.S., and Thompson, I.P. (2004). Temporal dynamics and degradation activity of an bacterial inoculum for treating waste metal-working fluid. Environ. Microbiol. *6*, 254–263.

Van Stempvoort, D., and Biggar, K. (2008). Potential for bioremediation of petroleum hydrocarbons in groundwater under cold climate conditions: A review. Cold Reg. Sci. Technol. *53*, 16–41.

Verce, M.F., Madrid, V.M., Gregory, S.D., Demir, Z., Singleton, M.J., Salazar, E.P., Jackson, P.J., Halden, R.U., and Verce, A. (2015). A long-term field study of in situ bioremediation in a fractured conglomerate trichloroethene source zone. Bioremediation J. *19*, 18–31.

Vogel, T.M. (1996). Bioaugmentation as a soil bioremediation approach. Curr. Opin. Biotechnol. *7*, 311–316.

White, C., Sharman, A.K., and Gadd, G.M. (1998). An integrated microbial process for the bioremediation of soil contaminated with toxic metals. Nat. Biotechnol. *16*, 572–575.

Whitehead, P., and Neal, C. (2005). Bioremediation of acid mine drainage: the Wheal Jane Mine wetlands project. Sci. Total Environ. *338*, 1–1.

Wick, L.Y., Shi, L., and Harms, H. (2007). Electro-bioremediation of hydrophobic organic soil-contaminants: A review of fundamental interactions. Electrochim. Acta *52*, 3441–3448.

Wong, J., Lim, C.H., and Nolen, G. (1997). Design of remediation systems (CRC, Boca Raton, FL, USA).

Zhang, C.L., Hughes, J.B., Nishino, S.F., and Spain, J.C. (2000). Slurry-phase biological treatment of 2,4-dinitrotoluene and 2,6-dinitrotoluene: Role of bioaugmentation and effects of high dinitrotoluene concentrations. Environ. Sci. Technol. *34*, 2810–2816.

# The Biofilm Concept from a Bioremediation Perspective

# 2

Benjamin Horemans, Pieter Albers and Dirk Springael

### Abstract

Biofilms have been extensively studied since they were identified as the primary growth mode of microbial life. In the clinical world and food processing industries, biofilms are considered a threat to human health, but biofilms also have beneficial properties as they are deployed in waste recycling and water treatment. Biofilm-based bioremediation systems such as biofilters, aerobic and anaerobic granular sludge reactors and rotating disk contactors are widely used nowadays. As with most microbial-based technologies, the creation of a robust and reliable biofilm-based remediation technology remains challenging. For this reason, an in-depth understanding of biofilm formation and of the specific processes that occur inside biofilms, is mandatory to improve bioremediation. In this chapter, we discuss the nature and role of biofilms from a bioremediation perspective. We first address the basics of biofilm formation and how this should be considered when using biofilms for bioremediation. The main stages in the biofilm life cycle, i.e. attachment, aggregation, biofilm maturation and dispersal are discussed. Afterwards, we address the role of biofilm properties such as the associated matrix of extracellular polymeric substances and spatial heterogeneity, and elaborate on important processes within the biofilm such as microbial interactions and gene exchange. Many of these properties and processes are unique for biofilms and provide creative opportunities to improve biofilm-based bioremediation.

## Introduction to biofilms

In 1978, Bill Costerton described microbial attachment and growth on surfaces embedded in a slimy layer as an important mode-of-life for bacteria. Since then, research has shown that biofilms are the predominant mode of microbial growth in various natural, industrial and clinical settings (Costerton *et al.*, 1978). Biofilms can be observed in a wide variety of natural environments including rivers and streams, as well as on man-made materials such as plastics and metal pipelines (Hall-Stoodley *et al.*, 2004). Various generic definitions of biofilms have been proposed (Costerton *et al.*, 1995; Davey and O'Toole, 2000; O'Toole *et al.*, 2000). Today biofilms are broadly defined as collections of microbial cells that are aggregated to themselves (flocs, sludge) or attached to (a)biotic surfaces at solid–liquid interfaces where they are commonly embedded in a strongly hydrated matrix of extracellular polymeric substances (EPS).

Biofilms in natural settings show great diversity harbouring microbial communities made up of all microbial prokaryotes and eukaryotes (bacteria, archaea, algae, yeasts, fungi and protists (Ley *et al.*, 2006; Schmeisser *et al.*, 2003). It is perhaps unsurprising therefore that biofilms display great functional diversity and elasticity in the biogeochemical cycling of nutrients including carbon, nitrogen and phosphorus. In the past few decades, this functionality has been exploited for the treatment of industrial, municipal and agricultural waste water streams and for the clean-up of contaminated natural water bodies such as in groundwater aquifers. Pollution such as the complex organic wastes, metals, nutrients and recalcitrant compounds with suspected toxicity are treated using reactor designs including trickling filters, rotating biological contactors, submerged bed biofilm reactors, membrane bioreactors, and fluidized bed reactors (as discussed in Chapter 1). Specific examples in waste water treatment are anaerobic and aerobic granular sludge technology (Bathe *et al.*, 2005), autotrophic nitrogen removal by rotating disk contactors (Pynaert *et al.*, 2003) and on-farm biopurification systems for the treatment of pesticide contaminated waste water at farms (Sniegowski *et al.*, 2011a).

Biofilms provide an ideal environment for microbial cells to adapt, survive and cooperate and offer protection during periods of stress, making biofilm-based bioremediation more robust and reliable compared to systems using only planktonic cells (Edwards and Kjellerup, 2013). Biofilms show great chemical heterogeneity compared to planktonic cultures. Plasmids carrying catabolic genes encoding for pollutant-degrading enzymes can be exchanged within the biofilm and different metabolic processes can take place simultaneously (Brune and Bayer, 2012). Moreover, biofilms provide an ideal environment for multi-species interactions (Elias and Banin, 2012; Little *et al.*, 2008). Owing to their complexity, studying multi-species biofilm communities in the environment is challenging but in the past years, valuable insights have been made in understanding both synergistic and antagonistic interactions in biofilms involved in bioremediation and the advent of new omics approaches are promising for future research on multispecies biofilm communities. In this introductory chapter, we elucidate the key features of biofilms and biofilm ecology, placing a strong emphasis on their relevance to biofilm-based bioremediation techniques. In addition, we will discuss how a profound understanding of biofilm ecology can contribute to the design or development of more robust and reliable biofilm-based bioremediation approaches.

## Biofilm formation

The biofilm life cycle consists of five distinct stages: (1) free-living planktonic state, (2) reversible and irreversible attachment, (3) aggregation and microcolony formation, (4) macrocolony development and maturation, (5) detachment and (re)dispersal (Fig. 2.1). The impact and relevance of the different stages of the biofilm life cycle have many implications for biofilm-based bioremediation.

### Reversible and irreversible attachment

Biofilm formation starts with the initial attachment of microbial cells to a surface. This stage happens in two phases: reversible and irreversible attachment. The initial reversible attachment is influenced by

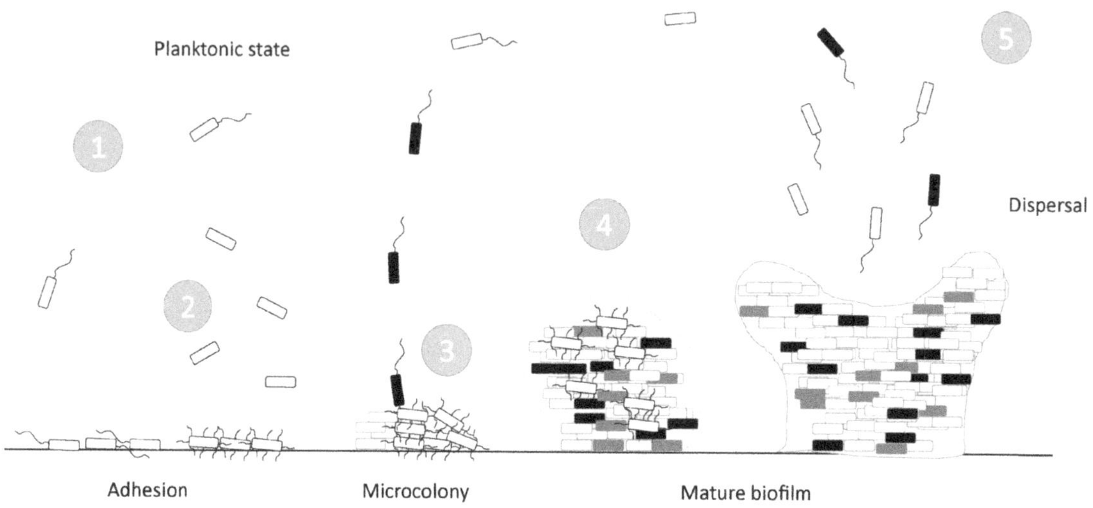


**Figure 2.1** Biofilm formation. 1. Pioneers: microbial cells collide with surfaces and reversibly attach. 2. Cells anchor to the surface by EPS, flagella and pili. 3. Microcolony formation through cell division, surface translocation and the attachment of new cells. 4. Microcolonies merge and expand to form macrocolonies which are encapsulated by a hydrated EPS matrix. 5. Degradation of EPS and cell lysis facilitates dispersal of planktonic cells from the mature biofilm. Cells disperse by swimming and swarming or by convection to find new environments to colonize.

both the cell surface properties of the microbial cells and the physicochemical characteristics of the (a)biotic surface. Once planktonic cells collide with a surface or other cells, they can adhere through a combination of attracting forces between the cell's surface and the surface of the abiotic/biotic substratum. The 'extended' Derjaguin–Landau–Verwey–Overbeek (DLVO) theory describes colloidal stability and is used to explain or predict cell adhesion qualitatively and quantitatively (Hermansson, 1999). The 'classical' DLVO theory explains bacterial adhesion as a combination of Lifschit–van der Waals interactions and electrostatic double layer interactions, which occur at medium to long range (> 100 nm). DLVO was later on extended with a term that includes hydrophilic/hydrophobic and osmotic interactions, which are close range (< 20 nm) (Van Oss, 1989). Osmotic interactions are neglected for microbial cells however (Hermansson, 1999).

Microbial cell surfaces have a negative charge at a neutral pH (Rijnaarts *et al.*, 1999) and this charge varies between microbial species (Matsumoto *et al.*, 2012). Since most (a)biotic substrata are also negatively charged, there will be an electrostatic repulsion between a cell and a substratum. The surface charge can be influenced by the ionic strength and the pH of the bulk liquid. At high ionic strength, ions are available to neutralize the negative charge and will facilitate cell attachment (van der Wal *et al.*, 1997). The closer the pH is to the isoelectric point of the cell, the better cells attach to a substratum (Husmark and Rönner, 1990).

Close range hydrophobic interactions between substratum and cell surface result from the strong hydrogen bonds between surrounding water molecules which forces hydrophobic groups together (Van Oss, 1991). The hydrophobic/hydrophilic nature of the cell surface is largely determined by the content of lipopolysaccharides in the cell membrane (de Weger *et al.*, 1989) and EPS positioned on the outside of the cell surface such as proteins, polysaccharides (Tsuneda *et al.*, 2003) and extracellular DNA (eDNA) (Regina *et al.*, 2014).

These initial binding forces are relatively weak and can be easily disturbed by shear forces invoked by fluid flow resulting in the loss of the cells into the bulk liquid. For biofilm formation, adhesion needs to become stronger and irreversible through specific cell-surface interactions. This involves a switch from a motile life style to a sessile life style which is triggered by sensing the surface, a process called mechanosensing. For a cell to sense surfaces, flagella serve as mechanosensors, although other mechanisms exist (Belas, 2014). Attached cells increase the production of EPS which is either embedded in the cell membrane or released into the bulk liquid (Vandevivere and Kirchman, 1993). Several extracellular biopolymers such as polysaccharides, proteins, DNA and amphiphilic molecules irreversibly anchor the cells to the surface (Flemming and Wingender, 2010). Also organelles such as fimbriae and pili covered with adhesins, improve attachment to surfaces (Axner *et al.*, 2011). For example, pili formation by *Bradyrhizobium japonicum* is observed to enhance cell attachment on biotic surfaces such as soybean roots (Vesper and Bauer, 1986).

Fluid flow is a major controlling factor in adhesion. A higher fluid flow results in an increased deposition of cells onto a surface but also invokes higher shear forces which hampers attachment or leads to detachment. Boks *et al.* (2008) distinguishes two critical shear rates. The first is the critical shear force to prevent adhesion and the second to detach adhering cells. However, hydrodynamic forces can alter cell surface characteristics. For instance, the fimbrial tip of *E. coli* is normally a rigid structure, but breaks open at high shear shaping the tip into a hook which allows anchoring to the surface at high fluid flow (Aprikian *et al.*, 2011). Also the numerous pili used by a cell to attach to a surface, are believed to evenly distribute the shear force exerted on the cell (Axner *et al.*, 2011).

Cell adhesion is or can be actively affected by surface modifications. Surface roughness and micro-topography can provide more hydrodynamic protection, increase cell attachment and improve biofilm formation. Nanoscale modifications of a glass substratum drastically increased the number of attached cells for different bacterial genera (Mitik-Dineva *et al.*, 2008). Hydrophilic materials such as plastics were more easily colonized compared to hydrophobic materials such as glass. Surfaces can also be 'primed' with a conditioning film to facilitate or hamper attachment. For instance, organic and inorganic molecules from the bulk liquid phase can form a conditioning film onto the surface, thereby altering surface properties such as hydrophobicity, surface charge and binding sites (Bar-Zeev *et al.*, 2012). Moreover, this layer can provide a source of nutrients to the cells, enhancing attachment and growth. In addition, conditioning of the substratum surface can be due to EPS left behind by microbial cells that were removed from the surface by shear forces, resulting in improved attachment of other cells. Otherwise, the role of so-called 'pioneer' bacteria have been suggested in facilitating the attachment of other bacteria. These primary colonizers adhere first to the surface to which other strains can successively attach (Rickard *et al.*, 2003) a process designated as co-aggregation. As will be discussed later in this chapter (see 'Social interactions between biofilm members', below), co-inoculation of strong aggregators with biodegrading strains can improve settlement and result in more successful bioaugmentation. Since the surface properties of substrata and of microbial cells can determine the success of biofilm formation and hence the performance of the engineered system, careful selection of the carrier or support material is important in the engineering of biofilm-based bioremediation systems. For instance, several fibrous materials such as carbon fibre, aromatic polyamide, preoxidized polyacrylonitrile and polyethylene, were tested as a biofilm support for a diverse range of bacterial strains for waste water treatment (Matsumoto *et al.*, 2012). The authors illustrated that the DLVO-theory predicted which fibrous material facilitated cell attachment. Adhesion to carbon fibres was feasible due to the lower negative zeta potential which allowed cells to surpass the energy barrier for adhesion, while for the other materials this energy barrier could not be overcome. If there is limited choice in carrier or support material, co-aggregation or surface conditioning can be a solution to facilitate adhesion of a target micro-organism.

## From microcolonies to macrocolonies

Microcolonies are small aggregates of surface-attached cells. Cells can cluster by migrating towards each other onto the surface by twitching and gliding motility as reported for pseudomonads (O'Toole and Kolter, 1998) or simply by cell division of individual attached cells (De Kievit *et al.*, 2001). However, gliding and twitching motility occurs mainly under static and low shear conditions. At higher fluid flow, surface translocation appears hampered and microcolony formation is primarily due to cell division (De Kievit *et al.*, 2001). Macrocolonies originate from microcolonies and are distinguishable by the EPS matrix which encapsulates a greater number of cells. For several freshwater bacteria, EPS is known to be important for aggregation of daughter cells within a surface microcolony. Mutants deficient in EPS production did attach but did not establish microcolonies (Allison and Sutherland, 1987). Macrocolonies also display greater cell differentiation. The number of cells in macrocolonies can be further increased by the surface translocation of entire cell clusters (e.g. S-type gliding motility), internal cell multiplication, the attachment of planktonic cells directly onto the colonies and the merging of microcolonies.

When macro-colonies are formed, biofilms reach maturation. Mature biofilms come in all shapes such as monolayers of cells (Horemans *et al.*, 2014), flat multilayer microbial mats of variable thickness (Reitner, 2011) and mushroom-shaped structures or streamers with interspersed water channels (Breugelmans *et al.*, 2008; Horemans *et al.*, 2014). Biofilm morphology depends in the first place on the environmental conditions. Hydrodynamics determine structure, mass transfer, EPS production and metabolism (Liu and Tay, 2002) and therefore take a prominent role in shaping biofilm morphology. Fluid hydrodynamics and diffusion control solute transport and therefore impact biofilm formation. A faster fluid flow around biofilms results in a greater influx of nutrient and energy sources from the bulk liquid. Cell growth is stimulated and biofilm biovolume increases. The thickness of the biofilm is controlled by the diffusion of solutes into the biofilm. Higher fluid flow, though, also involves greater shear stress and impacts morphology as biofilms change their viscoelastic properties and shape to resist shear stress (Stoodley *et al.*, 1999). Liu and Tay (2002) reported that dense biofilms, including aerobic and anaerobic granules, are formed only at higher hydrodynamic shear forces in reactors. Biofilm morphology is also affected by the availability of nutrients. For instance, the amount and the biodegradability of various sources of natural organic carbon can alter the biofilm structure and composition of pesticide-degrading biofilms (Horemans *et al.*, 2013b,c). Biofilm architecture and structure is further influenced by the resident microbial community composition. For instance, mono-species biofilms of Pseudomonads are in general mushroom shaped. Some species form a flat mat of cells under certain conditions, while others do not (Heydorn *et al.*, 2002). Biofilms can also be shaped to resist predation. A study on biofilm predation showed that the feeding behaviour of several protists resulted in either the formation of microcolonies or a change in biofilm porosity and surface area (Böhme *et al.*, 2009).

## Detachment and dispersal

From any point in the biofilm life cycle, cells can detach or disperse from the biofilm to colonize various and potentially more suitable micro-environments. This process is important as it guarantees the survival of species by allowing escape from conditions in which the biofilm is no longer suited and facilitating the colonization of new locations. Cells or cell clusters (clumps) can detach from the biofilm under unfavourable hydrodynamics of the bulk liquid. Depending on the viscoelasticity of the biofilms, shear stress can cause individual cells (erosion) or clusters of cells (sloughing) to break apart from the biofilm. The viscoelasticity of biofilms changes in response to hydrodynamics (Rupp *et al.*, 2005), depends on the microbial species in the biofilm, nutrient availability and can be altered by pH and surfactants in the bulk liquid (Rühs *et al.*, 2013). Also, active dispersion of cells from biofilms occurs and is triggered by environmental conditions such as nutrient starvation (Huynh *et al.*, 2012), nutrient excess (Sauer *et al.*, 2004), oxygen deprivation and oxidative and nitrosative stress (e.g. nitric oxide production under anaerobic conditions) (Barraud *et al.*, 2006) as well as by bacteriophage predation (Kunin *et al.*, 2008; Siringan *et al.*, 2011). Extracellular enzymes such as hydrolases and lyases which degrade EPS (Boyd and Chakrabarty, 1994), and surfactants such as rhamnolipids which reduce surface tension (Boles *et al.*, 2005) are produced to dissolve the surrounding biofilm matrix and facilitate cell detachment. Even lysis of subpopulations of sessile cells in the macrocolony has been reported, a process that provides nutrients for the planktonic cells (Webb *et al.*, 2003). For dispersion, cells within the mature biofilm undergo a physiological change as genes involved in the sessile growth mode (EPS formation, pili, etc.) are down-regulated and genes encoding for motility (flagella, chemotaxis, etc.) are up-regulated. This change from the sessile life style to the planktonic one and vice versa is in many bacterial species controlled by the universal bacterial second messenger cyclic dimeric-guanosine monophosphate (di-GMP) (Römling *et al.*, 2013).

After leaving the biofilm and adopting a planktonic state, the cells move through the bulk liquid or along wet surfaces in a passive or active manner and disperse to more favourable microenvironments. Passive movement occurs by random Brownian motion caused by the constant collision with particles and molecules in the liquid or by gravitation and convection of the bulk liquid. Active movement involves flagellar motion, A-type and S-type gliding motility and twitching motility. The flagellum allows movement through the bulk liquid, but is also involved in surface translocation. Other typical mechanisms for surface translocation are A-type and S-type gliding motility and twitching motility. They have been identified as an important mechanism of translocation in a wide variety of phylogenetic unrelated species. In contrast to flagellar motion, for gliding and twitching motility cells need to have contact with surfaces. A-type gliding motility occurs through movement by using various motor elements that are lined-up over the whole cell surface. These motor elements in combination with the adhesion sites on the outer membrane allows a gliding movement of cells over surfaces. S-type gliding motility and twitching motility use type IV pili. The difference between S-type gliding motility and twitching motility is that the latter is rather chaotic and jerky while the first occurs in a smooth and organized fashion. S-type gliding motility occurs when many cells are in close proximity and is perceived as a coordinated swarming of a group of cells along surfaces.

In some cases, this active movement is coupled with moving towards nutrients or away from toxic substances (chemotaxis), towards light (phototaxis) or along magnetic lines (magnetotaxis). Chemotaxis to various organic pollutants has been demonstrated such as for benzene, toluene, ethylbenzene and xylene (or BTEX) pollutants, 2,4-D and polyaromatic hydrocarbons (PAHs). Ortega-Calvo *et al.* (2003) suggested that chemotaxis towards PAHs can improve their bioavailability. Chemotaxis also enables bacteria to position themselves in non-toxic locations or form biofilms on the pollutant itself such as is the case for hydrocarbons (Klein *et al.*, 2010). As such, motility is an important bacterial characteristic to be considered in the remediation of heterogeneous pollutions such as PAHs in soil (Ortega-Calvo *et al.*, 2003), and dense non-aqueous phase liquids such as trichloroethene (Parales *et al.*, 2000) or hydrocarbons in sea water after oil spills (Mason *et al.*, 2012).

### The biofilm matrix

Biofilms are essentially built up of live and dead cells embedded in a matrix of extracellular polymeric substances (EPS). EPS are microbially produced biopolymers such as polysaccharides, proteins, glycoproteins, glycolipids and extracellular DNA and are for the largest part produced by the biofilm cells themselves. The EPS composition depends on the microorganisms that produce them and on the local conditions such as shear stress, temperature and nutrient availability. As such, the composition of EPS varies strongly between biofilms and within the same biofilm community over time. The EPS matrix directly affects mechanical stability, hydrodynamic properties such as porosity, density and water holding capacity and physico-chemical properties such as charge, sorption sites and hydrophobicity of the cells (Flemming and Wingender, 2010). The retention of nutrients, extracellular enzymes and lysed cell contents by EPS in combination with the close proximity of cells within the matrix, create optimal conditions for metabolic cooperation and improved cell-to-cell communication. This makes the biofilm a highly metabolically active entity. Major biopolymers found in EPS matrix are listed in Table 2.1 together with their functionality in the biofilm.

All biological material, such as biofilms, has the ability to sorb metals, radionuclides and organics to some extent. This phenomenon is called biosorption and is a non-metabolic, physico-chemical process acting through mechanisms such as absorption, adsorption, surface complexation and precipitation (Fomina and Gadd, 2014). Biosorption was postulated as a bioremediation technique for detoxifying various types of polluted waters. For instance, biofilms grown on activated carbon particles have been used to sorb Cr(VI) from polluted water (Lameiras *et al.*, 2008) and several heavy metal ions were observed to sorb to naturally formed biofilms on stones in a river (Ancion *et al.*, 2013; Kurniawan and Yamamoto, 2013). Several functional groups such as acetamide groups of chitin, structural polysaccharides, amino and phosphate groups of nucleic acids, amide, amine, sulfhydryl and carboxyl groups in proteins have the capacity to sequester metals (Ahluwalia and Goyal, 2006) including heavy metals (Mohanty *et al.*, 2006 and Pradhan *et al.*, 2007). However, while metals are predominantly sorbed onto microbial cells, organic pollutants sorb onto EPS (Späth *et al.*, 1998). Sorption onto biofilms involving hydrophobic interactions can also play a role in organic pollutant removal (Dong *et al.*, 2011; Headley *et al.*, 1998; Späth *et al.*, 1998; Writer *et al.*, 2011) and even improve their biodegradation in the biofilm (Wolfaardt *et al.*, 1995).

## Biofilm, a hotspot for horizontal gene transfer

Horizontal gene transfer (HGT) enables prokaryotes to exchange genetic information between members of the same species and eventually between members of different families and genera and is considered to be a driving force for evolution in prokaryotes. HGT can occur through transformation, conjugation and transduction.

In natural **transformation**, exogenous DNA is taken up by competent cells and then integrated into the genome. Biofilms might be hot spots for transformation since transformation rates were shown to be higher in biofilms compared to planktonic cells (Li *et al.*, 2001). The success of transformation depends on the amount of free DNA and the occurrence of competent bacterial cells in the biofilm. In monoculture biofilms of *Acinetobacter* sp. strain BD413, the transformation frequency was highest in actively growing biofilms exposed to a high amount of DNA (Hendrickx *et al.*, 2003). Interestingly, induction of natural competence in *Streptococcus pneumoniae* triggered the lysis and release of DNA from a subpopulation of cells albeit in a planktonic culture (Steinmoen *et al.*, 2002). Apart from the availability of DNA needed for transformation, also the competence of cells is a determining factor. The competence of cells in the natural environment depends on the growth stage and (partial) nutrient limitations. For instance, in the presence of soil extracts amended with phosphate, nitrogen and pyruvate, transformation of *Pseudomonas stutzeri* increased drastically when one of the added growth factors was limited (Lorenz and Wackernagel, 1991).

**Table 2.1** Major EPS constituents and their role in biofilms categorized by function. Adapted from Flemming and Wingender (2010)

| Function | Relevance for biofilms | EPS components |
|---|---|---|
| Adhesion | Initial colonization and long-term attachment on surfaces | Polysaccharides, proteins, DNA and amphiphilic molecules |
| Aggregation | Temporal immobilization, cell–cell recognition, cell bridging | Polysaccharides, proteins and DNA |
| Cohesion | Mechanical stability, biofilm architecture and cell–cell communication | Neutral and charged polysaccharides, proteins (lectins and amyloids), DNA |
| Water retention | Creates a hydrated environment and as such protects against desiccation | Hydrophilic polysaccharides, proteins |
| Protective barrier | Protection against antimicrobial agents, oxygen and grazing protozoa | Polysaccharides and proteins |
| Sorption of organic compounds | Accumulation of nutrients and bioaccumulation of xenobiotics | Charged or hydrophobic polysaccharides and proteins |
| Sorption of inorganic ions | Promotes polysaccharide gel formation, ion exchange, mineral formation and the accumulation of toxic metal ions | Charged polysaccharides and proteins including inorganic substituents such as phosphate and sulphate |
| Enzymatic activity | Digestion of exogenous macromolecules, degradation of structural EPS allowing cell dispersal | Proteins |
| Nutrient source | Source of carbon, nitrogen and phosphorus for biofilms cells | All EPS components |
| Exchange of genetic information | Facilitated horizontal gene transfer | DNA |
| Electron donor or acceptor | Redox activity in the biofilm | Proteins (pili and nanowires), humic substances |
| Export of cell components | Release of cellular material for metabolic turnover | Membrane vesicles containing nucleic acids, enzymes, lipopolysaccharides and phospholipids |
| Sink for excess energy | Storage of carbon excess | Polysaccharides |
| Binding of enzymes | Accumulation, retention, stabilization of enzymes through the interaction with polysaccharides | Polysaccharides and enzymes |

In **conjugation**, DNA is exchanged between cells through direct cell to cell contact. Formation of pili and plasmids play an essential role. In biofilms, cells meeting other cells occurs with lower frequency compared to planktonic cultures and hence it can be expected that HGT occurs at a lower frequency. However, the biofilm matrix appears to provide a stable and fixed environment which improves HGT (Hausner and Wuertz, 1999). It was observed that in a suspended culture of *E. coli*, plasmids were distributed gradually at a low rate until all cells were transconjugants. A sessile culture showed, however, a higher transfer rate but with only a limited amount of transconjugant cells (Licht *et al.*, 1999). In general, it is concluded that plasmid transfer occurs at higher efficiency in natural biofilms but also in biofilms in bioreactors and lab-scale systems (Beaudoin *et al.*, 1998; Björklöf *et al.*, 2000; Geisenberger *et al.*, 1999; Normander *et al.*, 1998; Roberts *et al.*, 1999; Springael *et al.*, 2002a). Conjugation transfer rates of plasmids from *E. coli* to *Cupriavidus metallidurans* was found to be independent of the nutrient conditions in biofilms (Hausner and Wuertz, 1999).

The last mechanism, **transduction**, involves bacteriophage-mediated DNA transfer and is recognized as an important mechanism of HGT in natural biofilms (Miller, 2001). Phage genes are highly up-regulated during biofilm development within species belonging to both Gram-positive and Gram-negative bacteria (Stanley *et al.*, 2003; Whiteley *et al.*, 2001). Webb *et al.* (2004) even observed that the filamentous phage Pf4 led to phenotypic variation within *P. aeruginosa* biofilms. Pf4-infected *P. aeruginosa* cells showed a high density of Pf4 filaments on the cell surface which enabled better attachment to surfaces and accelerated biofilm development.

Regarding bioremediation, HGT is of great importance for the distribution of catabolic genes and catabolic pathway evolution within a microbial community (Springael and Top, 2004). Catabolism of recalcitrant organic pollutants is often encoded by mobile elements like plasmids and as such biofilms might be hot spots for sharing catabolic genes by HGT and the expansion of catabolic pathways towards new recalcitrant pollutants within a community. Furthermore, the idea to use strains carrying catabolic mobile genetic elements (MGEs) for organic pollutant degradation in bioaugmentation to remediate contaminated soil, water, etc., has gained interest (Hendrickx *et al.*, 2003; Springael and Top, 2004; Top *et al.*, 2002). In that strategy, the survival of the inoculum strain is not mandatory since the catabolic genetic information can be distributed to resident bacteria (DiGiovanni *et al.*, 1996; Springael *et al.*, 2002b). For instance, Springael *et al.* (2002b) showed the out-competition of a 3-chlorobenzoate (3-CB)-degrading *Pseudomonas putida* strain used to inoculate a membrane-biofilm reactor treating 3-CB-contaminated water, by other 3-CB-mineralizing bacteria. These 3-CB-mineralizing bacteria had acquired the genes encoding 3-CB biodegradation originally carried by the inoculum strain (Springael *et al.*, 2002a).

## Chemical, physiological and genetic heterogeneity in biofilms

Biofilms show great chemical heterogeneity in concentrations and types of nutrients, microbially produced intermediates and signalling molecules as a result of diffusion processes and metabolic cell activity. These gradients create microniches inside biofilms and result in the diversification in the physiological state of cells, the acquisition of new genetic traits by affecting transformation competence and differences in local community composition. In turn, the physiological state of cells and the occurrence of different species creates micro-environments with new chemical signatures.

Planktonic cultures are generally grown in well-mixed, homogeneous conditions. Because of mixing, the major mode of solute transport is convection rather than the slower diffusion process. Diffusion occurs only in the very thin boundary layer around individual microbial cells. Diffusion is therefore rapid and can be neglected when considering the overall transport rate (Stewart, 2003). Convection of the bulk liquid brings solutes close to the outer layer of the biofilm or even inside the biofilm through channels. However, diffusion can no longer be neglected in biofilms. First, the boundary layer around a biofilm is thicker compared to planktonic cells and as such solutes need to diffuse over a longer distance to reach the outer cell layer in the biofilm. Once inside the biofilm matrix, solutes need to diffuse even further to reach cells deeper in the biofilm. Convection is limited in the biofilm matrix and as such diffusion becomes the main mode of solute transport within the biofilm matrix and impacts reaction rates of embedded biofilm cells (Stewart, 2003). By consuming substrates and releasing metabolites, cells influence their direct environment within the biofilm. Substrates in the bulk liquid will diffuse into the biofilm and be gradually consumed until being depleted. The outer cell layer of biofilms is therefore provided with the highest concentration of available substrate, while in the inner part of biofilms the substrate is depleted. On the other hand, the concentration of microbial produced intermediates or products will be highest within the biofilm and lowest in the bulk liquid (Stewart and Franklin, 2008). This local variation in nutrient and metabolite concentration creates micro-environments within a biofilm allowing different metabolic processes to occur simultaneously. For instance, oxygen penetrates only the upper layers of biofilms leaving the inside of the biofilm anoxic as observed in bacterial biofilm communities associated with suspended particles of organic and inorganic material in the marine environment, referred to as marine snow. Marine snow biofilms are involved in the degradation of particulate organic carbon in the marine pelagic environment. Anaerobic processes, like methanogenesis, nitrogen fixation and sulfide production have been detected in these particles despite the oxygenated bulk environment (Davey and O'Toole, 2000).

Antibacterial agents can be responsible for cell differentiation and a change in the physiological state of cells within a biofilm. For instance, in

*Pseudomonas aeruginosa* biofilms, a motile (type IV pili) population resistant to a membrane-targeting antibacterial agent formed a protective layer around a non-motile, non-resistant population (Haagensen *et al.*, 2007). Production of toxic product inside the biofilm such as nitric oxide, an increasing pH and oxidative stress can also induce dispersal from the biofilm. Bacteria can, furthermore, induce competitive and cooperative physiological responses in biofilms through quorum sensing (Zhang *et al.*, 2011) and physical contact (Shimoyama *et al.*, 2009). As such, the chemical environment inside the biofilm is created by gradients of local concentrations of nutrients, carbon and energy sources or anti-bacterial agents and signalling molecules, that diffuse throughout the biofilm and whose local concentrations depend on the (metabolic) activity of the micro-organisms residing inside the biofilm.

The chemical heterogeneity of the micro-environment in a biofilm further affects the genetic and phenotypic heterogeneity of a biofilm population. Genetic variability originates from adaptive mutations (Bjedov *et al.*, 2003; Oliver *et al.*, 2000; Taddei *et al.*, 1995), phase variation (Drenkard and Ausubel, 2002) and HGT (Hausner and Wuertz, 1999). As already described, HGT is an important and common mode for microorganisms to exchange genetic information and for microbial populations to evolve. The ability of *Cupriavidus metallidurans* cells to conjugate was shown to depend on the specific microenvironment inside the biofilm (Hausner and Wuertz, 1999). Similarly, the competence of *Acinetobacter* cells within biofilms depended on their position within the biofilm and transformation frequencies differed between locations (Hendrickx *et al.*, 2003). As such, chemical heterogeneity in biofilms can lead to genetic heterogeneity even among bacterial cells of the same species in biofilms.

## Social interactions between biofilm members

In natural biofilms, microbial communities often exist of various microbial species. The micro-organisms in such communities can interact in a beneficial or antagonistic way. While cooperation between micro-organisms often results in co-localization within a biofilm, antagonism results in separate growth (Molin *et al.*, 2004). These interactions differ, however, spatially inside the biofilm and can change as they are influenced by various factors, including the production of antibacterial agents, quorum sensing, motility, (co)aggregation, metabolic requirements, growth efficiency and environmental conditions (Moons *et al.*, 2009). Any change in one or more of these factors can dramatically impact the structure and dynamics of the biofilm community composition (Elias and Banin, 2012). Beneficial interactions are commonly known as symbiotic interactions and include mutualism and commensalism (Little *et al.*, 2008). Antagonistic interactions include the exploitative competition for nutrients and space and interference competition by inhibiting or killing organisms (Little *et al.*, 2008). Both beneficial interactions and competitive interactions can be crucial for the success of a bioremediation project and thus should be considered when choosing a bioremediation strategy.

## Synergy and its benefits for bioremediation

Many environmental processes are performed through a joint effort of different microbial species within a community. Together they achieve a result that is not attainable by the individual species (synergy). Degradation of organic xenobiotics in a complete and efficient manner without the formation of harmful metabolites is often due to the combined catabolic activities of and synergistic interactions between community members (Jacques *et al.*, 2008). Examples of synergism found in consortia involved in bioremediation are illustrated in Fig. 2.2 (mechanisms 1–4). During synergistic degradation, both partners (mutualism) or only one partner (commensalism) benefit from the association. Biofilms might provide an ideal environment for the evolution of efficient cooperation and synergistic dependencies (Bryant *et al.*, 1967). An obvious and important synergistic mechanism is syntrophy or cross-feeding in which metabolites and nutrients are exchanged between micro-organisms. The close proximity between cooperating micro-organisms optimizes syntrophic interactions as it facilitates the exchange of nutrients and metabolites (Costerton *et al.*, 1995; Davey and O'Toole, 2000). Therefore, syntrophic partners often co-localize in cell clusters in biofilms (Breugelmans *et al.*, 2008). Syntrophy was shown

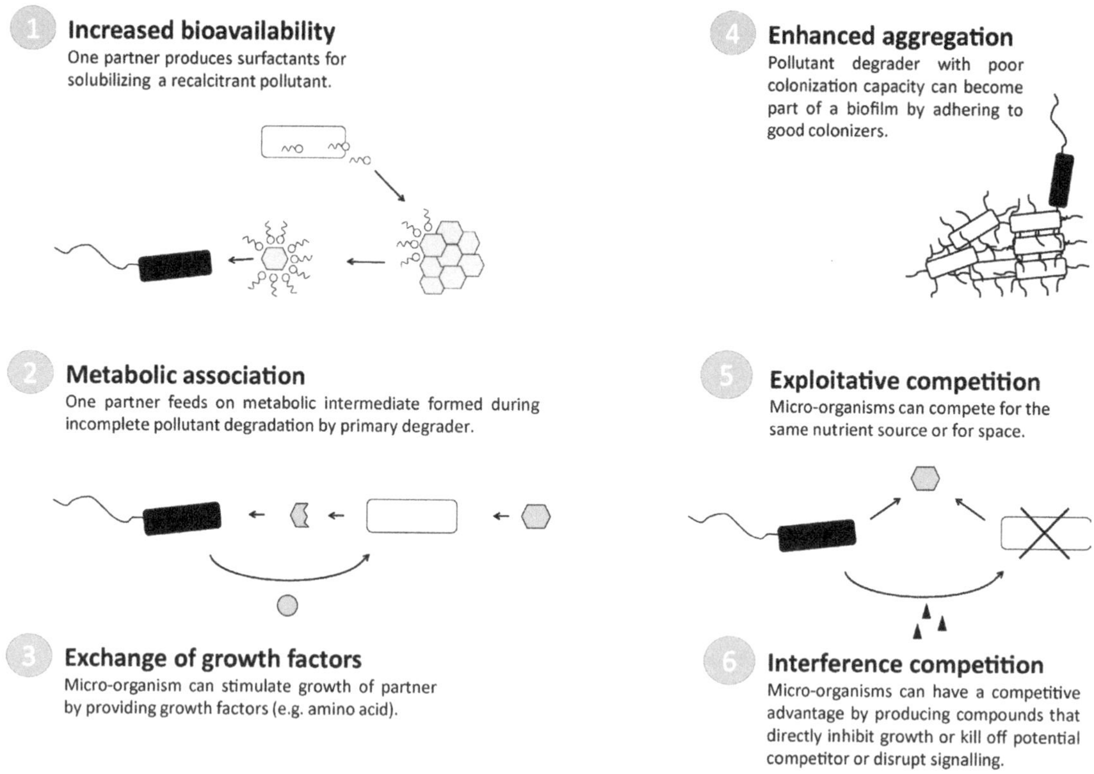


**Figure 2.2** Synergistic (1–4) and antagonistic (5–6) mechanisms that can take place between different species in biofilm communities.

for important microbial processes like methanogenesis (Schink, 1997), nitrification (Prosser *et al.*, 1990) and biodegradation of organic xenobiotics (Bradley, 2003; Feigel and Knackmuss, 1993; Katsuyama *et al.*, 2009). The most studied form of syntrophy is associated metabolism (Fig. 2.2, mechanism 2), in which the consortium members feed on metabolic pathway intermediates formed from incomplete pollutant degradation by other members (Sorensen *et al.*, 2002). Associated metabolism was shown for consortia degrading herbicides (Dejonghe *et al.*, 2003; Sørensen *et al.*, 2002), insecticides (Gilbert *et al.*, 2003; Katsuyama *et al.*, 2009), chlorobiphenyls (Nielsen *et al.*, 2000), crude oil hydrocarbons (Kasai *et al.*, 2005) and PAHs (Boonchan *et al.*, 2000; Kanaly *et al.*, 2002). Accumulation of metabolic pathway intermediates can be inhibitory for its producer and reduce overall pollutant degradation activity. In associated metabolism, cross-feeding between partners allows the continuous removal of the intermediate, contributing to efficient and complete degradation of the pollutant. For example, efficient $H_2$ transfer between community members is essential in the functioning of methanogenic communities in anaerobic digestion (see Box 2.1) and in bacterial organohalide respiration of halogenated pollutants. When the intermediate is an acid or a base, associated metabolism can prevent the acidification or alkalization of the environment as reported for degradation of the herbicide atrazine (de Souza *et al.*, 1998) and for the degradation of PAHs in wastewater (Qi *et al.*, 2007).

Other pollutant-degrading consortia include partners that are not directly involved in the metabolism of the pollutant, but indirectly improve biodegradation for instance by providing growth factors for the degraders (Fig. 2.2, mechanism 3) such as amino acids in the case of degradation of the phenylurea herbicide isoproturon (Sorensen *et al.*, 2002) and vitamins in the case of chlorobenzoate biodegradation (Holliger, 1999).

Aggregation and colonization capabilities of consortium partners can also enhance the performance of a bioremediation system (Fig. 2.2, mechanism 4). Yoshida *et al.* (2009) reported that

**Box 2.1 Synergistic methanogenesis through interspecies hydrogen transfer**

Methanogenesis is an important terminal electron-accepting process taking place during the fermentation of complex organic matter in anoxic environments. During methanogenesis a synergistic interaction between syntrophic fermentative bacteria and methanogenic bacteria/archaea is critical for processes like the anaerobic digestion of organic wastewater (Imachi *et al.*, 2000) and degradation of hydrocarbons in petrol contaminated fields (Kasai *et al.*, 2005). The degradation of the fermentation intermediates (ethanol, acetate, butyrate, alanine, propionate, etc.) is endergonic in standard, pure conditions and their conversion becomes only thermodynamically feasible when the $H_2$ partial pressure is kept sufficiently low (Ishii *et al.*, 2005; Overmann, 2006). In this example, these syntrophic bacteria can only grow when they are in close physical contact with $H_2$-consuming methanogens to ensure an efficient transfer of $H_2$. In anaerobic granules, this is achieved by organization of the community in a layered structure where fermentation occurs in the outer layers and methanogenesis in the inner layer of the granules (Satoh *et al.*, 2007). Since both partners benefit from interspecies hydrogen transfer, this association is mutualistic and clearly shows the importance of associated metabolism. Interspecies hydrogen transfer is also crucial in reductive dehalogenation of chlorinated pollutants by tight interactions between fermentative populations that produce hydrogen and organo halide respiring organisms that use the hydrogen for dehalogenation (Smidt and de Vos, 2004).

3-chlorobenzoate degradation by *Burkholderia* sp. NK8 was more efficient in a dual species biofilm than in monocultured biofilms because the presence of the second strain, i.e. *P. aeruginosa* PAO1 improved settlement of *Burkholderia* sp. NK8 on a solid surface. Gilbert *et al.* (2003) reported that efficient colonization of a flow cell surface by one strain promoted the adhesion of a parathion degrading strain and the development of a mixed-species biofilm. Bacterial strains that a have poor biofilm forming capacity in monocultures biofilms, can become well established and even dominant members of multi-species biofilms (Andersson *et al.*, 2008; Li *et al.*, 2008). In engineered systems such as bioreactors, introducing aggregators can drastically improve performance. Enhanced aerobic granulation and concomitant phenol degradation was observed when combining phenol-degraders with strong aggregators (Adav *et al.*, 2008) while Von Canstein *et al.* (2002) reported that mixing mercury reducers with strong biofilm formers prevented wash out in packed-bed bioreactors treating mercury polluted wastewater.

Consortium partners can further improve biodegradation by increasing the bioavailability of hydrophobic organic pollutants by the production of surfactants (Fig. 2.2, mechanism 1). Kanaly *et al.* (2002) documented that *Rhodanobacter* sp. BPC1 grows on metabolites within a benzo[*a*] pyrene mineralizing consortium and contributes to mineralization by solubilizing benzo[*a*]pyrene. Bioavailability of a pollutant can also be increased by enhanced mobilization of the microbial degraders. Enhanced bacterial degradation of PAHs was suggested to be caused by an improvement of the intrinsic motility of the bacteria by means of continuous (physiological) liquid films forming around fungal hyphae (Kohlmeier *et al.*, 2005; Wick *et al.*, 2007). Pollutant-degrading bacteria using these 'fungal highways' move up to 1 cm per day. This is particularly interesting in water-unsaturated soil where air pockets hinder flagella-driven motility (Kohlmeier *et al.*, 2005).

## Antagonistic interactions and implications for bioaugmentation

As mentioned above, competition between micro-organisms includes exploitative competition and interference competition (Little *et al.*, 2008). Exploitative competition for nutrients and space depends on several properties of the competing micro-organisms (Fig. 2.2, mechanism 5). Obviously, growth rate is an important factor during competition, but competition outcome is also dependent on nutrient concentration (Moons *et al.*, 2009). Motility and adhesion enable micro-organisms to move away from competitors, to efficiently colonize new sites and to rapidly occupy different adhesion sites avoiding colonization by other strains, a process called surface blanketing

(Rendueles and Ghigo, 2012). For instance, the co-immobilization of *Pseudomonas* sp. and *Rhodococcus* sp. on substratum used for augmentation of hydrocarbon-contaminated soil failed since *Rhodococcus* sp. cells outcompeted *Pseudomonas* sp. during the adhesion process on the porous matrix (Rivelli *et al.*, 2013). Interference competition involves the production of compounds that directly inhibit growth or kill off potential competitors (Fig. 2.2, mechanism 6). For example, soil-associated bacteria that produce phenazine derivatives have a strong growth inhibitory effect on *Phanerochaete chrysosporium*, a white rot fungus used for biodegradation of many recalcitrant pollutants (Radtke *et al.*, 1994). Next to growth inhibitors, micro-organisms produce a diverse range of microbial secondary metabolites inhibiting biofilm formation. Modes of action include blocking cell to cell communication and inhibition of adhesion (Rendueles and Ghigo, 2012).

In multispecies biofilms, competitive interactions can be less pronounced as compared to planktonic conditions thanks to the existence of microenvironments. Coexistence of two phenol degraders was not possible in activated sludge, but possible in aerobic granules due to the spatial isolation of the two competitors (Jiang *et al.*, 2007). Similarly, co-existence of two benzyl-alcohol degraders was not possible in a chemostat fed with benzyl-alcohol, whereas in biofilm the organisms established heterogeneous ecological niches enabling associated metabolism between the degraders (Christensen *et al.*, 2002). Furthermore, growth in biofilms can also reduce the permeability of inhibitory substances produced by competitors (Moons *et al.*, 2009). Since biofilm conditions allow coexistence of competitors, enrichment on solid support is used to isolate slow-growing degraders that otherwise are outcompeted in classical batch enrichment experiments (Breugelmans *et al.*, 2007; Verhagen *et al.*, 2011).

Competitive interactions also affect bioaugmentation, since bioaugmentation depends on the success of introducing a new strain in the endogenous community of the target environment. During bioaugmentation, the introduced strains occupy a metabolic niche that is usually not occupied by the endogenous community, i.e. the degradation of a pollutant. This increases the chance of successful invasion (Eisenhauer, 2013). Still, competitive factors are decisive for the outcome of bioaugmentation, often even more than abiotic factors (El Fantroussi and Agathos, 2005). To overcome ecological barriers, working with indigenous micro-organisms seems to be a successful bioaugmentation strategy (El Fantroussi and Agathos, 2005).

## Robustness and functional reliability of multispecies communities

Biological remediation techniques are often perceived as unpredictable and therefore as unreliable compared to physicochemical remediation techniques. Engineers are therefore challenged to develop robust and reliable bioremediation systems to cope with environmental perturbations. To achieve this, it is worth considering the diversity of the biodegrading community. A positive correlation between community diversity and robustness is not always found (Shade *et al.*, 2012), but in general it can be stated that the more diverse a community is, the more robust it will be against disturbances (Burmølle *et al.*, 2007). Moreover, multispecies communities are often more robust in biofilm conditions than in planktonic conditions.

First, multispecies communities are more robust against environmental fluctuations such as periods with limited availability of nutrient, carbon and energy sources (Brenner *et al.*, 2008). The combination of different species provides a greater diversity of metabolic processes and allows the exchange of metabolites within the community. For example, a herbicide-degrading triple-species consortium extends its synergy from the degradation of the herbicide to other natural carbon sources, which proves to be an interesting survival strategy during the absence of the herbicide as a carbon source (Horemans *et al.*, 2013b). On the other hand, the presence of other carbon sources often hinders biodegradation of the organic pollutant. In multispecies communities, this is, however, less pronounced than in monocultures as seen during degradation of the herbicide linuron when other carbon sources appear (Horemans *et al.*, 2013c; Horemans *et al.*, 2013a).

Multispecies communities are also more robust against toxic compounds. Sensitive strains profit from protection by more resistant populations because the latter detoxify the environment. This is more pronounced in biofilm conditions since the resistant populations can offer a more local protection through the creation of micro-environments of low toxicity (Moons *et al.*, 2009). For instance, a dual species biofilm was more resistant to shock loadings of 3,5-dinitrobenzoic acid than a monospecies biofilm of the 3,5-dinitrobenzoic acid degrader (Li *et al.*, 2008). Mixed species biofilms were also found to be more robust during mercury bioremediation than monospecies biofilms (Von Canstein *et al.*, 2002). Co-occurring pollutants *p*-cresol and 2-chloroethanol were more efficiently degraded by a dual species biofilm as the inhibitory effects of *p*-cresol on the 2-chloroethanol degrader were reduced by the surrounding *p*-cresol degrader (Cowan *et al.*, 2000). Multispecies biofilms can also reduce the permeation of toxic compounds into the biofilm because of interactions between matrix polymers of different species, making the EPS matrix more viscous (Burmolle *et al.*, 2006). Finally, multispecies communities are more robust against invasion by other 'unwanted' microbial species. This is because diverse communities exploit resources more efficiently, leaving less free niche space available to invaders (Eisenhauer *et al.*, 2013).

## Concluding remarks

Understanding the fundamental principles of biofilm formation can help us to develop more robust and reliable biofilm-based remediation techniques. The ability of micro-organisms to attach and form biofilms, or to flocculate and form granules in certain conditions, can mean the difference between failure or success of a bioremediation technique. Spatial heterogeneity and synergistic interactions between micro-organisms drastically increases the potential of biofilm-based remediation techniques. Complex waste streams and pollutants are tackled that would otherwise be impossible when relying on homogenous, planktonic systems with low ecological diversity. However, due to the complexity of multi-species biofilms or granules, investigating these systems is challenging and a profound knowledge should be acquired. This knowledge could prove invaluable for feedback purposes to improve and optimize biofilm-based remediation systems.

## References

Adav, S.S., Lee, D.J., and Lai, J.Y. (2008). Intergeneric coaggregation of strains isolated from phenol-degrading aerobic granules. Appl. Microbiol. Biotechnol. *79*, 657–661.

Allison, D.G., and Sutherland, I.W. (1987). The role of exopolysaccharides in adhesion of freshwater bacteria. J. Gen. Microbiol. *133*, 1319–1327.

Ancion, P.-Y., Lear, G., Dopheide, A., and Lewis, G.D. (2013). Metal concentrations in stream biofilm and sediments and their potential to explain biofilm microbial community structure. Environ. Pollut. *173*, 117–124.

Andersson, S., Kuttuva Rajarao, G., Land, C.J., and Dalhammar, G. (2008). Biofilm formation and interactions of bacterial strains found in wastewater treatment systems. FEMS Microbiol. Lett. *283*, 83–90.

Aprikian, P., Interlandi, G., Kidd, B.A., Le Trong, I., Tchesnokova, V., Yakovenko, O., Whitfield, M.J., Bullitt, E., Stenkamp, R.E., Thomas, W.E., *et al.* (2011). The bacterial fimbrial tip acts as a mechanical force sensor. PLoS Biol. *9*, e1000617.

Axner, O., Andersson, M., Björnham, O., Castelain, M., Klinth, J., Koutris, E., and Schedin, S. (2011). Assessing bacterial adhesion on an individual adhesin and single pili level using optical tweezers. In Bacterial Adhesion, D. Linke, and A. Goldman, eds. (Springer, Netherlands), pp. 301–313.

Bar-Zeev, E., Berman-Frank, I., Girshevitz, O., and Berman, T. (2012). Revised paradigm of aquatic biofilm formation facilitated by microgel transparent exopolymer particles. Proc. Natl. Acad. Sci. U.S.A. *109*, 9119–9124.

Barraud, N., Hassett, D.J., Hwang, S.-H., Rice, S.A., Kjelleberg, S., and Webb, J.S. (2006). Involvement of nitric oxide in biofilm dispersal of *Pseudomonas aeruginosa*. J. Bacteriol. *188*, 7344–7353.

Bathe, S., de Kreuk, M., McSwain, B., and Schwarzenbeck, N., eds. (2005). Aerobic Granular Sludge (IWA Publishing, London, UK).

Beaudoin, D.L., Bryers, J.D., Cunningham, A.B., and Peretti, S.W. (1998). Mobilization of broad host range plasmid from *Pseudomonas putida* to established biofilm of *Bacillus azotoformans*. II. Modeling. Biotechnol. Bioeng. *57*, 280–286.

Belas, R. (2014). Biofilms, flagella, and mechanosensing of surfaces by bacteria. Trends Microbiol. *22*, 517–527.

Bjedov, I., Tenaillon, O., Gérard, B., Souza, V., Denamur, E., Radman, M., Taddei, F., and Matic, I. (2003). Stress-induced mutagenesis in bacteria. Science *300*, 1404–1409.

Björklöf, K., Nurmiaho-Lassila, E.L., Klinger, N., Haahtela, K., and Romantschuk, M. (2000). Colonization strategies and conjugal gene transfer of inoculated *Pseudomonas syringae* on the leaf surface. J. Appl. Microbiol. *89*, 423–432.

Böhme, A., Risse-Buhl, U., and Küsel, K. (2009). Protists with different feeding modes change biofilm morphology. FEMS Microbiol. Ecol. *69*, 158–169.

Boks, N.P., Norde, W., van der Mei, H.C., and Busscher, H.J. (2008). Forces involved in bacterial adhesion to hydrophilic and hydrophobic surfaces. Microbiology *154*, 3122–3133.

Boles, B.R., Thoendel, M., and Singh, P.K. (2005). Rhamnolipids mediate detachment of *Pseudomonas aeruginosa* from biofilms. Mol. Microbiol. *57*, 1210–1223.

Boonchan, S., Britz, M.L., and Stanley, G.A. (2000). Degradation and mineralization of high-molecular-weight polycyclic aromatic hydrocarbons by defined fungal-bacterial cocultures. Appl. Environ. Microbiol. *66*, 1007–1019.

Boyd, A., and Chakrabarty, A.M. (1994). Role of alginate lyase in cell detachment of *Pseudomonas aeruginosa*. Appl Environ Microbiol *60*, 2355–2359.

Bradley, P.M. (2003). History and ecology of chloroethene biodegradation: A review. Bioremediat. J. *7*, 81–109.

Brenner, K., You, L., and Arnold, F.H. (2008). Engineering microbial consortia: a new frontier in synthetic biology. Trends Biotechnol. *26*, 483–489.

Breugelmans, P., D'Huys, P.J., De Mot, R., and Springael, D. (2007). Characterization of novel linuron-mineralizing bacterial consortia enriched from long-term linuron-treated agricultural soils. FEMS Microbiol. Ecol. *62*, 374–385.

Breugelmans, P., Barken, K.B., Tolker-Nielsen, T., Hofkens, J., Dejonghe, W., and Springael, D. (2008). Architecture and spatial organization in a triple-species bacterial biofilm synergistically degrading the phenylurea herbicide linuron. FEMS Microbiol. Ecol. *64*, 271–282.

Brune, K.D., and Bayer, T.S. (2012). Engineering microbial consortia to enhance biomining and bioremediation. Front. Microbiol. 3, 203.

Bryant, M.P., Wolin, E.A., Wolin, M.J., and Wolfe, R.S. (1967). *Methanobacillus omelianskii*, a symbiotic association of two species of bacteria. Arch. Microbiol. *59*, 20–31.

Burmolle, M., Webb, J.S., Rao, D., Hansen, L.H., Sorensen, S.J., and Kjelleberg, S. (2006). Enhanced biofilm formation and increased resistance to antimicrobial agents and bacterial invasion are caused by synergistic interactions in multispecies biofilms. Appl. Environ. Microbiol. *72*, 3916–3923.

Burmølle, M., Hansen, L., and Sørensen, S. (2007). Establishment and early succession of a multispecies biofilm composed of soil bacteria. Microb. Ecol. *54*, 352–362.

Christensen, B.B., Haagensen, J.A., Heydorn, A., and Molin, S. (2002). Metabolic commensalism and competition in a two-species microbial consortium. Appl. Environ. Microbiol. *68*, 2495–2502.

Costerton, J.W., Geesey, G., and Cheng, K. (1978). How bacteria stick. Sci. Am. *238*, 86–95.

Costerton, J.W., Lewandowski, D.E., Caldwell, D.E., Korber, D.R., and Lappin-Scott, H.M. (1995). Microbial biofilms. Annu. Rev. Microbiol. *49*, 711–145.

Cowan, S.E., Gilbert, E., Liepmann, D., and Keasling, J.D. (2000). Commensal interactions in a dual-species biofilm exposed to mixed organic compounds. Appl. Environ. Microbiol. *66*, 4481–4485.

Davey, M.E., and O'toole, G.A. (2000). Microbial biofilms: from ecology to molecular genetics. Microbiol. Mol. Biol. Rev. *64*, 847–867.

De Kievit, T.R., Gillis, R., Marx, S., Brown, C., and Iglewski, B.H. (2001). Quorum-sensing genes in *Pseudomonas aeruginosa* biofilms: Their role and expression patterns. Appl. Environ. Microbiol. *67*, 1865–1873.

de Souza, M.L., Newcombe, D., Alvey, S., Crowley, D.E., Hay, A., Sadowsky, M.J., and Wackett, L.P. (1998). Molecular basis of a bacterial consortium: interspecies catabolism of atrazine. Appl. Environ. Microbiol. *64*, 178–184.

de Weger, L.A., van Loosdrecht, M.C., Klaassen, H.E., and Lugtenberg, B. (1989). Mutational changes in physiochemical cell surface properties of plant-growth-stimulating *Pseudomonas* spp. do not influence the attachment properties of the cells. J. Bacteriol. *171*, 2756–2761.

Dejonghe, W., Berteloot, E., Goris, J., Boon, N., Crul, K., Maertens, S., Hofte, M., De Vos, P., Verstraete, W., and Top, E.M. (2003). Synergistic degradation of linuron by a bacterial consortium and isolation of a single linuron-degrading variovorax strain. Appl. Environ. Microbiol. *69*, 1532–1541.

DiGiovanni, G.D., Neilson, J.W., Pepper, I.L., and Sinclair, N.A. (1996). Gene transfer of *Alcaligenes eutrophus* JMP134 plasmid pJP4 to indigenous soil recipients. Appl. Environ. Microbiol. *62*, 2521–2526.

Dong, D., Guo, Z., Hua, X., Lan, Y., Zhou, J., Ding, X., and Qiao, Q. (2011). Sorption of DDTs on biofilms, suspended particles and river sediments: effects of heavy metals. Environ. Chem. Lett. *9*, 361–367.

Drenkard, E., and Ausubel, F.M. (2002). *Pseudomonas* biofilm formation and antibiotic resistance are linked to phenotypic variation. Nature *416*, 740–743.

Edwards, S.J., and Kjellerup, B.V. (2013). Applications of biofilms in bioremediation and biotransformation of persistent organic pollutants, pharmaceuticals/personal care products, and heavy metals. Appl. Microbiol. Biotechnol. *97*, 9909–9921.

Eisenhauer, N., Schulz, W., Scheu, S., and Jousset, A. (2013). Niche dimensionality links biodiversity and invasibility of microbial communities. Funct. Ecol. *27*, 282–288.

El Fantroussi, S., and Agathos, S.N. (2005). Is bioaugmentation a feasible strategy for pollutant removal and site remediation? Curr. Opin. Microbiol. *8*, 268–275.

Elias, S., and Banin, E. (2012). Multi-species biofilms: living with friendly neighbors. FEMS Microbiol. Rev. *36*, 990–1004.

Feigel, B.J., and Knackmuss, H.-J. (1993). Syntrophic interactions during degradation of 4-aminobenzenesulfonic acid by a two species bacterial culture. Arch. Microbiol. *159*, 124–130.

Flemming, H.-C., and Wingender, J. (2010). The biofilm matrix. Nat. Rev. Microbiol. *8*, 623–633.

Fomina, M., and Gadd, G.M. (2014). Biosorption: current perspectives on concept, definition and application. Bioresour. Technol. *160*, 3–14.

Geisenberger, O., Ammendola, A., Christensen, B.B., Molin, S.r., Schleifer, K.-H., and Eberl, L. (1999). Monitoring the conjugal transfer of plasmid RP4 in activated sludge and *in situ* identification of the transconjugants. FEMS Microbiol. Lett. *174*, 9–17.

Gilbert, E.S., Walker, A.W., and Keasling, J.D. (2003). A constructed microbial consortium for biodegradation of the organophosphorus insecticide parathion. Appl. Microbiol. Biotechnol. *61*, 77–81.

Haagensen, J.A.J., Klausen, M., Ernst, R.K., Miller, S.I., Folkesson, A., Tolker-Nielsen, T., and Molin, S. (2007). Differentiation and distribution of colistin- and sodium dodecyl sulfate-tolerant cells in *Pseudomonas aeruginosa* biofilms. J. Bacteriol. *189*, 28–37.

Hall-Stoodley, L., Costerton, J.W., and Stoodley, P. (2004). Bacterial biofilms: from the natural environment to infectious diseases. Nat. Rev. Microbiol. 2, 95–108.

Hausner, M., and Wuertz, S. (1999). High rates of conjugation in bacterial biofilms as determined by quantitative *in situ* analysis. Appl. Environ. Microbiol. 65, 3710–3713.

Headley, J.V., Gandrass, J., Kuballa, J., Peru, K.M., and Gong, Y. (1998). Rates of sorption and partitioning of contaminants in river biofilm. Environ. Sci. Technol. 32, 3968–3973.

Hendrickx, L., Hausner, M., and Wuertz, S. (2003). Natural genetic transformation in monoculture *Acinetobacter* sp. strain BD413 biofilms. Appl. Environ. Microbiol. *69*, 1721–1727.

Hermansson, M. (1999). The DLVO theory in microbial adhesion. Colloid. Surface. B *14*, 105–119.

Heydorn, A., Ersboll, B., Kato, J., Hentzer, M., Parsek, M.R., Tolker-Nielsen, T., Givskov, M., and Molin, S. (2002). Statistical analysis of *Pseudomonas aeruginosa* biofilm development: impact of mutations in genes involved in twitching motility, cell-to-cell signaling, and stationary-phase sigma factor expression. Appl. Environ. Microbiol. *68*, 2008–2017.

Horemans, B., Vandermaesen, J., Vanhaecke, L., Smolders, E., and Springael, D. (2013a). *Variovorax* sp.-mediated biodegradation of the phenyl urea herbicide linuron at micropollutant concentrations and effects of natural dissolved organic matter as supplementary carbon source. Appl. Microbiol. Biotechnol. *97*, 9837–9846.

Horemans, B., Breugelmans, P., Hofkens, J., Smolders, E., and Springael, D. (2013b). Environmental dissolved organic matter governs biofilm formation and subsequent linuron-degrading activity of a linuron-degrading bacterial consortium. Appl Environ Microbiol. 79, 4534–4542.

Horemans, B., Vandermaesen, J., Breugelmans, P., Hofkens, J., Smolders, E., and Springael, D. (2013c). The quantity and quality of dissolved organic matter as supplementary carbon source impacts the pesticide-degrading activity of a triple-species bacterial biofilm. Appl. Microbiol. Biotechnol. *98*, 43.

Horemans, B., Hofkens, J., Smolders, E., and Springael, D. (2014). Biofilm formation of a bacterial consortium on linuron at micropollutant concentrations in continuous flow chambers and the impact of dissolved organic matter. FEMS Microbiol. Ecol. *88*, 184–194.

Husmark, U., and Rönner, U. (1990). Forces involved in adhesion of *Bacillus cereus* spores to solid surfaces under different environmental conditions. J. Appl. Bacteriol. *69*, 557–562.

Huynh, T.T., McDougald, D., Klebensberger, J., Al Qarni, B., Barraud, N., Rice, S.A., Kjelleberg, S., and Schleheck, D. (2012). Glucose starvation-induced dispersal of *Pseudomonas aeruginosa* biofilms is cAMP and energy dependent. PLoS One 7, e42874.

Imachi, H., Sekiguchi, Y., Kamagata, Y., Ohashi, A., and Harada, H. (2000). Cultivation and *in situ* detection of a thermophilic bacterium capable of oxidizing propionate in syntrophic association with hydrogenotrophic methanogens in a thermophilic methanogenic granular sludge. Appl. Environ. Microbiol. 66, 3608–3615.

Ishii, S., Kosaka, T., Hori, K., Hotta, Y., and Watanabe, K. (2005). Coaggregation facilitates interspecies hydrogen transfer between Pelotomaculum thermopropionicum and Methanothermobacter thermautotrophicus. Appl. Environ. Microbiol. *71*, 7838–7845.

Jacques, R.J., Okeke, B.C., Bento, F.M., Teixeira, A.S., Peralba, M.C., and Camargo, F.A. (2008). Microbial consortium bioaugmentation of a polycyclic aromatic hydrocarbons contaminated soil. Bioresour. Technol. *99*, 2637–2643.

Jiang, H.L., Maszenan, A.M., and Tay, J.H. (2007). Bioaugmentation and coexistence of two functionally similar bacterial strains in aerobic granules. Appl. Microbiol. Biotechnol. 75, 1191–1200.

Kanaly, R.A., Harayama, S., and Watanabe, K. (2002). Rhodanobacter sp. strain BPC1 in a benzo[a]pyrene-mineralizing bacterial consortium. Appl. Environ. Microbiol. *68*, 5826–5833.

Kasai, Y., Takahata, Y., Hoaki, T., and Watanabe, K. (2005). Physiological and molecular characterization of a microbial community established in unsaturated, petroleum-contaminated soil. Environ. Microbiol. 7, 806–818.

Katsuyama, C., Nakaoka, S., Takeuchi, Y., Tago, K., Hayatsu, M., and Kato, K. (2009). Complementary cooperation between two syntrophic bacteria in pesticide degradation. J. Theor. Biol. *256*, 644–654.

Klein, B., Bouriat, P., Goulas, P., and Grimaud, R. (2010). Behavior of *Marinobacter hydrocarbonoclasticus* SP17 cells during initiation of biofilm formation at the alkane–water interface. Biotechnol. Bioeng. *105*, 461–468.

Kohlmeier, S., Smits, T.H., Ford, R.M., Keel, C., Harms, H., and Wick, L.Y. (2005). Taking the fungal highway: mobilization of pollutant-degrading bacteria by fungi. Environ. Sci. Technol. *39*, 4640–4646.

Kunin, V., He, S., Warnecke, F., Peterson, S.B., Garcia Martin, H., Haynes, M., Ivanova, N., Blackall, L.L., Breitbart, M., Rohwer, F., *et al.* (2008). A bacterial metapopulation adapts locally to phage predation despite global dispersal. Genome Res. *18*, 293–297.

Kurniawan, A., and Yamamoto, T. (2013). Biofilm polymer for biosorption of pollutant ions. Procedia Environ. Sci. *17*, 179–187.

Lameiras, S., Quintelas, C., and Tavares, T. (2008). Biosorption of Cr (VI) using a bacterial biofilm supported on granular activated carbon and on zeolite. Bioresour. Technol. 99, 801–806.

Ley, R.E., Harris, J.K., Wilcox, J., Spear, J.R., Miller, S.R., Bebout, B.M., Maresca, J.A., Bryant, D.A., Sogin, M.L., and Pace, N.R. (2006). Unexpected diversity and complexity of the Guerrero Negro hypersaline microbial mat. Appl. Environ. Microbiol. 72, 3685–3695.

Li, M., Peng, L., Ji, Z., Xu, J., and Li, S. (2008). Establishment and characterization of dual-species biofilms formed from a 3,5-dinitrobenzoic-degrading strain and bacteria with high biofilm-forming capabilities. FEMS Microbiol. Lett. *278*, 15–21.

Li, Y.-H., Lau, P.C.Y., Lee, J.H., Ellen, R.P., and Cvitkovitch, D.G. (2001). Natural genetic transformation of *Streptococcus mutans* growing in biofilms. J. Bacteriol. *183*, 897–908.

Licht, T.R., Christensen, B.B., Krogfelt, K.A., and Molin, S. (1999). Plasmid transfer in the animal intestine and other dynamic bacterial populations: the role of community structure and environment. Microbiology *145*, 2615–2622.

Little, A.E., Robinson, C.J., Peterson, S.B., Raffa, K.F., and Handelsman, J. (2008). Rules of engagement: interspecies interactions that regulate microbial communities. Annu. Rev. Microbiol. *62*, 375–401.

Liu, Y., and Tay, J.-H. (2002). The essential role of hydrodynamic shear force in the formation of biofilm and granular sludge. Water Res. *36*, 1653–1665.

Lorenz, M.G., and Wackernagel, W. (1991). High frequency of natural genetic transformation of *Pseudomonas stutzeri* in soil extract supplemented with a carbon/energy and phosphorus source. Appl. Environ. Microbiol. *57*, 1246–1251.

Mason, O.U., Hazen, T.C., Borglin, S., Chain, P.S.G., Dubinsky, E.A., Fortney, J.L., Han, J., Holman, H.-Y.N., Hultman, J., Lamendella, R., *et al.* (2012). Metagenome, metatranscriptome and single-cell sequencing reveal microbial response to Deepwater Horizon oil spill. ISME J. *6*, 1715–1727.

Matsumoto, S., Ohtaki, A., and Hori, K. (2012). Carbon fiber as an excellent support material for wastewater treatment biofilms. Environ. Sci. Technol. *46*, 10175–10181.

Miller, R. (2001). Environmental bacteriophage–host interactions: factors contribution to natural transduction. Antonie van Leeuwenhoek *79*, 141–147.

Mitik-Dineva, N., Wang, J., Stoddart, P.R., Crawford, R.J., and Ivanova, E.P. (2008). Nano-structured surfaces control bacterial attachment. Paper presented at: International Conference on Nanoscience and Nanotechnology, ICONN 2008.

Molin, S., Tolker-Nielsen, T., and Hansen, S.K. (2004). Microbial interactions in mixed-species biofilms. In Microbial Biofilms (ASM Press, Washington, DC, USA), pp. 206–221.

Moons, P., Michiels, C.W., and Aertsen, A. (2009). Bacterial interactions in biofilms. Crit. Rev. Microbiol. *35*, 157–168.

Nielsen, A.T., Tolker-Nielsen, T., Barken, K.B., and Molin, S. (2000). Role of commensal relationships on the spatial structure of a surface-attached microbial consortium. Environ. Microbiol. *2*, 59–68.

Normander, B., Christensen, B.B., Molin, S., and Kroer, N. (1998). Effect of bacterial distribution and activity on conjugal gene transfer on the phylloplane of the bush bean (*Phaseolus vulgaris*). Appl. Environ. Microbiol. *64*, 1902–1909.

O'Toole, G., Kaplan, H.B., and Kolter, R. (2000). Biofilm formation as microbial development. Annu. Rev. Microbiol. *54*, 49–79.

O'Toole, G.A., and Kolter, R. (1998). Flagellar and twitching motility are necessary for *Pseudomonas aeruginosa* biofilm development. Mol. Microbiol. *30*, 295–304.

Oliver, A., Cantón, R., Campo, P., Baquero, F., and Blázquez, J. (2000). High frequency of hypermutable *Pseudomonas aeruginosa* in cystic fibrosis lung infection. Science *288*, 1251–1253.

Ortega-Calvo, J.J., Marchenko, A.I., Vorobyov, A.V., and Borovick, R.V. (2003). Chemotaxis in polycyclic aromatic hydrocarbon-degrading bacteria isolated from coal-tar- and oil-polluted rhizospheres. FEMS Microbiol. Ecol. *44*, 373–381.

Overmann, J. (2006). Symbiosis between non-related bacteria in phototrophic consortia. Progr. Mol. Subcell. Biol. *41*, 21–37.

Parales, R.E., Ditty, J.L., and Harwood, C.S. (2000). Toluene-degrading bacteria are chemotactic towards the environmental pollutants benzene, toluene, and trichloroethylene. Appl. Environ. Microbiol. *66*, 4098–4104.

Prosser, J.I., Rose, A.H., and Tempest, D.W. (1990). Autotrophic nitrification in bacteria. In Advances in Microbial Physiology (Academic Press, Waltham, MA, USA), pp. 125–181.

Pynaert, K., Smets, B.F., Wyffels, S., Beheydt, D., Siciliano, S.D., and Verstraete, W. (2003). Characterization of an autotrophic nitrogen-removing biofilm from a highly loaded lab-scale rotating biological contactor. Appl. Environ. Microbiol. *69*, 3626–3635.

Qi, R., Yang, K., and Yu, Z.X. (2007). Treatment of coke plant wastewater by SND fixed biofilm hybrid system. J. Environ. Sci. (China) *19*, 153–159.

Radtke, C., Cook, W.S., and Anderson, A. (1994). Factors affecting antagonism of the growth of phanerochaete-chrysosporium by bacteria isolated from soils. Appl. Microbiol. Biotechnol. *41*, 274–280.

Regina, V.R., Lokanathan, A.R., Modrzyński, J.J., Sutherland, D.S., and Meyer, R.L. (2014). Surface physicochemistry and ionic strength affects eDNA's role in bacterial adhesion to abiotic surfaces. PLoS One *9*, e105033.

Reitner, J. (2011). Microbial mats. In Encyclopedia of Geobiology, J. Reitner, and V. Thiel, eds. (Springer, Netherlands), pp. 606–608.

Rendueles, O., and Ghigo, J.M. (2012). Multi-species biofilms: how to avoid unfriendly neighbors. FEMS Microbiol. Rev. *36*, 972–989.

Rickard, A.H., Gilbert, P., High, N.J., Kolenbrander, P.E., and Handley, P.S. (2003). Bacterial coaggregation: an integral process in the development of multi-species biofilms. Trends Microbiol. *11*, 94–100.

Rijnaarts, H.H.M., Norde, W., Lyklema, J., and Zehnder, A.J.B. (1999). DLVO and steric contributions to bacterial deposition in media of different ionic strengths. Colloid. Surface. B *14*, 179–195.

Rivelli, V., Franzetti, A., Gandolfi, I., Cordoni, S., and Bestetti, G. (2013). Persistence and degrading activity of free and immobilised allochthonous bacteria during bioremediation of hydrocarbon-contaminated soils. Biodegradation *24*, 1–11.

Roberts, A.P., Pratten, J., Wilson, M., and Mullany, P. (1999). Transfer of a conjugative transposon, Tn5397 in a model oral biofilm. FEMS Microbiol. Lett. *177*, 63–66.

Römling, U., Galperin, M.Y., and Gomelsky, M. (2013). Cyclic di-GMP: the first 25 years of a universal bacterial second messenger. Microbiol. Mol. Biol. Rev. *77*, 1–52.

Rühs, P.A., Böni, L., Fuller, G.G., Inglis, R.F., and Fischer, P. (2013). *In situ* quantification of the interfacial rheological response of bacterial biofilms to environmental stimuli. PLoS One *8*, e78524.

Rupp, C.J., Fux, C.A., and Stoodley, P. (2005). Viscoelasticity of *Staphylococcus aureus* biofilms in response to fluid shear allows resistance to detachment and facilitates rolling migration. Appl. Environ. Microbiol. *71*, 2175–2178.

Satoh, H., Miura, Y., Tsushima, I., and Okabe, S. (2007). Layered structure of bacterial and archaeal communities and their *in situ* activities in anaerobic granules. Appl. Environ. Microbiol. 73, 7300–7307.

Sauer, K., Cullen, M.C., Rickard, A.H., Zeef, L.A.H., Davies, D.G., and Gilbert, P. (2004). Characterization of nutrient-induced dispersion in *Pseudomonas aeruginosa* PAO1 biofilm. J. Bacteriol. *186*, 7312–7326.

Schink, B. (1997). Energetics of syntrophic cooperation in methanogenic degradation. Microbiol. Mol. Biol. Rev. *61*, 262–280.

Schmeisser, C., Stöckigt, C., Raasch, C., Wingender, J., Timmis, K.N., Wenderoth, D.F., Flemming, H.-C., Liesegang, H., Schmitz, R.A., Jaeger, K.-E., *et al.* (2003). Metagenome survey of biofilms in drinking-water networks. Appl. Environ. Microbiol. *69*, 7298–7309.

Shade, A., Peter, H., Allison, S.D., Baho, D.L., Berga, M., Burgmann, H., Huber, D.H., Langenheder, S., Lennon, J.T., Martiny, J.B., *et al.* (2012). Fundamentals of microbial community resistance and resilience. Front. Microbiol. 3, 417.

Shimoyama, T., Kato, S., Ishii, S., and Watanabe, K. (2009). Flagellum mediates symbiosis. Science *323*, 1574.

Siringan, P., Connerton, P.L., Payne, R.J.H., and Connerton, I.F. (2011). Bacteriophage-mediated dispersal of *Campylobacter jejuni* biofilms. Appl. Environ. Microbiol. *77*, 3320–3326.

Smidt, H., and de Vos, W.M. (2004). Anaerobic microbial dehalogenation. Annu. Rev. Microbiol. *58*, 43–73.

Sniegowski, K., Bers, K., Ryckeboer, J., Jaeken, P., Spanoghe, P., and Springael, D. (2011a). Robust linuron degradation in on-farm biopurification systems exposed to sequential environmental changes. Appl. Environ. Microbiol. 77, 6614–6621.

Sorensen, S.R., Ronen, Z., and Aamand, J. (2002). Growth in coculture stimulates metabolism of the phenylurea herbicide isoproturon by Sphingomonas sp. strain SRS2. Appl. Environ. Microbiol. *68*, 3478–3485.

Späth, R., Flemming, H.C., and Wuertz, S. (1998). Sorption properties of biofilms. Water Sci. Technol. 37, 207–210.

Springael, D., and Top, E.M. (2004). Horizontal gene transfer and microbial adaptation to xenobiotics: new types of mobile genetic elements and lessons from ecological studies. Trends Microbiol. *12*, 53–58.

Springael, D., Peys, K., Ryngaert, A., Roy, S.V., Hooyberghs, L., Ravatn, R., Heyndrickx, M., Meer, J.-R.v.d., Vandecasteele, C., Mergeay, M., *et al.* (2002a). Community shifts in a seeded 3-chlorobenzoate degrading membrane biofilm reactor: indications for involvement of *in situ* horizontal transfer of the clc-element from inoculum to contaminant bacteria. Environ. Microbiol. *4*, 70–80.

Springael, D., Peys, K., Ryngaert, A., Van Roy, S., Hooyberghs, L., Ravatn, R., van der Meer, J.R., Mergeay, M., and Diels, L. (2002b). Transfer of chlorocatechol degrading genes from initial inoculum to contaminant bacteria: Outcompetition of the initial inoculum in a 3-chlorobenzoaat degrading membrane reactor. Environ. Microbiol. *4*, 70–80.

Stanley, N.R., Britton, R.A., Grossman, A.D., and Lazazzera, B.A. (2003). Identification of catabolite repression as a physiological regulator of biofilm formation by *Bacillus subtilis* by use of DNA microarrays. J. Bacteriol. *185*, 1951–1957.

Steinmoen, H., Knutsen, E., and Håvarstein, L.S. (2002). Induction of natural competence in *Streptococcus pneumoniae* triggers lysis and DNA release from a subfraction of the cell population. Proc. Natl. Acad. Sci. U.S.A. *99*, 7681–7686.

Stewart, P.S. (2003). Diffusion in biofilms. J. Bacteriol. *185*, 1485–1491.

Stewart, P.S., and Franklin, M.J. (2008). Physiological heterogeneity in biofilms. Nat. Rev. Microbiol. *6*, 199–210.

Stoodley, P., Lewandowski, Z., Boyle, J.D., and Lappin-Scott, H.M. (1999). Structural deformation of bacterial biofilms caused by short-term fluctuations in fluid shear: An *in situ* investigation of biofilm rheology. Biotechnol. Bioeng. *65*, 83–92.

Taddei, F., Matic, I., and Radman, M. (1995). cAMP-dependent SOS induction and mutagenesis in resting bacterial populations. Proc. Natl. Acad. Sci. *92*, 11736–11740.

Top, E.M., Springael, D., and Boon, N. (2002). Catabolic mobile genetic elements and their potential use in bioaugmentation of polluted soils and waters. FEMS Microbiol. Ecol. *42*, 199–208.

Tsuneda, S., Aikawa, H., Hayashi, H., Yuasa, A., and Hirata, A. (2003). Extracellular polymeric substances responsible for bacterial adhesion onto solid surface. FEMS Microbiol. Lett. *223*, 287–292.

van der Wal, A., Norde, W., Zehnder, A.J.B., and Lyklema, J. (1997). Determination of the total charge in the cell walls of Gram-positive bacteria. Colloid. Surface. B *9*, 81–100.

Van Oss, C. (1989). Energetics of cell–cell and cell–biopolymer interactions. Cell Biophys. *14*, 1–16.

Van Oss, C.J. (1991). The forces involved in bioadhesion to flat surfaces and particles – their determination and relative roles. Biofouling *4*, 25–35.

Vandevivere, P., and Kirchman, D.L. (1993). Attachment stimulates exopolysaccharide synthesis by a bacterium. Appl. Environ. Microbiol. *59*, 3280–3286.

Verhagen, P., De Gelder, L., Hoefman, S., De Vos, P., and Boon, N. (2011). Planktonic versus biofilm catabolic communities: importance of the biofilm for species selection and pesticide degradation. Appl. Environ. Microbiol. *77*, 4728–4735.

Vesper, S.J., and Bauer, W.D. (1986). Role of pili (fimbriae) in attachment of *Bradyrhizobium japonicum* to soybean roots. Appl. Environ. Microbiol. *52*, 134–141.

Von Canstein, H., Kelly, S., Li, Y., and Wagner-Dobler, I. (2002). Species diversity improves the efficiency of mercury-reducing biofilms under changing environmental conditions. Appl. Environ. Microbiol. *68*, 2829–2837.

Webb, J.S., Thompson, L.S., James, S., Charlton, T., Tolker-Nielsen, T., Koch, B., Givskov, M., and Kjelleberg, S. (2003). Cell death in *Pseudomonas aeruginosa* biofilm development. J. Bacteriol. *185*, 4585–4592.

Webb, J.S., Lau, M., and Kjelleberg, S. (2004). Bacteriophage and phenotypic variation in *Pseudomonas aeruginosa* biofilm development. J. Bacteriol. *186*, 8066–8073.

Whiteley, M., Bangera, M.G., Bumgarner, R.E., Parsek, M.R., Teitzel, G.M., Lory, S., and Greenberg, E.P. (2001). Gene expression in *Pseudomonas aeruginosa* biofilms. Nature *413*, 860–864.

Wick, L.Y., Remer, R., Wurz, B., Reichenbach, J., Braun, S., Schafer, F., and Harms, H. (2007). Effect of fungal hyphae on the access of bacteria to phenanthrene in soil. Environ. Sci. Technol. *41*, 500–505.

Wolfaardt, G.M., Lawrence, J.R., Robarts, R.D., and Caldwell, D.E. (1995). Bioaccumulation of the herbicide diclofop in extracellular polymers and its utilization by a biofilm community during starvation. Appl. Environ. Microbiol. *61*, 152–158.

Writer, J.H., Ryan, J.N., and Barber, L.B. (2011). Role of biofilms in sorptive removal of steroidal hormones and 4-nonylphenol compounds from streams. Environ. Sci. Technol. *45*, 7275–7283.

Yoshida, S., Ogawa, N., Fujii, T., and Tsushima, S. (2009). Enhanced biofilm formation and 3-chlorobenzoate degrading activity by the bacterial consortium of Burkholderia sp. NK8 and Pseudomonas aeruginosa PAO1. J. Appl. Microbiol. *106*, 790–800.

Zhang, S.H., Yu, X., Guo, F., and Wu, Z.Y. (2011). Effect of interspecies quorum sensing on the formation of aerobic granular sludge. Water Sci. Technol. *64*, 1284–1290.

# Part II

# Methods and Monitoring

# Biofilm Survival Strategies in Polluted Environments

# 3

Marc A. Demeter, Joseph A. Lemire, Raymond J. Turner and Joe J. Harrison


## Abstract

Biofilm communities have the remarkable ability to withstand mechanical and toxic chemical perturbations that wipe out their planktonic cell counterparts. Many features of the biofilm mode-of-life protect microbial cells. These include an extracellular matrix, changes in cellular biochemistry, and social interactions that facilitate the exchange of metabolites, signal molecules and genetic material between cells. Here we provide a primer on the fundamental mechanical, physiological and ecological mechanisms that protect biofilm cells from antimicrobial substances and fluctuating shear forces. Biofilm research is not only advancing our understanding of these survival strategies, but also leading to the design of biofilm-based enrichment techniques that may harness new metabolic power for bioremediation.


## Introduction

Industrial wastes containing organics and metals can be toxic to nearly all forms of life, and these pollutants may devastate natural microbial communities (Chen *et al.*, 2014; Gans *et al.*, 2005; Giller *et al.*, 1998; Torsvik *et al.*, 1998). Harnessing the diverse metabolic biochemistries of microbes for sustainable remediation thus requires that the microbes can survive and thrive in the presence of the target pollutants. Biofilm-mediated processes may be advantageous to bioremediation because biofilm cells can withstand environmental stressors that eliminate planktonic microbes (Todhanakasem *et al.*, 2014). These stressors may include shear forces, toxic metal ions, biocides and extremes in pH and osmolarity as well as deadly fluxes in the pollutants themselves. Consequently, biofilm microbes may have a better chance of adapting and surviving in toxic surroundings.

While much has been learned about the survival strategies of infectious biofilms (Donlan and Costerton, 2002; Fux *et al.*, 2005; Hall-Stoodley and Stoodley, 2009; Mah and O'Toole, 2001; Stewart and Costerton, 2001), microbiologists and engineers have only just begun to unravel the resistance mechanisms for archaeal, bacterial and fungal biofilms found in highly polluted milieus. Studying biofilm stress resistance is difficult because it has numerous underlying causal factors (Harrison *et al.*, 2007). It depends on the environment as well as the organisms in the community, many of which have not yet been cultured in laboratories. Therefore, much of what we have learned is based on a limited number of model bacterial species. Overcoming these research challenges to understand the resistance mechanisms of environmental biofilm communities may be crucial for designing better treatment processes and may expand the repertoire of organisms recovered for use in bioremediation.

Here, we categorize the known phenomena that allow biofilms to withstand flushing forces and antimicrobial substances that may be present in polluted sites, soils, process waters and bioreactors. Although there are substantial gaps in our knowledge, it is becoming clear that many biofilms are protected by an extracellular matrix, active stress responses, ongoing genetic adaptation, and mutualistic interspecies interactions. The discussion that follows describes the most pertinent examples of these resistance mechanisms that have been substantiated using organisms that may be found in the environment.

## Cohesive and protective functions of the biofilm matrix

Although natural and laboratory biofilms formed by different species exhibit a variety of structural characteristics, most appear to be surrounded by a self-produced extracellular matrix (ECM) (Branda *et al.*, 2005). The biofilm matrix is a hydrogel (Hall-Stoodley *et al.*, 2004), and this matrix can make up more organic material in a biofilm than the microbial cells themselves. In fact, the ECM may be more than 90% of the dry mass of some biofilms (Zhang *et al.*, 1998).

The contents of the biofilm ECM vary from organism to organism and are comprised of a variety of extracellular polymeric substances (EPSs). These usually include polysaccharides (PSs), extracellular DNA (eDNA), and proteins (Branda *et al.*, 2005; Flemming *et al.*, 2007; Flemming and Wingender, 2010). The ECM also contains a variety of biomolecules that are packed in lipid membrane vesicles (MVs) (Choi *et al.*, 2011; Schooling and Beveridge, 2006). ECM polymers are held together by chemical interactions including hydrogen bonding, van der Waals interactions, and ionic attraction (Chen and Stewart, 2002). Ions within the matrix may strengthen electrostatic interactions by crosslinking charged polymers (Chen and Stewart, 2002; Grumbein *et al.*, 2014). Other ECM components may include metabolites and debris from dead and lysed cells (Asally *et al.*, 2012; Ma *et al.*, 2009) as well as biotic and abiotic molecules adsorbed from the external environment (Flemming and Wingender, 2010).

The biofilm ECM has many biological roles (Flemming *et al.*, 2007; Flemming and Wingender, 2010). Two important functions of the ECM are cohesion and protection. Many biofilm-specific mechanisms of tolerance to stressors may be attributable to these functions. These functions arise from the ability of the ECM to provide mechanical stability, resist external mass transfer and slow or restrict the penetration of toxicants.

### Mechanical stability

The retention of microbial biomass in high-flow conditions may be critical for the bioremediation of industrial wastewater (Lendenmann and Spain, 1998; Nicolella *et al.*, 2000). Biofilms are viscoelastic and can resist mechanical clearance (Klapper *et al.*, 2002; Stoodley *et al.*, 2002). Microbial EPS also affects the erosion process of sediments in water (Vignaga *et al.*, 2013). Therefore, EPS properties may help biofilms withstand shear forces in some treatment processes.

Biofilm viscoelasticity can be described using two physical parameters: the biofilm elastic modulus ($G$), which describes the reversible stretching of the biofilm under tension, and the biofilm failure strength ($\sigma_B$), which describes the overall strength of the biofilm (Stewart, 2014). Biofilm failure strengths can be determined using microcantilevers to measure the tensile stress experienced by an attached biofilm (Aggarwal *et al.*, 2010). As a mechanical tensile force ($T_F$) is applied to the biofilm, a biofilm will stretch reversibly to the extent determined by its elastic modulus ($G$), after which the biofilm will detach from the surface if the shear stress exceeds the failure strength (i.e. $T_F > \sigma_B$).

Matrix integrity is derived from the physicochemical interactions between polymers and is affected by polymer composition and concentration. The relationship of biofilm elastic modulus ($G$) to matrix polymer concentration ($C$) is approximately $G \approx C^{2.25}$; therefore, a doubling in polymer concentration results in a roughly 5-fold increase in biofilm elastic strength (Stewart, 2014). Studies examining multispecies biofilms grown from activated sludge or benthic sediment have correlated polysaccharide content from water-extracted ECM material with biofilm cohesiveness (Ahimou *et al.*, 2007; Lubarsky *et al.*, 2010).

In addition to extracellular polysaccharides, a variety of proteins are known to reinforce the biofilm matrix. In the soil bacterium *Pseudomonas aeruginosa*, for example, the extracellular adhesin CdrA provides structural integrity. Molecular modelling suggests that CdrA is a filamentous protein with putative haemagglutination and sugar binding domains. It has been hypothesized that CdrA reinforces the ECM by either directly or indirectly cross-linking PSL, an extracellular polysaccharide that co-immunoprecipitates with CdrA (Borlee *et al.*, 2010).

Other proteins important for adhesion and biofilm strength are the functional amyloids (Blanco *et al.*, 2012), which are thought to be abundant in environmental biofilms (Larsen *et al.*, 2007). Amyloid proteins have a distinct β-sheet-rich fold that forms fibres. These fibres spontaneously assemble and are highly resistant to protease digestion and denaturation. Examples of biochemically characterized biofilm amyloid proteins include FapC, which is highly conserved among *Pseudomonas* species (Dueholm *et al.*, 2010), and TasA, which is a major constituent of the *B. subtilis* biofilm matrix (Chai *et al.*, 2013; Romero *et al.*, 2010). Rheological measurements of synthetic biofilm matrices composed of PSs and recombinant amyloids, such as FapC, suggest that amyloid proteins may affect the elasticity and deformation characteristics of biofilms under shear stress (Lembre *et al.*, 2014).

Finally, in addition to biomolecules, inorganic ions adsorbed from the environment or antimicrobial treatments can have a significant influence on the mechanical properties of biofilms (Jones *et al.*, 2011; Koza *et al.*, 2009). The treatment of biofilms with metal cations, which might increase polymer cross-linking via the formation of salt bridges between anionic functional groups, can stiffen *P. aeruginosa* biofilms (Jones *et al.*, 2011). *B. subtilis* biofilms treated with $Cu^{2+}$, $Zn^{2+}$, $Fe^{2+}$, $Fe^{3+}$ or $Al^{3+}$ also become more rigid and are at least 100 times more resistant to mechanical clearance than in the absence of these metals (Grumbein *et al.*, 2014).

## External mass transfer resistance

Engineers have long recognized the potential functions of the ECM in mass transfer resistance. In biofilm microbiology, external mass transfer refers to the movement of solutes or gases from the bulk phase into the biofilm. External mass transfer resistance may arise from chemical properties of matrix biopolymers (Epstein *et al.*, 2011) as well as from fluid mechanics around biofilm structures via the formation of a concentration boundary layer (de Beer *et al.*, 1994b; Stewart, 2012).

Water repellent matrix polymers can create external mass transfer resistance (Epstein *et al.*, 2011). The best-characterized example of this is the *B. subtilus* biofilm surface layer (BslA) protein. This protein spontaneously polymerizes and localizes at the biofilm–air or biofilm–liquid interface (Hobley *et al.*, 2013; Kobayashi and Iwano, 2012). BslA monomers are amphiphilic and contain a hydrophobic cap that orients itself towards the external environment. In this fashion, BslA creates a non-wetting layer at the biofilm periphery. This layer is thought to surpass the water repellency of Teflon® (Epstein *et al.*, 2011). The BslA coating resists the penetration of liquids, such as ethanol (Fig. 3.1), as well as gasses (Epstein *et al.*, 2011; Hobley *et al.*, 2013; Kobayashi and Iwano, 2012). It has also been speculated that BslA may impair the penetration of toxic metals into the biofilm (Epstein *et al.*, 2011).

In addition to matrix polymers, theory in fluid mechanics suggests that a boundary layer could protect biofilms. Non-turbulent fluid motion, termed laminar flow, streamlines fluid velocity parallel to the biofilm surface. Under these conditions, the convective flow of bulk fluid and solutes into and

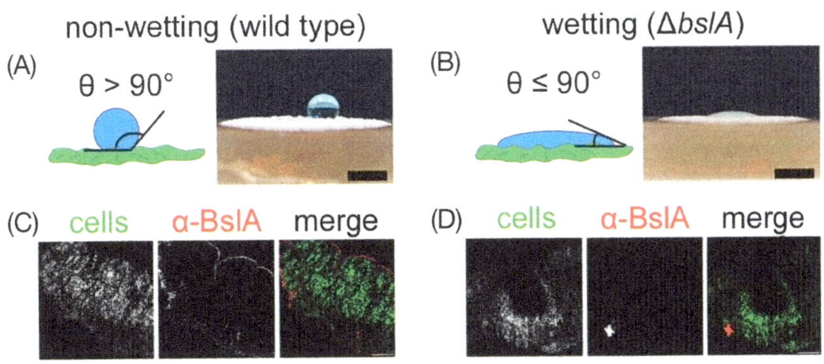


**Figure 3.1** *Bacillus subtilus* biofilms are coated with a water repellent biofilm surface layer (BslA) protein that creates external mass transfer resistance. The contact angle (θ) between a drop of water and a bacterial colony indicates whether it is non-wetting (θ > 90°) or wetting (θ ≤ 90°). (A) Wild type *B. subtilus* colony biofilms are non-wetting. (B) Biofilms lacking BslA are wetting. (C) Microscopy using fluorescent anti-BslA antibodies localizes BslA to layers at the biofilm periphery. (D) Biofilms lacking BslA do not have this water repellent layer. Black scale bars are 2 mm, white scale bars are 50 μM. Images in parts A and B are reproduced, with permission, from Kobayashi and Iwano (2012), © John Wiley & Sons Group Company. Images in parts C and D are reproduced, with permission, from Hobley *et al.* (2013), © National Academy of Sciences.

out of the biofilm may be limited. Limited diffusive transport occurs not only because fluid moves more slowly near the biofilm, but also because fluid layers in laminar flow do not efficiently mix. In this fashion, high solute concentrations that drive diffusion cannot be maintained at the biofilm surface. This creates a concentration boundary layer (Stewart, 2012; Zhang and Bishop, 1994). Direct measurements of the concentration boundary layer have been made with microelectrodes for many solutes, including $O_2$ (de Beer *et al.*, 1994b; de Beer *et al.*, 1993; Wasche *et al.*, 2002), $NH_4^+$ and $NO_2^-$ (de Beer *et al.*, 1993). Direct imaging of the $O_2$ boundary layer has been achieved using luminescent thin films coated onto glass coverslips (Staal *et al.*, 2011; Stewart, 2012). In principle, a concentration boundary layer might also hinder the penetration of toxic solutes into biofilms. For example, such a boundary layer has been measured for chlorine during disinfection of *P. aeruginosa* biofilms (de Beer *et al.*, 1994a).

## Slowed and restricted penetration

Diffusion, which is the passive movement of molecules from regions of higher to regions of lower concentration, is the transport mechanism responsible for solute penetration into biofilms. Diffusion of most solutes through biofilms is significantly slower than it is through water (Stewart, 2003). This reduced rate of diffusion may allow organisms an opportunity to activate stress responses before lethal concentrations of a toxicant accumulate in the depths of a biofilm (Tseng *et al.*, 2013). However, it is important to note that a reduced rate of diffusion alone does not restrict the penetration of toxicants into biofilms – it merely slows them down (Stewart, 2003). Given sufficient time, therefore, the concentrations of many antimicrobial agents are predicted to equilibrate across biofilms on time scales of seconds to minutes (Stewart, 1996).

By contrast, diffusion-reaction phenomena can restrict the penetration of antimicrobials into biofilms. Diffusion-reaction results from the reversible or irreversible reaction of solutes with functional groups in the biofilm matrix, on cell surfaces or within cells. Reactive functional groups in the ECM may include carboxylates, phosphates, thiols, amino and phenolic groups (Liu and Fang, 2002). The concentrations of these functional groups may be 20- to 30-fold higher in the ECM than on cell surfaces (Liu and Fang, 2002). Mathematical modelling predicts that the diffusion-reaction process will create a gradient distribution of the reactive solute within the biofilm (Stewart and Franklin, 2008), and moreover, that this will protect cells deep within the biofilm (Chambless *et al.*, 2006). Experimental measurements have so far substantiated this theory (Fig. 3.2). Reaction–diffusion phenomena have been measured for hypochlorite (Stewart *et al.*, 2001), chlorosulfamate (Stewart *et al.*, 2001) and monochloramine (Xue *et al.*, 2014), and have been well characterized for hydrogen peroxide ($H_2O_2$) (Stewart *et al.*, 2000).

Extracellular and periplasmic enzymes can mediate diffusion–reaction phenomena by catalysing reactions of toxic substrates that diffuse into the ECM. For instance, *P. aeruginosa* produces a periplasmic catalase (KatA) that is also packaged into outer membrane vesicles (Choi *et al.*, 2011). KatA degrades $H_2O_2$ into $H_2O$ and $O_2$. Direct measurements of $H_2O_2$ concentrations in biofilms using peroxide-sensing microelectrodes suggest that KatA consumes $H_2O_2$ and prevents it from completely penetrating into the biofilm, because $H_2O_2$ only penetrates into biofilms of strains that lack this enzyme (Fig. 3.2) (Stewart *et al.*, 2000). Biofilms formed by strains that lack *katA* are also much more susceptible to killing by $H_2O_2$ (Stewart *et al.*, 2001) (Fig. 3.2).

Reaction–diffusion phenomena may also be important for the protection of biofilms from toxic metal ions. Many different types of microbial EPS molecules possess functional groups that can bind to metal ions or their precipitates (Fang *et al.*, 2011; Flemming, 1995; Guibaud *et al.*, 2005, 2012; Jiao *et al.*, 2010; Mulcahy *et al.*, 2008; White and Gadd, 2000). Functional groups displayed on cell surfaces also may sequester certain metal species (Ahluwalia and Goyal, 2007; Langley and Beveridge, 1999; Mullen *et al.*, 1989). Coordination with these donor ligands might restrict the penetration of metals to the biofilm periphery (White and Gadd, 2000). Metals bound to cell surface sites may also serve as nucleation centres for mineralization, enabling reactions with cell metabolites to form metal precipitates (Flemming, 1995; White *et al.*, 1995). These observations seem to fit with reports that treatment of *P. aeruginosa* biofilms with toxic doses of $Cu^{2+}$ kills biofilm cells in the outermost regions of the biofilm (Teitzel and Parsek, 2003).

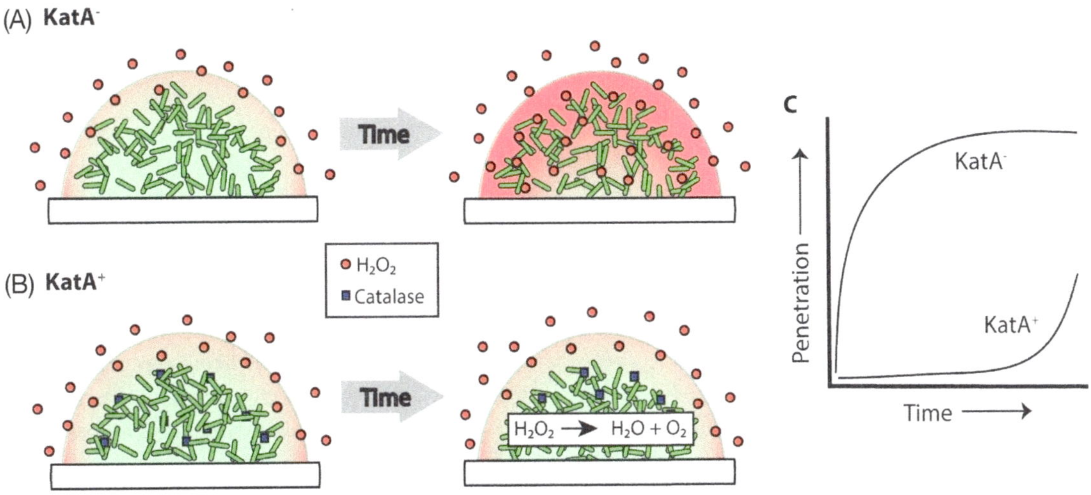


**Figure 3.2** Catalase-dependent diffusion-reaction restricts the penetration of deadly $H_2O_2$ into biofilms. (A) $H_2O_2$ rapidly penetrates into *P. aeruginosa* biofilms that lack catalase (KatA). (B) By contrast, the dismutation of $H_2O_2$ by KatA prevents this peroxide from penetrating into wild type biofilms. (C) Direct experimental measurements with $H_2O_2$-sensing microelectrodes suggest that $H_2O_2$ diffusion reaction prevents $H_2O_2$ from completely penetrating thick *P. aeruginosa* biofilms. The trends illustrated in (C) are based on previous studies of diffusion-reaction in *P. aeruginosa* biofilms (Stewart *et al.*, 2000).

Recently, $Fe^{2+}$ and $Cu^{2+}$ distributions have been imaged in living *Rhodobacter ferrooxidans* biofilms using metal-binding fluorescent probes and laser-scanning confocal microscopy (Hao *et al.*, 2013). This has co-localized Fe and Cu with biofilm matrix material. However, while the chemistry of metals and the observations summarized here suggest that diffusion-reaction processes may mediate biofilm-specific tolerance to some metal species, it is important to note that direct measurements of metal diffusion rates into biofilms have not been made.

## Active stress responses

Community proteomics of biofilms from toxic environments suggests that damage to biomolecules is a key challenge for microbial survival (Benndorf *et al.*, 2007; Ram *et al.*, 2005; Wu *et al.*, 2013). Indeed, some of the most abundant proteins expressed by these biofilm organisms are involved in protein folding and response to oxidative stress. The high level expression of these proteins may not be passive. A key challenge that microbiologists face is identifying active or anticipatory stress response pathways that protect biofilms. Two signalling pathways that are important for the ability of model biofilm organisms to withstand environmental stressors are cell-density dependent quorum sensing and the starvation-activated stringent response. We will briefly describe how these signal transduction pathways may help biofilms cope with stress.

### Quorum sensing

Quorum sensing (QS) is the regulation of gene expression in response to chemical signal molecules. These molecules, called autoinducers (AIs), can be self-recognized and stimulate changes in gene expression when they accumulate to a minimum threshold concentration (Miller and Bassler, 2001). Microorganisms secrete a wide variety of QS-signals that can be classified based on their structure (Camilli and Bassler, 2006). These QS communication circuits regulate a diverse array of physiological activities (Waters and Bassler, 2005), and can influence biofilm development in many ways for different bacterial species (Irie and Parsek, 2008).

In nature, many bacteria often reside within dense clusters or aggregates that contain less than $10^5$ cells (Connell *et al.*, 2010; Lock *et al.*, 1984; Morris *et al.*, 1998). The close association of cells and the slow diffusion of signals may result in the accumulation of autoinducers within these dense communities (McLean *et al.*, 1997; Parsek and Greenberg, 2005). Although flow rate can affect the amount of biomass required for QS induction (Kirisits *et al.*, 2007), recent findings indicate that the number of cells required for a quorum at high cell densities may be surprisingly small (Connell *et al.*, 2010).

One of the most studied types of QS is mediated by the acyl-homoserine lactones (AHLs) that are produced by Gram-negative bacteria. *P. aeruginosa* has been a paradigm species for studying AHL-mediated signalling among these organisms (Schuster *et al.*, 2013). One way *P. aeruginosa* QS-communication has been analysed is by using picoliter-sized porous cavities, called bacterial 'lobster traps,' to trap individual *P. aeruginosa* cells and track their growth and behaviour in real time (Wessel *et al.*, 2013). To assess whether *P. aeruginosa* populations in these traps communicate by QS, it is possible to use strains containing the gene encoding green fluorescent protein (GFP) under the control of a QS-responsive promoter. Such a reporter strain displays green fluorescence, for example, when the *P. aeruginosa* QS-signal 3-oxododecanoyl homoserine lactone (3OC12-HSL) reaches a critical concentration. To determine the minimum number of cells required for a quorum, it is possible to grow a reporter strain in differently sized traps, all of which have constant cell densities but different numbers of cells. These experiments have revealed that as few as few as 8500 *P. aeruginosa* cells may constitute a quorum when cells are trapped at these high densities ($\approx 10^{12}$ cells/ml) (Fig. 3.3) (Connell *et al.*, 2010). These observations suggest that QS-mediated gene regulation might be active in aggregates and cell clusters at sizes similar to those seen in bioreactors, polluted soils and wastewaters.

The *P. aeruginosa* QS-regulon is complex (Schuster and Greenberg, 2006). Many reports connect QS to the ability of *P. aeruginosa* to withstand environmental stressors. This includes heat and osmotic shock (Garcia-Contreras *et al.*, 2014), toxic doses of cadmium (Garcia-Contreras *et al.*, 2014) and copper (Thaden *et al.*, 2010), as well as oxidants (Garcia-Contreras *et al.*, 2014; Hassett *et al.*, 1999) and detergents (Davies *et al.*, 1998). Perhaps it is no surprise then that QS-deficient strains of *P. aeruginosa* form biofilms that are hypersensitive to many antimicrobial agents, including sodium dodecyl sulfate (SDS) (Davies *et al.*, 1998) and $H_2O_2$ (Bjarnsholt *et al.*, 2005). Similar observations of detergent hypersensitivity have been made for QS-deficient biofilms of *Burkholderia cenocepacia* (Tomlin *et al.*, 2005).

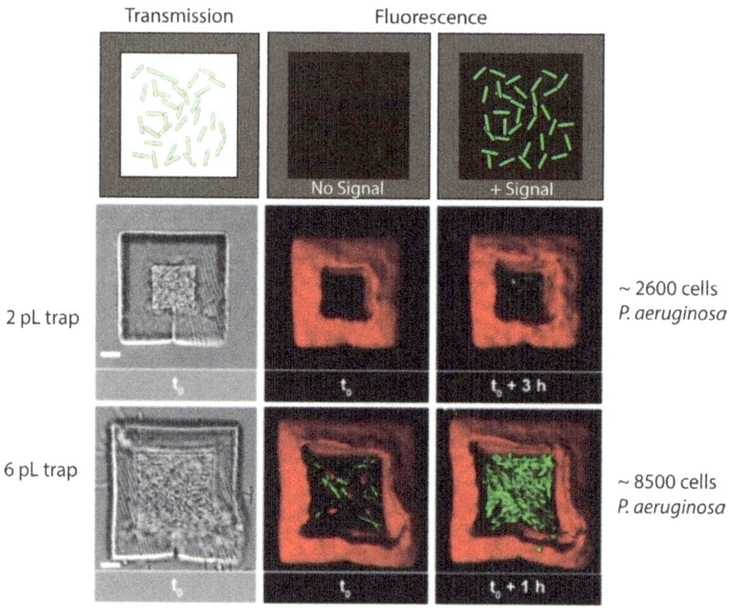


**Figure 3.3** Quorum-sensing-dependent gene expression is activated by small numbers of cells at a high densities. Bacteria in these 'lobster traps' express green fluorescence protein under the control of a promoter that responds to the signal 3OC12-HSL. Images are reproduced, with permission, from Connell *et al.* (2010) © The American Society for Microbiology.

### Stringent response

The starvation-signalling stringent response (SR) allows bacteria to adapt and respond to nutrient deprivation. Recent findings have identified that the stringent response may also protect nutrient-limited and biofilm bacteria from antimicrobial stress (Nguyen *et al.*, 2011).

Under amino acid starvation, uncharged tRNA molecules may cause the ribosome to stall during translation. This stimulates the activity of the ribosome-associated protein RelA (Dalebroux and Swanson, 2012; Potrykus and Cashel, 2008). When the ribosome stalls, RelA catalyses the pyrophosphorylation of GTP using ATP, which produces the second messenger molecule guanosine-3′,5′-bispyrophosphate (ppGpp). Synthesis of ppGpp can also be carried out by the bifunctional enzyme SpoT, which is activated by numerous environmental and nutrient stresses (Dalebroux and Swanson, 2012; Potrykus and Cashel, 2008). Although ppGpp acts as a signal in many cellular processes, in many bacteria ppGpp binds to the transcription factor DksA to redirect RNA polymerase to certain genes (Dalebroux and Swanson, 2012). The stringent regulon seems to vary considerably from organism to organism. However, in several bacterial species, activation of the stringent response down-regulates genes involved in cell replication and macromolecule production and up-regulates some genes involved in stress tolerance.

Many investigators have reported links between the stringent response and different aspects of biofilm formation. However, the work most relevant to environmental microbiology has been carried out using *P. aeruginosa*. The stringent response protects nutrient-limited *P. aeruginosa* by activating the production of catalase and superoxide dismutase (Nguyen *et al.*, 2011). Deletion mutations in *relA* and *spoT*, which inactivate the stringent response, make *P. aeruginosa* biofilms hypersensitive to $H_2O_2$ by preventing the expression of KatA (Khakimova *et al.*, 2013).

## Ecological and evolutionary processes

Biofilms are spatially structured environments with landscapes of selective pressures that drive not only ecological succession but also the evolution of microbial traits. These processes have been studied using a variety of laboratory models, ranging from drip-flow reactors to static microcosms. These investigations have revealed that genetic adaptation, mediated by horizontal gene transfer and adaptive radiation, as well as interspecies interactions, can profoundly influence biofilm fitness and survival.

### Horizontal gene transfer

Although horizontal gene transfer (HGT) may occur by natural transformation in biofilms (Hendrickx *et al.*, 2003), rates of conjugal transfer are estimated to be up to 1000-fold higher in biofilms than for comparable planktonic cell populations (Hausner and Wuertz, 1999). Perhaps this is expected given the high population density and close physical proximity of cells in a biofilm. However, a surprising discovery has been that the conjugation mechanism itself may promote biofilm formation and stabilize biofilm structure (Molin and Tolker-Nielsen, 2003).

The finding that conjugative plasmids enhance biofilm formation was made nearly 15 years ago – albeit with *Escherichia coli* K12 (Ghigo, 2001). Nevertheless, conjugative pili, which are encoded by all conjugative plasmids, may take the place of other factors in biofilm development and influence biofilm structure (Reisner *et al.*, 2003). Also, there is evidence that factors expressed from conjugative plasmids may allow planktonic bacteria to join biofilm communities (Ghigo, 2001). HGT may be a principal source of genetic diversity for selection, and conjugal plasmid transfer is known to occur in both natural biofilms (Normander *et al.*, 1998) and bioreactors (Angles *et al.*, 1993; Ehlers and Bouwer, 1999; Geisenberger *et al.*, 1999; Springael *et al.*, 2002). Moreover, in addition to genes for the conjugation machinery, some conjugative plasmids encode genes for the catabolism of target organics and metal resistance (DiGiovanni *et al.*, 1996). Therefore, HGT via conjugation may be of fundamental consequence for the fitness and evolution of microbial biofilm communities in polluted environments.

### Adaptive radiation

Microniches within biofilms present new challenges and resources to microorganisms. Adaptive radiation may allow an organism to rapidly diversify into new forms that can occupy these niches. A manifestation of this process in single-species laboratory biofilms is the appearance of colony morphology variants among bacteria isolated from the ageing population (Boles *et al.*, 2004; Davies *et al.*, 2007; Kirisits *et al.*, 2005; Penterman *et al.*, 2014; Poltak and Cooper, 2011; Rainey and Travisano, 1998; Spiers, 2014; Traverse *et al.*, 2013; Workentine *et al.*, 2010).

Many selective forces may drive evolution in biofilms. Competition for carbon sources and electron acceptors likely select for variants that overproduce EPS (Koza *et al.*, 2011; Xavier and Foster, 2007). The activation of lytic prophage may select for mutations in cell surface receptors, such as the type IV pilus, which subsequently alter motility (McElroy *et al.*, 2014). Fitness benefits may also arise from cross-feeding interactions that increase community productivity (Poltak and Cooper, 2011).

Adaptive laboratory evolution has revealed surprising mutational parallelism among independently evolved biofilm lineages (McElroy *et al.*, 2014). Moreover, these experiments suggest that interference competition may preserve genetic diversity in biofilms (Traverse *et al.*, 2013).

An interesting consequence of diversity is that it may increase the resistance of a community to environmental perturbation. The mechanistic basis of this stability is redundancy. Diversity may insure against a decline in community functioning because many genotypes provide greater guarantees that some will maintain functioning even if others fail. This ecological principle is known as the insurance hypothesis (Yachi and Loreau, 1999).

Microbial biofilms self-generate diversity that produces insurance effects. Diverse *P. aeruginosa* biofilm communities comprising multiple variants are many times more tolerant to $H_2O_2$ than biofilms comprising a single variant (Boles *et al.*, 2004). In fact, some biofilms composed of individual variants seem to be more resistant than the ancestor to certain antimicrobials, such as $H_2O_2$ or toxic metal ions (Davies *et al.*, 2007; Workentine *et al.*, 2010). However, in diverse biofilms, the frequency of these resistant variants does not seem to fix in the population, even after prolonged periods of selective antimicrobial pressure (Davies *et al.*, 2007; Workentine *et al.*, 2010). In this fashion, it is possible that intraspecific interactions may help to preserve some vulnerable genotypes during environmental perturbations – even if these genotypes remain at low frequencies.

## Interspecies interactions

Nearly all environments contain multiple coexisting species of prokaryotes and microbial eukaryotes. Yet our understanding of biofilm microbiology has been determined primarily by studying organisms that have been isolated and grown in pure culture. Emerging evidence suggests that biofilm stress resistance may be dictated by the numbers and types of microbial species in the community.

Biofilms have remarkable chemical heterogeneity and possess spatially segregated microniches (Stewart and Franklin, 2008). These microhabitats may be occupied by specific microbial species with metabolic requirements and functions that fit the ecological niche (Almstrand *et al.*, 2013). The different species in biofilms may interact through QS-mediated cell–cell communication, metabolic cooperation or competition (Burmolle *et al.*, 2014; Elias and Banin, 2012). Moreover, co-evolution in the biofilm environment may lead to metabolic interactions that increase community productivity (Hansen *et al.*, 2007). Altogether, these interactions drive community organization, and multispecies biofilms are often spatially stratified with different organisms (Elias and Banin, 2012).

Living in multispecies biofilms may provide a number of benefits to the community. One benefit may be protection against disinfectants (Burmolle *et al.*, 2006; Schwering *et al.*, 2013) and detergents (Lee *et al.*, 2014). For example, chlorine concentrations required to kill mixed species biofilms may be up to 300-fold higher than those concentrations required to kill biofilms comprised of the constituent species on their own (Schwering *et al.*, 2013). Similarly, the respiratory activity of multispecies biofilms comprising marine algal epibionts is unaffected by toxic doses of $H_2O_2$. By contrast, biofilms comprising individual epibiont species are significantly inhibited by the same dose of $H_2O_2$ (Fig. 3.4). In some cases, growth in multispecies communities seems to protect the most vulnerable species in the biofilm (Burmolle *et al.*, 2006; Lee *et al.*, 2014; Schwering *et al.*, 2013). Although the physiological mechanisms underlying this resilience remain a mystery, these studies illustrate that a foundation of biofilm community stability may lie in species diversity and interspecific interactions.

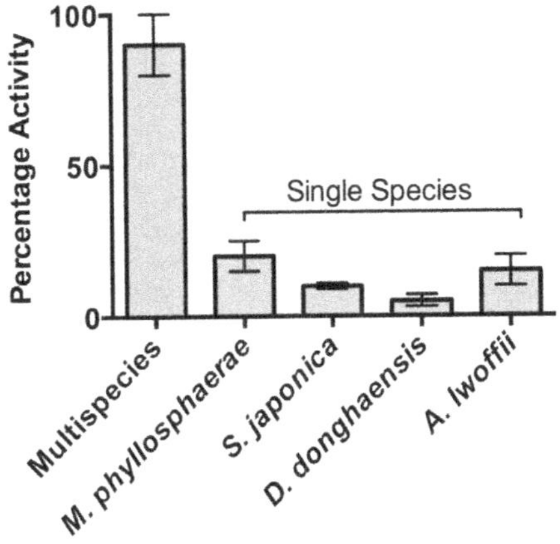


**Figure 3.4** Strength of biofilm communities to resist various stresses lies in differences, not in similarities among community members. Multispecies biofilms comprised of marine algal epibionts retain high levels of metabolic activity after exposure to toxic doses of $H_2O_2$. By contrast, biofilms comprised of the individual species from the community are vulnerable to $H_2O_2$. The data illustrated in this figure were derived from a previous study that used a tetrazolium salt as a respiratory indicator (Burmolle *et al.*, 2006).

## Summary and conclusions

Biofilms are well-protected, socially interactive and highly diverse communities that may readily adapt to changing environmental pressures. Although this chapter has focused on mechanisms that biofilms use to withstand shear forces and antimicrobial substances, many microorganisms also possess biofilm-specific means to resist predation (Matz *et al.*, 2005), desiccation (Chang *et al.*, 2007; Ronan *et al.*, 2013), and irradiation (Elasri and Miller, 1999; Pezzoni *et al.*, 2014). It should not be surprising then that many environmental stressors seem to induce EPS production and biofilm formation for a variety of bacterial species (Lapaglia and Hartzell, 1997; Priester *et al.*, 2006; Schmitt *et al.*, 1995; Sheng *et al.*, 2005).

Much of biofilm research has been motivated by the study of infectious disease. Consequently, many investigations have provided insight into the antibiotic resistance and tolerance of biofilm pathogens. However, general assumptions have been made that these protective mechanisms may be similar for other antimicrobials and biofilms of environmental organisms. This may not be the case. For instance, a large body of work has focused on multidrug tolerant 'persister' cells (Lewis, 2010; Maisonneuve and Gerdes, 2014), which can withstand lethal doses of antibiotics and may be present in high numbers in biofilms (Lewis, 2008). Other than linkage between the ability of *E. coli* biofilms to withstand some toxic metalloid oxyanions and a gene for high persistence (*hipA7*) (Harrison *et al.*, 2005), there is currently little if no evidence linking persister cells to the ability of biofilms to withstand any environmental stressors.

Finally, the means that microorganisms use to withstand toxic substances may pose a challenge to process engineers as the hardiest biofilms may not be the best biofilms for degrading pollutants. Hence, there may be trade-offs between certain mechanisms of resistance, such as barriers to external mass transfer, and metabolic capacity. As remediation methods and bioreactor designs progress, understanding the survival strategies of biofilm organisms may be crucial for engineering communities with the desired functions.

## References


Aggarwal, S., Poppele, E.H., and Hozalski, R.M. (2010). Development and testing of a novel microcantilever technique for measuring the cohesive strength of intact biofilms. Biotechnol. Bioeng. *105*, 924–934.

Ahimou, F., Semmens, M.J., Haugstad, G., and Novak, P.J. (2007). Effect of protein, polysaccharide, and oxygen concentration profiles on biofilm cohesiveness. Appl. Environ. Microbiol. *73*, 2905–2910.

Ahluwalia, S.S., and Goyal, D. (2007). Microbial and plant derived biomass for removal of heavy metals from wastewater. Bioresour. Technol. *98*, 2243–2257.

Almstrand, R., Daims, H., Persson, F., Sorensson, F., and Hermansson, M. (2013). New methods for analysis of spatial distribution and coaggregation of microbial populations in complex biofilms. Appl. Environ. Microbiol. *79*, 5978–5987.

Angles, M.L., Marshall, K.C., and Goodman, A.E. (1993). Plasmid transfer between marine bacteria in the aqueous phase and biofilms in reactor microcosms. Appl. Environ. Microbiol. *59*, 843–850.

Asally, M., Kittisopikul, M., Rue, P., Du, Y., Hu, Z., Cagatay, T., Robinson, A.B., Lu, H., Garcia-Ojalvo, J., and Suel, G.M. (2012). Localized cell death focuses mechanical forces during 3D patterning in a biofilm. Proc. Natl. Acad. Sci. U.S.A. *109*, 18891–18896.

Benndorf, D., Balcke, G.U., Harms, H., and von Bergen, M. (2007). Functional metaproteome analysis of protein extracts from contaminated soil and groundwater. ISME J. *1*, 224–234.

Bjarnsholt, T., Jensen, P., Burmolle, M., Hentzer, M., Haagensen, J.A., Hougen, H.P., Calum, H., Madsen, K., Moser, C., Molin, S., *et al.* (2005). *Pseudomonas aeruginosa* tolerance to tobramycin, hydrogen peroxide and polymorphonuclear leukocytes is quorum-sensing dependent. Microbiology *151*, 373–383.

Blanco, L.P., Evans, M.L., Smith, D.R., Badtke, M.P., and Chapman, M.R. (2012). Diversity, biogenesis and function of microbial amyloids. Trends Microbiol. *20*, 66–73.

Boles, B.R., Thoendel, M., and Singh, P.K. (2004). Self-generated diversity produces 'insurance effects' in biofilm communities. Proc. Natl. Acad. Sci. U.S.A. *101*, 16630–16635.

Borlee, B.R., Goldman, A.D., Murakami, K., Samudrala, R., Wozniak, D.J., and Parsek, M.R. (2010). *Pseudomonas aeruginosa* uses a cyclic-di-GMP-regulated adhesin to reinforce the biofilm extracellular matrix. Mol. Microbiol. *75*, 827–842.

Branda, S.S., Vik, S., Friedman, L., and Kolter, R. (2005). Biofilms: the matrix revisited. Trends Microbiol. *13*, 20–26.

Burmolle, M., Webb, J.S., Rao, D., Hansen, L.H., Sorensen, S.J., and Kjelleberg, S. (2006). Enhanced biofilm formation and increased resistance to antimicrobial agents and bacterial invasion are caused by synergistic interactions in multispecies biofilms. Appl. Environ. Microbiol. *72*, 3916–3923.

Burmolle, M., Ren, D., Bjarnsholt, T., and Sorensen, S.J. (2014). Interactions in multispecies biofilms: do they actually matter? Trends Microbiol. *22*, 84–91.

Camilli, A., and Bassler, B.L. (2006). Bacterial small-molecule signaling pathways. Science *311*, 1113–1116.

Chai, L., Romero, D., Kayatekin, C., Akabayov, B., Vlamakis, H., Losick, R., and Kolter, R. (2013). Isolation, characterization, and aggregation of a structured bacterial matrix precursor. J. Biol. Chem. *288*, 17559–17568.

Chambless, J.D., Hunt, S.M., and Stewart, P.S. (2006). A three-dimensional computer model of four hypothetical mechanisms protecting biofilms from antimicrobials. Appl. Environ. Microbiol. *72*, 2005–2013.

Chang, W.S., van de Mortel, M., Nielsen, L., Nino de Guzman, G., Li, X., and Halverson, L.J. (2007). Alginate production by *Pseudomonas putida* creates a hydrated microenvironment and contributes to biofilm architecture and stress tolerance under water-limiting conditions. J. Bacteriol. *189*, 8290–8299.

Chen, J., He, F., Zhang, X., Sun, X., Zheng, J., and Zheng, J. (2014). Heavy metal pollution decreases microbial abundance, diversity and activity within particle-size fractions of a paddy soil. FEMS Microbiol. Ecol. *87*, 164–181.

Chen, X., and Stewart, P.S. (2002). Role of electrostatic interactions in cohesion of bacterial biofilms. Appl. Microbiol. Biotechnol. *59*, 718–720.

Choi, D.S., Kim, D.K., Choi, S.J., Lee, J., Choi, J.P., Rho, S., Park, S.H., Kim, Y.K., Hwang, D., and Gho, Y.S. (2011). Proteomic analysis of outer membrane vesicles derived from *Pseudomonas aeruginosa*. Proteomics *11*, 3424–3429.

Connell, J.L., Wessel, A.K., Parsek, M.R., Ellington, A.D., Whiteley, M., and Shear, J.B. (2010). Probing prokaryotic social behaviors with bacterial 'lobster traps'. MBio *1*, e00202–00210.

Dalebroux, Z.D., and Swanson, M.S. (2012). ppGpp: magic beyond RNA polymerase. Nat. Rev. Microbiol. *10*, 203–212.

Davies, D.G., Parsek, M.R., Pearson, J.P., Iglewski, B.H., Costerton, J.W., and Greenberg, E.P. (1998). The involvement of cell-to-cell signals in the development of a bacterial biofilm. Science *280*, 295–298.

Davies, J.A., Harrison, J.J., Marques, L.L.R., Foglia, G.R., Stremick, C.A., Storey, D.G., Turner, R.J., Olson, M.E., and Ceri, H. (2007). The GacS sensor kinase controls phenotypic reversion of small colony variants isolated from biofilms of *Pseudomonas aeruginosa* PA14. FEMS Microbiol. Ecol. *59*, 32–46.

de Beer, D., van den Heuvel, J.C., and Ottengraf, S.P. (1993). Microelectrode measurements of the activity distribution in nitrifying bacterial aggregates. Appl. Environ. Microbiol. *59*, 573–579.

de Beer, D., Srinivasan, R., and Stewart, P.S. (1994a). Direct measurement of chlorine penetration into biofilms during disinfection. Appl. Environ. Microbiol. *60*, 4339–4344.

de Beer, D., Stoodley, P., Roe, F., and Lewandowski, Z. (1994b). Effects of biofilm structure on oxygen distribution and mass transport. Biotechnol. Bioeng. *43*, 1131–1138.

DiGiovanni, G.D., Neilson, J.W., Pepper, I.L., and Sinclair, N.A. (1996). Gene transfer of *Alcaligenes eutrophus* JMP134 plasmid pJP4 to indigenous soil recipients. Appl. Environ. Microbiol. *62*, 2521–2526.

Donlan, R.M., and Costerton, J.W. (2002). Biofilms: survival mechanisms of clinically relevant microorganisms. Clin. Microbiol. Rev. *15*, 167–193.

Dueholm, M.S., Petersen, S.V., Sonderkaer, M., Larsen, P., Christiansen, G., Hein, K.L., Enghild, J.J., Nielsen, J.L., Nielsen, K.L., Nielsen, P.H., *et al.* (2010). Functional amyloid in *Pseudomonas*. Mol. Microbiol. *77*, 1009–1020.

Ehlers, L.J., and Bouwer, E.J. (1999). RP4 plasmid transfer among species of *Pseudomonas* in a biofilm reactor. Water Sci. Technol. *39*, 163–171.

Elasri, M.O., and Miller, R.V. (1999). Study of the response of a biofilm bacterial community to UV radiation. Appl. Environ. Microbiol. *65*, 2025–2031.

Elias, S., and Banin, E. (2012). Multi-species biofilms: living with friendly neighbors. FEMS Microbiol. Rev. *36*, 990–1004.

Epstein, A.K., Pokroy, B., Seminara, A., and Aizenberg, J. (2011). Bacterial biofilm shows persistent resistance to liquid wetting and gas penetration. Proc. Natl. Acad. Sci. U.S.A. *108*, 995–1000.

Fang, L., Wei, X., Cai, P., Huang, Q., Chen, H., Liang, W., and Rong, X. (2011). Role of extracellular polymeric substances in Cu(II) adsorption on *Bacillus subtilis* and *Pseudomonas putida*. Bioresour. Technol. *102*, 1137–1141.

Flemming, H.C. (1995). Sorption sites in biofilms. Water Sci. Technol. *32*, 27–33.

Flemming, H.C., and Wingender, J. (2010). The biofilm matrix. Nat. Rev. Microbiol. *8*, 623–633.

Flemming, H.C., Neu, T.R., and Wozniak, D.J. (2007). The EPS matrix: the 'house of biofilm cells'. J. Bacteriol. *189*, 7945–7947.

Fux, C.A., Costerton, J.W., Stewart, P.S., and Stoodley, P. (2005). Survival strategies of infectious biofilms. Trends Microbiol. *13*, 34–40.

Gans, J., Wolinsky, M., and Dunbar, J. (2005). Computational improvements reveal great bacterial diversity and high metal toxicity in soil. Science *309*, 1387–1390.

Garcia-Contreras, R., Nunez-Lopez, L., Jasso-Chavez, R., Kwan, B.W., Belmont, J.A., Rangel-Vega, A., Maeda, T., and Wood, T.K. (2015). Quorum sensing enhancement of the stress response promotes resistance to quorum quenching and prevents social cheating. ISME J. *9*, 115–125.

Geisenberger, O., Ammendola, A., Christensen, B.B., Molin, S., Schleifer, K.H., and Eberl, L. (1999). Monitoring the conjugal transfer of plasmid RP4 in activated sludge and *in situ* identification of the transconjugants. FEMS Microbiol. Lett. *174*, 9–17.

Ghigo, J.M. (2001). Natural conjugative plasmids induce bacterial biofilm development. Nature *412*, 442–445.

Giller, K.E., Witter, E., and Mcgrath, S.P. (1998). Toxicity of heavy metals to microorganisms and microbial processes in agricultural soils: a review. Soil Biol. Biochem. *30*, 1389–1414.

Grumbein, S., Opitz, M., and Lieleg, O. (2014). Selected metal ions protect *Bacillus subtilis* biofilms from erosion. Metallomics *6*, 1441–1445.

Guibaud, G., Comte, S., Bordas, F., Dupuy, S., and Baudu, M. (2005). Comparison of the complexation potential of extracellular polymeric substances (EPS), extracted from activated sludges and produced by pure bacteria strains, for cadmium, lead and nickel. Chemosphere *59*, 629–638.

Guibaud, G., Bhatia, D., d'Abzac, P., Bourven, I., Bordas, F., van Hullebusch, E.D., and Lens, P.N.L. (2012). Cd(II) and Pb(II) sorption by extracellular polymeric substances (EPS) extracted from anaerobic granular biofilms: Evidence of a pH sorption-edge. J. Taiwan Inst. Chem. Eng. *43*, 444–449.

Hall-Stoodley, L., and Stoodley, P. (2009). Evolving concepts in biofilm infections. Cell. Microbiol. *11*, 1034–1043.

Hall-Stoodley, L., Costerton, J.W., and Stoodley, P. (2004). Bacterial biofilms: From the natural environment to infectious diseases. Nat. Rev. Microbiol. *2*, 95–108.

Hansen, S.K., Rainey, P.B., Haagensen, J.A., and Molin, S. (2007). Evolution of species interactions in a biofilm community. Nature *445*, 533–536.

Hao, L., Li, J., Kappler, A., and Obst, M. (2013). Mapping of heavy metal ion sorption to cell-extracellular polymeric substance-mineral aggregates by using metal-selective fluorescent probes and confocal laser scanning microscopy. Appl. Environ. Microbiol. *79*, 6524–6534.

Harrison, J.J., Ceri, H., Roper, N.J., Badry, E.A., Sproule, K.M., and Turner, R.J. (2005). Persister cells mediate tolerance to metal oxyanions in *Escherichia coli*. Microbiology *151*, 3181–3195.

Harrison, J.J., Ceri, H., and Turner, R.J. (2007). Multimetal resistance and tolerance in microbial biofilms. Nat. Rev. Microbiol. *5*, 928–938.

Hassett, D.J., J.F., M., Elkins, J.G., McDermott, T.R., Ochsner, U.A., West, S.E., Huang, C., Fredericks, J., Burnett, S., Stewart, P.S., *et al.* (1999). Quorum sensing in *Pseudomonas aeruginosa* controls expression of catalase and superoxide dismutase genes and mediates biofilm susceptibility to hydrogen peroxide. Mol. Microbiol. *34*, 1082–1093.

Hausner, M., and Wuertz, S. (1999). High rates of conjugation in bacterial biofilms as determined by quantitative *in situ* analysis. Appl. Environ. Microbiol. *65*, 3710–3713.

Hendrickx, L., Hausner, M., and Wuertz, S. (2003). Natural genetic transformation in monoculture *Acinetobacter* sp. strain BD413 biofilms. Appl. Environ. Microbiol. *69*, 1721–1727.

Hobley, L., Ostrowski, A., Rao, F.V., Bromley, K.M., Porter, M., Prescott, A.R., MacPhee, C.E., van Aalten, D.M., and Stanley-Wall, N.R. (2013). BslA is a self-assembling bacterial hydrophobin that coats the *Bacillus subtilis* biofilm. Proc. Natl. Acad. Sci. U.S.A. *110*, 13600–13605.

Irie, Y., and Parsek, M.R. (2008). Quorum sensing and microbial biofilms. Curr. Top. Microbiol. Immunol. *322*, 67–84.

Jiao, Y., Cody, G.D., Harding, A.K., Wilmes, P., Schrenk, M., Wheeler, K.E., Banfield, J.F., and Thelen, M.P. (2010). Characterization of extracellular polymeric substances from acidophilic microbial biofilms. Appl. Environ. Microbiol. *76*, 2916–2922.

Jones, W.L., Sutton, M.P., McKittrick, L., and Stewart, P.S. (2011). Chemical and antimicrobial treatments change the viscoelastic properties of bacterial biofilms. Biofouling *27*, 207–215.

Khakimova, M., Ahlgren, H.G., Harrison, J.J., English, A.M., and Nguyen, D. (2013). The stringent response controls catalases in *Pseudomonas aeruginosa* and is required for hydrogen peroxide and antibiotic tolerance. J. Bacteriol. *195*, 2011–2020.

Kirisits, M.J., Prost, L., Starkey, M., and Parsek, M.R. (2005). Characterization of colony morphology variants isolated from *Pseudomonas aeruginosa* biofilms. Appl. Environ. Microbiol. *71*, 4809–4821.

Kirisits, M.J., Margolis, J.J., Purevdorj-Gage, B.L., Vaughan, B., Chopp, D.L., Stoodley, P., and Parsek, M.R. (2007). Influence of the hydrodynamic environment on quorum sensing in *Pseudomonas aeruginosa* biofilms. J. Bacteriol. *189*, 8357–8360.

Klapper, I., Rupp, C.J., Cargo, R., Purvedorj, B., and Stoodley, P. (2002). Viscoelastic fluid description of bacterial biofilm material properties. Biotechnol. Bioeng. *80*, 289–296.

Kobayashi, K., and Iwano, M. (2012). BslA(YuaB) forms a hydrophobic layer on the surface of *Bacillus subtilis* biofilms. Mol. Microbiol. *85*, 51–66.

Koza, A., Hallett, P.D., Moon, C.D., and Spiers, A.J. (2009). Characterization of a novel air–liquid interface biofilm of Pseudomonas fluorescens SBW25. Microbiology *155*, 1397–1406.

Koza, A., Moshynets, O., Otten, W., and Spiers, A.J. (2011). Environmental modification and niche construction: developing $O_2$ gradients drive the evolution of the Wrinkly Spreader. ISME J. *5*, 665–673.

Langley, S., and Beveridge, T.J. (1999). Metal binding by *Pseudomonas aeruginosa* PAO1 is influenced by growth of the cells as a biofilm. Can. J. Microbiol. *45*, 616–622.

Lapaglia, C., and Hartzell, P.L. (1997). Stress-induced production of biofilm in the hyperthermophile

*Archaeoglobus fulgidus*. Appl. Environ. Microbiol. *63*, 3158–3163.

Larsen, P., Nielsen, J.L., Dueholm, M.S., Wetzel, R., Otzen, D., and Nielsen, P.H. (2007). Amyloid adhesins are abundant in natural biofilms. Environ. Microbiol. *9*, 3077–3090.

Lee, K.W., Periasamy, S., Mukherjee, M., Xie, C., Kjelleberg, S., and Rice, S.A. (2014). Biofilm development and enhanced stress resistance of a model, mixed-species community biofilm. ISME J. *8*, 894–907.

Lembre, P., Di Martino, P., and Vendrely, C. (2014). Amyloid peptides derived from CsgA and FapC modify the viscoelastic properties of biofilm model matrices. Biofouling *30*, 415–426.

Lendenmann, U., and Spain, J.C. (1998). Simultaneous biodegradation of 2,4-dinitrotoluene and 2,6-dinitrotoluene in an aerobic fluidized-bed biofilm reactor. Environ. Sci. Technol. 32, 82–87.

Lewis, K. (2008). Multidrug tolerance of biofilms and persister cells. Curr. Top. Microbiol. Immunol. *322*, 107–131.

Lewis, K. (2010). Persister cells. Annu. Rev. Microbiol. *64*, 357–372.

Liu, H., and Fang, H.P. (2002). Characterization of electrostatic binding sites of extracellular polymers by linear programming analysis of titration data. Biotechnol. Bioeng. *80*, 806–811.

Lock, M.A., Wallace, R.R., Costerton, J.W., Ventullo, R.M., and Charlton, S.E. (1984). River epilithon: toward a structural-functional model. Oikos *42*, 10–22.

Lubarsky, H.V., Hubas, C., Chocholek, M., Larson, F., Manz, W., Paterson, D.M., and Gerbersdorf, S.U. (2010). The stabilisation potential of individual and mixed assemblages of natural bacteria and microalgae. PLoS One *5*, e13794.

Ma, L., Conover, M., Lu, H., Parsek, M.R., Bayles, K., and Wozniak, D.J. (2009). Assembly and development of the *Pseudomonas aeruginosa* biofilm matrix. PLoS Path. *5*, e1000354.

McElroy, K.E., Hui, J.G., Woo, J.K., Luk, A.W., Webb, J.S., Kjelleberg, S., Rice, S.A., and Thomas, T. (2014). Strain-specific parallel evolution drives short-term diversification during *Pseudomonas aeruginosa* biofilm formation. Proc. Natl. Acad. Sci. U.S.A. *111*, E1419–1427.

McLean, R.J., Whiteley, M., Stickler, D.J., and Fuqua, W.C. (1997). Evidence of autoinducer activity in naturally occurring biofilms. FEMS Microbiol. Lett. *154*, 259–263.

Mah, T.F., and O'Toole, G.A. (2001). Mechanisms of biofilm resistance to antimicrobial agents. Trends Microbiol. *9*, 34–39.

Maisonneuve, E., and Gerdes, K. (2014). Molecular mechanisms underlying bacterial persisters. Cell *157*, 539–548.

Matz, C., McDougald, D., Moreno, A.M., Yung, P.Y., Yildiz, F.H., and Kjelleberg, S. (2005). Biofilm formation and phenotypic variation enhance predation-driven persistence of *Vibrio cholerae*. Proc. Natl. Acad. Sci. U.S.A. *102*, 16819–16824.

Miller, M.B., and Bassler, B.L. (2001). Quorum sensing in bacteria. Annu. Rev. Microbiol. *55*, 165–199.

Molin, S., and Tolker-Nielsen, T. (2003). Gene transfer occurs with enhanced efficiency in biofilms and induces enhanced stabilisation of the biofilm structure. Curr. Opin. Biotechnol. *14*, 255–261.

Morris, C.E., Monier, J.M., and Jacques, M.A. (1998). A technique to quantify the population size and composition of the biofilm component in communities of bacteria in the phyllosphere. Appl. Environ. Microbiol. *64*, 4789–4795.

Mulcahy, H., Charron-Mazenod, L., and Lewenza, S. (2008). Extracellular DNA chelates cations and induces antibiotic resistance in *Pseudomonas aeruginosa* biofilms. PLoS Path. *4*, e1000213.

Mullen, M.D., Wolf, D.C., Ferris, F.G., Beveridge, T.J., Flemming, C.A., and Bailey, G.W. (1989). Bacterial sorption of heavy metals. Appl. Environ. Microbiol. *55*, 3143–3149.

Nguyen, D., Joshi-Datar, A., Lepine, F., Bauerle, E., Olakanmi, O., Beer, K., McKay, G., Siehnel, R., Schafhauser, J., Wang, Y., *et al.* (2011). Active starvation responses mediate antibiotic tolerance in biofilms and nutrient-limited bacteria. Science *334*, 982–986.

Nicolella, C., van Loosdrecht, M.C., and Heijnen, J.J. (2000). Wastewater treatment with particulate biofilm reactors. J. Biotechnol. *80*, 1–33.

Normander, B., Christensen, B.B., Molin, S., and Kroer, N. (1998). Effect of bacterial distribution and activity on conjugal gene transfer on the phylloplane of the bush bean (*Phaseolus vulgaris*). Appl. Environ. Microbiol. *64*, 1902–1909.

Parsek, M.R., and Greenberg, E.P. (2005). Sociomicrobiology: the connections between quorum sensing and biofilms. Trends Microbiol. *13*, 27–33.

Penterman, J., Nguyen, D., Anderson, E., Staudinger, B.J., Greenberg, E.P., Lam, J.S., and Singh, P.K. (2014). Rapid evolution of culture-impaired bacteria during adaptation to biofilm growth. Cell Rep. *6*, 293–300.

Pezzoni, M., Pizarro, R.A., and Costa, C.S. (2014). Protective role of extracellular catalase (KatA) against UVA radiation in *Pseudomonas aeruginosa* biofilms. J. Photochem. Photobiol. B: Biol. *131*, 53–64.

Poltak, S.R., and Cooper, V.S. (2011). Ecological succession in long-term experimentally evolved biofilms produces synergistic communities. ISME J. *5*, 369–378.

Potrykus, K., and Cashel, M. (2008). (p)ppGpp: still magical? Annu. Rev. Microbiol. *62*, 35–51.

Priester, J.H., Olson, S.G., Webb, S.M., Neu, M.P., Hersman, L.E., and Holden, P.A. (2006). Enhanced exopolymer production and chromium stabilization in *Pseudomonas putida* unsaturated biofilms. Appl. Environ. Microbiol. *72*, 1988–1996.

Rainey, P.B., and Travisano, M. (1998). Adaptive radiation in a heterogeneous environment. Nature *394*, 69–72.

Ram, R.J., VerBerkmoes, N.C., Thelen, M.P., Tyson, G.W., Baker, B.J., Blake, R.C., Shah, M., Hettich, R.L., and Banfield, J.F. (2005). Community proteomics of a natural microbial biofilm. Science *308*, 1915–1920.

Reisner, A., Haagensen, J.A., Schembri, M.A., Zechner, E.L., and Molin, S. (2003). Development and maturation of *Escherichia coli* K-12 biofilms. Mol. Microbiol. *48*, 933–946.

Romero, D., Aguilar, C., Losick, R., and Kolter, R. (2010). Amyloid fibers provide structural integrity to *Bacillus subtilis* biofilms. Proc. Natl. Acad. Sci. U.S.A. *107*, 2230–2234.

Ronan, E., Yeung, C.W., Hausner, M., and Wolfaardt, G.M. (2013). Interspecies interaction extends bacterial survival at solid–air interfaces. Biofouling *29*, 1087–1096.

Schmitt, J., Nivens, D., White, D.C., and Flemming, H.-C. (1995). Changes of biofilm properties in response to sorbed substances – an FTIR-ATR study. Water Sci. Technol. *32*, 149–155.

Schooling, S.R., and Beveridge, T.J. (2006). Membrane vesicles: an overlooked component of the matrices of biofilms. J. Bacteriol. *188*, 5945–5957.

Schuster, M., and Greenberg, E.P. (2006). A network of networks: quorum-sensing gene regulation in *Pseudomonas aeruginosa*. Int. J. Med. Microbiol. *296*, 73–81.

Schuster, M., Sexton, D.J., Diggle, S.P., and Greenberg, E.P. (2013). Acyl-homoserine lactone quorum sensing: from evolution to application. Annu. Rev. Microbiol. *67*, 43–63.

Schwering, M., Song, J., Louie, M., Turner, R.J., and Ceri, H. (2013). Multi-species biofilms defined from drinking water microorganisms provide increased protection against chlorine disinfection. Biofouling *29*, 917–928.

Sheng, G.P., Yu, H.Q., and Yue, Z.B. (2005). Production of extracellular polymeric substances from *Rhodopseudomonas acidophila* in the presence of toxic substances. Appl. Microbiol. Biotechnol. *69*, 216–222.

Spiers, A.J. (2014). Getting Wrinkly Spreaders to demonstrate evolution in schools. Trends Microbiol. *22*, 301–303.

Springael, D., Peys, K., Ryngaert, A., Van Roy, S., Hooyberghs, L., Ravatn, R., Heyndrickx, M., van der Meer, J.R., Vandecasteele, C., Mergeay, M., *et al.* (2002). Community shifts in a seeded 3-chlorobenzoate degrading membrane biofilm reactor: indications for involvement of *in situ* horizontal transfer of the clc-element from inoculum to contaminant bacteria. Environ. Microbiol. *4*, 70–80.

Staal, M., Borisov, S.M., Rickelt, L.F., Klimant, I., and Kuhl, M. (2011). Ultrabright planar optodes for luminescence life-time based microscopic imaging of $O_2$ dynamics in biofilms. J. Microbiol. Methods *85*, 67–74.

Stewart, P.S. (1996). Theoretical aspects of antibiotic diffusion into microbial biofilms. Antimicrob. Agents Chemother. *40*, 2517–2522.

Stewart, P.S. (2003). Diffusion in biofilms. J. Bacteriol. *185*, 1485–1491.

Stewart, P.S. (2012). Mini-review: convection around biofilms. Biofouling *28*, 187–198.

Stewart, P.S. (2014). Biophysics of biofilm infection. Pathog. Dis. *70*, 212–218.

Stewart, P.S., and Costerton, J.W. (2001). The antibiotic resistance of bacteria in biofilms. Lancet *358*, 135–138.

Stewart, P.S., and Franklin, M.J. (2008). Physiological heterogeneity in biofilms. Nat. Rev. Microbiol. *6*, 199–210.

Stewart, P.S., Roe, F., Rayner, J., Elkins, J.G., Lewandowski, Z., Ochsner, A., and Hassett, D.J. (2000). Effect of catalase on hydrogen peroxide penetration into *Pseudmonas aeruginosa* biofilms. Appl. Environ. Microbiol. *66*, 836–838.

Stewart, P.S., Rayner, J., Roe, F., and Rees, W.M. (2001). Biofilm penetration and disinfection efficacy of alkaline hypochlorite and chlorosulfamates. J. Appl. Microbiol. *91*, 525–532.

Stoodley, P., Cargo, R., Rupp, C.J., Wilson, S., and Klapper, I. (2002). Biofilm material properties as related to shear-induced deformation and detachment phenomena. J. Ind. Microbiol. Biotechnol. *29*, 361–367.

Teitzel, G.M., and Parsek, M.R. (2003). Heavy metal resistance of biofilm and planktonic *Pseudomonas aeruginosa*. Appl. Environ. Microbiol. *69*, 2313–2320.

Thaden, J.T., Lory, S., and Gardner, T.S. (2010). Quorum-sensing regulation of a copper toxicity system in *Pseudomonas aeruginosa*. J. Bacteriol. *192*, 2557–2568.

Todhanakasem, T., Sangsutthiseree, A., Areerat, K., Young, G.M., and Thanonkeo, P. (2014). Biofilm production by *Zymomonas mobilis* enhances ethanol production and tolerance to toxic inhibitors from rice bran hydrolysate. N. Biotechnol. *31*, 451–459.

Tomlin, K.L., Malott, R.J., Ramage, G., Storey, D.G., Sokol, P.A., and Ceri, H. (2005). Quorum-sensing mutations affect attachment and stability of *Burkholderia cenocepacia* biofilms. Appl. Environ. Microbiol. *71*, 5208–5218.

Torsvik, V., Daae, F.L., Sandaa, R.A., and Ovreas, L. (1998). Novel techniques for analysing microbial diversity in natural and perturbed environments. J. Biotechnol. *64*, 53–62.

Traverse, C.C., Mayo-Smith, L.M., Poltak, S.R., and Cooper, V.S. (2013). Tangled bank of experimentally evolved *Burkholderia* biofilms reflects selection during chronic infections. Proc. Natl. Acad. Sci. U.S.A. *110*, E250–E259.

Tseng, B.S., Zhang, W., Harrison, J.J., Quach, T.P., Song, J.L., Penterman, J., Singh, P.K., Chopp, D.L., Packman, A.I., and Parsek, M.R. (2013). The extracellular matrix protects *Pseudomonas aeruginosa* biofilms by limiting the penetration of tobramycin. Environ. Microbiol. *15*, 2865–2878.

Vignaga, E., Sloan, D.M., Luo, X., Haynes, H., Phoenix, V.R., and Sloan, W.T. (2013). Erosion of biofilm-bound fluvial sediments. Nature Geoscience *6*, 770–774.

Wasche, S., Horn, H., and Hempel, D.C. (2002). Influence of growth conditions on biofilm development and mass transfer at the bulk/biofilm interface. Water Res. *36*, 4775–4784.

Waters, C.M., and Bassler, B.L. (2005). Quorum sensing: cell-to-cell communication in bacteria. Annu. Rev. Cell Dev. Biol. *21*, 319–346.

Wessel, A.K., Hmelo, L., Parsek, M.R., and Whiteley, M. (2013). Going local: technologies for exploring bacterial microenvironments. Nat. Rev. Microbiol. *11*, 337–348.

White, C., and Gadd, G.M. (2000). Copper accumulation by sulfate-reducing bacterial biofilms. FEMS Microbiol. Lett. *183*, 313–318.

White, C., Wilkinson, S.C., and Gadd, G.M. (1995). The role of microorganisms in biosorption of toxic metals and radionuclides. Int. Biodeterior. Biodegrad. *35*, 17–40.

Workentine, M.L., Harrison, J.J., Weljie, A.M., Tran, V.A., Stenroos, P.U., Tremaroli, V., Vogel, H.J., Ceri, H., and Turner, R.J. (2010). Phenotypic and metabolic profiling of colony morphology variants evolved from *Pseudomonas fluorescens* biofilms. Environ. Microbiol. *12*, 1565–1577.

Wu, J.H., Wu, F.Y., Chuang, H.P., Chen, W.Y., Huang, H.J., Chen, S.H., and Liu, W.T. (2013). Community and proteomic analysis of methanogenic consortia degrading terephthalate. Appl. Environ. Microbiol. *79*, 105–112.

Xavier, J.B., and Foster, K.R. (2007). Cooperation and conflict in microbial biofilms. Proc. Natl. Acad. Sci. U.S.A. *104*, 876–881.

Xue, Z., Lee, W.H., Coburn, K.M., and Seo, Y. (2014). Selective reactivity of monochloramine with extracellular matrix components affects the disinfection of biofilm and detached clusters. Environ. Sci. Technol. *48*, 3832–3839.

Yachi, S., and Loreau, M. (1999). Biodiversity and ecosystem productivity in a fluctuating environment: the insurance hypothesis. Proc. Natl. Acad. Sci. U.S.A. *96*, 1463–1468.

Zhang, T.C., and Bishop, P.L. (1994). Experimental determination of the dissolved oxygen boundary layer and mass transfer resistance near the fluid–biofilm interface. Water Sci. Technol. *30*, 47–58.

Zhang, X., Bishop, P.L., and Kupferle, M.J. (1998). Measurement of polysaccharides and proteins in biofilm extracellular polymers. Water Sci. Technol. *37*, 345–348.

# Tactic Responses of Bacteria to Pollutants: Implications for the Degradation Efficiency of Microbial Biofilms

4

Diana L. Vullo

## Abstract

In nature, microorganisms are mostly found attached to surfaces forming biofilm communities. The study of these biofilms is primarily focused on basic research to better understand their multicellular way of life, and negative implications for clinical cases and industrial processes. However, there is good reason to increase our knowledge regarding how to stimulate biofilm formation by bacteria, and particularly to exploit these biofilms for their efficient bioremediation of the wide spectrum of the environmental pollutants. Biofilm establishment and maintenance relies on a complex interaction of different mechanisms since bacterial movement and attachment is mediated by swimming, swarming and twitching motility, quorum-sensing mechanisms, biosurfactant secretion and the presence of the chemotactic responses. These chemotactic responses include reactions to inorganic species and xenobiotics that are commonly present in polluted aquatic or soil environments as a result of industrial processes. Microorganisms that display positive chemotactic responses are able to swim towards an adsorbed chemical and, following biofilm formation, can increase pollutant bioavailability by surfactant synthesis, further impacting the rate and extent of pollutant degradation or transformation. The manipulation of biofilm condition and chemotactic response may be managed in *ex situ* bioreactors to improve the bioremediation efficacy of a wide range of pollutants, and particularly for the treatment of metal contaminated media.

## Introduction

### Industrial activities and the environment

Beginning with the Industrial Revolution, the pace of technologic discovery and advancement has continued to increase. The conjunction of a growing world population and the search of a better lifestyle continues to fuel an ever greater demand for the world's natural resources. The intensive consumption of some of these resources, like oil and coal, will lead to their depletion whereas other resources, such as metals, have the potential for almost infinite recovery and reuse, although become more dispersed in our environment over time.

Organic pollutants of environmental concern include natural compounds such as the constituents of coal or crude oil. Alternatively they may be man-made, or anthropogenic compounds, usually termed xenobiotics. Both naturally derived and xenobiotic pollutants can be accidentally or deliberately released by processing industries, causing acute or chronic toxicity. Of particular concern, many anthropogenic substances are known to be highly recalcitrant. On the other hand, metal wastes originating either from material corrosion or discharges from metallurgy, leather tanning, electroplating, mining and pigment industries are not biodegradable once dispersed in the environment, although changes to metal speciation may impact pollutant toxicity and mobility. Therefore, metal emissions contribute to increased environmental concentrations of toxic metals that also have long residence times. Since metals cannot be degraded, this is frequently reflected by the huge remediation costs for metal contaminated soil and water.

The oil and gas industry generates more solid and liquid waste than all other categories of municipal, agricultural, mining, and industrial wastes combined, as has been monitored in the USA (O'Rourke and Connoly, 2003). The frequent oily wastes may be classified into two different categories depending on the ratio of water and solids as (a) simple oil or (b) sludge (Hu *et al.*, 2013). The former is highly viscous, being composed of a high percentage of solids and little water, whereas the latter is a stable water-oil emulsion. The hydrocarbons and other organic compounds in oily sludge can be generally classified into four fractions, including aliphatics, aromatics, nitrogen–sulphur–oxygen containing compounds, and asphaltenes. The aliphatic and aromatic fractions are represented by alkanes, cycloalkanes, benzene, toluene, xylenes, naphthalene, phenols and various polycyclic aromatic hydrocarbons (PAHs), such as phenanthrene, anthracene, chrysene, benzofluorene and pyrene. According to a report from the American Petroleum Institute (API), metals can also be found in oily sludge from petroleum refineries in concentrations around 7–80 mg/kg zinc, 0.001–0.12 mg/kg lead, 32–120 mg/kg copper, 17–25 mg/kg nickel, and 27–80 mg/kg chromium, and concentrations may be much higher in some cases (Hu *et al.*, 2013).

The industrial process of petroleum production consists of two operations: upstream and downstream processing. The first is related to oil exploration, drilling, extraction, transportation and storage. The second directly refers to the refining of the crude components to generate petroleum. In addition to operational leaks, oil spills may occur during extraction and transportation. Interestingly, only information about large oil spills is widely reported; smaller but cumulatively significant spills from shipping, pipelines and leaks often remain undocumented and thus remain hard to quantify (O'Rourke and Connoly, 2003).

Another key waste from the oil is an effluent known as 'produced water'. Produced water is extracted from the ground along with oil and is often reinjected into wells under high pressure to force more oil to the surface. Since a fraction of the produced water is usually not reinjected, its discharge into the environment is the main route for its disposal. Produced water is at least four times saltier than ocean water and often contains benzene, ethylbenzene, toluene and xylene (commonly referred to as BTEX pollutants). In addition, barium, arsenic, cadmium, chromium, and mercury have also been documented as being present in produced waters from oil extraction procedures.

The accidental, or purposeful release of hydrocarbons and contaminating metals into the environment has many implications. Most of the metals have a bio-cumulative effect and are particularly hazardous to both human and environmental health. In terms of hydrocarbons, polycyclic aromatic hydrocarbons (PAHs) are of major concern since they are genotoxic (Kanaly and Harayama, 2010). Oil industry wastes such as PAHs are significant contaminants of both marine and terrestrial environments, but can also migrate down through the soil profile and enter groundwater, thus causing serious adverse consequences in the sub-surface environment.

In addition to the widespread contamination of the environment by the oil processing industry there remain many other routes by which both organic and inorganic pollutants enter our environment. Regarding agricultural activities, the use of pesticides has constantly increased for the last forty years (Dyson, 2000), and in several countries, surface waters situated near rural or urban areas reveal a constant stream of pesticides presumably originating from soils related to agricultural uses (Hoffman *et al.*, 2000; Harman-Fetcho *et al.*, 2005). Although biological treatments may represent a more sustainable alternative to the sole use of agrochemicals, the use of chemicals remains a fundamental tool for controlling most crop diseases (Rice *et al.*, 2007).

## Bioavailability of xenobiotics and metals in soils and water bodies

The biodegradation of both xenobiotics and oil compounds is restricted by their hydrophobicity and low solubility in aquatic systems. Moreover, after remaining in the terrestrial environment for an extended period of time, the aged chemical residues may resist desorption and degradation because of their interaction with soil components, such as humin, fulvic acid, and humic acid. Their biodegradation either in natural or engineered environments is therefore often not as efficient as expected due to limited bioavailability. The bioavailable fraction of a pollutant can be defined as the fraction of the total amount of a chemical present in a specific environmental compartment that, within a given time span, is either available or potentially available for uptake by (micro)organisms (Ferreira *et al.*, 2013a). As a result of sorption to soils and sediments, xenobiotics and hydrocarbons often only exhibit weak chemical activity gradients that promote their uptake and transformation by cells. Accordingly, biodegradation rates may be dependant on restricted phase exchanges, and these compounds, together with their environmental risks, may persist for longer periods of time (Krell *et al.*, 2013).

To enhance hydrophobic compound bioavailability, microbial surfactants – biosurfactants – are often secreted. Such substances can be externally supplied to improve bioremediation processes, both *in situ* and *ex situ*. Table 4.1 provides experimental data of the surface tension (ST) of culture supernatants, expressed in dyne $(g\,cm/s^2)/cm$, derived from natural isolates grown in presence of different carbon sources (Barrionuevo and Vullo, 2013). A decrease in ST values implies the production of tensoactive compounds, such as biosurfactants. Depending on the carbon source, different results can be obtained: a hydrophilic substrate such as glucose induced tensoactive synthesis in the reference strain (*Pseudomonas aeruginosa* PA01) and in the natural isolate *Ralstonia taiwanensis* M2 (Vullo *et al.*, 2008), while hydrophobic substrates such as sunflower oil or the pesticide chlopyriphos made ST decrease in all cases.

Metals do not degrade in the environment, but their toxicity may be altered via microbial transformation. An accurate assessment of potential toxic effects that metals have on aquatic organisms must take into account three key factors: biogeochemical conditions that determine metal speciation, species-specific characteristics, and metal-specific properties. The geochemical and biochemical reactions that the metal must undergo in the environment include, for example, interconversions between various inorganic and organic species of the element in soil and water, leaching of soluble forms, immobilization by ion exchange and other adsorption processes on the solid soil phases, as well as biological uptake by micro- or macro-organisms. Thus, metal species in the environment are relevant to characterize not only the metal toxicity but also metal mobility between different environmental compartments. Toxicity is undoubtedly related to the metal chemical speciation: free cations and labile complexes are bioavailable because of their ability to cross biological membranes (Ferreira *et al.*, 2013a). In the case of strong complexes, metal–cell interactions are reduced. For example, the copper complexing capacity of different aquatic environments has been extensively studied (Santos Echeandía *et al.*, 2008) in cultures of heterotrophic bacteria (Gordon *et al.*, 2000) and on the roots of salt marsh plants (Mucha *et al.*, 2008). It has been observed that they excrete strong Cu-complexing ligands, most likely as a defence mechanism

**Table 4.1** Surface tension (ST) values of culture supernatants belonging to isolates from polluted soils and fresh waters. ST was determined from culture supernatants supplemented with different carbon sources: hydrophilic glucose and hydrophobic sunflower oil or chlorpyriphos

| Culture supernatants | Surface tension (dyne/cm) |
|---|---|
| Culture medium (sunflower oil, glucose) | 77.87 |
| Culture medium (chlorpyriphos) | 78.00 |
| *Pseudomonas aeruginosa* PA01, glucose | 35.31 |
| *Pseudomonas aeruginosa* PA01, sunflower oil | 38.93 |
| *Pseudomonas veronii* 2E, glucose | 76.22 |
| *Pseudomonas veronii* 2E, sunflower oil | 60.17 |
| *Ralstonia taiwanensis* M2, glucose | 50.77 |
| *Ralstonia taiwanensis* M2, sunflower oil | 55.04 |
| *Pseudomonas* sp. Hunvo2, glucose | 76.49 |
| *Pseudomonas* sp. Hunvo2, chlorpyriphos | 50.77 |
| *Pseudomonas* sp. Alnvo2, glucose | 77.41 |
| *Pseudomonas* sp. Alnvo2, chlorpyriphos | 64.11 |

against Cu toxicity. In both cases, Cu-complexing ligand concentrations and the stability constants were usually determined by cathodic stripping voltammetry using competitive ligand exchange (CLE) titrations, a technique well developed for this metal. An anodic stripping voltammetry (ASV) technique was also useful to evaluate moderate ligands present in ordinary culture media, able to complex Cd(II) (Ceretti *et al.*, 2006 and 2010). The labile metal, ligand concentrations and the stability constants ($K'$) can be calculated by ASV-titrations, in order to determine the complexing capacity (CC). Labile metal peak current ($Ip$) is plotted against each added metal concentration to obtain a titration curve (Fig. 4.1). When complexing effects are absent, a linear relation is obtained between both variables. However, when metal ligands are present in the solution, a deviation from such behaviour is expected and lower currents, or smaller labile metal concentrations, are obtained for similar metal additions. Deviation from a linear behaviour as shown in Fig. 4.1 (see arrows) is good evidence of a metal–ligand interaction that can be quantified from the experimental results. The information about $K'$ lies in the curved part of the titration graph. Careful attention must be paid to the slope ($S$) of the linear portion of the titration curve, which should match the linear behaviour obtained in the absence of ligands (i.e. in the buffer solution of the same pH). In fact, the value of $S$ is crucial in the transformation of Ip into a labile metal concentration ($[Me']$) since these two parameters are related by

$$[Me] = \frac{Ip}{S} \tag{4.1}$$

In order to calculate the values of ligand concentration ($L_t$) and the conditional stability constant ($K'$), all possible equilibriums should be considered. The following scheme is a representation of such a situation, where $L_j$ represents possible ligands in the solution being titrated. $L_j$ can also be involved in acid-base equilibriums (ionic charges are not included for the sake of simplicity):

$$\begin{array}{ccc} Me + L_j & \rightleftharpoons & MeLj \\ +nH \updownarrow & & \\ L_jH_n & & \end{array} \tag{4.1a}$$

Mass balances for the species involved are

$$[Me]_t = [Me'] + \Sigma\,[MeL_j]$$

$$[L_j]_t = [L'] + [MeL_j]$$

$K'$ for one of the ligands, $j$, is represented by

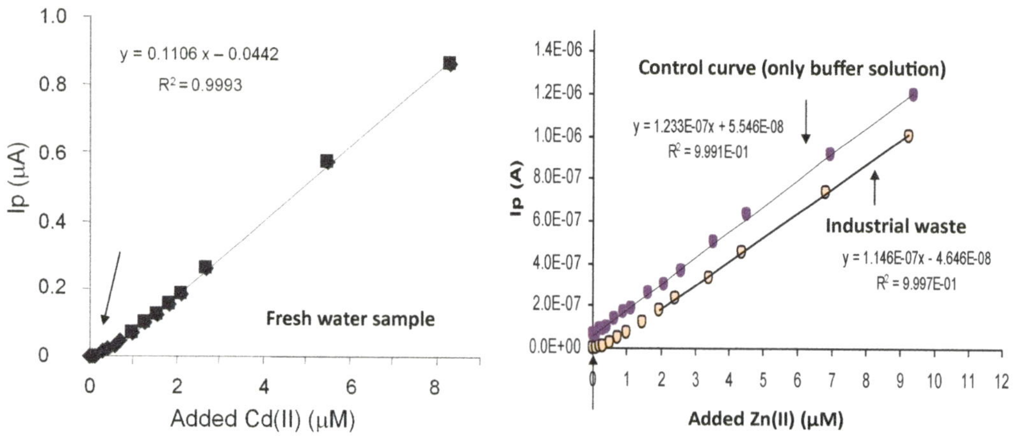


**Figure 4.1** Anodic stripping voltammetry (ASV) titration curves for the determination of Cd(II) complexing capacity in fresh water (left) and determination of Zn(II) complexing capacity in electroplating wastes (right). Arrows show the point at which complexation is detected.

$$K_j = \frac{MeL_j}{[Me]\,L_j} \quad (4.2)$$

where $Me'$ represents all labile forms of metal detected by the analytical technique; $L'$ includes all forms of $L$, except for $L$ associated to the metal. In this presentation only 1:1 complexes are considered. Based on the cases to be considered further, two possible scenarios are taken into account: (a) one metal–one ligand and (b) one metal–two ligands. As a result, generalized theoretical equations can be obtained for the titration curve with one ligand family:

$$[Me]_t = [Me] + \frac{K\,L_t\,[Me]}{1 + K\,[Me]} \quad (4.3)$$

Determination of $K'$ and $Lt$ from the titration curve can be accomplished by data linearization (Ruzic, 1982; Scatchard, 1949); more recently, computer programs can be applied to find the best $K'$ and $Lt$ values to fit the experimental titration curve.

Fig. 4.1 shows ASV titration curves ($Ip$, vs. added metal) for complexing capacity, or CC determinations, in fresh waters for Cd(II) (left), and in electroplating wastes for Zn(II) (right).

Data treatment through a linearization procedure has been performed using Ruzic's method (Ruzic, 1982).

$$\frac{[Me]}{[Me]_t - [Me]} = \frac{[Me]}{L_t} + \frac{1}{K\,L_t} \quad (4.4)$$

A plot of:

$$\frac{[Me]}{[Me]_t - [Me]} \quad (4.5)$$

as a function of $[Me']$ should be linear if the behaviour of the system is consistent with the presence of one family of ligands. $L_t$ and $K'$ can be obtained from the slope and intercept of Ruzic's plot. In both cases CC is related to the presence of moderate ligands ($K' = 10^5$–$10^7$) at a μM concentration (Ferreira *et al.*, 2013b; Alessandrello *et al.*, 2012).

## What can be done about environmental pollutants?

### Bacterial motility and biofilm establishment

In nature, microorganisms are mostly found attached to surfaces forming biofilms. These structures consist primarily of an extracellular matrix (EPS), composed of polysaccharides, proteins and nucleic acids, in which the bacterial cells are embedded. The study of biofilms initially focused on basic research to understand the multicellular way of life in these microbial communities or the negative effects of biofilms on clinical cases and industrial processes. Such negative impacts may occur as a consequence of the high level of adaptation and survival by biofilm-dwelling organisms, such as acquired antibiotic or host immune response resistances (Harmsen *et al.*, 2010). However, there is a good reason to increase our knowledge of how to stimulate biofilm formation in bacteria: to exploit biofilms for the efficient bioremediation of a wide spectrum of environmental pollutants.

Biofilm establishment and maintenance relies on different mechanisms as bacterial movement and attachment is variously mediated by flagella for swimming and swarming motility, by type IV pili for twitching motility and by mechanisms including, quorum sensing, biosurfactant and siderophore secretion and the presence and use of diverse chemotactic responses. The regulation and influence of such mechanisms strictly rely on the species involved in the biofilm development. A schematic description of the different steps of biofilm formation is shown in Fig. 4.2, and is also described in Chapter 2. The first stage consists of cell attachment mediated by physico-chemical interactions or extracellular matrix protein secretion forming a monolayer in the initial hours. Following this step, cell motility mechanisms, such as swimming, swarming, twitching and consequently chemotaxis, trigger biofilm maturation, increasing biomass and horizontal gene transfer processes. After several days, even weeks, cell development begins to decay and biodegradation phenomena occur, finally resulting in biofilm detachment, helped again by cell motility. This sequence is cyclic since these detached and motile cells provide the possibility of a new biofilm being established, but only if the

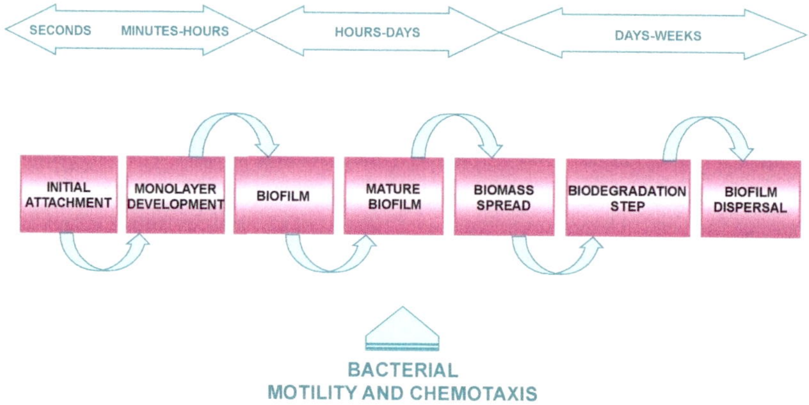


**Figure 4.2** Sequential steps of biofilm development.

environmental conditions allow it (Singh *et al.*, 2006).

For example, in *Pseudomonas aeruginosa* biofilms, motility caused by an active flagellum-driven transport is responsible for the bacteria meeting an appropriate surface for attachment. Biofilm development depends on a wide spectrum of bacterial components that includes flagella (swimming-swarming motilities), type IV *pili* (twitching motility), fimbriae (attachment), extracellular DNA and polysaccharide production (Harmsen *et al.*, 2010). *P. aeruginosa* produces at least two extracellular signals involved in cell-to-cell communication. The cell density-dependent expression of many secreted virulence factors could be also involved in the differentiation of *P. aeruginosa* biofilms. Such cell-to-cell signalling is, for example, shown to be involved in the development of specialized structures of 'fruiting' bacteria like *Myxococcus* (Davies *et al.*, 1998). In addition, heterogeneous biofilms are commonly found due to the coexistence of motile or non-motile subpopulations, dependant on siderophore (pyoverdine) synthesis and regulated by quorum sensing molecules under iron-limiting conditions. The Fe-pyoverdine produced by the stalk cells of the mushroom-shaped biofilms may be required for the settling of migrating bacteria on the 'cap', as iron limitation has been shown to induce twitching motility (Yang *et al.*, 2009). Additionally, biosurfactants may play roles in biofilm formation by facilitating swarming and twitching motilities (Harmsen *et al.*, 2010).

To conclude, as Stewart and Franklin (2008) states '... cells that are growing in biofilms are not only physiologically distinct from planktonic cells, but also vary from each other, both spatially and temporally, as biofilm development proceeds ... Elucidating the physiologies of biofilm-associated bacteria is necessary for our understanding of infection, ecological processes and bioreactor design, as well as other processes that are mediated by microorganisms'. In the following section, we consider the relationship of biofilm physiology with pollutant biodegradation.

## The tactic responses of biofilm bacteria to pollutants in contaminated environments

Since the origin of life on earth, microorganisms have evolved effective mechanisms that help them regulate their cellular functions in response to a changing environment. As shown in Fig. 4.3 on early Earth, when carbon and energy sources were abundant, but the population of microorganisms was low, no significant responses were needed for microbial survival. As a consequence of population increase, the resulting scarcity of carbon and energy sources forced the appearance of new bacterial behaviours such as aerotaxis, phototaxis and of course, chemotaxis. But in present times, new changes have been being introduced in the environment by anthropogenic activities. Exposures to pollutants such as xenobiotics, oil derivatives or inorganic species have triggered new microbial responses. Novel survival strategies have been developed, for example using new metabolic pathways that allow the biodegradation, biotransformation or immobilization of modern-day

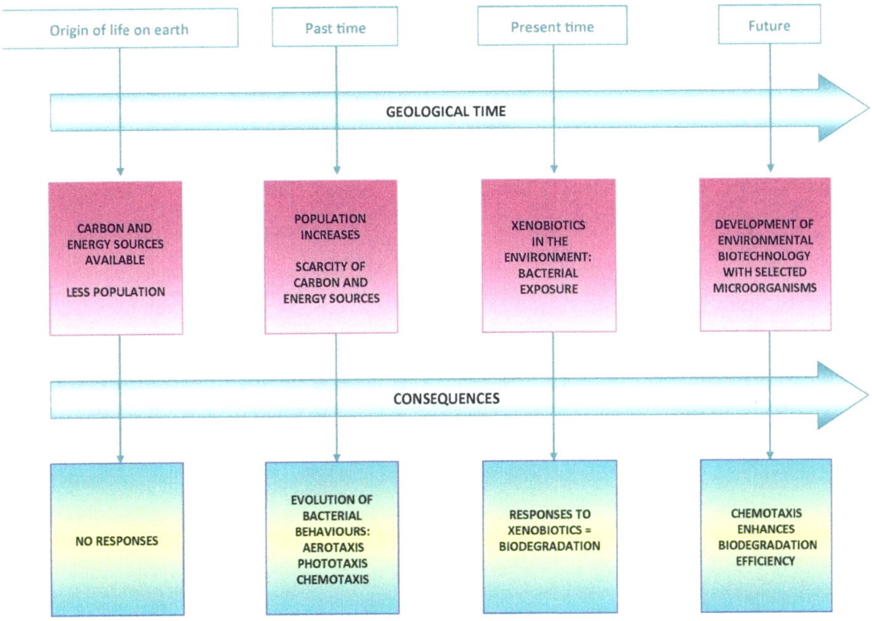


**Figure 4.3** Time sequence of bacterial responses for their survival in the environment. Adapted from Pandey and Jain (2002).

pollutants. These mechanisms remain the focus of studies in order to exploit microorganisms for environmental biotechnology processes, including bioremediation.

The chemotactic response of bacteria, that moderates the swimming of microorganisms towards or against the influence of a chemical gradient, was first studied for non-pollutant hydrophilic compounds as sugars, amino acids and aromatic acids. In the last decades, the study of chemotaxis towards environmental pollutants acquired more relevance because of its application in bioremediation processes. An increased number of chemoreceptors have been found in free-living bacteria, related to their chemoattraction to chemicals of environmental concern. Positive chemotaxis has been observed towards toluene, naphthalene, nitroaromatics, chloroaromatics, hydrocarbons, pesticides and metals as well as in a wide range of bacterial genera including *Pseudomonas*, *Ralstonia*, *Azospirillum* and *Burkholderia* (Iwaki *et al.*, 2007; Lacal *et al.*, 2011; Leungsakul *et al.*, 2005; Liu and Parales, 2009; Pandey *et al.*, 2012). This response is often associated with the fact that these compounds are useful in cell physiology. However, taking into account the toxic potential of most pollutants, it makes sense that bacteria have similarly evolved chemorepellent responses. Bacterial repellence has been reported, for example, to NaClO, $H_2O_2$, the PAHs anthracene and pyrene, some metals including Co(II), Ni(II), Zn(II), Cu(II), Cd(II) and silver nanoparticles (Krell *et al.*, 2013; Barrionuevo and Vullo, 2012a). Interestingly, some chemicals can act as chemoattractants for one bacterial species and be repellent to another. The bacterial response can also be dependent on the physical state of the chemical as was demonstrated for *Pseudomonas putida* G7 for which naphthalene was a chemorepellent in the vapour phase, but an attractant when the compound was dispersed in the aqueous phase (Hanzel *et al.*, 2010). The authors of this study termed this behaviour 'olfactory swimming', allowing bacteria to position themselves at a distance representing a compromise between an acceptable toxicity and a sufficient carbon source supply. Concluding, chemotactic responses can be a consequence of the action of several, potential antagonistic chemoreceptors that may differ in their sensitivity to a given compound.

The role of chemotaxis is a key factor that decides the environmental fate of a pollutant. Fig. 4.4 provides a scheme highlighting the differences between the two chemotactic responses to an organic compound. Positive chemotaxis is the first

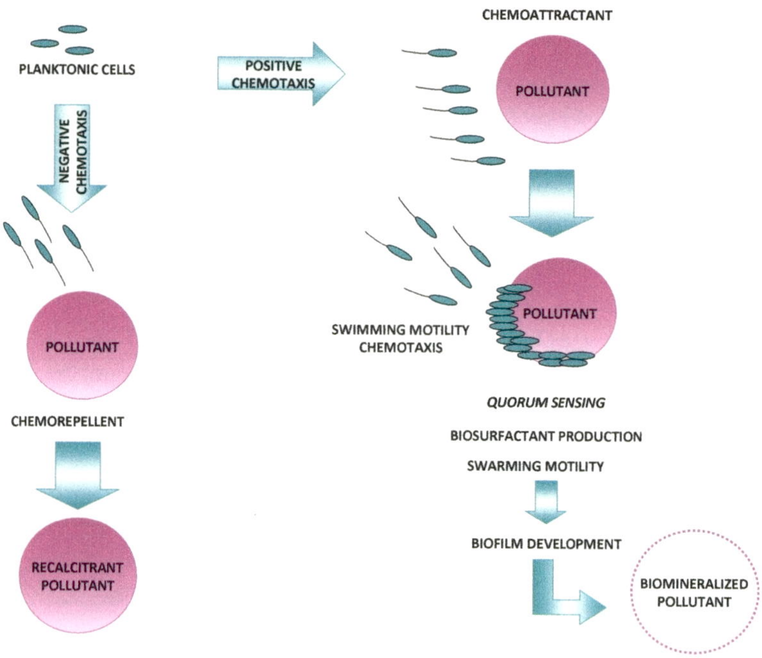


**Figure 4.4** Bacterial behaviour in polluted environments: schematic of the relationship between chemotactic responses and the biodegradation of pollutants, adapted from Chavez *et al.* (2006).

step that may lead to the biomineralization of a substrate. Several events occur immediately after the chemotactic response. Biosurfactant production may be directed to increase substrate bioavailability by emulsification or desorption from solid surfaces. Additionally, quorum sensing may impact biofilm development by determining swarming and twitching motilities. Once the microbial biofilm is installed, the biodegradation proceeds with the removal of the pollutant from this environment.

Some differences can be found if hydrophobic compounds are dispersed in an aqueous medium. Bacterial access to these compounds may be only possible with a direct adhesion to the non-aqueous phase liquid (NAPL)–water interface, a process that is facilitated by chemotaxis and biofilm formation (Singh *et al.*, 2006), as can be clearly visualized in Fig. 4.5. This process was shown in a study on naphthalene biodegradation by *Pseudomonas putida* G7: complete biodegradation was observed when the compound was present in a NAPL, using chemotactic strains; while non-chemotactic strains were not able to degrade the pollutant under such conditions (Krell *et al.*, 2013 and references therein).

Another strategy that enhances biodegradation rates, particularly at substrate–water interfaces, is the accumulation of cell density caused by hydrophobic interactions. For example, when cells were grown in presence of sunflower oil, a natural isolate *Ralstonia taiwanensis* M2 developed only at the water–oil interface where it accumulated at high density. Cell hydrophobicity (in %) can be calculated as:

$$\text{Hydrophobicity}\ (\%) = (1 - OD_f/OD_i) \times 100$$

where $OD_i$ is the optical density (600 nm) of the bacterial culture before its exposure to a non-polar solvent and $OD_f$ the optical density (600 nm) after the exposure. The results are presented in Fig. 4.6, where higher values were obtained when compared to glucose as carbon source (Barrionuevo and Vullo, 2012b; Rosenberg and Rosenberg, 1980) for *R. taiwanensis* M2. A different behaviour was observed for another natural isolate, *P. veronii* 2E (Vullo *et al.*, 2008): no hydrophobicity increase was registered while grown with sunflower oil despite tensoactive compound secretion (see Table 4.1). The biosurfactant secretion and the cell hydrophobicity

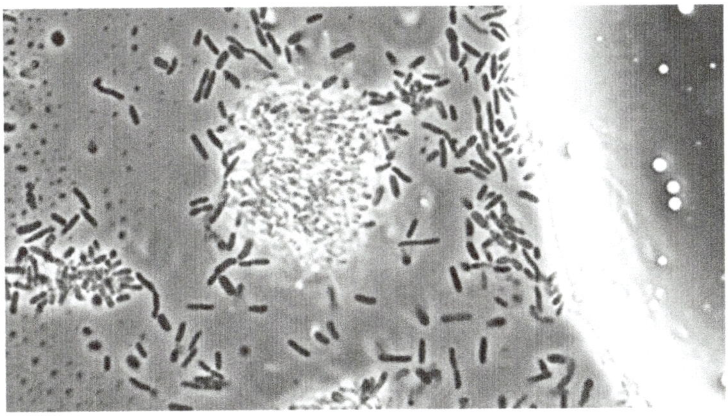

**Figure 4.5** Oil-consuming bacteria covering the oil–water interface of an oil globule viewed at 1500× magnification. Image is provided courtesy of Dr. Johannes Zedelius, MPI Bremen.

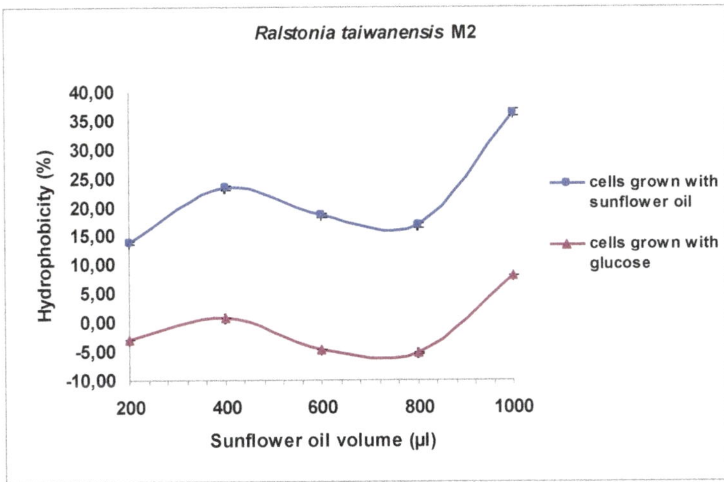


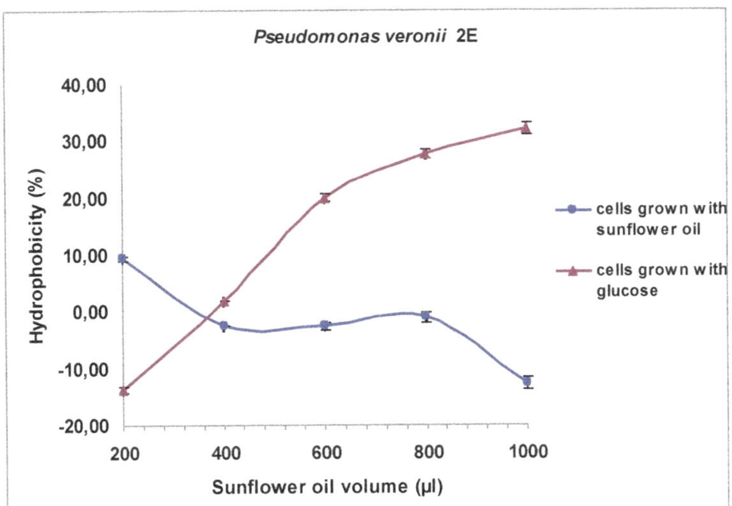


**Figure 4.6** Cell hydrophobicity determinations (as %) as a function of the non-polar solvent used for extraction (sunflower oil in this case).

increase could not be simultaneous responses to the presence of non-polar substrates. In many cases both events may be observed as different stages of biosurfactant production–secretion processes. *P. veronii* 2E was able to decrease ST in culture supernatants without any effect on cell hydrophobicity, while for *R. taiwanensis* M2 the opposite response was observed (Table 4.1 and Fig. 4.6).

In all cases negative chemotaxis will result in cells swimming away from the substrate, enhancing pollutant recalcitrance and increasing the environmental risks and consequences associated with the pollutants ongoing persistence in the environment.

Regarding metals and metalloids such as arsenic, positive or negative chemotaxis may influence their mobility in the environment. As is shown in Fig. 4.7, the interactions between microorganisms and metals depend on the metal itself and its speciation in the environment. Biosorption, bioaccumulation, biotransformation (redox mechanisms) and bioprecipitation by mineral formation are detoxifying strategies that can diminish metal toxicity in the cells immediate environment. Although the presence of efflux, redox mobilization or methylation mechanisms can be beneficial in removing contaminants from within microbial cells, they might not alter metal mobility or, might enhance it (Gadd, 2010).

A spectrum of responses has been observed in two species belonging to the genus *Pseudomonas* (Ferreira *et al.*, 2013a). *P. aeruginosa* PA01, a reference strain with great versatility in adaptive responses to variability in environmental conditions, showed changes in its swarming pattern promoted by cadmium or zinc. *P. aeruginosa* PA01 shows positive chemotaxis towards nutrients but this response becomes negative in the presence of Cd(II) and Zn(II) when concentrations surpassed 0.25 mM and 2 mM respectively. This highlights an overriding negative chemotactic response when a toxin is present (Fig. 4.8). Fig. 4.8 also shows that no changes in positive chemotaxis are observed when adding Cu(II) to the chemoattractant central plug, showing a selective chemotactic response dependent on individual pollutants and their concentrations.

*P. veronii* 2E, a natural isolate with metal-biosorption ability (Vullo *et al.*, 2007, 2008; Barrionuevo *et al.*, 2011; Mendez *et al.*, 2011), stimulated siderophore production when grown

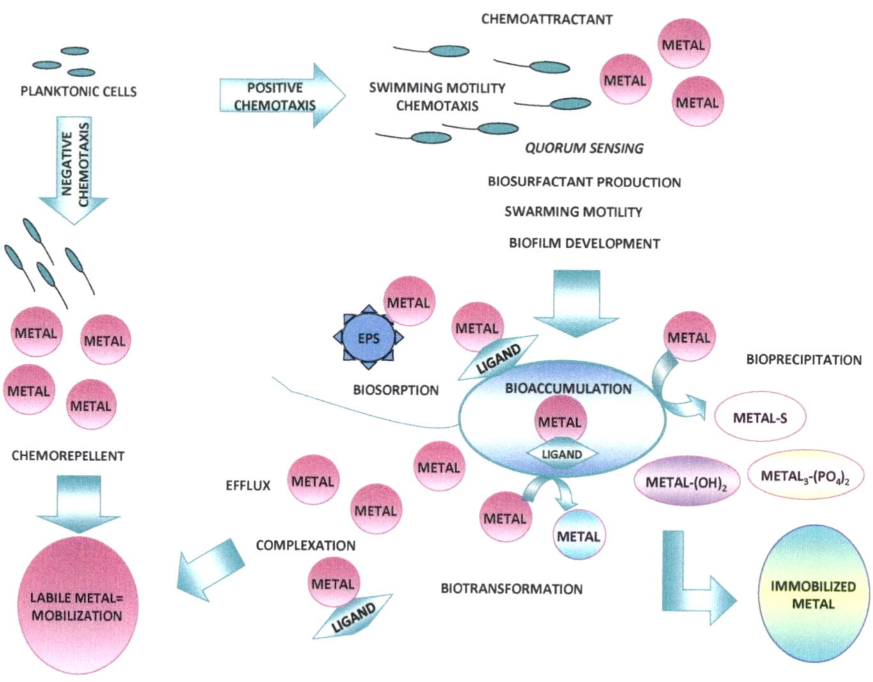


**Figure 4.7** Bacterial behaviour in polluted environments: relationship between chemotactic responses and metals.

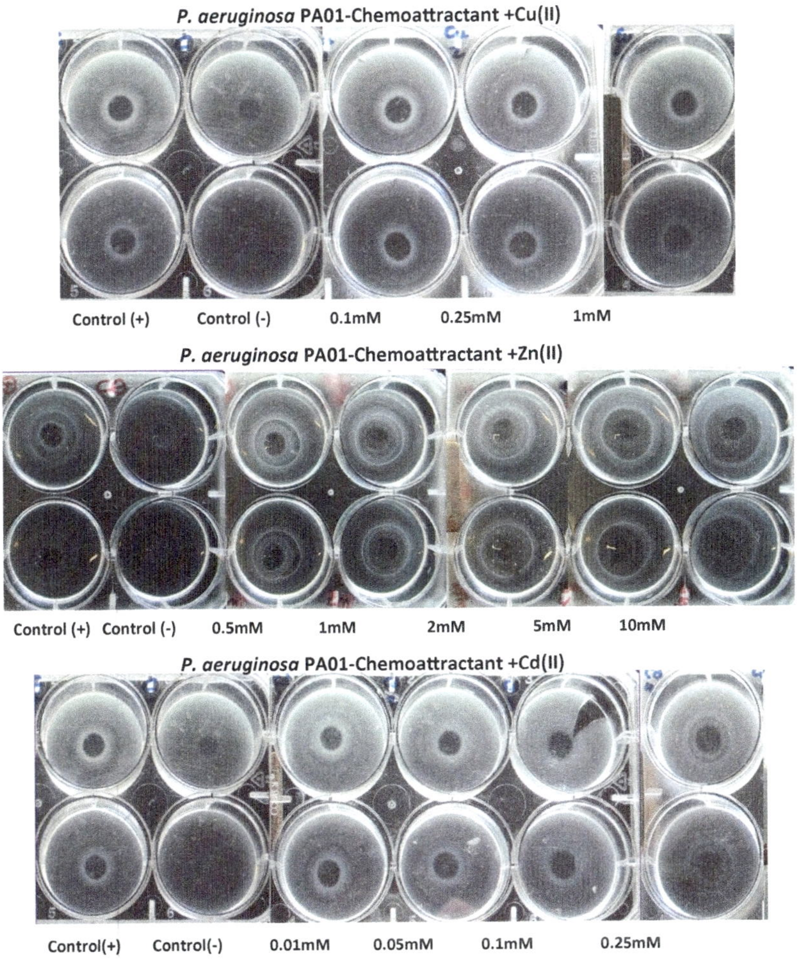


**Figure 4.8** Chemical in plug assays for chemotaxis detection. The central plug contains a chemoattractant compound with different concentrations of Cd(II), Zn(II) or Cu(II). A higher concentration of bacterial density around the plug indicates a positive chemotaxis, while a transparent halo between the plug and a bacterial density ring highlights a negative chemotactic response. Columns represent replicates of each assayed metal.

in an iron-free culture medium, or in a medium supplemented with Cd, Cu, Zn or Ni (0.25 mM). The absorption spectra of these metabolites and their metal-complexes are evidence of metal–siderophore interactions: typical absorbance peak displacements were observed when complexes were formed. Siderophore secretion is characterized by the compositional attributes of the siderophore and ability as a metal ligand. Interestingly, according to Barrionuevo and Vullo (2012a), negative chemotaxis was observed when exposing *P. veronii* 2E cells to metals in chemical in plug assays (Fig. 4.9). However, *Delftia acidovorans* AR, another bacteria isolated from polluted environments (Vullo *et al.*, 2008), exhibited a positive chemotactic response towards Cu(II), highlighting the differential responses of different species and strains to the same metals (Fig. 4.9).

In summary, positive chemotaxis can directly influence cell attachment to any substrate and impact subsequent biofilm development. The biofilm stability will impact the biodegradation of organic pollutants or immobilization of, metals or metalloids. However, the fact that chemotaxis assays are performed with cells in physiological stages more similar to exponential growth-phase planktonic cells than those present in natural biofilm cultures should not be disregarded. As a

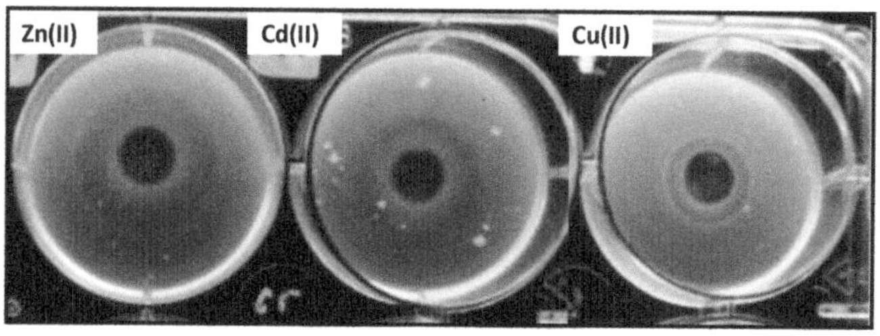


*Pseudomonas veronii* 2E

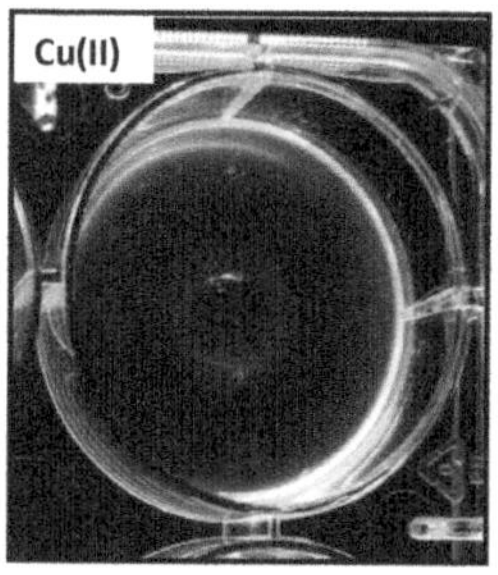


*Delftia acidovorans* AR

**Figure 4.9** Chemical plug assays for chemotactic response analysis of *Pseudomonas veronii* 2E (repellence) and *Delftia acidovorans* AR (attraction) towards metals.

consequence, it is likely that biofilms may establish in polluted environments, despite the negative chemotactic responses that may be observed *in vitro*.

A final influence to be considered is the accumulation of polyphosphates in bacteria. This natural phenomenon is a finely regulated process that depends on phosphate and energy source availability as well as the presence of $K^+$ and $Mg^{2+}$ ions. It is also dependent on additional nutritional or physicochemical factors (N or S deficiency or pH conditions, for example) (Nesmeyanova, 2000). Polyphosphates in bacteria act as phosphate donors, energy sources, as ligands for divalent cations and are involved in global regulatory systems. They are key for the survival for different bacteria, including pathogens under stress conditions (Kulaev and Kulakovskaya, 2000). Back to *P. aeruginosa*, the enzyme polyphosphate kinase is essential for its impact on bacterial motility, such as swimming, swarming and twitching and predictably for biofilm development, quorum sensing and production of virulence factors (Rashid and Kornberg, 2000; Rashid *et al.*, 2000). Polyphosphate-accumulating microorganisms are relevant in environmental biotechnology because of their impacts on metal immobilization (Renninger *et al.*, 2004) and their biotreatment potential for phosphate removal in wastewaters (McGrath and Quinn, 2003; Seviour *et al.*, 2003).

## Biofilms and bioremediation

With either *in situ* or *ex situ* bioremediation, the success of the process will depend not only on biofilm development and the chemotactic response of biofilm bacteria but on various interconnected factors that determine pollutant removal. Fig. 4.10 shows the relationship between microorganisms, pollutant and environmental conditions. Physicochemical conditions [temperature ($T$), pH, oxidation–reduction potential (ORP), water availability ($a_w$)], presence of P or N sources and electron acceptors or donors in cell surroundings will impact microbial activity and biofilm stability in addition to pollutant bioavailability. The biofilm maintenance will be ensured if optimal conditions are carefully maintained. In addition, the effectiveness of *in situ* bioremediation strategies will be dependant on the type of pollutant and the sites hydrogeochemical conditions: biodegradation rates must be faster than pollutant migration to prevent significant contaminant dispersal.

When bioremediation is conducted by biofilm-associated microorganisms, it constitutes a more

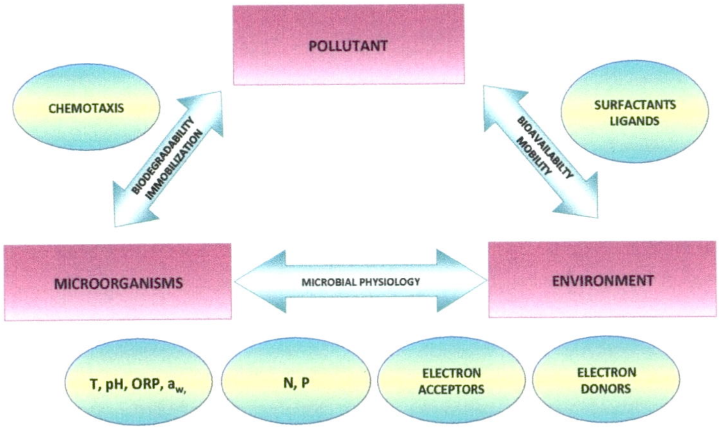


**Figure 4.10** Interconnections between microorganisms, pollutants and environment during bioremediation.

efficient alternative than when mediated by planktonic microorganisms. Cells in the biofilm have passed through different levels of adaptation that allow their survival (especially during periods of stress) and they have the additional protection generated by the matrix which may allow them to tolerate greater concentrations of the pollutant.

The efficiency of the process when using well-established biofilms relies on other limiting factors such as oxygen mass transfer, substrate solubility, toxicity, diffusion and effluent recovery. As an example, the diffusion and kinetic parameters for *Pseudomonas putida* biofilms were determined by Holden *et al.* (1997). Toluene diffusion and reaction in unsaturated biofilms allowed them to predict biodegradation rates. The diffusion coefficient for toluene biofilm was $1.3 \times 10^7\,cm^2/s$, two orders of magnitude lower than toluene diffusivity in water, being the bottleneck in toluene biomineralization in this system. To circumvent this limiting factor, bacterial motility via chemotaxis may be a solution, since it enhances the rate of substrate acquisition.

Xenobiotics- and hydrocarbon-polluted environments can be remediated by either *in situ* or *ex situ* strategies directly depending upon each system. The application of *ex situ* technologies is common way to ensure metal removal with safe and ecofriendly disposal. Biofilm-based reactors are commonly used for treating large volumes of dilute aqueous solutions such as industrial and municipal wastewaters (Fig. 4.11A). Fig. 4.11B provides a schematic representation of more innovative biofilm-based bioreactors composed by membrane reactors or extractive membrane reactors, requiring a two-phase treatment approach. Biofilm is developed in association with the aqueous phase, either on the inner or on the outer membrane surface depending on its location. The organic substrate is delivered through the membrane by diffusion and the product is continuously extracted through the membrane into the organic phase.

A different model appears with plastic composite supports used for biofilm establishment, releasing nutrients for cell growth. Other biofilm reactor configurations are air membrane surface bioreactors and rotating disc biofilm reactors, as well as electro-active biofilm reactors which can oxidize organic compounds present in wastewaters. This oxidation would lead to energy production by finally transferring electrons to a cathode. In all these cases biofilm growth should be controlled to prevent any blockage of the system. This is often achieved by regulating the agitation speed of the reactor fluids and their related shear forces.

Table 4.2 provides some examples of the biotreatment methods using biofilm-based reactors for the removal of xenobiotics or hydrocarbons (Singh *et al.*, 2006).

These biofilm-based bioreactors also represent a useful alternative for the bioremediation of metal loaded wastes. Bacterial growth in a biofilm enhances metal tolerance to lethal concentrations as compared to the corresponding planktonic cells, as has been described for *Pseudomonas* spp. and *Escherichia coli* biofilms (Workentine *et al.*, 2008, and references therein). As ligands, the matrix plays

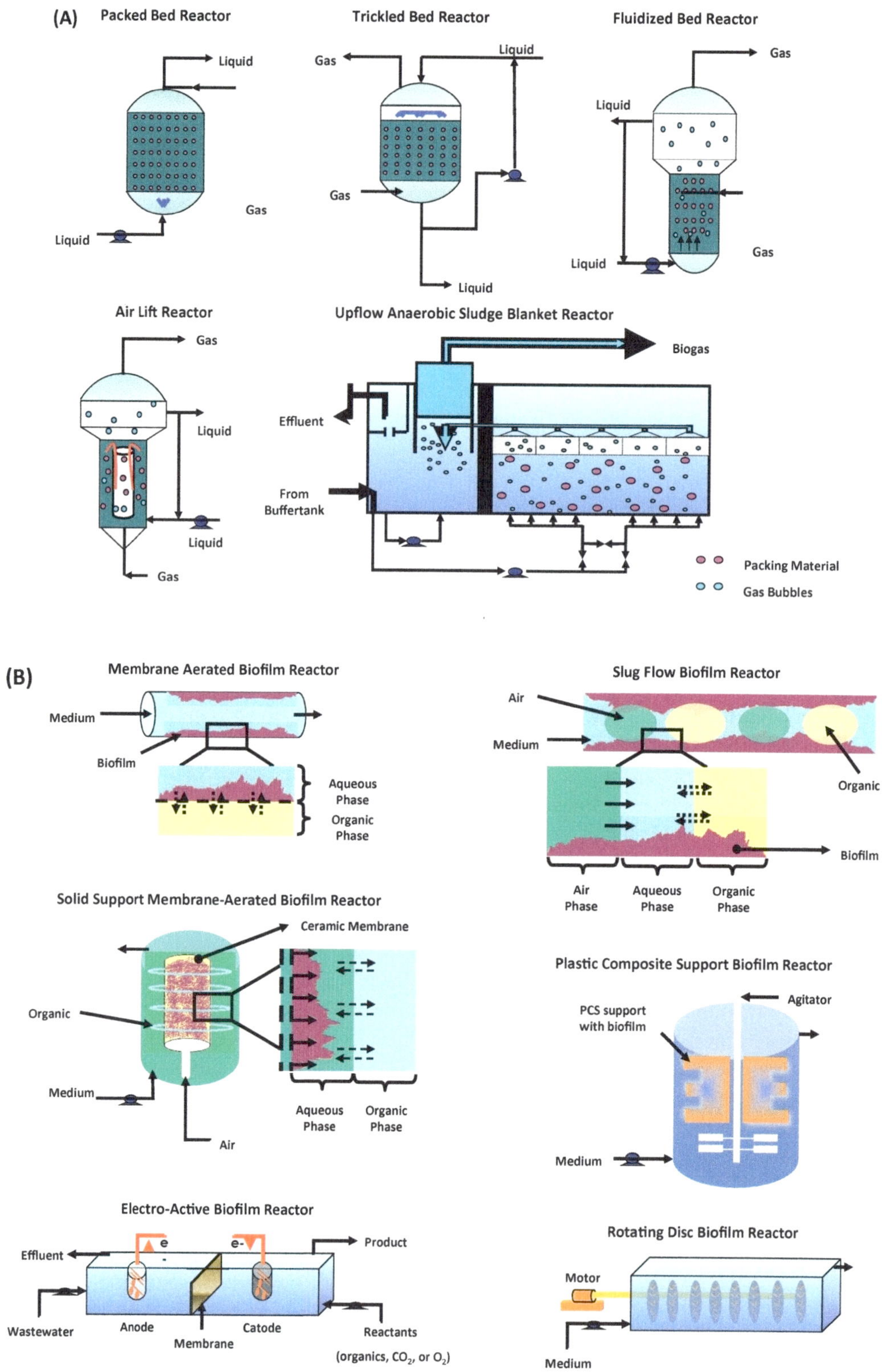


**Figure 4.11** Biofilm-based reactors for waste treatments. (A) Classical reactors and (B) innovative reactors (adapted from Halan *et al.*, 2012).

**Table 4.2** Examples of *ex situ* bioremediation processes developed for xenobiotics and hydrocarbons using biofilm-based bioreactors

| Compound | Reactor type | Microorganisms | Biodegradation efficiency |
|---|---|---|---|
| **Chlorophenols** | | | |
| 2-Chlorophenol | Silicone tube membrane bioreactor | Anaerobic sludge from a swine wastewater treatment plant | >90% |
| 4-Chlorophenol | Granular activated-carbon biofilm reactor | Bacterial consortium from rhizosphere of *Phragmites australis* | 70–100% |
| 2,4-Dichlorophenol | Rotating perforated tube biofilm reactor | *Pseudomonas putida* | ≈100% |
| 2,4,6-Trichlorophenol, 2,4-dichlorophenol, pentachlorophenol | Fluidized bed biofilm reactor | *Pseudomonas* sp., *Rhodococcus* sp. | ≈100% |
| Pyrene, phenantrene | Biofilm grown directly on liquid medium | *Polaromonas* sp., *Sphingomonas* sp. *Alcaligenes* sp., *Caulobacter and Variovorax* sp. | ≈50% (pyrene) ≈98% (phenantrene) |
| *o*-Cresol, naphthalene, phenol, 1,2,3-trimethylbenzene | Biofilm grown on NAPLs | *Pseudomonas fluorescens* | Not determined |
| n-alkanes | Rotating biological contactors | *Prototheca zopfii* | ≈65% |
| Carbon tetrachloride | Continuous-flow fixed biofilm reactor | *Providencia stuartii*, *Pseudomonas cepacia* | ≈100% |
| Toluene | Hollow-fibre membrane biofilter reactor; continuously fed biodrum reactor | Secondary sludge from wastewater treatment plant; aerobic and anaerobic bacteria | ≈84%; 65% |
| **Azo dyes** | | | |
| Acid orange 10, 14 | Laboratory-scale rotating drum biofilm reactor | *Methylosinus trichosporium* | ≈60% |
| Everzol Turquoise Blue G | Laboratory-scale activated sludge unit | *Coriolus versicolor* | ≈82% |
| Herbicides | | | |
| 2-(2-Methyl-4-chlorophenoxy) propionic acid, 2,4-dichlorophenoxyacetic acid | Granular activated-carbon biofilm reactor | Mixed culture of herbicide-degrading bacteria | Partial and complete respectively |

an important role in reducing metal toxicity by capturing cations and diminishing metal bioavailability. This binding capacity of EPS components may be attributed to the presence of functional groups including phosphate, sulphate, carboxylate, hydroxyl and amino groups, which can be evidenced by Fourier transform infrared (FT-IR) spectra. Fig. 4.12 is a typical FT-IR spectrum of the soluble EPS extracted from *P. veronii* 2E culture supernatants, where characteristic functional groups are evidenced (Ferreira *et al.*, 2013b) in each absorbance peak, a consequence of stretching or bending deformations. The binding capacity as CC was determined for Cd(II) by ASV titrations, showing preliminary $K'$ values of $3.78 \times 10^5$ and $4.9 \times 10^5$ and 67 and 83 μM ligand/gEPS at pH 7.5 and 8.1 respectively (Ferreira *et al.*, unpublished data).

EPS is commonly found in activated sludge, and its complexing capacity can be determined under different experimental conditions. The metal binding constants for EPS extracted from activated sludges varied over a wide range: for Cd(II) $-3.7 < \log K < 2.3$ and for Pb(II) $-1.4 < \log K < 4.1$ (Guibaud *et al.*, 2006, 2008, 2009). EPS is a biofilm component that enhances bacterial tolerance

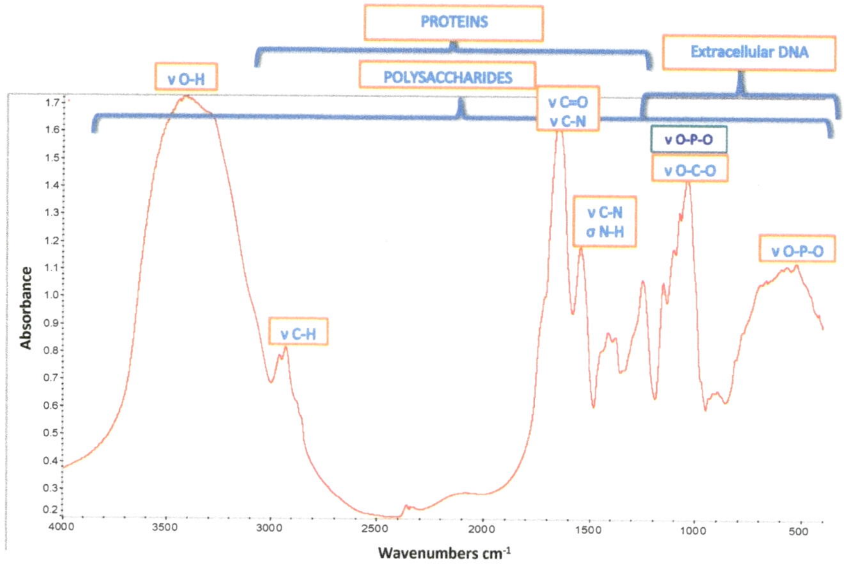


**Figure 4.12** FT-IR spectrum of *P. veronii* 2E soluble EPS. Absorbance peaks due to stretching (υ) or bending (σ) deformation in chemical bonds are characteristic of each functional group.

to changing environmental conditions and boosts metal removal from aquatic environments by promoting metal immobilization.

Considering that chemotaxis assays are often carried out with planktonic cells, some approximations have to be made when correlating biofilm stability to bacterial responses towards metals. As previously discussed by Barrionuevo and Vullo (2012a), for *P. veronii* 2E the presence of Cu(II) did not affect any type of motility despite its activity as a chemorepellent. This should not be a problem when considering *P. veronii* 2E as a living biomass for a reactor design. In the same way, Zn(II) did not alter any motility although this metal presented chemotactical activity as a repellent. Even more, the presence of Zn(II) enhanced chemotaxis towards a chemoattractant compound. These results indicate that Zn(II) would not potentially represent an obstacle for the implementation of bioremediation. A summary of final conclusions includes that despite the inhibition of swarming motility in the presence of Cd(II) and Cr(VI) and taking into account the tolerance to metals evidenced for *P. veronii* 2E, this strain seems to be suitable for fixed bed bioreactors without any adverse effect on an established biofilm. Now discussing *D. acidovorans* AR, neither Cd(II) nor Zn(II) produced a significant alteration on its chemotaxis, so it is likely that a biofilm remained well-established despite their presence. Nevertheless, the development of a new biofilm could be affected since the required mechanisms like swarming motility were altered by Cd(II) and Zn(II). So in this case, *D. acidovorans* AR could be a good candidate for its use in a bioreactor, only if the extraction of Cu(II) is the final goal. In addition, the capacity of this strain to tolerate the presence of chromium made it a good candidate to be used in new technologies designed for Cr(VI) extraction by biotransformation in fixed bed bioreactors. However, it does not seem suitable to be applied for Cu(II) removal; several effects produced by Cu(II) on this strain motility have been demonstrated (Barrionuevo and Vullo, 2012a).

## Conclusions

As a consequence of anthropogenic activities, the appearance of xenobiotic compounds, hydrocarbons, metals and metalloids in the environment have generated disequilibrium in different ecosystems. Microbial communities have shown the ability to restore them, providing an extremely important tool for the remediation of contaminated soil, sediment and water.

Biofilm development and pollutant biodegradation rely on the directed swimming towards or away from pollutants, a process termed chemotaxis. Microbial chemotactic responses are key for the success of bioremediation since they enhance the rate of substrate acquisition either by planktonic bacteria, or by encouraging a biofilm mode of life. In natural or engineered environments, biofilms provide microorganisms unique capacities that allow their exploitation for the removal of the wide spectrum of anthropogenic pollutants.

## References


Alessandrello, M., Barrionuevo, M., Daniel, M.A., Ferreira, M.L., Sosa, G., Ramirez, S.A., Zalts, A., and Vullo, D.L. (2012). Desarrollo de sistemas de biotratamiento de efluentes con contenido en metales. In SAProBio 2012, II Simposio Argentino de Procesos Biotecnológicos 'Prof. Dr. Rodolfo J.J. Ertola', C. Aguilar, S. Cavalitto, F. Luna, N. Rojas, eds. (Universidad Nacional de La Plata, La Plata, Argentina).

Barrionuevo, M., Daniel, M.A., Garavaglia, L., Lopez, N., Mendez, N., Sosa, G., Candal, R., Cerdeira, S., Ceretti, H.M., Ramirez, S.A., *et al.* (2011). Knowing more about metals, microbes and environment interactions: how to improve wastewater biotreatments. In Bioremediation: Biotechnology, Engineering and Environmental Management, A. Mason ed. (Nova Science Publishers, Inc., New York, NY, USA), pp. 383–404.

Barrionuevo, M.R., and Vullo, D.L. (2012a). Bacterial swimming, swarming and chemotactic response to heavy metal presence: which could be the influence on wastewater biotreatment efficiency? World J. Microbiol. Biotechnol. *28*, 2813–2825.

Barrionuevo, M.R., and Vullo, D.L. (2012b). Extracción y Caracterización Preliminar de Biosurfactantes Producidos por Bacterias Autóctonas de la Cuenca del Reconquista. In Ciencia y Tecnología Ambiental: un enfoque integrador, M. dos Santos Afonso and R.M. Torres Sanchez eds. (Asociación Argentina para el Progreso de las Ciencias, Buenos Aires, Argentina), pp. 701–706.

Barrionuevo, M.R., and Vullo, D.L. (2013). Screening of biosurfactant producer bacteria isolated from horticultural soils. IX Annual Meeting of the Argentinean Society for General Microbiology – SAMIGE. Abstract BF21.

Ceretti, H.M., Vullo, D.L., Zalts, A., and Ramirez, S.A.M. (2006). Cadmium complexation in culture media. Electroanalysis *18*, 493–498.

Ceretti, H.M., Vullo, D.L., Zalts, A., and Ramirez, S.A.M. (2010). Bacterial growth and complexing capacity in a culture medium supplemented with cadmium(II). World J. Microbiol. Biotechnol. *26*, 847–853.

Chavez, F.P., Gordillo, F., and Jerez, C.A. (2006). Adaptive responses and cellular behaviour of biphenyl-degrading bacteria toward polychlorinated biphenyls. Biotechnol. Adv. *24*, 309–320.

Davies, D.G., Parsek, M.R., Pearson, J.P., Iglewski, B.H., Costerton, J.W., and Greenberg, E.P. (1998). The Involvement of Cell-to-Cell Signals in the Development of a Bacterial Biofilm. Science *280*, 295–298.

Dyson, T. (2000). World food trends and prospects to 2025. Proc. Natl. Acad. Sci. U.S.A. *96*, 5929–5936.

Ferreira, M.L., Lazzarini Behrmann, I.C., Ceretti, H.M., Zalts, A., Vullo, D.L and. Ramirez, S.A. (2013a). Towards understanding the influence of cadmium on microbial behavior. In Cadmium: Characteristics, Sources of Exposure, Health and Environmental Effects, M. Hasanuzzaman and M. Fujita, eds. (Nova Science Publishers, Inc., New York, NY, USA), pp. 151–170.

Ferreira, M.L., Ramirez, S.A., and Vullo, D.L. (2013b). Characterization of EPS in *Pseudomonas veronii* 2E applicable in wastewater biotreatments. IX Annual Meeting of the Argentinean Society for General Microbiology – SAMIGE. Abstract BB002.

Gadd, G.M. (2010). Metals, minerals and microbes: geomicrobiology and bioremediation. Microbiology *156*, 609–643.

Gordon, A., Donat, J., Kango, R., Dyer, B., and Sturt, L. (2000). Dissolved copper-complexing ligands in cultures of marine bacteria and estuarine water. Marine Chem. *70*, 49–160.

Guibaud, G., van Hullebusch, E., and Bordas, F. (2006). Lead and cadmium biosorption by extracellular polymeric substances (EPS) extracted from activated sludges: pH-sorption edge tests and mathematical equilibrium modelling. Chemosphere *64*, 1955–1962.

Guibaud, G., Bordas, F., Saaid, A., d'Abzac, P., Van Hullebusch, E. (2008). Effect of pH on cadmium and lead binding by extracellular polymeric substances (EPS) extracted from environmental bacterial strains. Colloid. Surface. B *63*, 48–54.

Guibaud, G., van Hullebusch, E., Bordas, F., d'Abzac, P., and Joussein, E. (2009). Sorption of Cd (II) and Pb(II) by exopolymeric substances (EPS) extracted from activated sludges and pure bacterial strains: Modeling of the metal/ligand ratio effect an role of the mineral fraction. Bioresour. Technol. *100*, 2959–2968.

Halan, B., Buehler, K., and Schmid, A. (2012). Biofilms as living catalysts in continuous chemical syntheses. Trends Biotechnol. *30*, 453–465.

Hanzel, J., Harms, H., and Wick, L.Y. (2010). Bacterial chemotaxis along vapor phase gradients of naphthalene. Environ. Sci. Technol. *44*, 9304–9310.

Harman-Fetcho, J.A., Haperman, C.J., McConnel, L.L., Potter, T.L., Rice, C.P., Sadeghi, A.M., Smith, R.D., Bialek Kalinski, K., Sefton, K.A., and Schaffer, B.A. (2005). Pesticide occurence in selected South Florida canals and Biscayne Bay during high agricultural activity. J. Agric. Food Chem. *53*, 6040–6048.

Harmsen, M., Yang, L., Pamp, S.J., and Tolker-Nielsen, T. (2010). An update on *Pseudomonas aeruginosa* biofilm formation, tolerance, and dispersal. FEMS Immunol. Med. Microbiol. *59*, 253–268.

Hoffman, R.S., Capel, P.D., and Larson, S.J. (2000). Comparison of pesticides in eight U.S. urban streams. Environ. Toxicol. Chem. *19*, 2249–2258.

Holden, P.A., Hunt, J.R., and Firestone, M.K. (1997). Toluene Diffusion and Reaction in Unsaturated *Pseudomonas putida* Biofilms. Biotechnol. Bioeng. *56*, 656–670.

Hu, G., Li, J., and Zeng, G. (2013). Recent development in the treatment of oily sludge from petroleum industry: A review. J. Hazard. Mater. *261*, 470–490.

Iwaki, H., Muraki, T., Ishihara, S., Hasegawa, Y., Rankin, K.N., Sulea, T., Boyd, J., and Lau, P.C. (2007). Characterization of a pseudomonad 2- nitrobenzoate nitroreductase and its catabolic pathwayassociated 2-hydroxylaminobenzoate mutase and a chemoreceptor involved in 2-nitrobenzoate chemotaxis. J. Bacteriol. *189*, 3502–3514.

Kanaly, R.A., and Harayama, S. (2010). Advances in the field of high-molecular-weight polycyclic aromatic hydrocarbon biodegradation by bacteria. Microbiol. Biotechnol. *3*, 136–164.

Krell, T., Lacal, J., Reyes-Darias, J.A., Jimenez-Sanchez, C., Sungthong, R., and Ortega-Calvo, J.J. (2013). Bioavailability of pollutants and chemotaxis. Curr. Opin. Biotechnol. *24*, 451–456.

Kulaev, I., and Kulakovskaya, T. (2000). Polyphosphate and phosphate pump. Annu. Rev. Microbiol. *54*, 709–734.

Lacal, J., Munoz-Martinez, F., Reyes-Darias, J.A., Duque, E., Matilla, M., Segura, A., Ortega-Calvo, J.J., Jimenez-Sanchez, C., Krell, T., and Ramos, J.L. (2011). Bacterial chemotaxis towards aromatic hydrocarbons in *Pseudomonas*. Environ. Microbiol. *13*, 1733–1744.

Leungsakul, T., Keenan, B.G., Smets, B.F., and Wood, T.K. (2005). TNT and nitroaromatic compounds are chemoattractants for *Burkholderia cepacia* R34 and *Burkholderia* sp. strain DNT. Appl. Microbiol. Biotechnol. *69*, 321–325.

Liu, X., and Parales, R.E. (2009). Bacterial chemotaxis to atrazine and related s-triazines. Appl. Environ. Microbiol. *75*, 5481–5488.

McGrath, J.W., and Quinn, J.P. (2003). Microbial phosphate removal and polyphosphate production from wastewaters. Adv. Appl. Microbiol. *52*, 75–100.

Mendez, N., Ramirez, S.A.M., Ceretti, H.M., Zalts, A., Candal, R., and Vullo, D.L. (2011). *Pseudomonas veronii* 2E surface interactions with Zn(II) and Cd(II). Global J. Environ. Sci. Technol. *1*, 3.

Mucha, A., Almeida, C., Bordalo, A., and Vasconcelos, M. (2008). Salt marsh plants (*Juncus maritimus* and *Scirpus maritimus*) as sources of strong complexing ligands. Estuar. Coast. Shelf S. *77*, 104–112.

Nesmeyanova, M.A. (2000). Polyphosphates and enzymes of polyphosphate metabolism in *Escherichia coli*. Biochemistry (Moscow) *65*, 309–314.

O'Rourke, D., and Connolly, S. (2003). Just oil? The distribution of Environmental and social impacts of Oil production and consumption, Annu. Rev. Environ. Resour. *28*, 587–617.

Pandey, G., and Jain, R.K. (2002). Bacterial chemotaxis toward environmental pollutants: role in bioremediation. Appl. Environ. Microbiol. *68*, 5789–5795.

Pandey, J., Sharma, N.K., Khan, F., Ghosh, A., Oakeshott, J.G., Jain, R.K., and Pandey, G. (2012). Chemotaxis of *Burkholderia* sp. strain SJ98 towards chloronitroaromatic compounds that it can metabolise. BMC Microbiol. *12*, 19.

Rashid, M.H., and Kornberg, A. (2000). Inorganic polyphosphate is needed for swimming, swarming, and twitching motilities of *Pseudomonas aeruginosa*. Proc. Natl. Acad. Sci. U.S.A. *97*, 4885–4890.

Rashid, M.H., Rumbaugh, K., Passador, L., Davies, D.G., Hamood, A.N., Iglewski, B.H., and Kornberg, A. (2000). Polyphosphate kinase is essential for biofilm development, quorum sensing, and virulence of *Pseudomonas aeruginosa*. Proc. Natl. Acad. Sci. U.S.A. *97*, 9636–9641.

Renninger, N., Knopp, R., Nitsche, H., Clark, D.S., and Keasling, J.D. (2004). Uranyl precipitation by *Pseudomonas aeruginosa* via controlled polyphosphate metabolism. Appl. Environ. Microbiol. *70*, 7404–7412.

Rice, P.J., Rice, Patricia J., Arthur, E.L., and Barefoot, A.C. (2007). Advances in pesticide environmental fate and exposure assessments. J. Agric. Food Chem. *55*, 5367–5376.

Rosenberg, M., Gutnick, D., and Rosenberg, E. (1980). Adherence of bacteria to hydrocarbons: a simple method for measuring cell-surface hydrophobicity. FEMS Microbiol. Lett. *9*, 29–33.

Ruzic, I. (1982). Theoretical aspects of the direct titration of natural waters and its information yield for trace metal speciation. Anal. Chim. Acta *140*, 99–113.

Santos Echeandia, J., Laglera, L., Prego, R., and van den Berg, C. (2008). Copper speciation in estuarine waters by forward and reverse titrations. Marine Chem. *108*, 148–158.

Scatchard, G. (1949). The attractions of proteins for small molecules and ions. Ann. N.Y. Acad. Sci. *51*, 660–672.

Seviour, R.J., Mino, T., and Onuki, M. (2003). The microbiology of biological phosphorus removal in activated sludge systems. FEMS Microbiol. Rev. *27*, 99–127.

Singh, R., Paul, D., and Jain, R.K. (2006). Biofilms: implications in bioremediation. Trends Microbiol. *14*, 389–397.

Stewart, P.S., and Franklin, M.J. (2008). Physiological heterogeneity in biofilms, Nat. Rev. Microbiol. *6*, 199–210.

Vullo, D.L., Ceretti, H.M., Ramirez, S.A.M and Zalts, A. (2007). Heavy metals and microorganisms in the environment: taking advantage of reciprocal interactions for the development of a wastewater treatment. In Progress in Environmental Microbiology, Myung-Bo Kim eds. (Nova Science Publishers, Inc., New York, NY, USA), pp. 111–149.

Vullo, D.L., Ceretti, H.M., Daniel, M.A., Ramirez, S.A.M and Zalts, A. (2008). Cadmium, zinc and copper biosorption mediated by *Pseudomonas veronii* 2E. Bioresour. Technol. *99*, 5574–5581.

Workentine, M.L., Harrison, J.J., Stenroos, P.U., Ceri, H., and Turner, R.J. (2008). *Pseudomonas fluorescens'* view of the periodic table. Environ. Microbiol. *10*, 238–250.

Yang, L., Nilsson, M., Gjermansen, M., Givskov, M., and Tolker-Nielsen, T. (2009). Pyoverdine and PQS mediated subpopulation interactions involved in *Pseudomonas aeruginosa* biofilm formation. Mol. Microbiol. *74*, 1380–1392.

# Whole-cell Biosensors for Monitoring Bioremediation

# 5

Audrey S. Commault and Richard J. Weld

## Abstract

Whole-cell biosensors are sensing devices that typically use biofilm-dwelling microorganisms to detect specific physical or chemical aspects of environmental samples. Microbial detection produces a signal that is transformed into user accessible data which can be as simple as colour change on a paper strip or as complex as quantitative digital display. Whole-cells can be embedded on a transducer or used separately as part of a multi-step assay format. Both unmodified and genetically modified cells have been used in this way. Metabolic reporters are used to detect toxicity that inhibits cell metabolism while catabolic reporters can be used to detect specific contaminants. Biosensors can provide data on the bioavailability of contaminants and are very relevant to monitoring bioremediation. Several commercially available sensors have been developed and some of these have been widely tested and demonstrated to be effective at measuring environmental contaminants.

## Principle of whole-cell biosensors

A biosensor is an analytical device that uses a sensitive biological component combined with a transducer to detect the presence of an analyte of interest. The biological component (e.g. microorganisms, enzymes, aptamer, antibody, etc.) interacts by binding or simple recognition with the target; this interaction produces a signal (optical, electrochemical) read by the transducer, which then transforms it into a recordable signal (Fig. 5.1).

Compared to other biocomponents, intact living cells can be produced cheaply, are relatively easy to manipulate and have higher viability and stability (Byfield and Abuknesha, 1994). They also provide functional information relating to processes such as protein synthesis, apoptotic or necrotic cell death and the bioavailability of a target analyte, i.e. the quantity of the analyte that can pass the cellular membrane and be absorbed into the cell. The biocomponent of whole cell biosensors may include either prokaryotic or eukaryotic cells. Specific microbial biosensors can be achieved by adapting the microorganisms to an appropriate substrate of interest (target) through selective cultivation conditions (Garcia-Balboa and Lopez Rodas, 2012; Kumlanghan *et al.*, 2007) or exploiting specific metabolic capabilities (Zhang and Keasling, 2011).

Whole cell biosensors that use electrochemical techniques can be divided into amperometric, potentiometric, conductometric, voltammetric, and microbial fuel cell (MFC) sensors. An amperometric signal is a current generated by the reduction or oxidation of an electroactive metabolite on the surface of a working electrode poised at a given potential versus a reference electrode (Ding *et al.*, 2008). Conductometry is a measurement of electrolytic conductivity to monitor the production or consumption of ionic species related to the metabolic activity of the microorganisms (Shulga *et al.*, 1994). Potentiometric biosensors are based on the potential difference between a species-selective working electrode and a reference electrode. This potential is dependent on the concentration of the species in solution, which can then be calculated from the Nernst Equation (Torres *et al.*, 2008). In voltammetry, information about an analyte is obtained by measuring the current as the potential is varied. A peak of current is observed at the potential where the redox state of the analyte changes. The potential of the electrons transferred is related

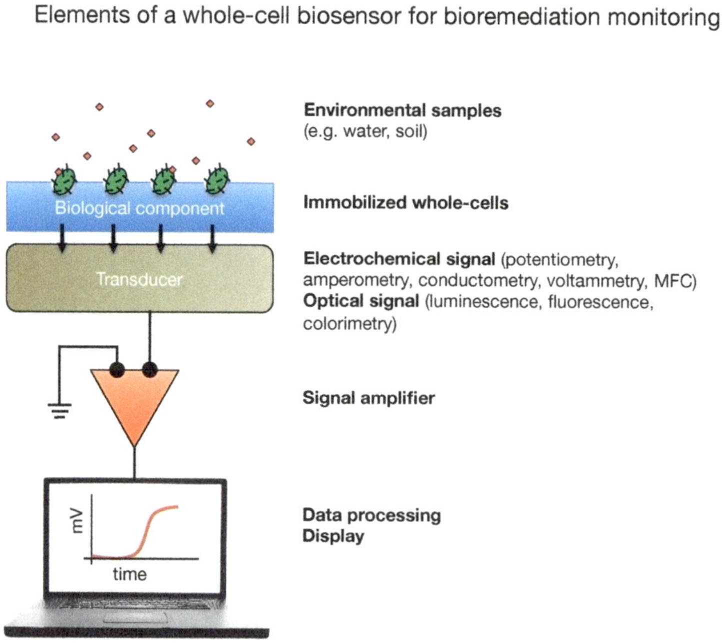


**Figure 5.1** Elements of a whole-cell biosensor for environmental sample monitoring.

to specific redox chemicals and the peak current density is proportional to their concentration in solution. Finally, MFCs are bioelectrochemical devices that convert chemical energy into electrical energy through microbial metabolism (Rabaey and Verstraete, 2005). In MFCs, the current is generated by the microorganisms and so increases with consumption of the analyte or decreases in the case of inhibition of metabolic pathways by toxic compounds. Therefore, MFCs allow the real-time monitoring of analyte concentrations. All these techniques are fast and sensitive, but their selectivity can vary. For instance, conductometry is less selective than potentiometry. Unlike the other methods, voltammetry allows the detection of multiple compounds with different peak potentials in a single scan.

Optical detection is also commonly used in microbial biosensors. The naturally bioluminescent bacteria, *Vibrio fischeri*, has been widely used as a bio-reporter for eco-toxicity (Farre *et al*., 2004) and forms the basis for several commercial assays. Toxicity that decreases metabolic activity results in a decrease in bioluminescence. Analyte-specific induction of microbial luminescence, fluorescence or colour change is possible with genetically engineered microorganisms having a reporter gene fused to an inducible promoter or gene. The activation of the inducible gene by the analyte consequently activates or represses the expression of the reporter gene, resulting in the production of an optical signal (Borisov and Wolfbeis, 2008).

This chapter discusses four different biosensing strategies using unmodified whole cell biosensors: (i) stimulation of microbial respiration by excess nutrients; (ii) inhibition of microbial metabolism due to environmental toxicity; (iii) specialist pathways naturally encoded by the microorganisms; and (iv) microbial degradation of toxicants to easily detectable products. Genetically modified catabolic and metabolic whole-cell bioreporters are also discussed.

## Whole-cell immobilization strategies

Immobilizing microorganisms on transducers is a crucial step in the fabrication of microbial biosensors to maintain long-term cell viability, to increase reproducibility of the response and to avoid spreading genetically modified cells into the environment, especially for field applications. Immobilization also offers a way to preserve the microorganisms without need for continuous cultivation. Whole cells are traditionally immobilized on transducers by physical methods (adsorption and encapsulation) or by chemical methods (covalent binding and cross-linking). More recently, there has been some interest in using natural biofilms.

## Artificial films

Artificial films are used to immobilize cell suspensions within a defined internal architecture by encapsulation, adsorption or covalent binding. Encapsulation consists of entrapping living cells in a matrix that provides a 3D environment where the cells remain viable but stop growing. An ideal immobilization matrix has to be functional at ambient temperature, assure a complete retention of the cells, and be non-toxic and permeable to nutrients and oxygen. The matrix should guaranty the viability of the cells over a long storage period. Organic (e.g. hydrogels like agar) and inorganic polymers (e.g. sol–gel) have shown good performance in cell encapsulation even though the opacity of the organic polymer often alters optical signal detection.

Hydrogels can be defined as organic polymers with a high water content providing a suitable environment for living cells. Agar is one example of a commonly used hydrogel. Immobilization of toxicity-sensing *E. coli* and *Salmonella typhimurium* in a Luria–Bertani agar matrix conferred on the biosensors a stable storage period at 4°C of up to 2 or 6 weeks respectively, without changing their dose-dependent response (Kim and Gu, 2003; Park *et al.*, 2005). An agar matrix was also successful at retaining the abilities of bioluminescent cyanobacterial reporter strains to measure phosphorus bioavailability in water (Schreiter *et al.*, 2001) and *Pseudomonas* strains sensing microbial volatile organic compounds (Ripp *et al.*, 2003). Alginate is another frequently used organic polymer, which can be cross-linked by divalent cations into a highly porous meshwork. Polyak *et al.* (2004) successfully immobilized a bioluminescent *E. coli* strain into a biotin–alginate conjugate for toxicity biosensing. Hydrogels based on polyvinyl alcohol (PVA) are also attractive for whole-cell biosensor applications (Lai Minh *et al.*, 2011; Wang *et al.*, 2006).

Sol-gel encapsulation consists of entrapping viable cells in a silica matrix. The sol–gel process leads to the fabrication of inorganic polymers by simple chemical reactions at close to ambient temperature. However, conventional methods starting from tetraethyl/methyl orthosilicate (TEOS/TMOS) produce ethanol/methanol as by-product, which can be lethal for the encapsulated cells. In the late eighties, Carturan and his co-workers (1989) pioneered the whole-cell immobilization field by encapsulating *Saccharomyces cerevisiae* in sol–gel silicates. *Saccharomyces cerevisiae*, known as bakers' or brewers' yeast, is widely used to convert sugars to carbon dioxide and alcohol, meaning that they can withstand the alcohol by-products of the alkoxide sol–gel processing. Since then, several bacterial strains have been successfully immobilized in silica gels (Carturan *et al.*, 2004). For instance, Yu *et al.* (2005) managed to encapsulate genetically engineered *Moraxella* spp. cells for the detection of organophosphate compounds using an all-aqueous and pH-neutral sol–gel encapsulation procedure. Premkumar *et al.* (2002) successfully encapsulated several recombinant bioluminescent *E. coli* reporter strains in sol–gel silicate to maintain cell viability and luminescence for more than 3 months at 4°C. The silicate film prevented leakage and proliferation of the entrapped bacteria, and protected the encapsulated culture from bacterial contamination. Sol-gel techniques offer the possibility to control the pore size of the silicate network and so control the mass transport of analytes (Bjerketorp *et al.*, 2006).

To avoid substrate limitations, Lee *et al.* (2010) encapsulated genetically modified *E. coli* in poly(ethylene glycol diacrylate) (PEGDA) micro-sized hydrogel microbeads; the diameter of the microbeads obtained was approximately 60 μM. After encapsulation, *E. coli* cells were still producing green and red fluorescent proteins, giving evidence of their viability. The high biocompatibility of the microbeads and the free diffusion of the substrate suggest possible applications of this encapsulation process in bioremediation and biosensing.

Inorganic polymers prevent the microorganisms from leaking out of the gel, while organic polymers also contribute to maintain the bioactivity of the immobilized microorganisms by providing a biocompatible environment (Dong *et al.*, 1996). In an attempt to combine the merits of organic and inorganic polymers, researchers developed hybrid materials (Desimone *et al.*, 2009). Georgieva *et al.* (2011) used silica hybrid materials containing tetramethyl siloxane (TMOS) as an inorganic precursor and apple pectin (AP) as an organic compound to entrap the filamentous yeast *Trichosporon cutaneum* for cadmium and copper ion removal from aqueous solutions. Liu *et al.* (2009) developed a long-term viable BOD sensor by entrapping the same yeast into an organic–inorganic hybrid material composed of silica and the grafting copolymer of poly(vinyl alcohol) and 4-vinylpyridine (PVA-g-P(4-VP)). This biochemical oxygen demand sensor retained 82% of its initial activity after 2 months of continuous measurement with a relative standard deviation (RSD) of the response being ± 8.37% (for three reduplicate measurements). The result indicated that the biocompatibility of hybrid material was good and promising for biosensor fabrication.

Conducting polymers (CP) (e.g. polyaniline, polypyrrole, polyacethylene, etc.) have also attracted much attention for immobilization of microorganisms in biosensors due to their unique electrochemical properties (Ahuja *et al.*, 2007). CPs have the ability to efficiently transfer electric charge produced by the biochemical reactions to an electronic circuit. CPs are commonly deposited on the anode surface by electropolymerisation. The potential of the electrode is cycled from a high to a low potential, and polymerization occurs at a discreet potential, allowing easy control of the thickness of the polymer by adjusting the number of cycles (Baskurt *et al.*, 2012). Baskurt *et al.* immobilized *Gluconobacter oxydans* cells in a conducting polymer with benzothiadiazole deposited on a graphite electrode. Glucose was used as the substrate and the respiratory activity of the cells was successfully recorded when the anode potential was set at −0.7 V versus Ag/AgCl. The biosensor response was directly proportional to the substrate concentration. Jha *et al.* (2009) also used electropolymerization to form a polystyrene sulphonate-polyaniline (PSS-PANI) conducting polymer on a platinum electrode. The bacterium *Brevibacterium ammoniagenes* was immobilized in the polymer to be used as a source of unpurified urease to develop a conductometric urea biosensor. The sensor response was linear over a range of 0–75 mM urea with a sensitivity of 0.125 mM. The sensor could be reused for 12–15 independent measurements and was stable for at least 7 days when stored at 4°C.

Whole-cell immobilization can also be achieved by attachment to a surface by adsorption or covalent binding. This method overcomes one of the major limitations of entrapment which is that the cells are not directly exposed to the analyte but must rely on substrate diffusion. Cells can be physically absorbed on filter-paper strips for fabrication of cheap and portable biosensors (Stocker *et al.*, 2003) with, however, the inconvenience of long incubation times and potential spreading of GMOs into the environment. Adsorption is affected by the surface charge and the hydrophobicity of the cell surface (Sharma and Rao, 2002). The latter plays a critical role in the adherence of bacteria to a wide variety of surfaces. In general, hydrophobic bacteria adhere on hydrophobic surfaces, whereas hydrophilic ones prefer hydrophilic surfaces (Sharma and Rao, 2002). Piccirillo *et al.* (2013) adsorbed selected bacterial strains onto natural hydroxyapatite ($Ca_{10}(PO_4)_6(OH)_2$, HAp) to act as heavy metal removers from aqueous streams. They observed that the more hydrophilic strain had the higher adherence to the HAp surface, since this material is hydrophilic under normal conditions. The same observation has been made by Guo *et al.* (2013) with electroactive microbes (e.g. *Geobacter*). They demonstrated that positively charged and hydrophilic surfaces were more selective for electroactive microbes and more conductive for electroactive biofilms. *Geobacter* sp. was the first organism found to have the ability to oxidize organic compounds and metals, including iron, radioactive metals and petroleum compounds in the environment while using iron oxide or other available metals as an electron acceptor. *Geobacter* is also capable of direct electron transfer to an electrode, making it a good candidate for biosensing applications (Logan, 2009).

Covalent binding of a cell on a surface requires surface modification that does not affect cell viability. Grafting carboxyl- or amine- groups onto electrode surfaces allows the immobilization of living cells (Kuzume *et al.*, 2013; Meyer *et al.*, 2010). The findings of Kuzume *et al.* (2013) indicate that carboxyl-group grafted gold electrodes provide anchoring for the outer-membrane cytochromes of *Geobacter sulfurreducens*, facilitating the electron transfer at the microorganism–electrode interface.

An important point to be considered when

immobilizing whole cells on a surface is the complex intercellular communication, called quorum sensing (QS) that influences cell behaviour (Mitchell *et al.*, 2011). At high cell density, microbes produce and respond to signalling molecules, 'auto-inducers', that regulate the expression of genes involved in genetic competence, biofilm formation and persistence, virulence, sporulation, bioluminescence, antibiotic production, and many other processes (Mitchell *et al.*, 2011). Cell surface immobilization produces high local cell densities that are likely to be QS up-regulated. It is still unclear what consequences this can have on the viability of immobilized cells, but it is certainly something to keep in mind when developing immobilized whole-cell biosensors.

All the immobilization methods described have their limitations. Covalent binding can greatly affect cell function and viability. Physical adsorption suffers from poor stability through microbe desorption. A major disadvantage of encapsulation is the additional diffusion resistance caused by the entrapment material, which would result in lower sensitivity of the biosensor. How to simply immobilize microbial cells without affecting their functionalities is a critical step in the fabrication of sensitive and cost-effective whole-cell biosensors (Su *et al.*, 2011).

## Cell microarrays

Cell microarrays are miniaturized whole-cell biosensor systems in which living cells are immobilized on platforms such as microchips, glass slides or optic fibres, for high-throughput analysis of many samples in parallel (Elad *et al.*, 2010).

To obtain fully automated environmental monitoring devices, it is critical to reproduce the natural cell environment on the solid support to overcome possible limitations due to cell preservation. The support, often called a 'chip', can be made of diverse substrates such as gold, silicon, glass, etc. The chip is spotted with varying biocompatible materials including oligonucleotides, antibodies, proteins, or lipids, which can interact with the cells, leading to their capture on specific spots (Ben-Yoav *et al.*, 2011). Simple adsorption of protein or oligopeptides is not suitable for long-term cell immobilization because many factors can affect binding, such as protein desorption or denaturation and uncontrolled sequence orientation on the surface. Covalent binding of peptide sequences on the support is thus necessary.

Some materials have the capacity to self-assemble as monolayers at the surface of solid supports, for example silicon and aluminium oxides, long-chain carboxylic acids on metal oxides and organosilane species on hydroxylated glass (Ulman, 1996). By modifying the end group of those monolayers, it is possible to modify their interfacial properties and so facilitate cell adhesion (Samanta and Sarkar, 2011). Choi *et al.* (2005) reported, for the first time, the self-assembly of a modified Arg-Gly-Asp peptide terminated with cysteine amino acid onto a gold surface. This thin film was used to immobilize *E. coli* O157:H7 for the detection of environmental pollutants such as phenol.

Different strategies have been investigated in the past two decades for the immobilization of reporter strains in microarrays for environmental toxicity monitoring. Nutrient agar was used on a chip or in a 384-well plate to capture 20 different biosensor strains of *E. coli* bacteria for the detection and classification of toxicity (Lee *et al.*, 2005). Kuang *et al.* (2004) crafted an *E. coli* genotoxin whole-cell biosensor, by inserting the cells into microwells on one end of an imaging fibre bundle with only one cell per microwell. The individual cellular responses were monitored simultaneously enabling sensing of a genotoxin with high sensitivity and an incubation time of 90 min. This sensor was reported to have an active sensing lifetime of more than six hours at ambient temperature and a shelf lifetime of two weeks when stored in minimal medium at 4°C. Hydrogel has also been used for bacterial cell immobilization on microchips. Polyacrylamide nanodrops containing an *E. coli* reporter strain responsive to arsenite deposited onto a glass support showed biosensing properties (Fesenko *et al.*, 2005). Following immobilization, bacterial viability was estimated to be approximately 60%, but storage performance was not described. Popovtzer *et al.* (2005) reported an electrochemical biochip, employing genetically engineered *E. coli* as sensor elements for toxicity detection in water. A current was produced within less than 10 min of exposure to various concentrations of toxicants with a direct correlation between the signal and the toxicant concentration.

Another state-of-the-art technique, is inkjet printing that 'prints-out' nano- to femtolitre volumes of cells onto surfaces such as glass slides, agar plates, or nitrocellulose membranes. This method does not affect the viability of bacterial cells as it has been reported that they retain 98.5% of their viability after being printed through a piezoelectric inkjet printer (Merrin *et al.*, 2007).

## Single-cell immobilization

A biosensor based on a cell population delivers a signal that is the average of the response of all individual cells. A mixed community will not be made of identical cells and, therefore, their responses will be different. The same problem occurs in clonal populations, even though the cells originated from one parent cell, stochasticity in gene expression will introduce genetic variability, which can also alter their response. Measurement at a single-cell level will, therefore, provide information that would be lost during analysis of whole populations (Cai *et al.*, 2006). A single bacterial cell biosensor that has been engineered to produce a measurable signal in the presence of one or more chemical targets, is a promising tool for analysis of microenvironment properties. Such a device can help us to understand the microphysiology of bacterial cells and its importance for macroscale processes like pollutant biodegradation, beneficial bacteria–eukaryote interactions, and infection (Tecon and van der Meer, 2006).

A few studies have reported the use of single-cell biosensors for the determination of microscale bioavailability of pollutants. By immobilizing an engineered *E. coli* sensor cells on the bottom glass plate of a microscope flow chamber, Jasper *et al.* (2001) monitored the diffusion process of octane in aqueous solution. The presence of mercury in samples was also monitored by immobilizing an *E. coli* biosensor strain on the face of an optical imaging fibre containing a high-density array of microwells. Each microwell accommodates a single bacterium whose response was measured when exposed to mercury. A response to 100 nM $Hg^{2+}$ could be detected after 1 h of incubation (Biran *et al.*, 2003).

## Natural biofilms

In nature, bacteria naturally assemble as groups of cells held together by an extracellular polymeric substance (EPS) matrix. Such groups of cells, referred to as biofilms, are ubiquitous on our planet and represent a common metabolic strategy in the microbial world (Costerton *et al.*, 1995). The dense extracellular matrix of the biofilm provides several advantages to the cells including protection against environmental stress, easy access to nutrients and resistance to detergents and antibiotics. Biofilms, besides being resistant to many stresses, have a reduced growth rate, facilitating their use as reliable biocomponent of biosensors. Additionally, the microbes within biofilms are able to compete, cooperate and freely communicate through quorum sensing, offering optimal conditions for cell immobilization. Even though microorganisms have non-specific metabolisms, highly selective microbial biosensors can be potentially achieved by adapting the microorganisms to a substrate of interest (target) through selective cultivation conditions. Therefore, it is potentially possible to engineer biofilm communities for different applications such as monitoring biological oxygen demand (BOD), the prevalence of antibiotics, or heavy metal sensing.

Microbial fuel cells (MFCs) have been used for several years as biosensors for measuring environmental parameters such as BOD and water toxicity. In such systems, the immobilization of the biofilm on the anode did not require the use of artificial techniques, the biofilms are simply selected using different inoculums and selective conditions. Modin and Wilen (2012) developed a bioelectrochemical BOD biosensor based on a mixed community biofilm selected from aerobic and anaerobic sludge from a wastewater treatment plant. The biofilm was engineered by poising the electrode at a fixed potential (0 V against Ag/AgCl reference electrode) while feeding it a nutrient medium with sodium acetate as sole carbon source. The current delivered by the biofilm showed a linear correlation for BOD concentrations ranging from 32 to 1280 mg/l in a reaction time of 20 h. Using the same approach, Davila *et al.* (2011) focused on the detection of toxic matter using a silicon-based MFC. The biocomponent was an anodic biofilm made of a pure culture of *Geobacter sulfurreducens* to operate the micro-device as a direct electron transfer MFC. A drop in the power production from the biofilm was observed when a toxic compound (0.1% formaldehyde) was added to the anode compartment.

These works illustrated the use of pure and mixed community biofilms as biocomponents of biosensors. Usually, BOD sensor biofilms are made of mixed communities selected from wastewater to assure a broader substrate usage than a pure culture biofilm (Peixoto *et al.*, 2011; Yang *et al.*, 2013; Zhang and Angelidaki, 2012). Moreover, mixed cultures are likely to be more stable as observed by Ishii *et al.* (2012), who determined that the dominant species changed with time over their year-long experiment, but the electricity generating performance and chemical oxidation rates remained consistent. Unlike mixed cultures, the stability of a pure culture may be challenged by regular introduction of competitors within test samples such as wastewater. In contrast, mixed communities are more likely to be self-optimizing when exposed to new carbon source fluctuations or competitors. This self-optimization is allowed only because the cells are not entrapped in a silicate matrix, adsorbed on the anode surface or covalently bound to it, and can therefore leak, grow and freely communicate. Developing specific mixed communities can also facilitate co-adhesion and synergistic interactions within the biofilm, making it more robust to environmental or substrate changes.

## Unmodified whole-cell biosensors

### Biological oxygen demand (BOD) sensing

One of the most ancient and universal environmental pollutants requiring remediation is not in itself an inhibitor of life but in fact a stimulant. Excess nutrients and organic carbon deliberately applied to land and water for agricultural purposes or emanating unintentionally as a consequence of industry, agriculture and the general business of human life, have the capacity to unbalance natural systems and tip them towards a state that is unfavourable for many multi-cellular beings. The basis of this environmental degradation is the stimulation of microbial growth, potentially causing blooms of toxin-producing microbes and/or oxygen depletion, both effects being detrimental to many of the forms of life that we view favourably. Organic pollution is one of the most widespread environmental problems (Zhang and Angelidaki, 2011). Wastewater with high organic loads is typically treated either prior to discharge by biological degradation in constructed systems or by natural refinement processes after discharge. If these are overwhelmed by the volume of waste, oxygenation can also be used to enhance natural refinement processes (Salanitro *et al.*, 2000). Organic carbon pollution can be measured as a function of biochemical oxygen demand (BOD); the theoretical amount of extra oxygen that would be consumed as a consequence of the aerobic microbial metabolism of that carbon.

BOD is typically measured by the $BOD_5$ assay, which is a 5-day incubation of initially air-saturated, diluted, aliquots of the sample with mixed communities of bacteria. The depletion of oxygen within the samples due to microbial metabolism is used as a measure of the BOD pollution originally contained in the samples. The $BOD_5$ assay is a relatively slow whole cell bioassay with its speed being one limitation for its utility and the drudgery of setting up multiple incubations in 300 ml bottles is also perceived negatively by those tasked with this chore. The assay is also constrained by the poor solubility of oxygen in water, which means that precisely diluted samples are needed to avoid the concentration of oxygen being limiting. Results of the $BOD_5$ assay can also depend on the temperature, the possible presence of toxins and the type of microorganism used. Efforts have been made over the years to speed up the measurement of biologically available nutrients using assays and biosensors that still retain the biological aspects of the $BOD_5$ assay. These efforts have generally focussed on: using shorter incubations, coupling oxygen sensors with microbial biofilms (Liu and Dong, 2013; Liu *et al.*, 2013; Torrents *et al.*, 2012); replacing oxygen, which is poorly soluble in water, with more soluble, chemical redox mediators such as potassium ferricyanide (Catterall *et al.*, 2001; Catterall *et al.*; Jordan *et al.*, 2013; Morris *et al.*, 2001; Pasco *et al.*, 2001, 2004); measuring $CO_2$ production rather than $O_2$ consumption (Chiappini *et al.*, 2010; Melidis *et al.*, 2008) or using electro-active microbial fuel cells (MFCs) and measuring electrical current from microbial respiration (Kumlanghan *et al.*, 2007; Modin and Wilen, 2012; Zhang and Angelidaki, 2011). All these methods essentially are using some form of measurement of microbial respiration as a measure of biologically available carbon. Several groups have developed rapid technology in this space, some of which is now commercially available (e.g. MB-DBO developed by the Spanish company Biosensores and SciTOX™ developed by Scitox Limited in New Zealand).

A problem for all rapid BOD technologies is that the 5-day $BOD_5$ assay is the incumbent standard against which all newcomers are compared. The five-day duration of the $BOD_5$ assay is an arbitrary choice and the assay is not free from variability. Compressing the $BOD_5$ assay from 5 days into an assay that takes only a few minutes or hours may seem a simple matter of speeding up the reaction by increasing the concentration of bacteria or electron acceptor. This assumption though, misjudges the complexity of the process of bacterial degradation (Pasco *et al.*, 2004; Zhou *et al.*, 1996). The strategy probably works reasonably well where only simple substrates are being metabolized. Where more complex substrates are present there may need to be a period of adjustment during which consortia of bacteria are formed allowing efficient degradation. Such consortia develop over time and whereas 5 days is a reasonable time to allow groups and biofilms to develop, a few hours are not. In fact, it would be reasonable to expect that a mixed community of bacteria incubated in the $BOD_5$ assay would be physiologically and genetically distinct after five days incubation, compared to the community at the start of the 5 days. There will be disproportionate growth of the best adapted community members, physiological adaptation to the changing growth conditions throughout the incubation, and associations formed between different members of the community. These adaptations cannot reasonably be expected to be mimicked in a rapid assay that takes a few hours. Population adjustments will take days not hours and even physiological adjustments would take up a significantly higher proportion of a short assay than they would of a long assay (Pasco *et al.*, 2004). The use of mature biofilms, such as enriched mixed consortium selected from anaerobic sludge over several months, could reduce the measuring time as their cellular machinery coupled with symbiosis phenomena allow them to degrade organic matter from the simplest to the most complex form. Studies showed reasonable reproducibility (variability lower than 10% between samples) of mixed consortium, but often a limited sensitivity (Kim *et al.*, 2003; Kumlanghan *et al.*, 2007; Sakai *et al.*, 1995).

The ideal BOD biosensor would be simple, fast, sensitive, reproducible, proportional to the $BOD_5$ assay, with a wide working range and achieving a high degree of substrate degradation (Jordan *et al.*, 2013). Any single biosensor approach may not be able to achieve all these criteria as there are trade-offs between each benefit (Jordan *et al.*, 2013). Variability in results within samples from different sources and even from the same source remains an issue for rapid BOD methods. Variability might be reduced to some extent by using a single species of bacteria with a broad substrate spectrum as the biocomponent in the sensor rather than using a mixed consortium of bacteria, for example a community taken from activated sludge (Liu and Mattiasson, 2002; Raud and Kikas, 2013). However, single bacteria are unlikely to have the same breadth of substrate metabolism as multi-species populations and so are less likely to match the results of the $BOD_5$ assay that exploits mixed bacterial communities (Chiappini *et al.*, 2010; Liu and Mattiasson, 2002). One promising method that attempts to overcome this conundrum is the use of 'bioelectronic tongues' – arrays of biosensors, each with a biocomponent that is semi-specific to a different substrate, with data output assembled by multivariate analysis (Raud and Kikas, 2013).

## Toxicity measurement through inhibition of cellular metabolism

Whole-cell microbial biosensors for measuring environmental BOD can equally be used to detect and measure environmental toxicity. Rather than using metabolic activity as a proxy for measuring nutrients, toxicants can be quantified by measuring metabolic inhibition (Davila *et al.*, 2011). For example, the SciTOX™ technology that is currently being commercialized for rapid BOD analysis has also been successfully demonstrated for use as a toxicity assay (Pasco *et al.*, 2008). There are also some examples where unique metabolic pathways have been exploited to quantify metabolic inhibition.

The naturally bioluminescent bacterium *Vibrio fischeri* has been widely used as a metabolic bioreporter for environmental toxicity assessment (Farre *et al.*, 2004; Fulladosa *et al.*, 2007). Several products based on the rapid fading of *V. fischeri* luminescence in response to inhibiting metabolism are currently commercially available as toxicity assays for use in monitoring soil and water remediation and industrial discharges, including the widely used Microtox® assay. Multi-laboratory studies have demonstrated the practical application of these assays as rapid and cost effective methods for general toxicity assessment in real complex samples (Farre *et al.*, 2004). However, assays based on naturally luminescent *V. fisheri* may not be as effective for assessment of contaminated sites as assays using contaminant-degrading bacteria, or near relatives, that are genetically modified to be luminescent (Bhattacharyya *et al.*, 2005).

Soil toxicity assessments using luminescent bacteria usually require diluted environmental samples to be incubated with the bacteria for short periods (seconds or minutes), the bacteria removed from the samples and bacterial light production measured with a photometer. Comparison with controls indicates the level of inhibition and hence toxicity. Because contaminants, such as heavy metals, bind to soil particles, liquid phase assays may underestimate the sample toxicity and solid-phase assays are preferable. Some of the luminescent bioreporter assays can be used as solid-phase assays where the bacteria are in contact with soil particles in a slurry.

A similar metabolic biosensor approach is to use sulphur-oxidizing bacteria (Gurung and Oh, 2013). These bacteria oxidize sulphur to sulphuric acid, releasing protons that increase the electrical conductivity of their medium and decrease pH. In the presence of toxic substances, the activity of the sulphur-oxidizing bacteria is inhibited and there is no change in conductivity and pH of the medium. Metabolic inhibition of the bacteria can thereby be easily measured. This system has been used to detect toxic substances in wastewater and hexavalent chromium ions in contaminated soil samples. In preliminary experiments the sensor was found to be effective at detecting chromium down to 5 ppm.

This biosensor approach of general metabolic inhibition allows the detection of general toxicity but not the identification or quantification of specific toxicants. To quantify specific toxicants with a whole cell biosensor approach, it is necessary to either genetically engineer specific signalling pathways into the whole cell component or to use microorganisms with unique and useful metabolic pathways. Several different microorganisms with novel metabolic pathways have been exploited for detecting specific environmental toxicants (Sadik *et al.*, 2004). The pathways include those that allow the detection of heavy metals, pesticides, anionic surfactants, linear alkylbenzene sulfonates, pathogens, endocrine disrupters, polycyclic aromatic hydrocarbons, acrylamide, chlorophenols and cyanide (D'Souza, 2001; Lei *et al.*, 2007; Nomura *et al.*, 1998; Rasooly and Herold, 2006; Rodriguez-Mozaz *et al.*, 2004; Sadik *et al.*, 2004; Taranova *et al.*, 2004; Verma and Singh, 2005; Wanekaya *et al.*, 2008).

### Specific metabolic pathways

Specific metabolic inhibition can be used to detect some specific substances. For example, herbicides that inhibit photosynthesis can be detected by whole-cell biosensors that incorporate a photosynthetic organism as the sensor biocomponent (Campas *et al.*, 2008; Husu *et al.*, 2013; Preuss and Hall, 1995; Shing *et al.*, 2013; Touloupakis *et al.*, 2012). Such a system, using immobilized *Synechococcus* sp. in a calcium alginate matrix has been used to detect the herbicides, atrazine and diuron (Preuss and Hall, 1995). The biosensor relies on herbicide inhibition of photosystem II electron transfer. A redox mediator is used to transfer electrons from the photosystem to an electrode on which the cells are immobilized. Once photosynthesis is inhibited, electron transfer to the electrode is also inhibited. Detection was as low as 1 ppb. Such a sensor is only analyte-specific to the extent that there are no other inhibitors of photosynthesis present. Other inhibitors of photosynthesis including heavy metals and co-contamination of sites is probably more usual than contamination with a single substance (Diplock *et al.*, 2010). Consequently, similar photosynthetic biosensors have been promoted as sensors for more general toxicity including that from heavy metals and pesticides than as herbicide-specific biosensors (Shing *et al.*, 2013).

Many toxic substances that are considered to be environmental contaminants are a source of nutrition for some microorganisms that have acquired specific pathways to metabolize them. Microorganisms that have evolved to utilize such compounds are potentially useful as biocomponents in whole cell biosensors as the presence of the specific compound will stimulate their metabolism and this can be used to quantify those substances.

Benzene, toluene, ethylbenzene and xylene (BTEX) petroleum compounds are carcinogens that are widespread in the environment because of their extensive industrial and domestic applications. These compounds have become a focus of concern because of their mutagenic potential even at low concentrations (Wang and Bartha, 1994). While chromatographic methods are regularly used to evaluate the extent of soil BTEX contamination before and after remediation, these methods do not necessarily indicate bioavailability, something whole-cell biosensors should intrinsically do. Certain bacteria such as *Pseudomonas putida* mt2, *P. putida* F1, *P. mendocina* and *Burkholderia cepacia* GE can specifically catabolize BTEX, and this can be measured by the rate of resazurin dye reduction to a fluorescent product. The detection range of this method is 0.2 to 1.5 mM, which is considerably higher than the detection limits for chromatographic methods stipulated by the USA Environmental Protection Agency (Tizzard *et al.*, 2006).

Benzene degradation has been monitored in the laboratory using a biosensor approach that measures the respiration activity of the benzene-degrading microorganism, *Pseudomonas putida* ML2. Lanyon *et al.* (2006) immobilized *P. putida* cells on a screen-printed gold electrode poised at a fixed potential. The cells were sandwiched between the gold electrode and a cellulose acetate membrane. As oxygen diffuses across the membrane it is reduced at the gold electrode, producing a current. In the presence of benzene, aerobic bacterial catabolism of benzene reduces the concentration of oxygen, and therefore the current at the electrode, in proportion to the benzene concentration. The system was designed to be portable and to operate in a range between 0.01 and 0.1 mM benzene. However, the sensor has a short lifetime and was suitable only as a single-use device.

Using an electrode to measure the depletion of oxygen resulting from substrate-specific bacterial degradation has been used for several surfactants. Surfactants with aromatic rings have high stability in the environment and can act to solubilize other pollutants, concentrating them in the water phase. Nonylphenol is a surfactant used in the wood-pulp industry and nonylphenol ethoxylate (NPEO) can be degraded by the bacterium *Comamonas testosterone* TI. A biosensor for detecting NPEO was made by trapping *Comamonas testosterone* TI in 20% agar gel, and immobilizing them with a nylon net on a Clark oxygen electrode (Taranova *et al.*, 2004). This system had an NPEO detection limit of 0.25 mg/l and was stable for 10 days of continuous measurements. This same strategy, but using the NPEO degrading bacteria, *P. alcaligenes*, has also been used to monitor microbial degradation of NPEO in wastewater (Plekhanova *et al.*, 2011). Several strains of *Pseudomonas* and *Achromobacter* have also shown promise for use in similar whole-cell biosensors for surfactants, with target analytes including sodium dodecyl sulphate, volgnate, toluene sulphonate and alcylbenzene sulphonate (Reshetilov, 2005; Semenchuk *et al.*, 2000; Taranova *et al.*, 2002).

## Substrate degradation products

Some toxic chemicals, such as organophosphate pesticides, are degraded by naturally occurring bacterial enzymes to electro-active products that are facile to detect. Organophosphate pesticides (OPs) are widely used neurotoxic chemicals that can be persistent in the environment (Cheke *et al.*, 2013). Whole-cell biosensors have been devised whereby microorganisms break down pesticides such as parathion releasing the electro-active compound p-nitrophenol which can be detected by electrochemical methods that are rapid, simple and require only inexpensive equipment and reagents (Brogan and Walt, 2005; Schoning *et al.*, 2003). Bacteria that can naturally convert organophosphate pesticides to p-nitrophenol include *Pseudomonas* and *Flavactenium* species. Through genetic manipulation, bacteria that over-express the enzymes that catalyse the conversion have been created and used as more sensitive biosensor strains. These include biosensors with genetically modified *Moraxella* sp., *Pseudomonas putida* and *Escherichia coli* with surface-exposed organophosphate hydrolysing enzymes as the biological recognition element (Guliy *et al.*, 2003; Kumar *et al.*, 2006; Mulchandani *et al.*, 2006; Rainina *et al.*, 1996). Hybrid approaches have also been developed that use either organophosphate hydrolysing enzyme or the pesticide target enzyme cholinesterase, combined in a two-step process with whole-cell metabolism (Lei *et al.*, 2006; Stoytcheva *et al.*, 2009).

In recent years, development of biosensors for the detection of specific compounds has to a large extent moved away from exploiting unmodified whole cells. The greatest research effort now appears to be directed into developing biosensors that use: isolated components of cells such as enzymes; synthetic constructions that bind specific substrates; and whole-cell sensors that use genetically modified metabolic pathways.

## Genetically engineered whole-cell biosensors

### Metabolic bioreporters

Genetically engineered bacteria have been widely developed for environmental biosensor applications and there are also examples where genetically modified yeast, protozoa, algae, plant cells and even human cells have been used (Buonasera *et al.*, 2010; Close *et al.*, 2012; Giardi *et al.*, 2009; Tillner *et al.*, 1996, 1998). Genetically modified cells can be used: to detect and quantify specific compounds; to indicate the presence and activity of specific toxin-degrading microbes introduced into the environment; and as biosensors for general toxicity. Genetically modified bioreporters are generally classified as either catabolic reporters or metabolic reporters depending on whether the reporter gene is constitutively or inducibly expressed.

Metabolic bioreporter cells have a reporter gene, encoding an easily measurable phenotype such as light production, which is constitutively expressed. Constitutively expressed reporter genes, such as those encoding bioluminescence, can be used to monitor the presence of the transgenic microorganism in the environment and to gauge the cells' general metabolic condition. Constitutively expressed reporter genes are used to measure general eco-toxicity as conditions that suppress metabolic activity will reduce the signal from the reporter gene (Heitzer *et al.*, 1998). Bioluminescence encoded by the *luxCDABE* gene cluster is most often used for this application as it is particularly sensitive to the cell's metabolic state as the production of light requires numerous reactions that consume metabolic energy (del Busto-Ramos *et al.*, 2008; Heitzer *et al.*, 1998; Porteous *et al.*, 2000).

Through genetic engineering it is possible to use environmentally relevant bacteria that are not naturally luminescent as environmental bioreporters. Bacteria engineered to constitutively express *luxCDABE* include environmental isolates of bacteria such as *Pseudomonas* and *Acinetobacter* as well as common laboratory strains of bacteria including *Escherichia coli*. Such bioreporters have been used to assess toxicity in contaminated soil and water samples in numerous studies demonstrating their utility and superiority to naturally luminescent bioreporters (Bhattacharyya *et al.*, 2005; Kurvet *et al.*, 2011; Xu *et al.*, 2013).

Cell metabolism can also be re-engineered to enhance an existing metabolic pathway or providing a new pathway that allows the cell to specifically utilize a target substrate or which increases the cells resistance or sensitivity to a substrate. This approach has been used to develop an amperometric biosensor for environmental determination of organophosphate nerve agents which increased the respiration rate of genetically engineered *Moraxella* (Mulchandani *et al.*, 2006). Similarly, engineered *Chlamydomonas* strains with amino acid modifications in the D1 subunit of photosystem II conferring increased sensitivity or resistance towards a particular herbicide class have been developed as bioreporters for environmental herbicide detection (Buonasera *et al.*, 2010; Giardi *et al.*, 2009; Scognamiglio *et al.*, 2009). Chlorophyll fluorescence was used to measure the extent of inhibition of photosynthesis in each mutant after incubation with environmental samples.

### Catabolic bioreporters

While metabolic bioreporters measure environmentally induced cytotoxicity, catabolic bioreporters can be used to more specifically detect and quantify compounds or groups of compounds. Catabolic bioreporters are generally engineered by placing a reporter gene under the control of an inducible promoter expressed in response to the target analyte. Whole-cell catabolic bioreporters have been used to measure specific contaminants in complex environments with very low detection limits (Eldridge *et al.*, 2007). The degree of specificity of the bioreporter can be modified by fusing the reporter gene to promoters that are induced by a group of compounds, rather than a specific compound, or that are induced in response to activation of a particular signal transduction pathway that responds to multiple inputs or more general conditions such as cellular redox state. For example, a fluorescent reporter gene has been placed under the control of the yeast stress-sensing pathway to induce reporter gene expression in response to environmental stress (Radhika *et al.*, 2005). Similarly, a panel of biosensors has been constructed by fusing different stress-inducible promoters to a bioluminescent reporter allowing different classes of antibiotics to be discriminated depending on the stress response that they induce (Shapiro and Baneyx, 2007).

Commonly used reporter genes include the *E. coli lacZ* gene encoding β-galactosidase as a chromogenic marker (Tag *et al.*, 2007), genes encoding fluorescent proteins such as GFP (Axtell and Beattie, 2002; Fiorentino *et al.*, 2009) and its wavelength-shifted variants, and luciferase genes such as the eukaryotic *lucF* gene, and the bacterial *luxCDABE* (Kong, 2006; Paton *et al.*, 2006; Tom-Petersen *et al.*, 2001) gene cluster for bioluminescence. The *Vibrio fischeri luxCDABE* gene cassette has been extensively used as it is a sensitive reporter that does not require any exogenous substrates (Eldridge *et al.*, 2007; Li *et al.*, 2008). Eukaryotic luciferases have high light production and exert a lower metabolic burden on the host cell than bacterial luciferases but they require exogenous luciferin which complicates assays (Wilson and Hastings, 1998). As with GFP variants, wavelength-shifted variants of *LucF* have now been developed allowing multiplex analysis (Branchini *et al.*, 2007; Shapiro *et al.*, 2005). The bioluminescence reporter has a fast response time and short-term sensitivity in comparison with GFP, which is more stable. GFP fluorescence can be detected long after reporter gene induction has ceased and even after cell death, which is generally a disadvantage for biosensor applications (Li *et al.*, 2008).

Using techniques such as laser-induced fluorescence confocal spectroscopy, fluorescent reporters can be as sensitive as bioluminescent reporters (Kohlmeier *et al.*, 2007). In contrast to luminescence, fluorescent proteins allow interrogation of single-cell responses. Fluorescent reporter proteins that fluoresce at different wavelengths (e.g. green fluorescent protein and red fluorescent protein) allow the opportunity for using multiple fluorescent reporters in the same cell. It is possible to use dual fluorescence reporting where one colour is repressed by an analyte and another colour is induced (Martineau *et al.*, 2010). This may increase the sensitivity of the assay but it is not a strategy that has been widely adopted. Neither luminescence nor fluorescence biosensors are suitable for use in situations where soil particles or pigments can attenuate light transmission.

Engineered whole-cell biosensors can either use planktonic cells (Horry *et al.*, 2004, 2007) or cells immobilized on the surface of a solid substrate such as an optical fibre, a membrane or an electrode (Smutok *et al.*, 2007). Signal transduction can be by detection of fluorescence or luminescence, colour change, or electrical current production depending on the reporter gene used. Colorimetric biosensors allow simple visual inspection for determination of the presence/absence of a specific contaminant above a certain threshold (Freire-Picos and Lamas-Maceiras, 2006; Pasco *et al.*, 2004). Amperometric detection is possible where the reporter gene encodes a redox protein (Smutok *et al.*, 2007), where the reporter enzyme catalyses the production of a redox species (Lei *et al.*, 2007; Paitan *et al.*, 2004) or where the reporter gene increases cellular respiration (Lehmann *et al.*, 2000; Song *et al.*, 2012, 2013; Tag *et al.*, 2007).

Induced changes in cellular respiration by a reporter phenotype has been achieved by using a gene involved in respiration such as *nuoA* as the reporter gene or less directly through the induced expression of β-galactosidase (*lacZ*) (Lehmann *et al.*, 2000; Song *et al.*, 2012, 2013; Tag *et al.*, 2007). Increased respiration with the *lacZ* gene results from the enzymatic cleavage of lactose into metabolisable sugars. Amperometric detection of β-galactosidase conversion of the electro-inactive substrate p-aminophenyl-β-D-galactopyranoside (PAPG) to the electro-active product PAP has also been achieved (Paitan *et al.*, 2004).

## Summary

There is a vast array of different technologies that have been developed for environmental biosensing. Whole-cell, unmodified bioreporters for general toxicity monitoring have been developed for many years and assay kits based on this technology are widely commercially available. Genetic modification of whole-cells has increased the utility, sensitivity and specificity of these biosensors. Whole-cell biosensors allow economical quantification of biologically active contaminants and should be more widely used in place of, or as a complementary addition to, more expensive and complex chemical analyses. Published research into biosensors appears to show a relative increase in interest in biosensors that use genetically modified cells and also towards biosensors that incorporate biocomponents such as enzymes and DNA rather than unmodified whole cells. The future trend for all biosensors appears to be towards miniaturization, high throughput systems and increased integration, enabling the development of single devices with multiple channels, simpler support systems and more rapid response time, at lower cost.

## References


Ahuja, T., Mir, I.A., Kumar, D., and Rajesh (2007). Biomolecular immobilization on conducting polymers for biosensing applications. Biomaterials *28*, 791–805.

Axtell, C.A., and Beattie, G.A. (2002). Construction and characterization of a proU-gfp transcriptional fusion that measures water availability in a microbial habitat. Appl. Environ. Microbiol. *68*, 4604–4612.

Baskurt, E., Ekiz, F., Demirkol, D.O., Timur, S., and Toppare, L. (2012). A conducting polymer with benzothiadiazole unit: Cell based biosensing applications and adhesion properties. Colloid. Surface. B *97*, 13–18.

Ben-Yoav, H., Melamed, S., Freeman, A., Shacham-Diamand, Y., and Belkin, S. (2011). Whole-cell biochips for biosensing: integration of live cells and inanimate surfaces. Crit. Rev. Biotechnol. *31*, 337–353.

Bhattacharyya, J., Read, D., Amos, S., Dooley, S., Killham, K., and Paton, G.I. (2005). Biosensor-based diagnostics of contaminated groundwater: assessment and remediation strategy. Environ. Pollut. *134*, 485–492.

Biran, I., Rissin, D.M., Ron, E.Z., and Walt, D.R. (2003). Optical imaging fiber-based live bacterial cell array biosensor. Anal. Biochem. *315*, 106–113.

Bjerketorp, J., Hakansson, S., Belkin, S., and Jansson, J.K. (2006). Advances in preservation methods: keeping biosensor microorganisms alive and active. Curr. Opin. Biotechnol. *17*, 43–49.

Borisov, S.M., and Wolfbeis, O.S. (2008). Optical biosensors. Chem. Rev. *108*, 423–461.

Branchini, B.R., Ablamsky, D.M., Rosenman, J.M., Uzasci, L., Southworth, T.L., and Zimmer, M. (2007). Synergistic mutations produce blue-shifted bioluminescence in firefly luciferase. Biochemistry *46*, 13847–13855.

Brogan, K.L., and Walt, D.R. (2005). Optical fiber-based sensors: application to chemical biology. Curr. Opin. Chem. Biol. *9*, 494–500.

Buonasera, K., Pezzotti, G., Scognamiglio, V., Tibuzzi, A., and Giardi, M.T. (2010). New platform of biosensors for prescreening of pesticide residues to support laboratory analyses. J. Agr. Food Chem. *58*, 5982–5990.

Byfield, M.P., and Abuknesha, R.A. (1994). Biochemical aspects of biosensors. Biosens. Bioelectron. *9*, 373–400.

Cai, L., Friedman, N., and Xie, X.S. (2006). Stochastic protein expression in individual cells at the single molecule level. Nature *440*, 358–362.

Campas, M., Carpentier, R., and Rouillon, R. (2008). Plant tissue- and photosynthesis-based biosensors. Biotechnol. Adv. *26*, 370–378.

Carturan, G., Campostrini, R., Dire, S., Scardi, V., and Dealteriis, E. (1989). Inorganic gels for immobilization of biocatalysts – inclusion of invertase active whole cells of yeast (*Saccharomyces cerevisiae*) into thin layers of $SiO_2$ gel deposited on glass sheets. J. Mol. Catal. *57*, L13–L16.

Carturan, G., Dal Toso, R., Boninsegna, S., and Dal Monte, R. (2004). Encapsulation of functional cells by sol–gel silica: actual progress and perspectives for cell therapy. J. Mater. Chem. *14*, 2087–2098.

Catterall, K., Morris, K., Gladman, C., Zhao, H., Pasco, N., and John, R. (2001). The use of microorganisms with broad range substrate utilisation for the ferricyanide-mediated rapid determination of biochemical oxygen demand. Talanta *55*, 1187–1194.

Catterall, K., Zhao, H., Pasco, N., and John, R. (2013). Development of a rapid ferricyanide-mediated assay for biochemical oxygen demand using a mixed microbial consortium. Anal. Chem. *75*, 2584–2590.

Cheke, R.A., Adranyi, E., Cox, J.R., Farman, D.I., Magoma, R.N., Mbereki, C., McWilliam, A.N., Mtobesya, B.N., and van der Walte, E. (2013). Soil contamination and persistence of pollutants following organophosphate sprays and explosions to control red-billed quelea (*Quelea quelea*). Pest Manage. Sci. *69*, 386–396.

Chiappini, S.A., Kormes, D.J., Bonetto, M.C., Sacco, N., and Corton, E. (2010). A new microbial biosensor for organic water pollution based on measurement of carbon dioxide production. Sensor. Actuat. B-Chem. *148*, 103–109.

Choi, J.W., Park, K.W., Lee, D.B., Lee, W., and Lee, W.H. (2005). Cell immobilization using self-assembled synthetic oligopeptide and its application to biological toxicity detection using surface plasmon resonance. Biosens. Bioelectron. *20*, 2300–2305.

Close, D., Webb, J., Ripp, S., Patterson, S., and Sayler, G. (2012). Remote detection of human toxicants in real time using a human optimized bioluminescent bacterial luciferase gene cassette bioreporter. In Sensing Technologies for Global Health, Military Medicine, Disaster Response, and Environmental Monitoring II: And Biometric Technology for Human Identification Ix, A.H.J. Kolk, S.O. Southern, K.N. Montgomery, C.W. Taylor, B. Kumar, S. Prabhakar, and A.A. Ross, eds. (SPIE, Baltimore, MD, USA).

Costerton, J.W., Lewandowski, Z., Caldwell, D.E., Korber, D.R., and Lappinscott, H.M. (1995). Microbial biofilms. Annu. Rev. Microbiol. *49*, 711–745.

D'Souza, S.F. (2001). Microbial biosensors. Biosens. Bioelectron. *16*, 337–353.

Davila, D., Esquivel, J.P., Sabate, N., and Mas, J. (2011). Silicon-based microfabricated microbial fuel cell toxicity sensor. Biosens. Bioelectron. *26*, 2426–2430.

del Busto-Ramos, M., Budzik, M., Corvalan, C., Morgan, M., Turco, R., Nivens, D., and Applegate, B. (2008). Development of an online biosensor for *in situ* monitoring of chlorine dioxide gas disinfection efficacy. Appl. Microbiol. Biotechnol. *78*, 573–580.

Desimone, M.F., Alvarez, G.S., Foglia, M.L., and Diaz, L.E. (2009). Development of sol–gel hybrid materials for whole cell immobilization. Recent Pat. Biotechnol. *3*, 55–60.

Ding, L., Du, D., Zhang, X., and Ju, H. (2008). Trends in cell-based electrochemical biosensors. Curr. Med. Chem. *15*, 3160–3170.

Diplock, E.E., Alhadrami, H.A., and Paton, G.I. (2010). Application of microbial bioreporters in environmental microbiology and bioremediation. In Whole Cell Sensing Systems II: Applications, S. Belkin, and M.B. Gu, eds. (Springer-Verlag, Berlin, Germany), pp. 189–209.

Dong, S., Cheng, G., Li, B., Kou, W., and Zhang, J. (1996). Construction of enzyme electrodes based on PVA-g-PVP hydrogel. Chin. Pat. CN 961235284.

Elad, T., Lee, J.H., Gu, M.B., and Belkin, S. (2010). Microbial cell arrays. In Whole Cell Sensing Systems I: Reporter Cells and Devices, S. Belkin, and M.B. Gu, eds. (Springer-Verlag, Berlin, Germany), pp. 85–108.

Eldridge, M.L., Sanseverino, J., Layton, A.C., Easter, J.P., Schultz, T.W., and Sayler, G.S. (2007). Saccharomyces cerevisiae BLYAS, a new Bioluminescent bioreporter for detection of androgenic compounds. Appl. Environ. Microbiol. 73, 6012–6018.

Farre, M., Arranz, F., Ribo, J., and Barcelo, D. (2004). Interlaboratory study of the bioluminescence inhibition tests for rapid wastewater toxicity assessment. Talanta *62*, 549–558.

Fesenko, D.O., Nasedkina, T.V., Prokopenko, D.V., and Mirzabekov, A.D. (2005). Biosensing and monitoring of cell populations using the hydrogel bacterial microchip. Biosens. Bioelectron. *20*, 1860–1865.

Fiorentino, G., Ronca, R., and Bartolucci, S. (2009). A novel E-coli biosensor for detecting aromatic aldehydes based on a responsive inducible archaeal promoter fused to the green fluorescent protein. Appl. Microbiol. Biotechnol. *82*, 67–77.

Freire-Picos, M.A., and Lamas-Maceiras, M. (2006). KlHIS4-modulated histidine auxotrophy in yeast and characterisation of a KlHIS4-lacZ-based cadmium and copper biosensor. Enzym. Microb. Tech. *40*, 67–70.

Fulladosa, E., Murat, J.-C., Bollinger, J.-C., and Villaescusa, I. (2007). Adverse effects of organic arsenical compounds towards Vibrio fischeri bacteria. Sci. Total Environ. *377*, 207–213.

Garcia-Balboa C., C.E., and Lopez Rodas, V. (2012). Microalgae biosensors for the detection of environmental contaminants: a review. Rev. Comp. Sci. Vet. *6*, 51–67.

Georgieva, N., Rangelova, N., Peshev, D., and Nenkova, S. (2011). Novel pectin–silica hybrids used for immobilization of trichosporon cutaneum cells efficient in removal of cadmium and copper ions from waste water. Cr. Acad. Bulg. Sci. *64*, 1421–1428.

Giardi, M.T., Scognamiglio, V., Rea, G., Rodio, G., Antonacci, A., Lambreva, M., Pezzotti, G., and Johanningmeier, U. (2009). Optical biosensors for environmental monitoring based on computational and biotechnological tools for engineering the photosynthetic D1 protein of *Chlamydomonas reinhardtii*. Biosens. Bioelectron. 25, 294–300.

Guliy, O.I., Ignatov, O.V., Makarov, O.E., and Ignatov, V.V. (2003). Determination of organophosphorus aromatic nitro insecticides and p-nitrophenol by microbial-cell respiratory activity. Biosens. Bioelectron. *18*, 1005–1013.

Guo, K., Freguia, S., Dennis, P.G., Chen, X., Donose, B.C., Keller, J., Gooding, J.J., and Rabaey, K. (2013). Effects of surface charge and hydrophobicity on anodic biofilm formation, community composition, and current generation in bioelectrochemical systems. Environ. Sci. Technol. *47*, 7563–7570.

Gurung, A., and Oh, S.-E. (2013). Use of sulfur-oxidizing bacteria for assessment of chromium-contaminated soil. Environ. Earth Sci. *70*, 139–143.

Heitzer, A., Applegate, B., Kehrmeyer, S., Pinkart, H., Webb, O.F., Phelps, T.J., White, D.C., and Sayler, G.S. (1998). Physiological considerations of environmental applications of lux reporter fusions. J. Microbiol. Meth. *33*, 45–57.

Horry, H., Durand, M.J., Picart, P., Bendriaa, L., Daniel, P., and Thouand, U. (2004). Development of a biosensor for the detection of tributyltin. Environ. Toxicol. *19*, 342–345.

Horry, H., Maul, A., and Thouand, G. (2007). Optimization of a bacterial bioluminescent biosensor through experimental design. Sensor. Actuat. B-Chem. *127*, 649–657.

Husu, I., Rodio, G., Touloupakis, E., Lambreva, M.D., Buonasera, K., Litescu, S.C., Giardi, M.T., and Rea, G. (2013). Insights into photo-electrochemical sensing of herbicides driven by *Chlamydomonas reinhardtii* cells. Sensor. Actuat. B-Chem. *185*, 321–330.

Ishii, S., Suzuki, S., Norden-Krichmar, T.M., Nealson, K.H., Sekiguchi, Y., Gorby, Y.A., and Bretschger, O. (2012). Functionally stable and phylogenetically diverse microbial enrichments from microbial fuel cells during wastewater treatment. PLoS One 7, e30495.

Jaspers, M.C.M., Meier, C., Zehnder, A.J.B., Harms, H., and van der Meer, J.R. (2001). Measuring mass transfer processes of octane with the help of an alkS-alkB: gfp-tagged Escherichia coli. Environ. Microbiol. 3, 512–524.

Jha, S.K., Kanungo, M., Nath, A., and D'Souza, S.F. (2009). Entrapment of live microbial cells in electropolymerized polyaniline and their use as urea biosensor. Biosens. Bioelectron. *24*, 2637–2642.

Jordan, M.A., Welsh, D.T., John, R., Catterall, K., and Teasdale, P.R. (2013). A sensitive ferricyanide-mediated biochemical oxygen demand assay for analysis of wastewater treatment plant influents and treated effluents. Water Res. *47*, 841–849.

Kim, B.C., and Gu, M.B. (2003). A bioluminescent sensor for high throughput toxicity classification. Biosens. Bioelectron. *18*, 1015–1021.

Kim, B.H., Chang, I.S., Gil, G.C., Park, H.S., and Kim, H.J. (2003). Novel BOD (biological oxygen demand) sensor using mediator-less microbial fuel cell. Biotechnol. Lett. *25*, 541–545.

Kohlmeier, S., Mancuso, M., Tecon, R., Harms, H., van der Meer, J.R., and Wells, M. (2007). Bioreporters: gfp versus lux revisited and single-cell response. Biosens. Bioelectron. *22*, 1578–1585.

Kong, I.C. (2006). An optimization of a bioassay for toluene analogs using bioluminescence reporter strain KG1206. Soil Sediment. Contam. *15*, 231–239.

Kuang, Y., Biran, I., and Walt, D.R. (2004). Living bacterial cell array for genotoxin monitoring. Anal. Chem. *76*, 2902–2909.

Kumar, J., Jha, S.K., and D'Souza, S.F. (2006). Optical microbial biosensor for detection of methyl parathion pesticide using Flavobacterium sp whole cells adsorbed on glass fiber filters as disposable biocomponent. Biosens. Bioelectron. *21*, 2100–2105.

Kumlanghan, A., Liu, J., Thavarungkul, P., Kanatharana, P., and Mattiasson, B. (2007). Microbial fuel cell-based biosensor for fast analysis of biodegradable organic matter. Biosens. Bioelectron. *22*, 2939–2944.

Kurvet, I., Ivask, A., Bondarenko, O., Sihtmaee, M., and Kahru, A. (2011). LuxCDABE-transformed constitutively bioluminescent *Escherichia coli* for toxicity screening: comparison with naturally luminous *Vibrio fischeri*. Sensors *11*, 7865–7878.

Kuzume, A., Zhumaev, U., Li, J.F., Fu, Y.C., Fueg, M., Esteve-Nunez, A., and Wandlowski, T. (2013). An *in situ* surface electrochemistry approach toward whole-cell studies: Charge transfer between *Geobacter sulfurreducens* and electrified metal/electrolyte interfaces through linker molecules. Electrochim. Acta. *112*, 933–942.

Lai Minh, Q., Do Phuong, K., Hira, D., Fujii, T., and Furukawa, K. (2011). Reject water treatment by improvement of whole cell anammox entrapment using polyvinyl alcohol/alginate gel. Biodegradation *22*, 1155–1167.

Lanyon, Y.H., Tothill, I.E., and Mascini, M. (2006). An amperometric bacterial biosensor based on gold screen-printed electrodes for the detection of benzene. Anal. Lett. 39, 1669–1681.

Lee, J.H., Mitchell, R.J., Kim, B.C., Cullen, D.C., and Gu, M.B. (2005). A cell array biosensor for environmental toxicity analysis. Biosens. Bioelectron. *21*, 500–507.

Lee, K.G., Park, T.J., Soo, S.Y., Wang, K.W., Kim, B.H., Park, J.H., Lee, C.S., Kim, D.H., and Lee, S.J. (2010). Synthesis and utilization of *E. coli*-encapsulated PEG-based microdroplet using a microfluidic chip for biological application. Biotechnol. Bioeng. *107*, 747–751.

Lehmann, M., Riedel, K., Adler, K., and Kunze, G. (2000). Amperometric measurement of copper ions with a deputy substrate using a novel *Saccharomyces cerevisiae* sensor. Biosens. Bioelectron. *15*, 211–219.

Lei, Y., Mulchandani, P., Chen, W., and Mulchandani, A. (2006). Biosensor for direct determination of fenitrothion and EPN using recombinant *Pseudomonas putida* JS444 with surface expressed organophosphorus hydrolase. 1. Modified Clark oxygen electrode. Sensors *6*, 466–472.

Lei, Y., Mulchandani, P., Chen, W., and Mulchandani, A. (2007). Biosensor for direct determination of fenitrothion and EPN using recombinant *Pseudomonas putida* JS444 with surface-expressed organophosphorous hydrolase. 2. Modified carbon paste electrode. Appl. Biochem. Biotech. *136*, 243–250.

Li, Y.F., Li, F.Y., Ho, C.L., and Liao, V.H.C. (2008). Construction and comparison of fluorescence and bioluminescence bacterial biosensors for the detection of bioavailable toluene and related compounds. Environ. Pollut. *152*, 123–129.

Liu, C., and Dong, S. (2013). Field application of a biofilm reactor based BOD prototype in Taihu Lake, China. Talanta *109*, 147–151.

Liu, C., Zhao, H., Gao, S., Jia, J., Zhao, L., Yong, D., and Dong, S. (2013). A reagent-free tubular biofilm reactor for on-line determination of biochemical oxygen demand. Biosens. Bioelectron. *45*, 213–218.

Liu, J., and Mattiasson, B. (2002). Microbial BOD sensors for wastewater analysis. Water Res. *36*, 3786–3802.

Liu, L., Shang, L., Guo, S.J., Li, D., Liu, C.Y., Qi, L., and Dong, S.J. (2009). Organic–inorganic hybrid material for the cells immobilization: Long-term viability mechanism and application in BOD sensors. Biosens. Bioelectron. *25*, 523–526.

Logan, B.E. (2009). Exoelectrogenic bacteria that power microbial fuel cells. Nat. Rev. Microbiol. *7*, 375–381.

Martineau, R.L., Stout, V., and Towe, B.C. (2010). Optical tracking of a stress-responsive gene amplifier applied to cell-based biosensing and the study of synthetic architectures. Biosens. Bioelectron. *25*, 1881–1888.

Melidis, P., Vaiopoulou, E., and Aivasidis, A. (2008). Development and implementation of microbial sensors for efficient process control in wastewater treatment plants. Bioproc. Biosyst. Eng. *31*, 277–282.

Merrin, J., Leibler, S., and Chuang, J.S. (2007). Printing multistrain bacterial patterns with a piezoelectric inkjet printer. PLoS One *2*, e663.

Meyer, R.L., Zhou, X.F., Tang, L.N., Arpanaei, A., Kingshott, P., and Besenbacher, F. (2010). Immobilisation of living bacteria for AFM imaging under physiological conditions. Ultramicroscopy *110*, 1349–1357.

Mitchell, R.J., Lee, S.K., Kim, T., and Ghim, C.-M. (2011). Microbial linguistics: perspectives and applications of microbial cell-to-cell communication. BMB Rep. *44*, 1–10.

Modin, O., and Wilen, B.-M. (2012). A novel bioelectrochemical BOD sensor operating with voltage input. Water Res. *46*, 6113–6120.

Morris, K., Catterall, K., Zhao, H., Pasco, N., and John, R. (2001). Ferricyanide mediated biochemical oxygen demand – development of a rapid biochemical oxygen demand assay. Anal. Chim. Acta. *442*, 129–139.

Mulchandani, P., Chen, W., and Mulchandani, A. (2006). Microbial biosensor for direct determination of nitrophenyl-substituted organophosphate nerve agents using genetically engineered *Moraxella* sp. Anal. Chim. Acta. *568*, 217–221.

Nomura, Y., Ikebukuro, K., Yokoyama, K., Takeuchi, T., Arikawa, Y., Ohno, S., and Karube, I. (1998). Application of a linear alkylbenzene sulfonate biosensor to river water monitoring. Biosens. Bioelectron. *13*, 1047–1053.

Paitan, Y., Biran, I., Shechter, N., Biran, D., Rishpon, J., and Ron, E.Z. (2004). Monitoring aromatic hydrocarbons by whole cell electrochemical biosensors. Anal. Biochem. 335, 175–183.

Park, K.S., Baumstark-Khan, C., Rettberg, P., Horneck, G., Rabbow, E., and Gu, M.B. (2005). Immobilization as a technical possibility for long-term storage of bacterial biosensors. Radiat. Environ. Bioph. *44*, 69–71.

Pasco, N., Hay, J., and Webber, J. (2001). Biosensors: MICREDOX – a new biosensor technique for rapid measurement of BOD and toxicity. Biomarkers *6*, 83–89.

Pasco, N., Baronian, K., Jeffries, C., Webber, J., and Hay, J. (2004). MICREDOX – development of a ferricyanide-mediated rapid biochemical oxygen demand method using an immobilised *Proteus vulgaris* biocomponent. Biosens. Bioelectron. *20*, 524–532.

Pasco, N.F., Goonerate, R., Daniel, R.M., Czollner, A., and Scott, A.J. (2008). Toxicity assessment of chlorophenols using a mediated microbial toxicity assay. Int. J. Environ. Anal. Chem. *88*, 1063–1075.

Paton, G.T., Cheewasedtham, W., Marr, I.L., and Dawson, J.J.C. (2006). Degradation and toxicity of phenyltin compounds in soil. Environ. Pollut. *144*, 746–751.

Peixoto, L., Min, B., Martins, G., Brito, A.G., Kroff, P., Parpot, P., Angelidaki, I., and Nogueira, R. (2011). In situ microbial fuel cell-based biosensor for organic carbon. Bioelectrochemistry *81*, 99–103.

Piccirillo, C., Pereira, S.I.A., Marques, A., Pullar, R.C., Tobaldi, D.M., Pintado, M.E., and Castro, P.M.L. (2013). Bacteria immobilisation on hydroxyapatite surface for heavy metals removal. J. Environ. Manage. *121*, 87–95.

Plekhanova, Y.V., Reshetilov, A.N., Manolov, T.V., and Taranova, L.A. (2011). Biosensor monitoring of microbial treatment of wastewater from nonylphenol polyethoxylates under flow-through conditions. Appl. Biochem. Microbiol. *47*, 846–851.

Polyak, B., Geresh, S., and Marks, R.S. (2004). Synthesis and characterization of a biotin–alginate conjugate and its application in a biosensor construction. Biomacromolecules *5*, 389–396.

Popovtzer, R., Neufeld, T., Biran, N., Ron, E.Z., Rishpon, J., and Shacham-Diamand, Y. (2005). Novel integrated electrochemical nano-biochip for toxicity detection in water. Nano. Lett. *5*, 1023–1027.

Porteous, F., Killham, K., and Meharg, A. (2000). Use of a lux-marked rhizobacterium as a biosensor to assess changes in rhizosphere C flow due to pollutant stress. Chemosphere *41*, 1549–1554.

Premkumar, J.R., Sagi, E., Rozen, R., Belkin, S., Modestov, A.D., and Lev, O. (2002). Fluorescent bacteria encapsulated in sol–gel derived silicate films. Chem. Mater. *14*, 2676–2686.

Preuss, M., and Hall, E.A.H. (1995). Mediated herbicide inhibition in a pet biosensor. Anal. Chem. *67*, 1940–1949.

Rabaey, K., and Verstraete, W. (2005). Microbial fuel cells: novel biotechnology for energy generation. Trends Biotechnol. *23*, 291–298.

Radhika, V., Milkevitch, M., Audige, V., Proikas-Cezanne, T., and Dhanasekaran, N. (2005). Engineered Saccharomyces cerevisiae strain BioS-1, for the detection of water-borne toxic metal contaminants. Biotechnol. Bioeng. *90*, 29–35.

Rainina, E.I., Efremenco, E.N., Varfolomeyev, S.D., Simonian, A.L., and Wild, J.R. (1996). The development of a new biosensor based on recombinant E-coli for the direct detection of organophosphorus neurotoxins. Biosens. Bioelectron. *11*, 991–1000.

Rasooly, A., and Herold, K.E. (2006). Biosensors for the analysis of food- and waterborne pathogens and their toxins. J. Aoac. Int. *89*, 873–883.

Raud, M., and Kikas, T. (2013). Bioelectronic tongue and multivariate analysis: A next step in BOD measurements. Water Res. *47*, 2555–2562.

Reshetilov, A.N. (2005). Microbial, enzymatic, and immune biosensors for ecological monitoring and control of biotechnological processes. Appl. Biochem. Microbiol. *41*, 442–449.

Ripp, S., Daumer, K.A., McKnight, T., Levine, L.H., Garland, J.L., Simpson, M.L., and Sayler, G.S. (2003). Bioluminescent bioreporter integrated-circuit sensing of microbial volatile organic compounds. J. Ind. Microbiol. Biot. *30*, 636–642.

Rodriguez-Mozaz, S., Marco, M.P., de Alda, M.J.L., and Barcelo, D. (2004). Biosensors for environmental monitoring of endocrine disruptors: a review article. Anal. Bioanal. Chem. *378*, 588–598.

Sadik, O.A., Wanekaya, A.K., and Andreescu, S. (2004). Advances in analytical technologies for environmental protection and public safety. J. Environ. Monitor. *6*, 513–522.

Sakai, Y., Abe, N., Takeuchi, S., and Takahashi, F. (1995). BOD sensor using magnetic activated sludge. J. Ferment. Bioeng. *80*, 300–303.

Salanitro, J.P., Johnson, P.C., Spinnler, G.E., Maner, P.M., Wisniewski, H.L., and Bruce, C. (2000). Field scale demonstration of enhanced MTBE bioremediation through aquifer bioaugmentation and oxygenation. Environ. Sci. Technol. *34*, 4152–4162.

Samanta, D., and Sarkar, A. (2011). Immobilization of bio-macromolecules on self-assembled monolayers: methods and sensor applications. Chem. Soc. Rev. *40*, 2567–2592.

Schoning, M.J., Arzdorf, M., Mulchandani, P., Chen, W., and Mulchandani, A. (2003). A capacitive field-effect sensor for the direct determination of organophosphorus pesticides. Sensor. Actuat. B-Chem. *91*, 92–97.

Schreiter, P.P.Y., Gillor, O., Post, A., Belkin, S., Schmid, R.D., and Bachmann, T.T. (2001). Monitoring of phosphorus bioavailability in water by an immobilized luminescent cyanobacterial reporter strain. Biosens. Bioelectron. *16*, 811–818.

Scognamiglio, V., Raffi, D., Lambreva, M., Rea, G., Tibuzzi, A., Pezzotti, G., Johanningmeier, U., and Giardi, M.T. (2009). *Chlamydomonas reinhardtii* genetic variants as probes for fluorescence sensing system in detection of pollutants. Anal. Bioanal. Chem. *394*, 1081–1087.

Semenchuk, I.N., Taranova, L.A., Kalenyuk, A.A., Il'yasov, P.V., and Reshetilov, A.N. (2000). Effect of various methods of immobilization on the stability of a microbial biosensor for surfactants based on *Pseudomonas rathonis* T. Appl. Biochem. Microbiol. *36*, 69–72.

Shapiro, E., and Baneyx, F. (2007). Stress-activated bioluminescent *Escherichia coli* sensors for antimicrobial agents detection. J. Biotechnol. *132*, 487–493.

Shapiro, E., Lu, C., and Baneyx, F. (2005). A set of multicolored *Photinus pyralis* luciferase mutants for *in vivo* bioluminescence applications. Protein Eng. Des. Sel. *18*, 581–587.

Sharma, P.K., and Rao, K.H. (2002). Analysis of different approaches for evaluation of surface energy of microbial cells by contact angle goniometry. Adv. Colloid Interfac. *98*, 341–463.

Shing, W.L., Heng, L.Y., and Surif, S. (2013). Performance of a cyanobacteria whole cell-based fluorescence biosensor for heavy metal and pesticide detection. Sensors *13*, 6394–6404.

Shulga, A.A., Soldatkin, A.P., Elskaya, A.V., Dzyadevich, S.V., Patskovsky, S.V., and Strikha, V.I. (1994). Thin-film conductometric biosensors for glucose and urea determination. Biosens. Bioelectron. *9*, 217–223.

Smutok, O., Dmytruk, K., Gonchar, M., Sibirny, A., and Schuhmann, W. (2007). Permeabilized cells of flavocytochrome b(2) over-producing recombinant yeast *Hansenula polymorpha* as biological recognition element in amperometric lactate biosensors. Biosens. Bioelectron. *23*, 599–605.

Song, W., Pasco, N., Gooneratne, R., and Weld, R.J. (2012). Comparison of three genetically modified *Escherichia coli* biosensor strains for amperometric tetracycline measurement. Biosens. Bioelectron. *35*, 69–74.

Song, W., Pasco, N., Gooneratne, R., and Weld, R.J. (2013). An improved genetically modified *Escherichia coli* biosensor for amperometric tetracycline measurement. Appl. Microbiol. Biotechnol. *97*, 9081–9086.

Stocker, J., Balluch, D., Gsell, M., Harms, H., Feliciano, J., Daunert, S., Malik, K.A., and Van der Meer, J.R. (2003). Development of a set of simple bacterial biosensors for quantitative and rapid measurements of arsenite and arsenate in potable water. Environ. Sci. Technol. *37*, 4743–4750.

Stoytcheva, M., Zlatev, R., Velkova, Z., Valdez, B., Ovalle, M., and Petkov, L. (2009). Hybrid electrochemical biosensor for organophosphorus pesticides quantification. Electrochim. Acta. *54*, 1721–1727.

Su, L., Jia, W., Hou, C., and Lei, Y. (2011). Microbial biosensors: A review. Biosens. Bioelectron. *26*, 1788–1799.

Tag, K., Riedel, K., Bauer, H.J., Hanke, G., Baronian, K.H.R., and Kunze, G. (2007). Amperometric detection of Cu2+ by yeast biosensors using flow injection analysis (FIA). Sensor. Actuat. B-Chem. *122*, 403–409.

Taranova, L., Semenchuk, I., Manolov, T., Iliasov, P., and Reshetilov, A. (2002). Bacteria-degraders as the base of an amperometric biosensor for detection of anionic surfactants. Biosens. Bioelectron. *17*, 635–640.

Taranova, L.A., Fesay, A.P., Ivashchenko, G.V., Reshetilov, A.N., Winther-Nielsen, A., and Emneus, J. (2004). *Comamonas testosteroni* strain TI as a potential base for a microbial sensor detecting surfactants. Appl. Biochem. Microbiol. *40*, 404–408.

Tecon, R., and van der Meer, J.R. (2006). Information from single-cell bacterial biosensors: what is it good for? Curr. Opin. Biotech. *17*, 4–10.

Tillner, J., Winckler, T., and Dingermann, T. (1996). Developmentally regulated promoters from *Dictyostelium discoideum* as molecular markers for testing potential teratogens. Pharmazie *51*, 902–906.

Tillner, J., Nau, H., Winckler, T., and Dingermann, T. (1998). Evaluation of the teratogenic potential of valproic acid analogues in transgenic *Dictyostelium discoideum* strains. Toxicol. In Vitro *12*, 463–469.

Tizzard, A.C., Bergsma, J.H., and Lloyd-Jones, G. (2006). A resazurin-based biosensor for organic pollutants. Biosens. Bioelectron. *22*, 759–763.

Tom-Petersen, A., Hosbond, C., and Nybroe, O. (2001). Identification of copper-induced genes in *Pseudomonas fluorescens* and use of a reporter strain to monitor bioavailable copper in soil. FEMS Microbiol. Ecol. *38*, 59–67.

Torrents, A., Mas, J., Xavier Munoz, F., and Javier del Campo, F. (2012). Design of a microfluidic respirometer for semi-continuous amperometric short time biochemical oxygen demand (BODst) analysis. Biochem. Eng. J. *66*, 27–37.

Torres, C.I., Marcus, A.K., Parameswaran, P., and Rittmann, B.E. (2008). Kinetic experiments for evaluating the Nernst-Monod model for anode-respiring bacteria (ARB) in a biofilm anode. Environ. Sci. Technol. *42*, 6593–6597.

Touloupakis, E., Boutopoulos, C., Buonasera, K., Zergioti, I., and Giardi, M.T. (2012). A photosynthetic biosensor with enhanced electron transfer generation realized by laser printing technology. Anal. Bioanal. Chem. *402*, 3237–3244.

Ulman, A. (1996). Formation and structure of self-assembled monolayers. Chem. Rev. *96*, 1533–1554.

Verma, N., and Singh, M. (2005). Biosensors for heavy metals. BioMetals *18*, 121–129.

Wanekaya, A.K., Chen, W., and Mulchandani, A. (2008). Recent biosensing developments in environmental security. J. Environ. Monitor. *10*, 703–712.

Wang, X.P., and Bartha, R. (1994). Effects of bioremediation on toxicity, mutagenesis, and microbiota in hydrocarbon-polluted soils. In Remediation of Hazardous Waste Contaminated Soils, Wise, D.L., and Trantolo, D.J., eds. (Marcel Dekker, Inc., New York, NY, USA), pp. 175–198.

Wang, Y., Yang, X., Li, H., and Tu, W. (2006). Immobilization of *Acidithiobacillus ferrooxidans* with complex of PVA and sodium alginate. Polym. Degrad. Stabil. *91*, 2408–2414.

Wilson, T., and Hastings, J.W. (1998). Bioluminescence. Annu. Rev. Cell Dev. Biol. *14*, 197–230.

Xu, T., Close, D.M., Sayler, G.S., and Ripp, S. (2013). Genetically modified whole-cell bioreporters for environmental assessment. Ecol. Indic. *28*, 125–141.

Yang, G.-X., Sun, Y.-M., Kong, X.-Y., Zhen, F., Li, Y., Li, L.-H., Lei, T.-Z., Yuan, Z.-H., and Chen, G.-Y. (2013). Factors affecting the performance of a single-chamber microbial fuel cell-type biological oxygen demand sensor. Water Sci. Technol. *68*, 1914–1919.

Yu, D., Volponi, J., Chhabra, S., Brinker, C.J., Mulchandani, A., and Singh, A.K. (2005). Aqueous sol-gel encapsulation of genetically engineered Moraxella spp. cells for the detection of organophosphates. Biosens. Bioelectron. *20*, 1433–1437.

Zhang, F., and Keasling, J. (2011). Biosensors and their applications in microbial metabolic engineering. Trends Microbiol. *19*, 323–329.

Zhang, Y., and Angelidaki, I. (2011). Submersible microbial fuel cell sensor for monitoring microbial activity and BOD in groundwater: focusing on impact of anodic biofilm on sensor applicability. Biotechnol. Bioeng. *108*, 2339–2347.

Zhang, Y.F., and Angelidaki, I. (2012). A simple and rapid method for monitoring dissolved oxygen in water with a submersible microbial fuel cell (SBMFC). Biosens. Bioelectron. *38*, 189–194.

Zhou, J., Bruns, M.A., and Tiedje, J.M. (1996). DNA recovery from soils of diverse composition. Appl. Environ. Microbiol. *62*, 316–322.

# Modern Methods in Microscopy for the Assessment of Biofilms and Bioremediation

6

Gunaratna K. Rajarao

**Abstract**

Imaging techniques have an important and well-accepted role in many fields including bioremediation. New advanced methods are enhancing quantitative *in situ* analyses of the structure, composition, processes and dynamics of complex biofilm microbial communities that perform key roles in pollutant degradation and immobilization. Biofilm matrices consist of water or solvent (97%) bound to microbial cells and the physical properties are determined by the solutes present in it. Biofilm systems are suitable for bioremediation due to their enhanced persistence, complexity, mechanisms to increase pollutant availability for degradation and also because of enhanced by gene transfer among the associated organisms. Several microscopy-based techniques have been developed to better understand the role of microbial communities and associated features such as quorum sensing for community structure and functioning. However, there are impediments to characterizing biofilms *in situ*. This chapter gives an overview of microscopic methods for biofilm bacterial community analysis and mainly focuses on recent advancements in microscopic methods for the assessment of extracellular polymeric substances (EPS) that form a key component of the biofilm matrix.

## Introduction

In the past few decades there has been substantial growth in chemical industries and associated toxic waste effluents that lead to environmental contamination (Paul *et al.*, 2005; Pandey and Jain, 2002). Bioremediation may be used as an *in situ* technology to clean up environmental pollutants using microorganisms, and may score higher than chemical and physical processes in terms of economy and contaminant removal efficiency (Paul *et al.*, 2005). Both organic pollutants and heavy metal contaminants can be metabolized and oxidized by indigenous bacterial communities. However the efficiency of these bacterial communities may be limited by their low abundance and activity, as well as difficulties in accessing nutrients in their environment which may reduce the speed and efficacy of bioremediation. Microbial biofilms are often more proficient and safer to use than planktonic cells since organisms within the complex biofilm matrix have a better survival rate and adaptability to their surrounding environment (Decho, 2000).

As described in Chapter 2, microorganisms anchor to substrates (biotic or abiotic) surrounded by extracellular polymeric substances (EPS) to form biofilms. In nature, biofilms are a complex gathering of multiple microbial species and EPS with heterogeneous architectures and biochemistry (Hall-Stoodley *et al.*, 2004; Branda *et al.*, 2005). The matrix of the biofilm is a combination of complex material such as polymers, lipids, phospholipids, proteins, polysaccharides, DNA, RNA, absorbed nutrients and metabolites, as well as cell lysis products and particulate material from detritus in the immediate surrounding environment (Flemming, 1995). The transportation of solutes, including contaminants, is driven by the biofilm structure, which includes pores and water channels, as well as by diffusion (Horn and Morgenroth, 2006).

Biofilms can be used as a biomarker for ecosystem monitoring (e.g. stream water quality; Washington *et al.*, 2013) and bioremediation (Edwards and Kjellerup, 2013). In the early 1980s, Akinson reported water and wastewater treatment using biofilms (Akinson, 1981). Although wastewater treatment processes involve a biological process, the use of biofilm reactors has been brought into focus recently. Singh *et al.* (2006) discusses the applications of biofilm in bioremediation processes and the importance of gene transfer for the regulation of biofilms. Microbial communities have the potential to exchange genetic information through gene transfer to protect themselves from the environment (chemicals and shear stress). Indeed, microorganisms in biofilms (activated sludge) have a higher tolerance against changes in nutrient concentrations, predation and toxic chemicals, as well as environmental stresses such as pH, temperature and salt concentrations (Heipieper *et al.*, 1991; Beviridge *et al.*, 1997; Hall-Stoodley *et al.*, 2004). As an example of this, a biofilm may require 100–1000 times higher concentration of antibiotics than planktonic bacteria for it to respond to treatment in the human body (Mah *et al.*, 2003). For this reason, microbial cells growing in highly contaminated sites tend to grow as biofilms, allowing them to tolerate harsh environment conditions. These natural characteristics of microbial communities can be utilized to implement strategies for bioremediation (Gross *et al.*, 2007; Edwards and Kjellerup, 2013). In a review article by Pal and Paul (2008) the benefits of EPS for the bioremediation of heavy metals by specific bacteria is described. Researchers are giving much importance to EPS characterization and highlighting the availability of genetic engineering tools to facilitate more creative ways for biofilm produced EPS to be used for bioremediation in the future. Biofilms found in rivers are important in biogeochemical cycling and for the biodegradation of pollutants within riverine systems that significantly contribute to self-purification (Lawrence *et al.*, 2001; Chenier *et al.*, 2003). Tien *et al.* (2013) reported the biodegradation of carbamate pesticides (methonyl, carbaryl, carbofuran) by natural river biofilms and their effects on the structure of the biofilm community. Hence it is essential to understand biofilm community structure, functions and relationships for the future use of biofilm in bioremediation and ecosystem monitoring. A main aim of this chapter is to discuss the role of advances in microscopy for developing a better understanding of the effective use of biofilms in remediation, towards the optimization of bioremediation procedures. A summary of some of the main advantages and disadvantages of a variety of key microscopic methods for biofilm assessment is provided in Table 6.1.

## Biofilm community analysis

The characterization and quantification of microbial cells/EPS fragments during the formation/detachment of biofilm is a challenging task. Traditional methods like viable cell counting, density measurements, and direct counts are tedious and often dependent on obtaining accurate dilution ratios. Moreover, since biofilms are complex 3-dimensional structures of cells and EPS, direct cell counting is a major challenge. Techniques used to characterize EPS/cell fractions are similarly time-consuming. Though fluorescent staining techniques and confocal laser scanning microscopy (Cook *et al.*, 2000) were developed as an improvement on the direct cell count method, it still requires skilled personnel and is expensive and time consuming, typically using flow cytometry alone or in combination with PCR (Lebaron *et al.*, 2002; Vives Rego *et al.*, 2000). At the same time, indirect measurements of cell population density have been developed using molecular immunological techniques and genotyping methods like real-time quantitative PCR and RT-PCR (Bleve *et al.*, 2003; Van der Vossen and Hofstra, 1996; Zhang *et al.*, 2009), as described in Chapter 7. These molecular techniques are able to distinguish viable and dead or non-culturable cells. However, these techniques provide little information about the nature and role of EPS.

Basic bacterial culturing techniques used to quantify cell abundance requires the desorption and dispersion of bacterial cells onto an agar plate, which influences the results (Hannig *et al.*, 2007). Most of the bacteria are not culturable, hence alternative methods are being developed for the direct visualization and quantification of bacteria in adherent states using fluorescent dyes so they can be visualized by fluorescent microscopy (Aas *et al.*, 2005; Thurnheer *et al.*, 2001). Fluorescence *in situ* hybridization (FISH) marks bacterial species with fluorescent probes and is also a technique developed to stain live/dead cells (Hannig *et al.*, 2010). There are various drawbacks that come along with the fluorescent-based detection techniques. These drawbacks include differences in the target area, dye penetration and, in most cases, a maximum of only seven different dyes can be used at the same time.

**Table 6.1** Advantages and limitations of some key microscopic methods

| Techniques | Advantages | Limitations |
|---|---|---|
| Laser scanning microscopy (LSM) | Microscopy of samples *in situ*<br>Use of hydrated samples<br>Probe dependent, allowing visualization of specific cells/features<br>3-D analysis of macromolecules, cells and communities<br>Multichannel analysis allows use of multiple probes<br>Flexible in terms of sample staining and mounting | Requires laser penetration<br>Only limited probes can be used at once<br>Only targeted compounds will be detected |
| Magnetic resonance imaging (MRI) | Microscopy of samples *in situ*<br>Use of hydrated samples<br>Probe independent<br>Allows chemical species analysis<br>Suited for analysis of structure, flow and transport within biofilms | Slow with ferromagnetic samples or not possible<br>Requires experience for measuring biological<br>Samples and data analysis |
| Scanning transmission X-ray microscopy (STXM) | Microscopy of samples *in situ* chemical analysis<br>Use of hydrated samples<br>Probe independent, no staining or fixation of sample required<br>Analysis of organic and inorganic substances<br>Speciation of molecular structures<br>Tomographic imaging<br>Resolution to less than 50 nm | Sample damage due to radiation<br>Suited to analysis of very thin samples (<200 nm) |
| Atomic force microscopy (AFM) | Allows analysis under near physiological conditions<br>Simple sample preparation<br>Informs on nanostructure and nanomechanics of cell sample<br>Provides surface topology information<br>Suited for analyses of cell adhesion and cellular elasticity | Small scanning area and slow scan speed<br>Image artefacts produced if approach not appropriately optimized<br>Probe tips cannot easily scan the topology of steep walls and overhangs |
| Raman microscopy (RM) | Non-destructive sample analysis<br>Fingerprints of bacteria chemical composition<br>Data mining extracts biochemical information<br>Simple sample preparation | As the Raman effect is very weak, the approach must be highly optimized<br>Fluorescence of impurities or of the sample itself can hide the Raman spectrum<br>Sample heating by radiation can destroy the sample or cover the Raman spectrum |
| Surface enhance raman spectroscopy (SERS) | Enhanced Raman scattering<br>Provides chemical spectra (composition) of living cells | As the Raman effect is very weak, the approach must be highly optimized<br>Fluorescence of impurities or of the sample itself can hide the Raman spectrum<br>Photodegradation of sample |
| Environmental scanning electron microscopy (ESEM) | Used to visualize poorly conductive hydrolysed biological samples<br>Spatial resolution 10 nm or less<br>Minimum processing time | May provide only a small field of view |
| Scanning acoustic microscopy (SAM) | Can be used to quantify sample thickness and mechanoelastic properties | Slow sample processing time<br>Requires contrast agents linked with targeted ligands |

## Characterization of extracellular polymeric substances (EPS)

Generally EPS has to be purified from the biofilm and the separated cellular compounds analysed with destructive methods (Nielsen and Jahn, 1999) to overcome the weak intermolecular interactions such as electrostatic, van der Waals forces and hydrogen bonds that otherwise maintain the structure of the biofilm matrix (Garny *et al.*, 2010). For this reason, the techniques developed to characterize EPS in aquatic samples require EPS extraction which often causes cell lysis (Frolund *et al.*, 1996). Physical and chemical EPS extraction methods have both been used. However, boiling and alkaline treatments are avoided since they lead to drastic structural damage of both cells and EPS.

There are several techniques that have been used to characterize EPS after extraction such as colorimetry, spectrometry/chromatography and microscopy (Sheng *et al.*, 2010). For example, EPS fractions can be extracted by EDTA and components such as carbohydrate, protein, uronic acid and DNA determined by photocolorimetric methods. Alternatively, chromatography/spectrometry based methods are used to obtain detailed information about saccharides and amino acid composition or FTIR spectrophotometry used to analyse the multicomponent EPS fractions without separation (Castro *et al.*, 2014). These techniques will not provide information on spatial and temporal variations of EPS production and their interactions since the structure of the biofilm is destroyed by the process of analysis. Lectin-tagged fluorophores can be used to bind the exopolymeric material for confocal laser scanning microscopy (CSLM) or epifluorescence microscopy (Flemming and Wingender, 2010; Sheng *et al.*, 2010), which allows visualization of intact biofilm structures. In 1999, Neu and Lawrence (1999) used such an approach to report the spatiotemporal development of EPS glycoconjugates with fluorescence-conjugated lectins. However, the small number of fluorescent labels that be used in CLSM limits the applications of the technique. There are additional difficulties that arise when observing biofilms under a microscope in hydrated conditions due to the densely packed nature of the microorganisms. Therefore new methods need to be developed in order to characterize cells and in biofilm without disturbance. To address these biological questions, a variety of microscopic techniques have developed and optimized to investigate biofilms including the use of gene reporters to inform on real-time variations in biofilm physiology.

## Microscopic methods

Recent advancement in confocal laser scanning microscopy (CLSM) and other methods have enabled us to document biofilm morphology and physiology under *in situ* conditions. Additionally, the continuous development of instrumentation has provided new microscopic techniques while also continuing to generate 'new applications of older techniques' (Palmer and Sternberg, 1999).

To identify the microorganisms present in biofilm, fluorescently labelled probes, such as those used in FISH, multicolour fluorescence microscopy or CSLM (Caldwell *et al.*, 1992; Gu *et al.*, 2005; Yun *et al.*, 2006) may be utilized. Such approaches provide *in situ* non-destructive studies of biofilm in hydrated forms thus giving information on the physical structures of the biofilm (Beyenal *et al.*, 2004). On the other hand, it provides images containing out of focus sections, hence preventing the acquisition of a large part of the fluorescent emission. The illumination by high light intensities to excite the probe fluorophores may also bleach the fluorescent molecules. Moreover, the fluorochromes can be expensive and confocal microscopes may not be available in many research laboratories. Carla *et al.* (2007) explored the possibility of obtaining 3-D images of biofilm using optical microscopy. In this technique, when the bright field of transmitted light passes through the biofilm, the presence of cells in the X-Y plane decreases proportionally to the light passed in the Z axis (perpendicular). It is possible to determine the thickness by finding the relation between the amount of light observed in 2D images and the number of cells.

### Laser scanning microscopy

Laser scanning microscopy (LSM) is used as a tool for imaging microbial communities in *situ* (Lawrence *et al.*, 1991). It is used based on the visualization of multiple features, detecting multichannel signals in four dimensions (Neu *et al.*, 2010) from the fluorescence signal of probes to enable analyses of the structure, composition, activity and volume of complex biofilm communities. Confocal laser scanning microscopy (CLSM) creates a thin plane of highly focused laser light to scan samples, thereby providing deeply penetrating excitation energy. Photomultiplier tubes detect the excitation of specific the fluorophores (e.g. GFP, chlorophyll) or exogenous probes (fluorescently labelled antibodies or nucleotides) in each sample. A point-scanning device offers high lateral and axial resolution however but with delayed image collection. The single-photon CLSM is used mainly for biological specimens that are stained with fluorescent probes. In combination with the use of conventional microscopy and cryo-embedding microtome sectioning of multilayered biofilm, it is possible to visualize the biofilm. The main concern is fixation and dehydration of the sample that is required for sectioning since this affects the three-dimensional structure of biofilm. Hence modern embedding techniques have improved CLSM processing. Agarose embedded biofilms are good and simple though the agarose solution is fragile at low concentrations and must be kept warm. Acrylamide (20%) was shown to be a good embedding medium for the analysis of GFP expression or 16S rRNA gene probe analysis (Christensen *et al.*, 1999; Moller *et al.*, 1998). However, the biofilm needs to be fixed, dehydrated or in still in living conditions. The use of CSLM has generated many insights into biofilm structure and physiology. For example, Singleton *et al.* (1997) showed a multidisciplinary approach of using confocal data to model the flow of nutrients within biofilms, whereas Davies *et al.* (1998) revealed the importance of the signalling molecule homoserine in regulating the architecture of biofilms.

As our understanding of the complexity of biofilms continues to increase, the requirement for visual data on the 3-dimensional organization of biofilms and their relationship with EPS and their substratum becomes more significant. To distinguish among the similar morphologies of the cells in biofilm, they can be observed using species selective fluorescently labelled oligonucleotide probes. Fluorescent-based techniques have both advantages and disadvantages. The positive side is that it is a powerful tool used in identifying the spatial relationship of taxa or genes within a biofilm. The hybridization steps involve polysaccharide stabilization of biofilms to stabilize the three-dimensional structure and therefore the cells will not be withdrawn during the washing steps and can be stored to be viewed later. The disadvantage is that it requires a dehydration process and the physiological conditions may not be studied. Hence, other techniques need to be combined with the hybridization techniques to study the physiological interactions. A study by Schramn *et al.* (1996) revealed how single cell-levels of microcolonies can be correlated with microelectrode profiles related to changes in concentrations of soluble chemical species of nitrate/nitrite and oxygen in biofilms in wastewater treatment. The activity of genes in biofilm can be assessed using reporter strains expressing the lacZ enzyme system since active production of the enzyme luciferase produces light that can be visualized by cameras (Palmer and White, 1999). Using a slightly modified approach, this system has been used with a genetically engineered *Pseudomonas putida* biofilm to show the heterogeneity of toluene degradation (Palmer and Sternberg, 1999), as the bacteria was engineered to produce light as toluene degradation genes were expressed. Confocal microscopy can be used to visualize GFP expression in the three-dimensional biofilm structure. When taken together, confocal and epifluorescence microscope techniques cannot visualize un-stained material such as exopolymeric substances. Hence, the structure and functional attributes of these exopolymers may need to be explored using other techniques (Wolfaardt *et al.*, 1998).

## Electron microscopy

Transmission electron microscopy (TEM) and scanning electron microscopy (SEM) provide high resolution images and require sample penetration and dehydration of the sample to disrupt the structure of the biofilm (Carla *et al.*, 2007). SEM can reveal bacterial adhesion properties, which are influenced by the substrate, liquid medium, carbon source, pH and the hydrodynamics and flow rate for biofilm development and distribution (Bragadeeswaran *et al.*, 2010; Herald and Zottola, 1988; Hamadi *et al.*, 2005). SEM can help in elucidating the surface structure of biofilms (Motta *et al.*, 1994) in order to understand the ecology and physiology of microbial systems since it is shown that bacterial cells are packaged in a dense glycocalyx (Blenkinsopp *et al.*, 1991; Fassel *et al.*, 1991). It can also be used to visualize the thickness of the biofilm by cross sectioning the freeze-dried samples as shown by Ganderton *et al.* (1992). Thus SEM allows a higher resolution image of biofilm topology (3.5 nm), measuring quantity data in three dimensions. In order to visualize samples in SEM, the sample first has to be fixed chemically, dehydrated and coated with conductive material. Many sample preparation methods have been developed to withstand high vacuum conditions such as glutaraldehyde fixation, negative staining, sputter cryo technology and coating with gold or osmium to enhance the contrast and reduce damage to the sample (Allan-Wojtas *et al.*, 2008; Hassan *et al.*, 2003; Lamed *et al.*, 1987).

To visualize biofilm under TEM, samples need to be embedded in resin followed by taking ultrathin sections, mounting on copper grids and staining with uranyl acetate or lead acetate. Bonaventura *et al.* (2004) showed that biofilm formation by *Stenotrophomonas maltophilia* after 24 h of growth visualized by TEM as cotton like structures surrounded by bacteria where the former are suggested to regulate the adherence onto polystyrene surfaces. Cells treated with the antibiotic mixofloaxin (100 mg/ml) showed the degradation of glycocalyx and detachment from the surface. Biofilm formed by a microbial community was used to study copper removal from groundwater in sub-merged fixed biofilter. TEM together with energy-dispersive X-ray micrograph (EDX) showed the accumulation of copper onto EPS by binding with phosphate or silicates (Vilchez *et al.*, 2007).

## Combinations of microscopy techniques

Combinations of different analytical methods can enhance our understanding of the structure, functions and relationship of a biofilm [e.g. by combining CLSM and Raman microscopy (Ivleva *et al.*, 2009; Wagner *et al.*, 2009) and CLSM and magnetic resonance microscopy (Majors *et al.*, 2005; McLean *et al.*, 2008; Wind *et al.*, 2002) and many more]. CLSM is often the first choice of method used in characterizing biofilm structures, detecting different macromolecules in biofilms of up to 100 μm in thickness or following cryosectioning. However, the detection of various macromolecules will be limited using this approach due to the usage of fluorescent stains, their low specificity and difficulties in quantifying the biofilm matrix. For this reason, the use of a combination of techniques is often desirable to allow the thorough characterization of EPS.

### Raman microscopy (RM)

Raman microscopy (RM) helps to detect the chemical composition of the biofilm matrix without using a staining method. The spectroscopic fingerprints of living cells and bacteria are identified by nondestructive RM methods based on their chemical composition and molecular structure. Raman scattering signals classify the chemical constituents of biofilms (polysaccharides, proteins). It provides the spatial resolution in μm ranges and fingerprint spectra of biofilm matrices, which allows the characterization and identification of different substances in single bacteria without needing to alter the structure of the biofilm (Harz *et al.*, 2005; Pätzold *et al.*, 2006; Rösch *et al.*, 2005; Sandt *et al.*, 2007; Wagner *et al.*, 2009). RM can be used to differentiate *Bacillus anthracis, B. thuringiensis* and *B. atrophaeus* spores and *B. cereus* vegetative cells following principal component analysis (PCA) of the spectroscopic fingerprints to differentiate biological threats (Guicheteau *et al.*, 2008). However, RM with single spot spectra is time consuming (100 s exposure time). Subsequent research in the future should focus on Raman mapping to compare biofilm matrices and reduce the time taken for RM biofilm characterization. Wagner *et al.* (2009) compared the CLSM staining method with RM to show the chemical composition of different parts of the biofilm and cellular morphologies. Raman microscopy measures any small shifts in photon energy in visible or infrared lasers (Palmer and Sternberg, 1999). Several commercial confocal Raman microscopies are available to investigate the chemical nature of biofilms in axial and lateral dimensions.

### Surface-enhanced Raman spectroscopy (SERS)

Surface-enhanced Raman spectroscopy (SERS) can be used to characterize biological molecules such as DNA, protein, lipids, cells, tissues, bacteria and yeast using different nanostructured surfaces (Efeoglu and Culha, 2013). Wagner *et al.* (2009) revealed the comparison of SERS and CLSM and observed that SERS performance was superior to CLSM. The chemical changes of bacterial (*Escherichia coli*) cell wall were detected when exposed to antibiotics using chitosan-coated silver nanoparticles as a substrate for SERS analysis. The peaks that appeared in the SERS revealed that the response might be caused by DNA or RNA released from bacterial cells due to attraction of negative charges of the substrate.

## Scanning transmission X-ray microscopy (STXM)

Scanning transmission X-ray microscopy (STXM) uses near edge X-ray absorption fine structure/spectroscopy (NEXAFS) to collect images with a range of energy profiles for quantitative mapping of chemical species in the biofilm. NEXAFS provides information based on the bonding structures of chemical species (Kirz *et al.*, 1995; Ade and Hitchcock, 2008). STXM can be used in hydrated samples by utilizing 'soft' X-rays to penetrate water (Lawrence *et al.*, 2003). Gilbert *et al.* (1999), demonstrated the efficiency of soft X-ray microscopy for imaging the *Pseudomonas putida* biofilm using STXM with synchrotron radiation. STXM can be used for chemical analysis of hydrated samples either in mineral water or at the mineral–microbe interface. STXM is capable of analysing and addressing questions on the nature and distribution of proteins, carbohydrates, lipids and nucleic acids in the EPS. Quantitative mapping of organic macromolecules within a biofilm using intrinsic absorption spectra was demonstrated (Lawrence *et al.*, 2003). This could allow users to distinguish cells that are rich in polysaccharide lipids or proteins. STXM can also detect and provide information on metals like nickel, copper, manganese and iron. For example, Lawrence *et al.* (2012) demonstrated the sorption of copper nanoparticles on EPS and nickel sorption of microbial biofilm by manganese oxides has similarly been reported (Dynes, 2006).

Detailed information on STXM and other microscopic techniques was reviewed by Neu *et al.* (2010). The spatial distribution of Fe(II) and Fe(III) in *P. aeruginosa* biofilm was analysed by STXM and it showed that the Fe(II) was located in the EPS and Fe(III) was on the cell surface (Hunter *et al.*, 2008). Miot *et al.* (2009) revealed the organic substances associated with iron biomineralization by the anaerobic nitrate-dependent iron-oxidizing bacterium *Acidovorax* sp. strain BoFeN1 using STXM and mapped the proteins present at the periphery of the cell poles. The detected carbon showed the presence of exopolysaccharides and Fe content in the cells as extracellular precipitates. STXM also demonstrated that lipids and polysaccharides of polymeric fibres act as templates for mineral precipitation (Miot *et al.*, 2009). The distribution of carbon and iron species in *Shewanella oneidensis* MR-1 cells was identified by a combination of STXM and X-ray magnetic circular dichroism (XMCD) techniques. Lam *et al.* (2010) demonstrated the magnetosomes produced by *Vibrio* sp. using XMCD in STXM. Proteins and lipids produced by bacteria are associated with dissimilatory Fe(III) reduction and are revealed as magnetite (Coker *et al.*, 2012). Arsenic immobilization by freshwater Fe(II)-oxidizing bacteria was demonstrated using low-energy X-ray fluorescence in STXM (Hitchcock *et al.*, 2012).

## Atomic force microscopy (AFM)

EPS and other components such as calcium fill the space between cells and bring them together to influence the cohesive strength of the biofilm (Mayer *et al.*, 1999; Patrauchan *et al.*, 2005). Atomic force microscopy (AFM) provides molecular resolution of three-dimensional surface ultra-structures with minimal sample preparation using chemically functionalized AFM probes. Cell surface hydrophobicity and charges have also been detected using this approach (Ahimou *et al.*, 2002; Towler *et al.*, 2003). AFM can also be used to measure bacterial turgor pressure, surface polymers and cell elasticity (Arnoldi *et al.*, 2000; Velegol *et al.*, 2002). The advantages of AFM in studying the bacterial adhesion to biomaterial surfaces and image processing for quantification of biofilm structure was reviewed and discussed (Nehal *et al.*, 2013). Ahimou *et al.* (2007) reported the applications of AFM at the nanoscale level of measuring *in situ* cell/EPS and EPS/EPS interactions of biofilm under defined volume. It was shown that the biofilm cohesion energy was increased in the presence of calcium. The presence of salty bridges between cations and exopolymer functional group of anions (carboxyl, phosphate, sulfate, glycerate, pyruvate and succinate groups) has also been identified using this approach (Ahimou *et al.*, 2007). The affinities of multivalent ions are $Ca^{2+}$, $Cu^{2+}$, $Mg^{2+}$ and $Fe^{2+}$ to anionic ligands. This can be confirmed by AFM where the intact biofilms can be visualized. This technique helps to provide information on thickness and sloughing of the biofilm needed for various applications.

The biochemical and biomechanical properties of *P. putida* (KT2440) living cells using combined AFM and RM under physiological conditions was revealed by Zhou *et al.* (2010). The adhesion force of flagella is higher than the cell wall (20–36%) as revealed by AFM and the chemical components of EPS were determined by RM. This combined microscopy method simplified the method of sample preparation, since biofilm properties can be measured directly in the aqueous phase.

## Additional fluorescence microscopic methods

Different fluorescent microscopic methods, such as fluorescence recovery photobleaching (FRAP), fluorescence correlation spectroscopy (FCS) and fluorescence life time imaging (FLIM), for diffusion reactions of macromolecules (antibiotics) in biofilms have been studied (Oubekka *et al.*, 2012). The time lapse approach method has the possibility to monitor the penetration of fluorescent macromolecules in all parts of a biofilm over a period of time and allows the measurement of the diffusion reaction when the macromolecules reach the equilibrium inside the biofilm matrix. The research has previously focused on the penetration of fluorescently labelled vancomycin in the biofilm of *Staphyloccus aureus* human isolates (Oubekka *et al.*, 2012).

## Environmental scanning electron microscopy (ESEM)

Environmental scanning electron microscopy (ESEM) can be used to visualize poorly conductive hydrolysed biological samples under low-vacuum (Danilatos, 1993). Trinidad *et al.* (2010) gave a detailed description and explained the principle of ESEM using a gaseous detection device. ESEM helps to visualize highly hydrated glycocalyx and the sample can be stored in a refrigerator before examination, providing a spatial resolution of 10 nm or less. EPS volume may be calculated from the EPS surface and imaged as a semi-transparent coating of bacteria (Costeston 1999). Samples do not require fixing and staining, thus preserving native morphologies and minimizing biofilm dehydration (Walker *et al.*, 2001), shrinkage and artifice. Thus, ESEM could provide complimentary information of biological samples, e.g. cell structure can be visualized in SEM, whereas cells surrounded by external polymers are visible in ESEM (Doucot *et al.*, 2005) to provides the exact topography and architecture of intact hydrated biofilms. In order to enhance the contrast resolution of the ESEM method for imaging hydrated samples of biofilm to discriminate EPS, specific staining and of the targeted sample is required and use of the system with increasing electron density (Priester *et al.*, 2007). Investigation of biofilms often recommend to use different techniques in order to gain valuable information in different aspects. For example, ESEM is well suited to study the topography and structure of glycocalyx whereas standard SEM may only provide the rough structure of dehydrated samples. Since CSLM is well suited for the analysis of live cells, a combination of ESEM and CLSM can be used to provide comprehensive investigations of biofilm biology.

## Scanning probe microscopy (SPM)

A useful technique to view the topography of biofilm is scanning probe microscopy (SPM) which moves a cantilever probe across the surface. It can be used in hydrated samples (Kuznetsov *et al.*, 1997). Force microscopy provides information at the atomic level by measuring attraction/repulsion of the probe tip. In order to get different resolutions, a range of different approaches are used including phase detection microscopy, electrostatic force microscopy, magnetic force microscopy and force modulation microscopy. The relative strength of interactions between the cells and the substratum as well as the influence of lipopolysaccharide in aqueous solution has been quantified using force microscopy (Razatos *et al.*, 1998).

## Scanning acoustic microscopy (SAM)

Mechanical and structural frequency scanning acoustic microscopy of biofilms may be assessed by high-frequency scanning acoustic microscopy (SAM) to quantify the thickness and mechanoelastic properties of biofilm. Ultrasound contrast agents (UCAs) are encapsulated gas bubbles that can be used to enhance biofilm detection by providing unique acoustic scattering signature. The outer membrane of UCAs can be linked with a ligand to bind to the tissue. A combination of time resolved high frequency SAM is ideal for quantifying mechanical and elastic properties of biofilms and can be used as a non-invasive setting for clinical diagnostic applications (Anastasiadis *et al.*, 2014).

## Summary

Research today focuses on bioremediation techniques as it is an important tool for controlling environmental pollutants. In order to understand the mechanisms of microorganisms in bioremediation, a thorough knowledge of biofilm matrices is necessary. To obtain an overall comprehension of the different aspects of biofilms, combinations of *in situ* analytical techniques are crucial. New promising microscopy-based methods and approaches are being developed, such as those mentioned above, to meet the new challenges and opportunities provided by the detailed characterization of microbial biofilms.


## References


Aas, J.A., Paster, B.J., Stokes, L.N., Olsen, I., and Dewhirst, F.E. (2005). Defining the normal bacterial flora of the oral cavity. J. Clin. Microbiol. *43*, 5721–5732.

Ade, H., and Hitchcock, A.P. (2008). NEXAFS microscopy and resonant scattering composition and orientation probed in real and reciprocal space. Polymer *49*, 643–675.

Ahimou, F., Denis, F.A., Touhami, A., and Dufrene, Y.F. (2002). Probing microbial cell surface charges by atomic force microscopy. Langmuir *18*, 9937–9941.

Ahimou, F., Semmens, M.J., Novak, P.J., and Haugstad, G. (2007). Biofilm cohesiveness measurement using a novel atomic force microscopy methodology. Appl. Environ. Microbiol. *73*, 2897–2904.

Akinson, B. (1981). Immobilized biomass – a basis for process development in wastewater treatment. In Biological Fluidized Bed Treatment of Water and Wastewater, Cooper, P.E., and Akinson, B., eds. (Ellis Horword, Chichester, UK), pp. 22–34.

Allan-Wojtas, P., Hansen, L.T., and Paulson, A.T. (2008). Microstructural studies of probiotic bacteria loaded alginate microcapsules using standard electron microscopy techniques and anhydrous fixation. LWT-Food Sci. Technol. *41*, 101–108.

Anastasiadis, P., Mojica, K.D.A., Allen, J.S., and Matter, M.L. (2014). Detection and quantification of bacterial biofilms combining high-frequency acoustic microscopy and targeted lipid microparticles. J. Nanobiotechnol. *12*, 24.

Arnoldi, M., Fritz, M., Bauerlein, E., Radmacher, M., Sackmann, E., and Boulbitch, A. (2000). Bacterial turgor pressure can be measured by atomic force microscopy. Phy. Rev. *61*, 1034–1044.

Beveridge, T.J., Makin, S.A., Kadurugamuwa, J.L., and Li, Z. (1997). Interactions between biofilms and the environment. FEMS Microbiol. Rev. *20*, 291–303.

Blenkinsopp, A.S., and Costerton, J.W. (1991). Understanding bacterial biofilms. Trends Biotechnol. *9*, 138–143.

Bleve, G., Rizzotti, L., Dellaglio, F., and Torriani, S. (2003). Development of reverse transcription (RT)-PCR and real time RT-PCR assays for rapid detection and quantification of viable yeasts and molds contaminating yoghurts and pasteurized food products. Appl. Environ. Microbiol. *69*, 4116–4122.

Bonaventura, G.D., Spedicato, I., D'Antonio, D., Robuffo, I., and Piccolomini, R. (2004). Biofilm formation by *Stenotrophomonas maltophilia*: modulation by quinolones, trimethoprim-sulfamethoxazole and ceftazidime. Antimicrobiol. Agents Chemother. *48*, 151–160.

Bragadeeswaran, S., Balasubramanian, S.T., Raffi, S.M., Rani Sophia, S. (2010). Scanning electron microscopy elemental studies of primary film. World Appl. Sci. J. *10*, 169–172.

Branda, S.S., Vik, Å., Freidman, L., and Kolter, R. (2005). Biofilms: the matrix revisited. Trends Microbiol. *13*, 20–26.

Caldwell, D.E., Korber, D.R., and Lawrence, J.R. (1992). Confocal laser microscopy and digital image analysis in microbial ecology. Adv. Microb. Ecol. *12*, 1–67.

de Carvalho, C.C., and da Fonseca, M.M. (2007). Assessment of three-dimensional biofilm structure using an optical microscope. Biotechnology *42*, 616, 618–620.

Castro, L., Zhang, R., Munoz, J.A., Gonzalez, F., Blazquez, L., Sand, W., and Ballester, A. (2014). Characterization of exopolymeric substances (EPS) produced by *Aeromonas hydrophila* under reducing conditions. Biofouling *30*, 501–511.

Chenier, M.R., Beaumier, R.D., Roy, R., Driscoll, B.T., Lawrence, J.R., and Greer, C.W. (2003). Impact of seasonal variations and nutrient inputs on the cycling of nitrogen and the degradation of hexadecane by replicated river biofilms. Appl. Environ. Microbiol. *13*, 589–603.

Christensen, B.B., Sternberg, C., Andersen, L., Palmer, R. Jr, Nielsen, A., Givskov, M., and Molin, S. (1999). Molecular tools for study of biofilm physiology. Met. Enzymol. *310*, 20–42.

Coker, V.S., Byrne, J.M., Telling, N.D., VanDerlaan, G., Lloyd, J.R., Hitchcock, A.P., Wang, J., and Pattrick, A.D. (2012). Characterization of the dissimilatory reduction of Fe(III) oxyhydroxide at the microbe–mineral interface: the application of STXM-XMCD. Geobiology *10*, 347–354.

Cook, G., Costerton, J.W., and Darouiche, R.O. (2000). Direct confocal microscopy studies of the bacterial colonization *in vitro* of a silver-coated heart valve sewing cuff. Int. J. Antimicrob. Agents *13*, 169–173.

Costerton, J.W. (1999). Introduction to biofilm. Int. J. Antimicrob. Agents *11*, 217–221.

Danilatos, G.D. (1993). Introduction to the ESEM instrument. Microsci. Res. Tech. *25*, 354–361.

Davies, D., Parsek, M., Pearson, J., Iglewski, B., Costerton, J., and Greenberg, E. (1998). The involvement of cell-to-cell signals in the development of a bacterial biofilms. Science *280*, 295–298.

Decho, A.W. (2000). Microbial biofilms in intertidal systems: an overview. Cont. Shelf Res. *20*, 1257–1273.

DeLong, E.F., Wickham, G.S., and Pace, N.R. (1989). Phylogenetic strains: ribosomal RNA-based probes for the identification of single cells. Science *243*, 1360–1363.

De Soete, G. (1973). A least-square algorithm for fitting additive trees to proximity data. Psychometrika *48*, 621–626.

Doucet, F.J., Maguire, L., Achterberg, E.P., and Millward, G.E. (2005). Visualization of natural aquatic colloids and particles – a comparison of conventional high vacuum and environmental scanning electron microscopy. J. Environ. Monit. *7*, 115–121.

Dynes, J.J., Tyliszczak, T., Araki, T., Lawrence, J.R., Swerhone, G.D.W., Leppard, G.G., and Hitchcock, A.P. (2006). Speciation and quantitative mapping of metal species in microbial biofilms using scaning transmission X-ray microscopy. Environ. Sci. Technol. *40*, 1556–1565.

Efeoglu, E., and Culha, M. (2013). Surface-enhanced raman scattering for biofilm characterization. Spectroscopy *28*, 36–41.

Erwards, S.J., and Kjellerup, B.V. (2013). Applications of biofilms in bioremediation and biotransformation of persistent organic pollutants, pharmaceuticals/personal are products and heavy metals. Appl. Microbiol. Biotechnol. *97*, 9909–9921.

Fassel, T.A., Van Over, J.E., Hauser, C.C., Edmiston, C.E., and Sanger, J.R. (1991). Adhesion of Staphylococci to breast prosthesis biomaterials: an electron microscopic evaluation. Cells Materials *1*, 199–208.

Flemming, H.C. (1995). Sorption sites in biofilms. Water Sci. Technol. *32*, 27–33.

Flemming, H., and Wingender, J. (2010). The biofilm matrix. Nat. Rev. Microbiol. *8*, 623–633.

Frolund, B., Palmgren, R., Keidding, K., and Nielsen, P.H. (1996). Extraction of extracellular polymers from activated sludge using a cation exchange resin. Water Res. *30*, 1749–1758.

Ganderton, L., Chawla, J., Winters, C., Wimpenny, J., and Stickler, D. (1992). Scanning electron microscopy of bacterial biofilms on indwelling bladder catheters. Eur. J. Clin. Microbiol. Infect. Dis. *11*, 789–796.

Garny, K., Neu, T.R., Horn, H., Volke, F., and Manz, B. (2010). Combined application of $^{13}C$ NMR spectroscopy and confocal laser scanning microscopy-investigation on biofilm structure and physic-chemical properties. Chem. Eng. Sci. *65*, 4691–4700.

Gilbert, E.S., Khelbnikov, A., Meyer-Ilse, W., and Keasling, J.D. (1999). Use of soft X-ray microscopy for the analysis of early stage biofilm formation. Water Sci. Technol. *39*, 269–272.

Gross, R., Hauer, B., Otto, K., and Schmid, A. (2007). Microbial biofilms: new catalysts for maximizing productivity of long-term biotransformations. Biotechnol. Bioeng. *98*, 1123–1134.

Gu, F., Lux, R., Du-Thumm, L., Stokes, I., Kreth, J., Anderson, M.H., Wong, D.T., Wolinsky, L., Sullivan, R., and Shi, W. (2005). In situ and non-invasive detection of specific bacterial species in oral biofilms using fluorescently labeled monoclonal antibodies. J. Microbiol. Met. *62*, 145–160.

Guicheteau, J., Atgue, L., Emge, D., Hyre, A., Jacobson, M., and Christensen, S. (2008). Bacillus spore classification via surface-enhanced raman spectroscopy and prinicipal component analysis. Appl. Spectrosc. *62*, 267–272.

Hall-Stoodley, L., Costerton, J.W., and Stoodley, P. (2004). Bacterial biofilms: from the natural environment to infectious diseases. Nat. Rev. *2*, 95–108.

Hamadi, F., Latrache, H., Mabrrouki, M., Elghmari, A., Outzourhit, A., Ellouali, M., and Chtaini, A. (2005). Effect of pH on distribution and adhesion of *Staphylococcus aureus* to glass. J. Adhesion. Sci. Technol. *19*, 73–85.

Hannig, C., Hannig, M., Rehmer, O., Braun, G., Hellwig, E., and Al-Ahmed, A. (2007). Fluorescence microscopic visualization and quantification of initial bacterial colonization on enamel *in situ*. Arch. Oral Biol. *51*, 1048–1056.

Hannig, C., Follo, M., Hellwig, E., and Al-Ahmed, A. (2010). Visualization of adherent microorganisms using different techniques. J. Med. Microbiol. *59*, 1–7.

Harz, M., Rösch, P., Peschke, K.D., Ronneberger, O., Burkhardt, H., and Popp, J. (2005). Micro-Raman spectroscopic identification of bacterial cells of the genus *Staphylococcus* and dependence on their cultivation conditions. Analyst *130*, 1543–1550.

Hassan, A.N., Frank, J.F., and Elsoda, M. (2003). Observation of bacterial exopolysaccharide in dairy products using cryo-scanning electron microscopy. Inter. Dairy J. *13*, 755–762.

Heipieper, H.J., Keweloh, H., and Rehm, H.J. (1991). Influence of phenols on growth and membrane permeability of free and immobilized *Escherichia coli*. Appl. Environ. Microbiol. *57*, 1213–1217.

Herald, P.J., and Zottola, E.A. (1988). Scanning electron microscopic examination of *Yersinia enterocolitica* attached to stainless steel at selected temperatures and pH values. J. Food Sci. *51*, 445–448.

Hitchcock, A. (2001). Chemical mapping with soft X-ray spectromicroscopy. Am. Lab. *33*, 30–36.

Hitchcock, A.P., Obst, M., Wang Lu, Y.S., and Tyliszczak, T. (2012). Advances in the detection of As in environmental samples using low energy X-ray fluorescence in a scanning transmission X-ray microscopic: arsenic immobilization by an Fe(II)-oxidizing freshwater bacteria. Environ. Sci. Technol. *46*, 2821–2829.

Horn, H., and Morgenroth, E. (2006). Transport of oxygen, sodium chloride and sodium nitrate in biofilms. Chem. Eng. Sci. *61*, 1347–1356.

Hunter, R.C., Hitchcock, A.P., Dynes, J.J., Obst, M., and Beveridge, T.J. (2008). Mapping the speciation of iron in *Pseudomonas aeruginosa* biofilms using scanning transmission X-ray microscopy. Environ. Sci. Technol. *42*, 8766–8772.

Ivleva, N.P., Wagner, M., Horn, H., Niessner, R., and Haisch, C. (2009). Towards a non destructive chemical characterization of biofilm matrix by raman microscopy. Anal. Bioanal. Chem. *393*, 197–206.

Kirz, J., Jacobsen, C., and Howells, M. (1995). Soft X-ray microscopes and their biological applications. Q. Rev. Biophys. *28*, 33–130.

Kuznetsov, Y., Malkin, A., and McPherson, A. (1997). Atomic force microscopy studies of living cells: visualization of motility, division, aggregation, transformation and apoptosis. J. Struct. Biol. *120*, 180–191.

Lam, K.P., Hitchcock, A.P., Obst, M., Lawrence, J.R., Swerhone, G.D.W., Leppard, G.G., Tyliszczak, T., Karunakaran, C., Wang, J., Kaznatcheev, K., *et al.* (2010). Characterizing magnetism of individual magnetosomes by X-ray magnetic circular dichroism in a scanning transmission X-ray microscope. Chem. Geol. *270*, 110–116.

Lamed, R., Naimark, J., Morgenstern, E., and Bayer, E.A. (1987). Scanning elelctron microscopic delineation of bacterial surface topology using cationinzed ferritin. J. Microbiol. Met. *17*, 233–240.

Lawrence, J.R., Korber, D., Hoyle, B.D., Costerton, J.W., and Caldwell, D. (1991). Optical sectioning of microbial biofilms. J. Bacteriol. *173*, 6558–6567.

Lawrence, J.R., Kopf, G., Headley, J.V., and Neu, T.R. (2001). Sorption and metabolism of selected herbicides in river biofilm communities. Can. J. Microbiol. *47*, 634–641.

Lawrence, J.R., Swerhone, G.D.W., Leppard, G.G., Araki, T., Zhang, X., West, M.M., and Hitchcock, A.P. (2003). Scanning transmission X-ray laser scanning, and transmission electron microscopy mapping of the exopolymeric matrix of microbial biofilms. Appl. Environ. Microbiol. *69*, 5543–5554.

Lawrence, J.R., Dynes, J.J., Korber, D.R., Swerhone, G.D.W., Leppard, G.G., and Hitchcock, A.P. (2012). Monitoring the fate of copper nanoparticles in river biofilms using scanning transmission X-ray microscopy (STXM). Chem. Geol. *329*, 18–25.

Lebaron, P., Servias, P., Baudoux, A., Bourrain, M., Courties, C., and Parthuisot, N. (2002). Variations of bacterial-specific activity with cell size and nucleic acid content assessed by flow cytometry. Aquat. Microb. Ecol. *28*, 131–140.

Little, B., Wagner, P., Ray, R., Pope, R., and Scheetz, R. (1991). Biofilms: an ESEM evaluation of artifacts introduced during SEM preparation. J. Ind. Microbiol. *8*, 213–222.

McLean, S.J., Ona, O.N., and Majors, P.D. (2008). Correlated biofilm imaging, transport and metabolism measurements via combined nuclear magnetic resonance and confocal microscopy. ISME J. *2*, 121–131.

Mah, T.F., Pitts, B., Pellock, B., Walker, G.C., Stewart, P.S., and O'Toole, G.A. (2003). A genetic basis of a biofilm community developing at the liquid–liquid interface between polychlorinated bisphenyls and water. Appl. Environ. Microbiol. *71*, 7301–7309.

Majors, P.D., McLean, J.S., Pinchuk, G.E., Gorby, F.Y.A., Minard, K.R., and Wind, R.A. (2005). NMR methods for *in situ* biofilm metabolism studies. J. Microbiol. Methods *62*, 337–344.

Mayer, C., Moritz, R., Kirschner, C., Borchard, W., Maibaum, R., Wingender, J., and Flemming, H.C. (1999). The role of intermolecular interactions studies on model systems for bacterial biofilms. Int. J. Biol. Macromol. *26*, 3–16.

Mayer, C., Lattner, D., and Schurks, N. (2001). 13C nuclear magnetic resonance studies on selectively labeled bacterial biofilms. J. Ind. Microbial. Biotechnol. *26*, 62–69.

Miot, J., Benzarara, K., Morin, G., Kappler, A., Bernard, S., Obst, M., Ferard, C., Skouri-Panet, F., Guigner, J.-M., Posth, N., *et al.* (2009). Iron biomineralization by anaerobic neurophilic iron-oxidizing bacteria. Geochim. Cosmochim. Acta. *73*, 696–711.

Miot, J., Benzerara, K., Obst, M., Kappler, A., Hegler, F., Schädler, S., Bouchez, C., Guyot, F., and Morin, G. (2009b). Extracellular iron biomineralization by photoautotrophic iron-oxidizing bacteria. Appl. Environ. Microbiol. *75*, 5586–5591.

Moller, S., Sternberg, C., Andersen, J., Christensen, B.B., and Molin, S. (1998). In situ gene expression in mixed culture biofilms: evidence of metabolic interaction between community members. Appl. Environ. Microbiol. *64*, 721–732.

Motta, P.M., Makabe, S., Naguro, T., and Correr, S. (1994). Oocyte follicle cells association during development of human ovarian follicle. A study by high resolution scanning and transmission electron microscopy. Arch. Histol. Cytol. *57*, 369–394.

Nehal, I., Lail, A., and Beyenal, H. (2013). Characterization of bacteria–biomaterial interaction from a single cell to biofilms. In Characterization of Biomaterials, A. Bandyopadhyay, S. Bose, eds. (Elsevier, Amsterdam, the Netherlands), pp. 207–253.

Neu, T.R., and Lawrence, J.R. (1999). In situ characterization of extracellular polymeric substanes (EPS) in biofilm systems. In Microbial Extracellular Polymeric Substances; Characterization, Structure and Function, J. Winglender, T.R. Neu, and H.C. Flemming, eds. (Springer-Verlag, Berlin, Germany), pp. 21–47.

Neu, T.R., Manz, B., Volke, F., Dynes, J.J., Hitchcock, A.P., and Lawrence, J.R. (2010). Advanced imaging techniques for assessment of structure composition and function in biofilm systems. FEMS Microb. Ecol. *72*, 1–21.

Nielsen, A.T., and Jahn, A. (1999). Extraction of EPS. In Microbial Extracellular Polymeric Substances; Characterization, Structure and Function, J. Winglender, T.R. Neu, and H.C. Flemming, eds. (Springer-Verlag, Berlin, Germany), pp. 49–72.

Oubekka, S.D., Braindet, R., Fontaine-Aupart, M.P., and Steenkeste, K. (2012). Correlative time-resolved fluorescence microscopy to assess antibiotic diffusion-reaction in biofilms. Antimicriob. Agents Chemother. *56*, 3349–3358.

Pal, A., and Paul, A.K. (2008). Microbial extracellular polymeric substances: central elements in heavy metal bioremediation. Indian J. Microbiol. *48*, 11289–11298.

Palmer, R. Jr., and White, D. (1999). Spatially resolved, quantitative determination of luciferase activity by photon-counting microscopy. Met. Enzymol. *310*, 152–160.

Palmer, R. Jr., and Sternberg, C. (1999). Modern microscopy in biofilm research: confocal microscopy and other approaches. Curr. Opin. Biotechnol. *10*, 263–268.

Pandey, G., and Jain, R.K. (2002). Bacterial chemotaxis toward environmental pollutants: role in bioremediation. Appl. Environ. Microbiol. *68*, 5789–5795.

Patrauchan, M.A., Sarkisova, S., Sauer, K., and Franklin, M.J. (2005). Calcium influences cellular and extracellular product formation during biofilm associated growth of a marine *Pseudomonas* sp. Microbiology *151*, 2885–2897.

Pätzold, R., Keuntje, M., and Anders-von Ahlften, A. (2006). A new approach to non-destructive analysis of biofilms by confocal Raman Spectroscopy. Anal. Bioanal. Chem. *386*, 286–292.

Paul, D., Pandey, G., Pandey, J., and Jain, R.K. (2005). Accessing microbial diversity for bioremediation and environmental restoration. Trends Biotechnol. *23*, 135–142.

Priester, J.H., Horst, A.M., Van, D.E., Werfhorst, L.C., Saleta, J.L., Mertes, L.A.K., and Holden, P.A. (2007).

Enhanced visualization of microbial biofilms by staining and environmental scanning electron microscopy. J. Microbiol. Met. *68*, 577–587.

Razatos, A., Ong, Y.-L., Sharma, M., and Georgiou, G. (1998). Molecular determinants of bacterial adhesion monitored by atomic force microscopy. Proc. Natl. Acad. Sci. U.S.A. *95*, 11059–11064.

Rösch, P., Harz, M., Schmitt, M., Peschke, K.D., Ronneberger, O., Burkhardt, H., Motzkus, H.W., Lankers, M., Lofers, S., Thiele, H., and Popp, J. (2005). Chemotaxonomic identification of single bacteria by micro-Raman Spectroscopy: application to clean-room-relevant biological contaminations. Appl. Environ. Microbiol. *71*, 1626–1637.

Sandt, C., Smith-Palmer, T., Pink, J., Brennan, L., and Pink, D. (2007). Confocal Raman microspectroscopy as a tool for studying the chemical heterogeneities of biofilms *in situ*. J. Appl. Microbiol. *103*, 1808–1820.

Schramn, A., Larsen, L., Revsbech, N., Ramsing, N., Amann, R., and Schleifer, K.H. (1996). Structure and function of a nitrifying biofilm as determined *in situ* hybridization and the use of microelectrodes. Appl. Environ. Microbiol. *62*, 4641–4647.

Sheng, G., Yu, H., and Li, X. (2010). Extracellular polymeric substances (EPS) of microbial aggregates in biological wastewater treatment systems. A review. Biotechnol. Adv. *28*, 882–894.

Singh, R., Paul, D., and Jain, R.K. (2006). Biofilms: implications in bioremediation, Trends Microbiol. *14*, 389–397.

Singleton, S., Treolar, R., Warren, R., Watson, G., Hodgson, R., and Allison, C. (1997). Methods for microscopic characterization of oral biofilms: analysis of colonization, microstructure and molecular transport phenomena. Adv. Dent. Res. *11*, 133–149.

Stewart, P.S., Murga, R., Srinivasan, R., and de Beer, D. (1995). Biofilm structural heterogeneity visualized by three microscopic methods. Water Res. *29*, 2006–2009.

Thumheer, T., Gmur, R., Giertsen, E., and Guggenheim, B. (2001). Automated fluorescent *in situ* hybridization for the specific detection and quantification of oral Streptococci in dental plaque. J. Microbiol. Methods *44*, 39–47.

Tien, C.-J., Lin, M.-C., Chiu, W.-H., and Chen, C.S. (2013). Biodegradation of carbomate pesticides by natural river biofilms in different seasons and their effects on biofilm community structure. Environ. Pollut. *179*, 95–104.

Towler, B.W., Rupp, C.J., Cunningham, A.B., and Stoodley, P. (2003). Viscoelastic properties of a mixed culture biofilm from rheometer creep analysis. Biofouling *19*, 279–285.

Trinidad, A., Ibanez, A., Gomez, D., Garcia-Berrocal, J.R., and Ramirez-Camacho, R. (2010). Application of environmental scanning electron microscopy for study of biofilms in medical devices. Microscopy: Science, Technology, Application and Education, A. Mendez-Vilas and J. Diaz, eds. (Formatex Research Centre, Badajoz, Spain), pp. 204–210.

Van der Vossen, J.M., and Hofstra, H. (1996). DNA based typing, identification and detection systems for food spoilage microorganisms: development and implementation. Int. J. Food Microbiol. *33*, 35–49.

Velegol, S.B., and Logan, B.E. (2002). Contributions of bacterial surface polymers, eleastostatics, and cell elasticity to the shape of AFM force curves. Langmuir *18*, 5256–5262.

Vilchez, R., Pozo, C., Gomez, M.A., Rodelas, B., and Gonzalez-Lopez, J. (2007). Dominance of Sphingomonads in a copper exposed biofilm community for groundwater treatment. Microbiology *153*, 325–337.

Vives Rego, J., Lebaron, O., and Nebe von Caron, G. (2000). Current and future applications of flow cytometry in aquatic microbiology. FEMS Microbiol. Rev. *24*, 429–448.

Wagner, M., Ivleva, N.P., Haisch, C., Neissner, R., and Horn, H. (2009). Combined use of confocal laser scanning microscopy (CLSM) and Raman microscopy (RM): investigation on EPS-matrix. Water Res. *43*, 63–76.

Walker, J.T., Verran, J., Boyd, R.D., and Percival, S. (2001). Microscopy methods to investigate structure of potable water biofilms. Met. Enzymol. *337*, 243–255.

Washington, V.J., Lear, G., Neale, M.W., and Lewis, G.D. (2013). Environmental effects on biofilm bacterial communities: a comparison of natural and anthropogenic factors in New Zealand streams. Freshwater Biol. *58*, 2277–2286.

Wind, R.A., Majors, P.D., Minard, K.R., Ackerman, E.J., Daly, D.S., Holton, G.R., Thrall, G.R., and Weber, T.J. (2002). Combined confocal and magnetic resonance microscopy. Appl. Magn. Reson. *22*, 145–158.

Wolfaardt, G., Lawrence, J., Robarts, R., and Caldwell, D. (1998). In situ characterization of biofilm exopolymers involved in the accumulation of chlorinated organics. Microb. Ecol. *35*, 213–223.

Yamamoto, Y., and Shinhora, K. (2002). Application of X-ray microscopy in analysis of living hydrated cells. Anant. Rec. (New Anat.) *269*, 217–223.

Yun, M.A., Yeon, K.M., Park, J.S., Lee, C.H., Chun, J., and Lim, D.J. (2006). Characterization of biofilm structure and its effect on membrane permeability in MBR for dye wastewater treatment. Water Res. *40*, 45–52.

Zhang, W., Sturm, B.S.M., Knapp, C.W., and Graham, D.W. (2009). Accumulation of tetracycline resistance genes in aquatic biofilms due to periodic waste loadings from swine lagoons. Environ. Sci. Technol. *43*, 7643–7650.

Zhang, W., McLamore, E.S., Garland, N.T., Chavez Leon, J.V., and Banks, M.K. (2013). A simple method for quantifying biomass cell and polymer distribution in biofilms. J. Microbiol. Met. *94*, 367–374.

Zhou, A.H., McEwen, G.D., and Wu, Y.Z. (2010). Combined AFM/RM microspectroscopy for characterization of living cells in near physiological conditions. In Microscopy: Science, Technology, Application and Education, A. Mendez-Vilas and J. Diaz, eds. (Formatex Research Centre, Badajoz, Spain), pp. 515–522.

# Molecular Methods for the Assessment of Microbial Biofilms in Bioremediation

# 7

Gavin Lear

## Abstract

Advances in molecular methods are providing new insights into the diversity and organization of biofilm microbial communities as well as the genetic basis of biofilm-mediated contaminant degradation. In the first part of this chapter, we consider the various methods available to inform on the taxonomic and genetic composition of microbial biofilms and to visualize the location of microorganisms genes within complex biofilm communities. Methods directed towards the analysis of functional subsets of these communities, including contaminant degraders are also described. The advantages and disadvantages of a variety of contemporary DNA fingerprinting and sequencing methodologies are each considered. In the second part of this chapter, case studies focusing on the use of molecular techniques in bioremediation assessments are presented. We highlight how the labelling of bacterial taxa with gene specific probes can reveal the complex structural organization of natural biofilms as well as the impact of pollutants on their composition and organization. Finally, we detail how the DNA of contaminant-degrading bacteria can be isolated to identify even low abundances of pollutant degrading taxa within diverse and active biofilm communities. The research opportunities provided by modern molecular methods continue to rapidly advance our understanding, and application of, biofilms in bioremediation.

## Part 1: molecular methods to investigate the role of biofilms in bioremediation

A wide variety of molecular techniques have been used to characterize microbial community attributes in natural and bioreactor environments following their use for the bioremediation of organic and inorganic pollutants. These include approaches to both identify and quantify the presence of bacterial taxa and their functional genes in the environment, methods to visualize the location of taxa and genes within complex biofilm communities and methods to investigate the phylogenetic relatedness of individual taxa and functional genes. In the following section, a variety of the most common molecular techniques used to explore the role of biofilms in bioremediation are described and contrasted (as summarized in Table 7.1). In Part 2 of this chapter, we finally consider the recent application of a variety of such methods to increase our understanding of the role biofilms play in contaminant remediation.

### An overview of common PCR and hybridization techniques

#### The polymerase chain reaction

The polymerase chain reaction, or PCR, is a common and almost indispensable technique in biology and medical research. This method is used to amplify single pieces of DNA by many orders of magnitude to generate thousands or millions of copies. The amplified DNA may then be visualized following gel electrophoresis, and used for downstream applications including community DNA fingerprinting and DNA sequence analysis. PCR relies on the repeated heating and cooling of DNA in the presence of a polymerase enzyme which acts to exponentially amplify the copy numbers of DNA. The target sequence of DNA amplification is directed by the use of short DNA fragments, or primers, which are complementary only to the

**Table 7.1** Brief description of molecular methods used for the investigation of biofilms and bioremediation

| Technique | General approach | Reference |
|---|---|---|
| **Identification of taxa/genes by microscopy** | | |
| Fluorescence *in situ* hybridization (FISH) | Fluorescent probe hybridized to organisms containing corresponding DNA sequence. Used to track presence and abundance of microbial taxa or functional genes | Nielsen *et al.* (2000) |
| Use of fluorescent protein labelled strains | Strains engineered to contain a fluorescent protein that may be used to track their presence, and even an organisms interaction with specific pollutants | Wouters *et al.* (2010) |
| **Direct quantification of microbial taxa/gene abundance** | | |
| Flow cytometry | To count the total number of microbial cells, or labelled cells (e.g. labelled using FISH) | Shah *et al.* (2006) |
| DNA fingerprint analysis | | |
| Automated ribosomal intergenic spacer analysis (ARISA) | Profiles microbial community composition based on variation in the length of intergenic DNA regions | Lear *et al.* (2009a) |
| Denaturing gradient gel electrophoresis (DGGE) | Profiles microbial community composition based on the GC content of the amplified DNA | Cattony *et al.* (2005) |
| Terminal-restriction fragment length polymorphism (T-RFLP) | Profiles microbial community composition based on the presence of specific DNA sequences in the amplified fragments | Bordenave *et al.* (2007) |
| **DNA sequence analysis** | | |
| Sanger sequencing | A traditional approach to sequence DNA fragments. Requires cloning step to isolate DNA from mixed communities and offers a low to medium sequencing throughput | Lear *et al.* (2009b) |
| Roche 454 | High-throughput generation of DNA sequence reads. Generally provides longer sequence reads than Illumina and Ion Torrent | Gomez-Alvarez *et al.* (2013) |
| Illumina | High throughput generation of DNA sequence reads. Generates short reads compared with 454 and Ion Torrent platforms but provides very high throughput analysis | Krohn-Moult *et al.* (2013) |
| Ion Torrent | High-throughput generation of DNA sequence reads. Generates relatively short reads. | Yergeau *et al.* (2013) |
| **Analysis of community function/activity** | | |
| Stable isotope probing | Used to track the fate of labelled substrate into microbial DNA. Subsequent DNA sequencing analysis can be used to identify substrate degraders | Manefield *et al.* (2002) |
| Functional microarrays | Screen for the presence of a wide variety of genes involved in key processes such as nutrient cycling, contaminant degradation and resistance. | Liang *et al.* (2014) |
| Shotgun metagenomics | Used to characterize the functional and phylogenetic attributes of entire communities in combination with DNA sequencing strategies | Tyson *et al.* (2004) |

DNA region of interest. While our choice of primer selection can be used to amplify DNA from any part of the microbial genome, genes encoding for the rRNA of small ribosomal subunits (e.g. 16S rRNA genes for bacterial and archaeal cells, and 18S rRNA genes from eukaryotes) are commonly targeted. The DNA sequence of these particular molecules are chosen for analysis since these ribosomes are necessary for bacterial survival, cannot tolerate large mutations and are little affected by horizontal gene transfer. As a consequence, they are well suited to explore microbial evolution, or the relatedness of individuals since the extent of genetic variability is characteristically greater for more distantly related organisms. This region of DNA is commonly exploited by a variety of community DNA fingerprinting methods and by DNA sequence analysis to describe variation in the composition and diversity of complex microbial communities.

One major drawback of traditional, or endpoint PCR, is that it cannot be reliably used to quantify the abundance of genes within a sample. This is because after a period of rapid amplification, available quantities of critical components required for the PCR, or their respective activities (e.g. polymerase enzyme activity) decline during the PCR process. Consequently, the amount of final PCR product does not always reflect the amount of target DNA present in the original sample. To quantify the abundance of any target gene, it is necessary to witness the amplification of PCR products during their exponential amplification, rather than when the availability or activity of key PCR reagents has expired. The quantification of concentrations of target DNA can be achieved in this way using an approach termed real-time, or quantitative PCR (qPCR). Here, intercalating dyes are added to the PCR reaction. Crucially, these dyes are able to record the generation of double stranded DNA as the PCR progresses. Specially designed thermocyclers monitor the amount of PCR product generated. This means that the extent of DNA amplification may be calculated during exponential amplification, rather than after rates of amplification have declined or halted. qPCR has been widely used to track the abundance of specific organisms within complex biofilm communities [such as populations of the *Geobacteraceae*, a commonly studied group of organisms associated with metal cycling and particularly Fe(III) reduction in subsurface biofilms (Lear *et al.*, 2010)]. An additional strength in the use of this process is its use to monitor changes in the abundance of functional genes within environmental samples, such as genes encoding for pollutant degradation and transformation. For example, the abundance of *ars* genes has been used to quantify numbers of arsenic-metabolizing bacterial present in a pulp mill biomass bioreactor (Sun *et al.*, 2004). Similarly, Beller *et al.* (2002) used a qPCR approach to report on the abundance of anaerobic, hydrocarbon degrading bacteria in BTEX contaminated aquifer sediment by quantifying the numbers of genes encoding for the large alpha subunit of benzylsuccinate synthase. While the abundance of DNA may not necessarily be proportional to differences in gene expression and protein production, increases in the abundance of degradative genes are nevertheless widely correlated with increased rates of pollutant degradation (Monard *et al.*, 2012). Alternatively, quantitative real-time PCR (or qRT-PCR) can be used to report levels of gene expression (rather than gene abundance) among biofilm-dwelling microbial taxa (Yergeau *et al.*, 2010).

### Microarrays

Standard PCR and qPCR methods are commonly used to record the presence or abundance of single genes in a community. In contrast, DNA microarrays have been developed to provide a comprehensive overview of the microbial species richness or functional potential in environmental samples via the simultaneous analysis of hundreds or thousands of individual genes that can be expected to be present in small volumes of natural biofilm communities. DNA microarrays are solid slides, usually made of glass or silicon, onto which DNA has been added in an organized grid fashion. In standard arrays, DNA probes are synthesized and attached via covalent bonds to spot locations on the support matrix. Depending on the procedure, most microarrays will contain hundreds to tens of thousands of probes and the location of each probe is carefully recorded so that, following analysis, the genetic 'identity' of each spot is known. Each probe consists of a small fragment of single-stranded DNA encoding for a gene of interest. The utility of the microarray then relies on the hybridization that occurs between complementary DNA strands located in the sample and those printed onto the solid surface of the array. Fluorescently labelled sample DNA is washed over the microarray. Where DNA strands in the extract sample are complementary to the DNA probes spotted to the surface of the slide, hybridization can occur. After any non-bound DNA is washed from the probe, fluorescence at individual spot locations on the array highlights the likely presence of that particular DNA sequence within the original environmental sample.

Two types of microarray commonly used in microbial ecology include 16S rRNA gene arrays to provide information on the presence/absence of microbial taxa and functional gene arrays to provide an indication of the metabolic potential of microbial communities. A number of different 16S microarrays are available for use, the most common being the PhyloChip which was developed for the rapid identification of bacterial and archaeal cells from complex environmental samples (Brodie *et al.*, 2006). The chip contains tens of thousands of probes targeting 16S rRNA genes related to a diversity of cultured microbial taxa as well as candidate divisions of uncultured strains. Handley *et al.* (2012) used deep community sequencing with Phylochip to explore microbial taxonomic variability among samples associated with uranium bioremediation experiments. This approach provided a detailed profile of community composition following acetate enrichment and provided phylogenetic evidence for increased sulphur cycling and taxa capable of uranium reduction (*Shewanallaceae*, *Desulfovibrionaceae* and *Peptococcaceae*) following addition of this labile carbon source. Such detailed profiles of the presence and abundance of microbial taxa can provide valuable information on bacterial community composition in response to both nutrient enrichment and contamination events. An alternative approach to the use of PhyloChips is to explore the functional potential of microbial communities using gene arrays specially designed to identify diverse functional genes present within environmental samples. The GeoChip array includes tens of thousands of genes involved in metabolic processes including carbon, nitrogen, sulphur and phosphorus cycling, energy metabolism, antibiotic and metal resistance, organic contaminant degradation, as well as phylogenetic marker genes (He *et al.*, 2010). These microarrays are used for diverse purposes including for analysing the functional attributes of microorganisms in contaminated and biofilm-associated communities (detailed later in this chapter). However, while microarrays allow the high-throughput screening of samples to monitor for the presence of phylogenetic and functional gene markers within complex environmental communities there remain numerous limitations for their use. The hybridization signal intensity of different probes may be considered to be proportional to the abundance of target organisms or genes, but the rapid saturation of probes by excess sample DNA mean that such methods are generally appropriate only to describe their presence, and not abundance. Additionally, cross hybridization, or the binding of sample DNA to probes that are intended to detect other DNA fragments is a major limitation. Finally, microarrays are poorly suited for the detection of novel microbial taxa and genes for which probes have not yet been developed. Nevertheless, microarrays remain useful for the high-throughput screening of the presence of microbial taxa and associated functional genes within complex biofilm communities.

## Methods to identify the presence and location of microbial taxa or genes within biofilm communities

In combination with microscopy, the location of individual taxa or genes can be recorded from within complex biofilm communities. This may be achieved either by using fluorescent probes to target DNA of interest, or via the use of microbial taxa that are genetically engineered to fluoresce, or luminesce, to reveal their location within the community.

### Fluorescence *in situ* hybridization (FISH)

Fluorescence *in situ* hybridization is one of the most direct ways to observe microorganisms in their natural environment. Here, fluorescently labelled primers (or probes) are selected that are specific for the microbe of interest (Fig. 7.1). Before hybridization can occur, the cells must first be fixed to preserve the cell morphology while simultaneously permeating cells for the entry of labelled primer(s). The most commonly used fixative is a 1–4% solution of formaldehyde, followed by a series of dehydration steps in increasing concentrations of ethanol to permeabilize the cell wall. Subsequently, a fluorescent primer, in hybridization buffer, is added to the fixed sample and incubated to hybridize the probe to complementary nucleotide sequences within the microbial cells. Excess primer is removed by gentle washing and the remaining fluorescent tag then reveals the location of organisms of interest. An epifluorescent microscope, fitted with filters to excite the specific fluorescent tag is used to visualize probe fluorescence related to individual microbial cells. Used in combination with a confocal laser scanning microscope, or CLSM, FISH allows the location of specific microorganisms (e.g. based on their 16S rRNA gene sequence) or their functional attributes (e.g. the presence of ammonia oxidation

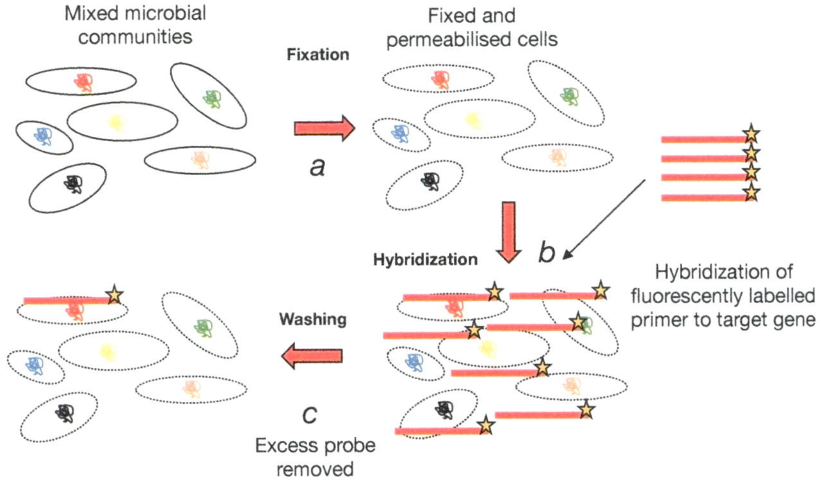


**Figure 7.1** Principles of FISH. (a) A mixed community of microbial cells is fixed and permeabilized, usually using a combination of formamide and alcohol before (b) a fluorescently tagged primer is added to the sample. The primer hybridizes to corresponding sequences in the genome of the fixed cells and (c) any excess primer is washed away such that the target cells are identifiable via their fluorescent signal.

genes) to be visualized within the three-dimensional structure of complex biofilms. As an example of this, Saha *et al.* (2012) used a custom-designed primer to identify group I pseudomonads present within biofilm communities associated with metalworking fluids (MWF) even though prior approaches had revealed the low recovery of culturable cells of this genera. Use of the FISH approach revealed significant variation in the architecture of biofilms inoculated with different *Pseudomonas* strains, possibly correlating with the response of each strain to the various disinfection methods used to control biofilm development within MWF. Catalysed reporter deposition-FISH, or CARD-FISH, provides an improvement on the standard FISH procedure. It is particularly suited for use in aquatic habitats or communities with a prevalence of small, slow-growing or nutrient starved bacteria that may contain relatively few ribosomes. Here, an enzyme is attached to the primer, usually horseradish peroxidase (HRP). After probe hybridization, tyramide is added to the samples which amplifies the fluorescence by HRP such that single hybridized primers are capable of generating a bright fluorescent signal. Despite the many advantages of FISH, numerous difficulties remain associated with its use, including autofluorescence, photobleaching and variable probe uptake among different bacterial cells (Hogardt *et al.*, 2000).

## Tracking the fate of taxa labelled with fluorescent proteins

An alternative approach to DNA hybridization methods such as FISH is to track the fate of microbial strains that have been genetically engineered to fluoresce. This allows their fate to be monitored *in situ*, but also in real time. A green fluorescent protein (GFP) purified from the jellyfish *Aequorea victoria* (Shimomura *et al.*, 1962) has been engineered into a wide variety of organisms, including many bacteria. This GFP, and other fluorescent proteins, have become very useful for monitoring the presence and abundance of the labelled strains within complex biofilm communities as well as for detecting levels of gene expression *in situ* (Nielsen *et al.*, 2000; Andrews *et al.*, 2006).

Like FISH, gfp can be used to track the presence and abundance of microbial taxa within complex biofilm communities. For example, Wouters *et al.* (2010) incubated a triple species combination which included a *Mycobacterium* sp., gfp-tagged *Sphingomonas* sp. and rfp-tagged *Pseudomonas* sp. to show the dominance of the former two strains when grown as biofilms in the presence of the polycyclic aromatic hydrocarbon (PAH) phenanthrene. A remarkable layering was observed whereby the *Sphingomonas* sp. is seen to grow trapped between two layers of the *Mycobacterium* sp. thereby being deprived of direct contact with the source of phenanthrene. These observations support the existence of highly structured commensal relationships between biofilm strains, which may facilitate the organized transport of carbon and help to explain why certain genera commonly colocate within PAH-contaminated environments, for example. While the exact mechanisms involved in regulating the species organization of such biofilm communities remains largely unknown, evidence derived from the use of gfp-labelled strains reveal that the micro-scale organization of organisms in their environment is not only dependent on the physiological characteristics of each individual organism, but also by complex multi-species interaction phenomena.

A particular strength associated with the use of organisms tagged with *gfp* and similar proteins is their use as whole-cell bioreporters to monitor the activity of specific metabolic processes, including contaminant degradation pathways. These bioreporter strains are often genetically engineered such that the expression of the fluorescent protein is linked to the expression of a gene involved in contaminant degradation or resistance under the control of a single operon. Such biosensors have become important tools to complement the chemical analysis of polluted environments (Andrews *et al.*, 2006) offering the advantage that they are mainly affected by the bioavailable fractions of contaminants in soil and water (Chinalia *et al.*, 2008). To demonstrate the utility of *gfp* bioreporters for studying degradation in mixed species biofilms Andrews *et al.* (2006) constructed artificial biofilm communities using *Acinetobacter* sp., C6 and *Pseudomonas putida* :: *gfp*. The *P. putida* has a chromosomally expressed cassette which encodes a *gfp* downstream of a promoter that controls the degradation of benzoate, produced by the *Acinetobacter* as it degrades benzyl alcohol. Therefore the interaction of these two strains in the degradation of benzyl alcohol and its benzene degradation product should be evidenced by the production of green fluorescent protein. Microscopic observations using a confocal laser scanning microscope confirmed the expression of gfp, but also indicated a preferred internal organization of bacteria in the biofilm, with *P. putida* in the bottom layer, and *Acinetobacter* on top. Such observations provide valuable information on optimal inoculation conditions, and order, to enhance the rates and extents of contaminant degradation in soils and biofilm reactors. Consequently, a wide variety of microbial bioreporters have now been developed to signal the presence of both organic and inorganic pollutants in the environment (Tecon *et al.*, 2006; Branco *et al.*, 2013).

An alternative type of biosensor uses the *lux*-CDABE gene cassette to produce a light emitting reaction, catalysed by the enzyme luciferase in the presence of oxygen. When constitutively expressed in wild-type strains of the bacterium *Vibrio fischeri*, or in genetically engineered organisms, the decrease in light resulting from the toxicological effects of various contaminants can be used to monitor the toxic impacts of low levels of single, or mixed solutions, of pollutant compounds. This 'lights-off' approach offers an important tool to rapidly screen environmental samples for the presence of any inhibitory biological compounds (Fig. 7.2). For example, Chinalia *et al.* (2008) monitored the effects of zinc and copper toxicity on the relative light intensity of a *lux*-marked *Burkholderia* strain and related the toxic impacts of these metals to 2,4-D degradation rates by the organism. While such approaches can be used to monitor the health and activity the labelled taxa, an alternative, 'lights-on' approach is often used to examine the expression of genes involved in specific metabolic pathways for degradation. For example, Dorn *et al.* (2005) used *Pseudomonas putida* RB 1353 which includes genes for naphthalene catabolism, fused with the light producing *lux*CDABE gene cassette. This genetic modification then encodes for light production in the presence of naphthalene. The author's injected naphthalene into packed media columns with thick biofilm communities where the *P. putida* strain had been allowed to grow. The media columns were spiked with fibre-optic cables so that the expression of the naphthalene degradation gene, or rather the light produced by the activation of the associated *luxCDABE* could be monitored. This remarkable combination of techniques allowed Dorn *et al.* (2005) to observe naphthalene catabolism *in situ*. Monitoring the location of naphthalene degrading bacteria over time was used to show how the bioactive zone moved from the site of inoculation towards the contaminant plume. Such approaches allow us to better predict the mobilization of bioactive zones in complex subsurface biofilm communities to inform on optimal strategies for inoculation. For

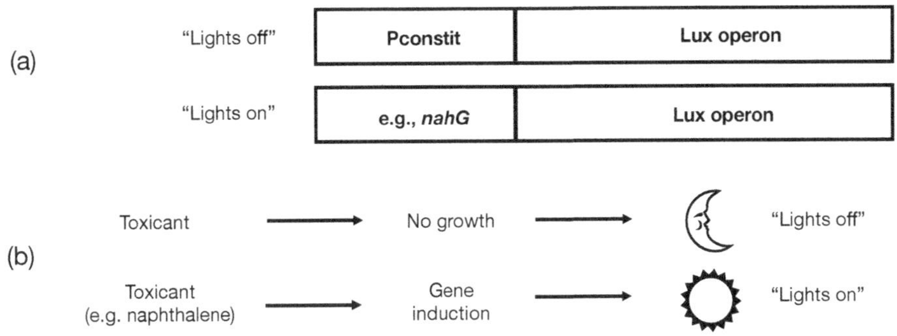


**Figure 7.2** Use of two bioreporter organisms to signal the presence of bioavailable contaminants in the environment showing (a) the configuration of genes with respect to the lux operon and (b) the effect of toxicant on light production. In the 'lights off' system, light is normally produced by the organism (the lux operon is constitutively expressed) but light production is reduced by any toxin or stressful condition. In the 'lights on' system an inducible gene for the pathway of interest (e.g. naphthalene degradation, *nahG*) is fused to the lux operon such that light is produced when the promoter is expressed.

such purposes, a wide range of bioluminescent bacterial biosensors have been developed to be induced by organic compounds including naphthalene (Dorn *et al.*, 2005), BTEX compounds (Applegate *et al.*, 1998), 2,4-dichlorophenol (Hay *et al.*, 2000) and phenols (Shingler and Moore 1994); biosensors for a wide variety of metals and metalloids are similarly available (Jouanneau *et al.*, 2012). However, despite the enormous potential offered by artificial bioreporter strains, a major limitation to their widespread use remains regulatory issues concerned with the release of genetically modified microorganisms into the environment.

## Quantifying the abundance of labelled strains of bacteria by flow cytometry

So far, we have described how the labelling of organisms with fluorescent or luminescent markers can be used to identify their location within complex communities or media. In many cases when using labelled strains of bacteria it is desirable to quantify the abundance of these organisms in their environment. Crude measures of bacterial abundance can be gained by counting cells using microscopic techniques, or by measuring the total output of light using fluorimetric assays as Dorn *et al.* (2005). However, an alternative and generally more accurate approach to quantify the number of cells tagged with a fluorescent protein (e.g. gfp) or stained with a fluorescent dye (e.g. FISH) is to count their number with a flow cytometer. Here, sample material is injected into a fluid stream forcing one cell at a time to pass through a fine laser beam. A detector then counts the number of particles excited by a laser beam of wavelength appropriate to the fluorescent label. We are thus able to count the number of labelled cells among a substantial background population of non-fluorescent cells and debris, at a rate of many thousands of cells per second. An extension of this approach is the use of a fluorescently activated cell sorter (FACS) where cells are electrostatically charged as they pass through the detector. This charge is used to guide cells into different collection vessels (e.g. separating fluorescent and non-fluorescent stains), counting and sorting individual cells at speed. The separation of fluorescent cells from the background population has various downstream applications. For example, cells which are hybridized by a particular probe (e.g. FISH of naphthalene degradation genes) can be separated and the DNA of this subset of the population sequenced to confirm the abundance, identity and diversity of organisms possessing this particular genetic attribute. A comparable approach was used by Shah *et al.* (2006) to distinguish among cells of a single species (*E. coli*) that developed different physiological forms as either non- or slow-growing persistor cells when grown as biofilms. However, despite the many advantages provided by flow cytometric methods, and particularly by fluorescence-activated cell sorting, the analysis of biofilm communities, and particularly biofilm communities growing on complex media remains challenging. To accurately count the number of microbial cells present in a biofilm, the biofilm must first be homogenized and cells separated into individual units for counting. A number of methods, such as sonication and the use of surfactants are suggested for use (Yarwood *et al.*, 2004). However, the strong clumping of biofilm-associated cells mean this issue remains a challenge. Additionally, larger particles associated with soils and microbial exudates can block the liquid sample stream, requiring that microbial taxa be separated from soil particles and organic detritus before counting, a process which itself may create significant sample bias (Whiteley *et al.*, 2003).

## Approaches to assess the composition and diversity of biofilm bacterial communities from their nucleic acids

Methods to identify variability in the composition and structure of microbial communities can be broadly grouped into two main categories: (i) DNA-sequencing methods and (ii) community fingerprinting methods. DNA fingerprinting methods are generally used to profile 'differences' in microbial community data among samples from their DNA. These methods do not normally provide information on the identity of individual taxa and are commonly used as a 'screening approach' to identify groups of samples, or sample outliers that may be studied further, including by DNA sequencing. The key advantage of DNA sequencing is that comparison of the DNA sequence data to the sequence of known genes allows the identity of individual taxa, or genes, to be revealed. Until recently, however, the DNA sequence analysis of large volumes of DNA has been prohibitively expensive, especially where the analysis of multiple samples is deemed necessary. In the following section, we review the main community fingerprinting and DNA sequencing methods used to characterize microbial communities, as well as their advantages and limitations.

### Commonly used community fingerprinting methods

Community fingerprinting refers to a set of methods used to rapidly profile differences in community composition between samples. In general, these approaches target a particular gene or DNA region for analysis that varies among the study group of organisms to provide information on variability in community composition or genetic diversity among samples. Community fingerprinting is most commonly employed to track changes in community composition over space or time, for example to work out in which samples microbial communities are responding most strongly to environmental perturbation (Lear *et al.*, 2009b; Jechalke *et al.*, 2010). Unlike DNA sequencing, community fingerprinting is not normally used to reveal the identity of any individual microbial taxa since multiple microbial species are normally capable of producing similar fingerprint profiles. Additionally, it is well recognized that fingerprinting methods grossly underestimate the true diversity of complex environmental communities, excluding rare taxa in particular (Bent *et al.*, 2007). Nevertheless, community DNA fingerprinting approaches remain widely used since analysis can be performed relatively cheaply and, in some instances can accommodate the simultaneous analysis of large sample numbers.

#### Denaturing gradient gel electrophoresis (DGGE)

In denaturing gradient gel electrophoresis (DGGE), DNA is amplified using primers targeting a specific molecular marker, such as the bacterial 16S rRNA gene. The DGGE approach relies on the fact that while the amplified PCR targets are about the same size, the nucleotide sequence of the target DNA varies among taxa (Fig. 7.3). Variation in DNA fragment sequence dictates the threshold at which individual strands of DNA will denature, or separate to form single strands of DNA. The most important determinant of this denaturing threshold is the G + C content of each DNA fragment, since the paired nucleotide bases of guanine (G) and cysteine (C) are joined by triple hydrogen bonds in comparison with the double hydrogen bonds that typically join the bases adenosine (A) and thymine (T). An important aspect of the DGGE PCR is the addition of a GC-rich tail to one of the primers required for amplification. This generates a GC-clamp which strongly binds one end of the newly amplified double-stranded template together. Following DGGE-PCR, amplified DNA fragments are loaded onto an acrylamide gel which contains a gradient from being poorly denaturing (low concentrations of urea and formamide) to being highly denaturing (high concentrations of urea and formamide). As the DNA is run through the gel, each double stranded DNA sequence will reach

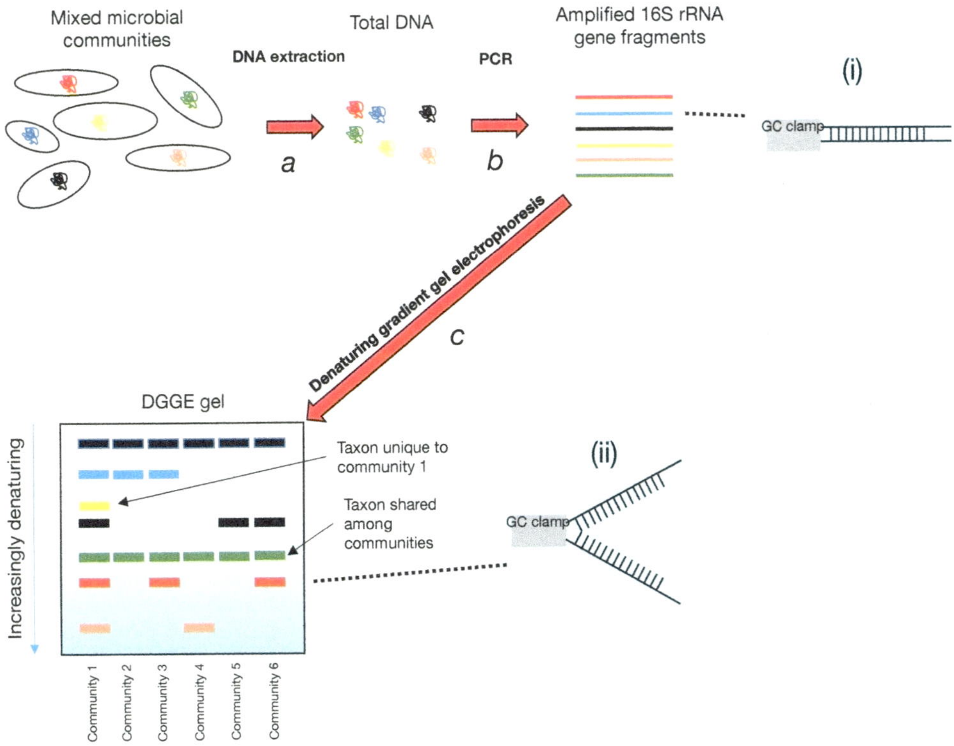


**Figure 7.3.** Principles of denaturing gradient gel electrophoresis (DGGE). (a) The total DNA of the biofilm communities are extracted, then (b) DNA fragments of interest (usually the bacterial 16S rRNA gene fragment) is amplified by PCR. One of the primers will have a G-C rich sequence that binds the double-stranded sequence firmly at one end (see inset i). (c) The amplified PCR product is loaded onto a denaturing gradient gel which typically contains higher concentrations of the denaturants urea and formamide furthest away from the loading wells. As electrophoresis occurs, the DNA travels into increasingly denaturing zones of the gel. Sequences with a higher G-C content travel further along the gel before denaturing. As fragments denature, they remain bound at one end by the primer G-C clamp. The partly denatured DNA can no longer migrate easily through the gel, and a 2-D gel fingerprint is created based on the sequence identity of every DNA fragment in each mixed community.

its melting point (threshold of enough denaturant for the double strands of DNA to separate). At this location the double stranded DNA will begin to separate and its movement slows within the gel. Since the nucleotides G and C bind more strongly to each other than the nucleotides A and T, DNA fragments originating from bacteria with a higher GC content in that specific sequence of DNA will travel further down the gel. Using this approach we are able to create a 2-D fingerprint of bacterial community composition, separating bacterial DNA by nucleotide sequence attributes. DGGE is most commonly used to profile differences in the composition of total bacterial communities (Lear *et al.*, 2004), but is an approach that can also be used to characterize the diversity and relative abundance of various bacterial genes. For example, Cattony *et al.* (2005) used DGGE-profiling techniques to individually examine variation among communities of bacteria, archaea and sulphate-reducing bacteria (SRB) in biofilm communities subject to toluene exposure. DGGE-profiling confirmed that communities of bacteria and SRB were impacted by toluene and ethanol amendments increasing both bacterial diversity and abundance.

### Terminal restriction fragment length polymorphism (T-RFLP)

A common alternative for rapid microbial community analysis is terminal restriction fragment length polymorphism, or T-RFLP (Fig. 7.4). In this approach, a target gene, which is often an rRNA gene, is amplified by PCR using a primer set in which one or both of the primers is tagged with

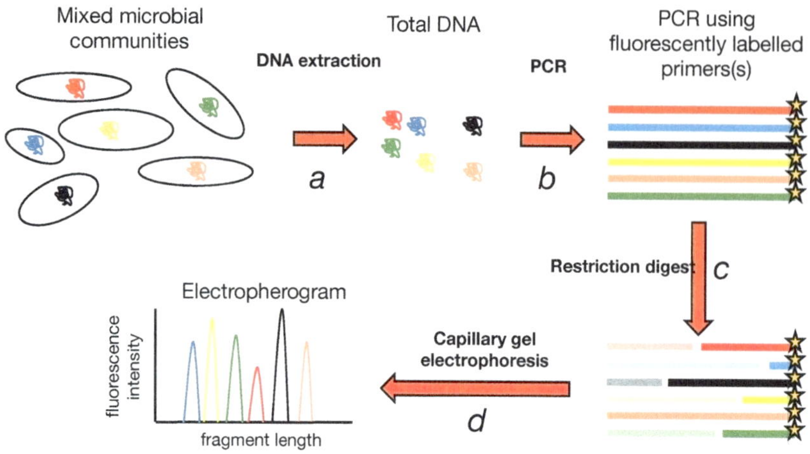


**Figure 7.4** Principles of terminal restriction fragment length polymorphism (T-RFLP). (a) The total DNA of the biofilm community is extracted, then (b) DNA fragments of interest (often from rRNA genes) are amplified by PCR using one (or two) primers with a fluorescent tag added to the 5′ end. (c) PCR products are purified and then digested using one or more restriction enzymes which cleave recognized sequences in the dsDNA fragments. (d) The digested fragments are run through capillary gel electrophoresis which records the length of the terminal fragments, with fluorescent dye attached. Electropherogram profiles are then used to plot the relative abundance of each differently sized terminal fragment, creating a crude profile of taxa community composition.

fluorescent dye. The PCR products, which should be of similar fragment length, are then cut with one or more restriction enzymes which recognize and cleave specific DNA sequences. This process generates a series of DNA fragments of varying length, depending on how many restriction sites are present in the target DNA. The digested PCR products are then separated and sized using an automated DNA sequencer which records the length of only the fluorescently labelled terminal fragments [such as those tagged with 6-carboxyhexachlorofluorescein (HEX) or 6-carboxyfluorescein (FAM)]. The diversity of any bacterial community, or set of genes, is therefore characterized by the length of each DNA terminal sequence (i.e. with fluorescently labelled primer attached) following digestion. While T-RFLP remains popular as a DNA-fingerprinting methodology, one particular disadvantage is that unlike DGGE or ARISA (later described) the amplified DNA can no longer be used for downstream applications (e.g. DNA sequencing) since the integrity of the DNA is destroyed by the process of digestion. Nevertheless, T-RFLP is widely used to compare microbial community composition among reference and contaminated biofilm communities (Bordenave *et al.*, 2007) and to monitor the community composition of abundant biofilm taxa within both bioreactors and field studies (Mariner *et al.*, 2008; Chiellini *et al.*, 2013). T-RFLP is most commonly used to assess the diversity of rRNA gene sequences among various groups of organisms including bacteria (Turpeinen *et al.*, 2004), archaea (Tischer *et al.*, 2013), fungi (Rieder and Frey, 2013) and protozoa (Dopheide *et al.*, 2009). However, the same approach can be used to assess the diversity of a wide variety of functional genes, including those involved in microbial bioremediation. For example, De la Iglesia *et al.* (2010) used a T-RFLP approach to assess the impact of copper pollution on the genetic composition of copper-resistance determinants using biofilm samples from a non-impacted intertidal shore as reference material. Yao *et al.* (2011) and Huang *et al.* (2011) used this method to characterize the diversity of ammonia oxidizing and nitrogen fixing bacteria, respectively. Hence T-RFLP offers significant opportunities to explore the diversity of a wide range of genes involved in the biodegradation process.

### Automated ribosomal intergenic spacer analysis (ARISA)

Similar to T-RFLP, automated ribosomal intergenic spacer analysis, or ARISA, also capitalizes on the use of a capillary DNA sequencer to differentiate bacterial DNA based on the length of sequence fragments (Fig. 7.5). During bacterial ARISA, primers are typically used to target highly conserved regions of bacterial 16S rRNA and 23S rRNA genes. Between these two genes lies a non-coding region of DNA, termed an internal transcribed spacer (ITS) region, which varies widely in length from about 50 bp to over 1050 bp (Ranjard *et al.*, 2001). Variation in the fragment length of the ITS region is analysed by amplifying this section of DNA with oligonucleotide primers, one of which is labelled at the 5′-end with a fluorescent tag. Use of this fluorescent tag then enables the automated analysis of DNA fragment lengths, as is also used in T-RFLP. In similar with T-RFLP, the main output of this approach is an electropherogram showing the relative abundance of differently sized PCR fragments. Advantages of ARISA are that, in comparison with DGGE, the use of a capillary genetic sequencer mean that hundreds of samples can be run, and compared, with relative ease (Ranjard *et al.*, 2013). Additionally ARISA can have a higher resolution compared to T-RFLP and thus is a rapid and relatively sensitive method to gauge variability in bacterial community composition from large sample numbers. For this reason, ARISA has been widely used to characterize variability in the community composition of bacteria (Lear *et al.*, 2013; Yamamura *et al.*, 2013), archaea (Ciesielski *et al.*, 2013) and fungi (Ranjard *et al.*, 2001).

## DNA sequencing

The technology of DNA sequencing is undergoing a revolution, with new methods being developed and advanced at an astonishing rate, leading to exponential increases in the amount of DNA sequence data that can be analysed for a given cost (Shendure and Lieberman Aiden, 2012). A key feature of DNA sequencing is that analysis of the data can provide valuable insights into the taxonomic identity of organisms within a community, as well as providing information on community structure or diversity. In the following section, we consider the selection of DNA sequencing methodologies available to the wider scientific community at the time of publication, and consider the advantages and disadvantages associated with each.

### Sanger sequencing

The advent of DNA sequencing, which began in the 1970s has significantly accelerated biological research and understanding. The first widely used DNA sequencing approach, the Sanger DNA sequencing method remains the most common DNA sequencing approach used today and represents the first generation of DNA sequencing

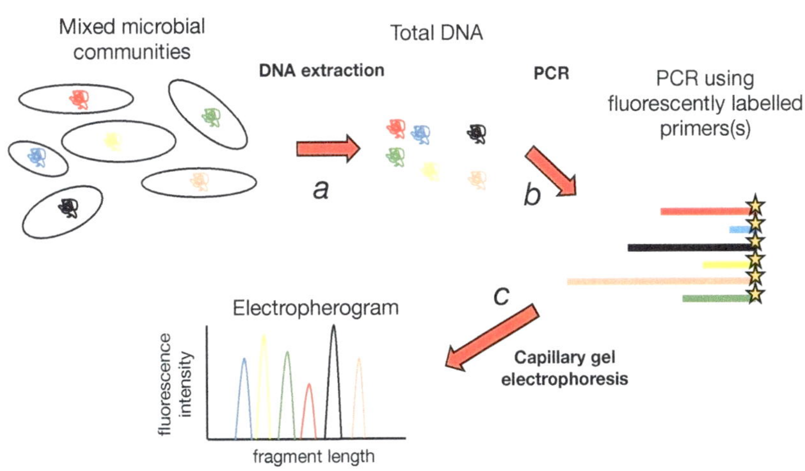


**Figure 7.5** Principles of automated ribosomal intergenic spacer analysis (ARISA). (a) The total DNA of the biofilm community is extracted, then (b) DNA fragments of interest (often the intergenic region located between rRNA genes which is known to vary in length among taxa) are amplified by PCR. One of primers used in this process is marked with a fluorescent tag (e.g. FAM, HEX). (c) Amplified PCR products are then purified and run through capillary gel electrophoresis which effectively records the base pair length of each fragment. Electropherogram profiles are then used to plot the relative abundance of each differently sized fragment, creating a crude profile of taxa community composition.

methodologies. The Sanger approach requires the synthesis of new strands of DNA using PCR-based amplification. The DNA strand to be sequenced is used as a template and an appropriate DNA primer is selected as the starting point for amplification. A DNA polymerase amplifies the single stranded DNA template by adding either normal dNTPs, or modified nucleotides (dideoxy-NTPs, or ddNTPs). The ddNTPs differ from the standard dNTPs in that the 3′ carbon lacks a hydroxyl group. Under such conditions, DNA synthesis will take place until a ddNTP is added to the growing chain, rather than a dNTP, causing DNA synthesis to stop. For this reason the Sanger approach to DNA sequencing is frequently referred to as a chain termination DNA sequencing method. This method therefore provides multiple amplified DNA sequence fragments, each of different length depending on when a ddNTP was incorporated. Since each of the four ddNTP chain terminators is labelled with a different fluorescent dye, the colour signal associated with each fragment provides the identity of the nucleotide base (A, C, G, T) at the site of termination. These fragments can thus be separated by capillary gel electrophoresis whereby a laser beam detects the size of individual DNA fragments by the order in which they exit the gel column. The size of the DNA fragment, and the colour of the fluorescent dye at the termination site is then used to generate a chromatogram for each sample, which can be interpreted to provide the actual sequence read of each DNA sample. To date, the Sanger sequencing approach remains the method of choice for the analysis of small numbers of DNA fragments. However, a major drawback for the use of this approach is that it is only suitable for the analysis of DNA from pure cultures. This has significant limitations for the analysis of microbial community composition and diversity in natural microbial communities which are often highly diverse. To circumvent problems associated with the analysis multi-species communities, the cloning of DNA fragments into recipient cells of *E. coli* is widely practised. Using such an approach it has been possible to produce pure cultures of *E. coli*, each containing DNA cloned from a single original DNA fragment that may each be analysed in separate sequencing reactions. This approach has provided enormous insight into the diversity of biofilm-dwelling bacterial communities, and the vital functions that they perform. However, the cost of sample preparation remains inhibitively high for the analysis of large sample numbers. In the following section we review three 'next-generation' technologies that have recently allowed us to generate and process enormous volumes of DNA sequence data, largely by bypassing the rate-limiting step of DNA cloning prior to sequence analysis.

### Advanced DNA sequencing technologies

So called 'next-generation' technologies involve attaching short pieces of DNA to a solid substrate using a specific oligonucleotide sequence called an adapter. An adapter sequence at the other end of the sequence anneals to a primer used to initiate the PCR. During PCR, all of the surface attached DNA is then sequenced simultaneously, such that these methods are also commonly referred to as 'massively parallel sequencing' strategies. Importantly, the use of DNA adapters has also allowed the indexing or 'barcoding' of samples so that many DNA libraries can be mixed (or multiplexed) together on one sequencing lane. An example of a barcoded DNA primer is shown below. Here the Illumina TruSeq universal adaptor sequence, underlined, allows the single stranded DNA to bind to the solid surface of a flow cell. The region comprised of 'N's refers to a sequence of any six nucleotides, providing a nucleic acid sequence which can be readily identified as unique to a single DNA library. The remaining sequence is the template DNA specific PCR primer sequence.

5′- GATCGGAAGAGCACACGTC
TGAACTCCAGTCAC-**NNNNNN**-
<u>ATCTCGTATGCCGTC TTCTGCTTG</u>- 3′

A different 'barcoded' primer is annealed to the DNA amplified from each individual sample. As DNA sequencing steps are performed, the sequence information provided by the unique barcode can then be used to associate each individual DNA read with the sample that it originated from. Such methods are in routine use for the parallel analysis of hundreds of DNA samples within a single sequence run.

#### *DNA pyrosequencing*

DNA pyrosequencing or 454 sequencing, was the first next-generation method to gain widespread use. The DNA fragments to be analysed are attached to individual microscopic beads and then amplified in an emulsion PCR until each bead is covered with perhaps a million copies of each DNA fragment. The DNA fragments are denatured and an amplification procedure initiated by a DNA polymerase. One dNTP nucleotide (i.e. dATP, dCTP, dGTP or dTTP) is then added at a time in the presence of ATP sulfurylase. When a dNTP is incorporated, released phosphorylate reacts with the ATP sulfurylase to generate a flash of light. A high resolution camera, linked to computer imaging software can thus track the growth of millions of newly synthesized DNA strands, *in situ*, one base at a time.

To date, the 454 DNA sequencing platform remains the most commonly used and published next-generation technology, generating substantial insights into the microbial community composition and function of biofilms during contamination and bioremediation events (Gomez-Alvarez *et al.*, 2013, 2014; Dalahmeh *et al.*, 2014).

### *Illumina*

In the Illumina method, short nucleotide adaptor sequences are annealed to target DNA to facilitate it binding to the glass surface of the flow cell. A process of solid-phase amplification then generates hundreds or thousands of copies of each single template molecule in close proximity to each other, using a process often known as 'bridge PCR' (Bentley *et al.*, 2008). Once these clusters of DNA sequences are obtained, a 'sequencing by synthesis approach' is used to add fluorescently labelled reversible terminator bases in the presence of a DNA polymerase. Non-incorporated bases are washed away and sequential imaging of the location of the different dyes (one for each nucleotide base) reveals the sequence of each DNA cluster. Between each amplification step non-incorporated nucleotides are washed away before the next step of the amplification cycle begins. The end result allows a robust base calling procedure, including for repetitive sequence regions and within homopolymers.

Currently, the Illumina sequencing platform is gaining popularity over the more traditional DNA pyrosequencing approach, largely because it is capable of generating far more reads per sample run. The Illumina 'HiSeq' approach is currently capable of producing > 8 billion 125 base-paired ends reads, while the smaller-scale 'MiSeq' approach will generate approximately 50 million 300 base-paired end reads. Interestingly, in studies which have used the HiSeq approach to produce order of magnitude increases in DNA sequence data, little difference in measures such as beta diversity are noted between the outputs of both HiSeq and MiSeq DNA sequence analysis (Caporaso *et al.*, 2012). This suggests that, depending on the scientific question being posed, the decreased costs of sequencing could more effectively be used to sequence a larger number of samples, rather than to sequencing individual samples to a greater depth. Nevertheless, the substantial sequence data provided by the HiSeq platform present valuable opportunities for the detection of rare taxa, and also for use in genomic and metagenomic studies (later described) where the provision of multiple sequence read data for DNA regions is of great importance.

### *Ion Torrent approach*

In contrast to DNA pyrosequencing and dye-labelled terminator approaches (e.g. Sanger and Illumina methods) Ion Torrent, or ion semiconductor sequencing, is based on the detection of hydrogen ions released during the polymerization of DNA. The Ion Torrent sequencing chip consists of a dense microwell array, in which each well of the array acts as an individual PCR chamber for a single DNA fragment. Located underneath the layer of microwells is an ion sensitive layer and field-effect transistor, aligned with the array microwells. Sample preparation for Ion Torrent analysis is similar to that required for DNA pyrosequencing. A library of short DNA fragments is created, flanked by Ion Torrent sequence adapters. The library fragments are then amplified using an emulsion PCR process before these sample beads, containing the cloned DNA are applied to the Ion sequencing chip. The microwells, each now containing many copies of a single-stranded template DNA molecule, are then sequentially flooded with unmodified dNTPs, adding one nucleotide at a time (i.e. either dATP, dCTP, dGTP or dTTP). If a complementary dNTP is incorporated, a hydrogen ion (and pyrophosphate) is released. The release of this hydrogen ion changes the pH of the microwell solution which is detected by an ion-sensitive field-effect transistor (ISFET) to confirm that a nucleotide (A, G, C, T) has been incorporated into the newly forming double-stranded DNA sequence. Any unattached dNTPs are washed away and the process is repeated.

The series of electrical pulses produced from each well can then be translated into DNA sequence data, avoiding the need to use any modified nucleotides, or optical measurements. This means that the cost per Ion Torrent run tends to be far less than alternative sequencing methods. The perceived limitations of this approach are the relatively short read length obtained compared to other sequencing strategies and DNA sequencing errors caused by homopolymer repeats of the same nucleotide (e.g. CCCCCC).

In a recent study by Yergeau *et al.* (2012), an Ion Torrent sequencing approach was used to investigate the impact of naturally occurring bituminous compounds on biofilm bacterial community composition. Sediments collected from a river known to contain naturally occurring bitumen at various concentrations were circulated through rotating biofilm reactors. Biofilms exposed to high or low concentrations of the bitumen were explored after 50 days of incubation. Biofilm samples in each reactor were taken, DNA extracted, and 16S rRNA genes amplified to inform the composition and phylogenetic relatedness of bacterial taxa within each sample. DNA sequences from each of the thirty six individual samples were generated using an Ion Torrent platform to provide a total of approximately 2000 DNA sequence reads per sample. The DNA sequence data revealed a reduction in the relative abundance of *Cyanobacteria*, *Bacteriodetes* and *Firmicutes* in biofilms exposed to high concentrations of bitumen, suggesting that bitumen has inhibitory effects on the presence of photosynthetic organisms. Nevertheless specific microbial taxa and their respective functional genes were suggested to be stimulated, providing insight into taxa (e.g. *Alphaproteobacteria* and *Betaproteobacteria*) that may utilize the carbon present in bituminous compounds. While there exist a number of advantages and disadvantages associated with the use of different next-generation DNA sequencing platforms (see Table 7.1), an earlier study conducted by the authors on the same site was used to extensively validate the use of Ion Torrent sequencing in comparison with a 454 sequencing platform approach. Both platforms yielded similar results despite obtaining longer sequence reads from the 454 approach (Yergeau *et al.*, 2012).

## Approaches to assess the activity of biofilm bacterial communities from their nucleic acids

In many cases, researchers may not require information on the composition, diversity and taxa identity of individuals within whole biofilm communities. In bioremediation studies in particular, a key aim is often to identify only the organisms capable of contaminant degradation or transformation. This can be particularly challenging to achieve when the population of these organisms is small compared to background numbers of microbial taxa. This is often the case for studies exploring the degradation of low concentrations of organic compounds where the carbon/energy yields from contaminant degradation may be insufficient for degrading organisms to become dominant members of the community.

A variety of methods have been developed to track the fate of contaminants into the biomass of degrading taxa. Such methods normally require the use of substrates labelled with isotopes (e.g. $^{13}C$, $^{14}C$) to monitor incorporation of the substrate into microbial biomass (e.g. DNA, proteins, or fatty acids). A method commonly used in molecular microbiology is to identify the DNA of taxa incorporating labelled substrates into their nucleic acids. The stable-isotope probing of microbial DNA and RNA is described in the following section.

### Stable isotope probing

Stable isotope probing (SIP) is a molecular tool that can be used to link a specific metabolic process to specific members of a microbial community. SIP may be used to track the fate of $^{13}C$-labelled substrates (Manefield *et al.*, 2002; Lear *et al.*, 2007; Hery *et al.*, 2008), or less commonly $^{15}N$- (Bell *et al.*, 2011) and $^{18}O$-labelled substrates (Woods *et al.*, 2011) into microbial biomass. Using this approach, a labelled substrate (typically with $^{13}C$) is provided to the microbial community which may use this carbon source as a substrate for growth, and in doing so will incorporate the labelled DNA into its cellular biomass. Stable isotope probing was first developed using phospholipid-derived fatty acids (PLFA) (Boschker *et al.*, 1998) but later the labelling of both DNA (Radajewski *et al.*, 2000) and rRNA (Manefield *et al.*, 2002) was utilized, which may be collectively referred to as nucleic acid stable-isotope-probing, or NA-SIP. Using DNA-SIP as our example, let us consider how one would identify the bacterial taxa responsible for the degradation of phenanthrene within a diverse biofilm community (Fig. 7.6). First, $^{13}$C-labelled substrate would be added to the community and allowed to incubate for a suitable period of time before the total DNA of that community is extracted. This DNA would then be separated using density gradient centrifugation. During centrifugation heavier molecules will be forced to reside at a lower position in the centrifuge tube than lighter molecules. Since the DNA of organisms that have incorporated the $^{13}$C will be heavier than $^{12}$C-dominated DNA, we are then able to identify the components of the community that have used our labelled substrate. The 'heavy fraction' of DNA can then be sequenced to identify which bacteria have degraded the labelled substrate presented to them. NA-SIP is ideal for identifying which microbes are capable of degrading a substrate among a mixed community, and is particularly useful for identifying the bacteria responsible for the degradation of low concentrations of toxic substances, from which simple observations of the total community may not be sufficient. However, the use of NA-SIP also provides a number of drawbacks, in addition to the often long periods of centrifugation and sample processing. A key drawback in the use of SIP remains the process of cross-feeding (Neufeld *et al.*, 2007). Here the death of the primary degraders, or loss otherwise of their cellular constituents will allow the label to be taken up by secondary organisms, which may be incapable of degrading the original substrate. This may lead to 'false positives' increasing the likelihood that potential phenanthrene degraders, would be incorrectly identified. Strategies suggested to minimize this effect of cross-feeding include shortening incubation times, conducting time-course experiments (Leigh *et al.*, 2007) or combining the results of both RNA-SIP and DNA-SIP approaches (Lear *et al.*, 2007).

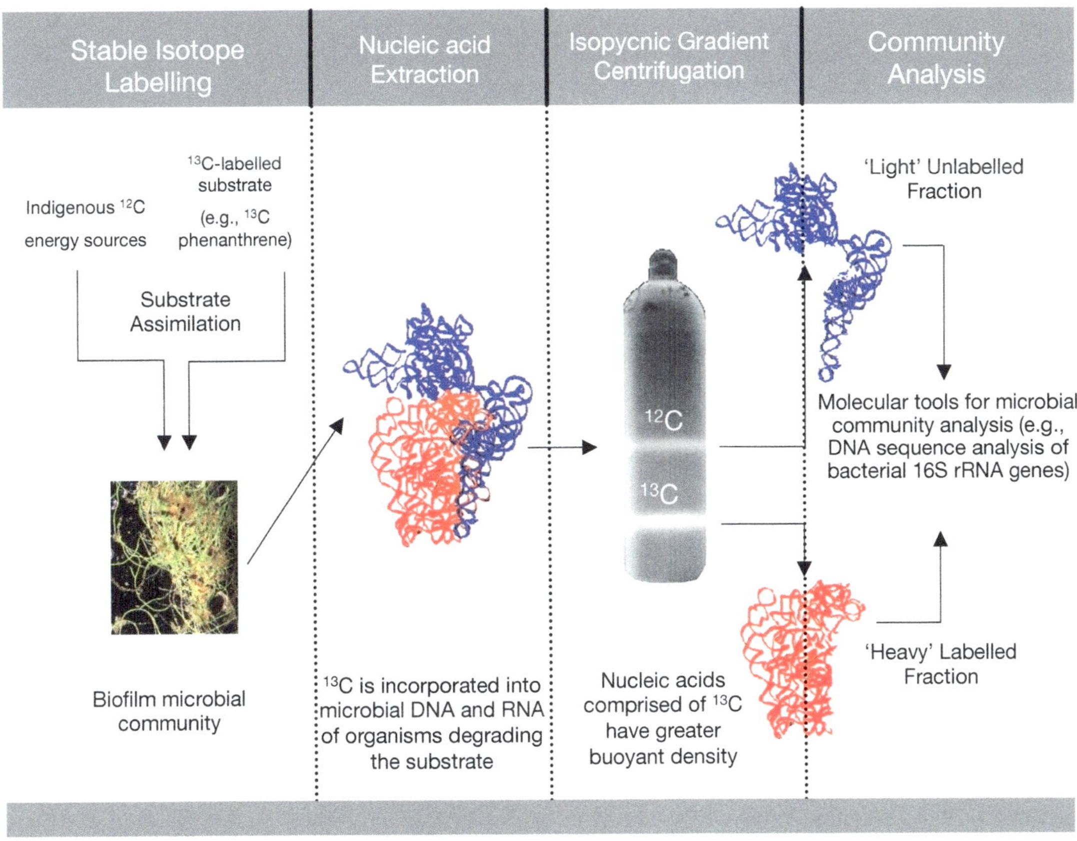


**Figure 7.6** Principles of nucleic acid stable isotope probing.

## Approaches to assess the functional attributes of biofilm bacterial communities from their nucleic acids

In many studies exploring the degradation of environmental pollutants, it is desirable to catalogue the presence/abundance/diversity of genes involved in the contaminant degradation, rather than to explore the taxonomic identity of these communities. Thus a variety of taxonomic approaches have been developed that now allow researchers to investigate the various functional genes present within contaminated environments. One of the most simple approaches to achieve this is to undertake PCR, or qPCR, using primers specific to the functional gene of interest, an approach discussed earlier in this chapter. Alternatively, more complex methodologies including functional gene microarrays and shotgun metagenomics may be used to screen samples for the presence of a much wider array of genes, including those with unknown or unproven functions.

### Functional microarrays

A number of different functional gene arrays have been developed and used in recent years (Brazma *et al.*, 2001; Steward *et al.*, 2004; He *et al.*, 2007). However, perhaps the most commonly used functional microarray at the present time remains the GeoChip microarray which is currently available in its fourth iteration, Geochip 4 (Tu *et al.*, 2014). This array contains approximately 82,000 probes encoding for 420 functional gene families related to microbial carbon, nitrogen, sulphur and phosphorus cycling, energy metabolism, antibiotic resistance, metal reduction/resistance, organic remediation, stress responses, bacteriophage and virulence. The hybridization of sample DNA to probes on the chip can be used to inform on variability in the presence and richness of genes encoding for key ecosystem processes among samples. To date, the Geochip approach has been used to characterize functional variability in biofilm microbial communities and their response to contaminants in both natural and engineered systems. For example, Dopheide *et al* (2015) surveyed the functional diversity and composition of stream biofilm communities using the GeoChip 3.0. Biofilm communities were compared from both native forest streams and in heavily impacted urban and rural catchments. A majority of the gene families were found in similar proportions among all samples from contrasting streams. This suggests the existence of significant functional gene redundancy within different stream biofilm communities, since they exhibit quite similar microbial functional potential, despite contrasting taxonomic composition. However, this study which was restricted to the analysis of only a small number of samples, provides valuable baseline data characterizing the functional gene signature of stream biofilm communities. Liang *et al.* (2014) used a more advanced version of the microarray to explore the phylogenetic and functional attributes of a biofilm community capable of the bioelectrochemical reduction of nitrobenzene to the less toxic aniline. This occurred during a carbon source switchover from organic glucose to inorganic bicarbonate. Use of the functional microarray revealed a proportionately higher intensity of multi-haem cytochrome c and carbon fixation genes in the bicarbonate-fed system, presumably to meet the requirement of energy conservation after organic glucose was removed. Associated with the biocathode of the electrochemical system, genes encoding for extracellular pilin were more abundant when supplied with glucose. This finding suggests that the pilin genes detected in this study may contribute to the transfer of extracellular electrons during biochemical nitrobenzene reduction, supported by the addition of glucose as a growth carbon source. The abundance of cytochrome P450 reductase genes increased following the switch from glucose to bicarbonate addition, indicating that these may be involved in electron transfer during nitrobenzene reduction by the biocathodic biofilm. The findings of this study provide valuable insights into the functional mechanisms that allow the biocathodic reduction of nitrobenzene to be maintained under various carbon source supply conditions. A wide range of research has now proven the utility of the GeoChip microarray, and similar systems, for investigation of microbial functional attributes in biofilm and bioremediation related studies.

### Shotgun metagenomics

A limitation of functional gene microarrays is that for the most part they can only be used to observe the presence of a select number of known genes. In comparison, shotgun metagenomic methods allow researchers to interrogate the entire genetic material within any sample. Indeed, a major goal of metagenomics is to generate genomic data from uncultured microbial taxa, allowing previously unparalleled investigations into the functional capability of unculturable bacteria taxa. Handelsman *et al.* (1998) described an early process to explore the metagenome of soil microbial communities. The approach that they describe involves the extraction of the genomic DNA from all organisms inhabiting that environment, and the fragmentation of this DNA into pieces small enough to allow their cloning and subsequent DNA sequencing and analysis. The utility of Handelsman's original approach was restricted by the number of clones that could be screened for DNA sequence analysis. However, this processing bottleneck has been superseded by more recent developments in high throughput DNA sequencing. Regardless of the approach used, metagenomics can be broadly separated into approaches that are either sequence driven or function driven.

In sequence-driven metagenomics the desired outcome of the study is usually to catalogue the diversity of all genes present by the analysis of as much DNA as is possible. The total environmental DNA is extracted from the environment and fragmented into smaller pieces using processes such as sonication, hydrodynamic shearing and enzymatic digestion. The DNA fragments can then be sequenced and compared against the sequences of known genes to predict their putative function. Such an approach was used by Tyson *et al.* (2004) to provide novel insights into the functional attributes of biofilm communities associated with acid mine drainage (AMD). The reduced diversity of organisms present in the AMD environment enabled the full genetic information (10× genome coverage) to be generated for both a *Leptospirillum* group II species, and an uncultured *Ferroplasma* species. This allowed the environmental role of these biofilm dwelling organisms to be elucidated from their genes, without the need for culturing. For example, the genome of a *Leptospirillum* revealed its role in iron oxidation, likely resistance mechanisms to elevated Cu, Zn, Cd and As, and also indicated that this organism required fixed N, being incapable of fixing N from $N_2$.

In many cases it is not necessary to understand the total genetic component of an environment. An example of this is when we wish to explore the presence or genetic diversity of organisms responsible for only specific functions. Suppose then that we wish to identify the genes used by a community of biofilm bacteria to degrade a particular contaminant such as pentachlorophenol within a bioreactor system. An approach that we can use is to extract the total DNA from the biofilm community, fragment the DNA, and then incorporate individual DNA fragments into a host such as *E. coli* using a vector plasmid. These host bacteria, containing the cloned DNA fragments can then be grown on a media enriched in pentachlorophenol. Degradation assays provide evidence of host bacteria expressing pentachlorophenol resistance or degradation attributes. The cloned DNA of host strains exhibiting these attributes can then be DNA sequenced to identify genes for PCP degradation that were present in our original biofilm community, including those that were previously held by unculturable taxa.

At the present time, the utility of shotgun metagenomics remains limited by the huge amount of DNA sequence data that may be required to make inferences from complex and diverse microbial communities. However, continued exponential increases in the amount of DNA sequence data produced per dollar mean that large-scale metagenomic investigations will likely become a prominent feature of future research in biofilms and bioremediation.

## Part 2: applications of molecular techniques in bioremediation assessments

Molecular methods have been widely used to explore the diversity of microbial biofilm communities, as well as their interactions with contaminants during bioremediation. In the remainder of this chapter we consider how molecular methods have advanced our understanding of biofilm formation, microbial diversity and organization, as well as being used to elucidate the specific taxa responsible for contaminant degradation in mixed communities. As we will discover, many of the recent advances in identifying and encouraging the process of contaminant degradation in the environment following major pollution events, as well as in *ex situ* treatment bioreactors, have been made via molecular observations of microbial community DNA.

## Monitoring biofilm formation

A wide variety of both molecular and microscopy-based techniques have provided valuable insights into the establishment and structural organization of biofilm communities. As outlined in Chapter 2, there exists a general model of biofilm formation which starts with a pre-conditioning phase where the physical and chemical characteristics of a sterile surface are altered as polymers and colloids in the overlying water adhere to the surface to form a nutrient-rich 'conditioning film'. As bacterial cells in the overlying water encounter this film, they may begin to adhere to this conditioned surface, a process which in the early stages, is considered to be reversible. After a period of time (e.g. several hours) members of the attached biofilm community may become irreversibly attached via strong dipole–dipole interactions, hydrophobic and ionic binding. The developing structure of the mature biofilm is then dictated by factors including nutrient and oxygen availability (Lawrence *et al.*, 2004), cell death and lysis (Bayles, 2007), and bacterial quorum sensing responses to their environment (Bassler and Losick, 2006).

One of the key approaches used to explore the spatial composition of complex biofilm communities is fluorescence *in situ* hybridization, or FISH. Wilmes *et al.* (2009) used a combination of FISH, epifluorescent microscopy and transmission electron microscopy, or TEM, to explore the organismal and functional organization of biofilm communities at the air–solution interface of underground acid mine drainage (AMD) streams and pools in the USA. These particular AMD biofilms were chosen since their acidity (pH $\approx$1) and moderately high temperature (40°C) provides a system with relatively low species richness. This same reasoning has led to AMD samples being used as model systems for exploratory metagenomic studies, as later described. Biofilm pellicles were carefully removed from a number of locations across sampling transects which included both the newest, thinnest zones of biofilm, to the oldest, thickest regions. These samples were preserved before the spatial organization of the biofilm microbial communities was monitored on lateral sections of the material. FISH probes were used to highlight the presence of bacteria, archaea and eukarya as well as more specific taxa, using probes targeting the *Thermoplasmatales* (a group of acidophilic archaea) and *Leptospirillum* groups II and III (a genus of iron-oxidizing bacteria). Additionally, an immunohistological assay for cytochrome $Cyt_{579}$ was used, since this protein undertakes a central role in iron oxidation which plays an important role in the formation of AMD. At one site, the authors describe the dominance of *Leptospirillum* group II in biofilms thought to be at an early stage of development. In contrast, with increasing maturity, a greater abundance of archaea (mainly belonging to the Thermoplasmatales) as well as eukarya (mainly fungal filaments) were observed. The presence of fungi and protozoa only in biofilms at a later stage of development was suggested to indicate a threshold in the availability of carbon required to support the abundance of these organisms. At another biofilm location, distinct strata were detected within mature biofilm, being dominated by a densely packed layer of *Leptospirillum* group II at the base, and an upper layer containing an abundance of archaea. Since $Cyt_{579}$ was also localized at the base of the biofilm, the authors suggest that the majority of iron-oxidization, catalysed by $Cyt_{579}$, occurs at the biofilm/surface interface. Such studies, combining molecular and microscopic methods have been vital for increasing our understanding of the structural organization of individual taxa within the complex 3-D structure of biofilm communities, and implications for the varied functional processes that they undertake.

A recent study by Bar-Zeev *et al.* (2012) used a different combination of confocal laser scanning microscopy and cell staining techniques to provide a revised paradigm of biofilm formation. Using a simple filtration technique to remove particles $>0.7$ µm, they explored the formation of biofilm by communities that either did, or did not, include transparent exopolymer particles (TEPs) or other large organic microgels. These microgels may be generated by algal mucous, or the gelatinous material that surround many algae. This gel may represent an important source of nutrients and also provide a substantial surface area for microbial interaction. The settling of TEPs also provides a mechanism for the transport of nutrients and microbial cells to the floor of aquatic environments. Epifluorescent microscopy observations of stained cells revealed substantial differences in biofilm development, with 300-fold greater biofilm coverage of surfaces in experiments where large TEPs were not removed.

This provides evidence that microgel particles such as TEPs are important for their role in conditioning pristine surfaces, providing a favourable substrate for subsequent microbial attachment. Additionally, the microbial taxa that are already associated with the TEPs could provide a 'jump-start' for the early development of biofilm communities. The labelling of individual microbial taxa, or genes, with fluorescent and luminescent markers has provided valuable new insights into biofilm community dynamics. This has revealed biofilms to most frequently be highly structured communities, with taxa being well organized to make optimal use of available resources and physicochemical conditions. Improved knowledge of their organization continues to enhance our ability to understand and promote the remediation of pollutants by biofilm communities.

## Assessing the composition and diversity of biofilm microbiota

In addition to studies which have sought to record the diversity and composition of natural biofilm communities (Lear and Lewis, 2009; Lear *et al.*, 2008, 2012a, 2013), the response of biofilms to environmental stressors, including pollutants, is being used to track both the impacts of environmental contamination and ecosystem responses to bioremediation practices.

As an example of the use of innovative methods to monitor the response of biofilm communities to contaminants, Yergeau *et al.* (2010) used an anonymous DNA microarray to explore the impact of pharmaceuticals on the expression of genes by freshwater bacterial communities. Water was collected from two rivers and exposed to environmentally relevant concentrations of erythromycin (ER), gemfibrozil (GM), sulfamethazine (SN) and sulfamethoxazole (SL) in rotating annular reactors designed to allow biofilm development. Total DNA and RNA were extracted from samples of biofilm pooled from each treatment, fragmented and cloned fragments of the DNA printed onto glass slides. RNA (or rather cDNA) from each reactor treatment was then allowed to hybridize with corresponding DNA sequences present on the microarray grid. Statistical correspondence analysis could then be used to identify DNA probes for which more hybridization occurred in some treatments than others, signalling the differential expression of DNA among treatment samples. The DNA sequence of these differentially expressed probes was then confirmed by Sanger sequencing. This novel approach allowed both positive and negative transcriptional responses caused by low concentrations of pharmaceuticals to be identified. For example, the addition of 0.5 μg/l sulfamethoxazole caused changes in gene expression related to the cell envelope and outer membrane, which could affect biofilm formation. The different biofilm response observed by both the sulfamethoxazole and sulfamethazine antibiotics highlight the specificity of microbial responses to antibiotics. Such results demonstrate that low concentrations of small molecules can induce transcriptional changes in complex microbial communities, which may have significant implications for aquatic ecosystems in receipt of treated sewage effluent. A later approach by the same lead author (Yergeau *et al.*, 2014) was to explore microbial expression profiles in the rhizosphere of willows grown in contaminated soil. Plants under stress have a tendency to exude more nutrients via their roots (Naik *et al.*, 2009), and in doing so stimulate the activity of microorganisms present in the rhizosphere. This increased microbial activity may contribute the degradation and transformation of organic contaminants in the soil, by so-called 'phytoremediation'. For this reason Yergeau *et al.* (2014) sought to understand what microbial genes are up-regulated in the rhizosphere of willow grown in contaminated soil. Gene expression profiles were compared following metatranscriptomic analysis using an Illumina platform. Genes related to hydrocarbon, and particularly PAH degradation were generally up-regulated in contaminated soils where initial PAH concentrations were ≈ 60 mg/kg, helping to validate the strength of relationship between biodegradation and gene expression. The presence of the willow tree also had a significant impact on microbial gene expression, including of various hydrocarbon-degrading genes, even though concentrations of hydrocarbons did not differ significantly among planted and unplanted soil samples. This study provided the first extensive metatranscriptomic view of microbial gene expression in the rhizosphere. It also provides information which could be of relevance to stimulate the degradation of hydrocarbons by biofilm-dwelling organisms frequenting the root zone of higher plants.

If the response of biofilm-dwelling microbial communities to pollutant stresses is predictable, then biofilm communities could provide an informative indicator of environmental health (Lear *et al.*, 2012a; Washington *et al.*, 2013; Ancion *et al.*, 2014) and recovery after bioremediation (Lear *et al.*, 2009a). For example, Lear *et al.* (2009a) observed distinct bacterial communities present in streams located in either neutral, acidic, or very acidic stream water that had been impacted by acid mine drainage. Interesting, it would appear that despite being impacted by a variety of contaminants, the addition of lime to a formally acid stream led to the bacterial community within the site resembling that of an unimpacted stream. This observation confirms that there may have been a partial recovery of the site following pH adjustment for remediation purposes. Such findings highlight the utility of microbial community data to not only record the impacts of pollution events, but also as a tool to monitor the success of remediation activities.

## Identification, isolation and characterization of contaminant-degrading bacteria

Advances in a variety of molecular methods have vastly improved our ability to identify contaminant-degrading organisms even when they may be present in low abundance, or are unculturable using traditional methods. Molecular methods have helped to overcome these issues by allowing the high-throughput analysis of large volumes of nucleic acid data, such as in metagenomic studies, or by tracking the fate of substrate-derived carbon in the bacterial genome, as in stable isotope probing.

Stable isotope probing (SIP) has been particularly useful to identify degraders of pollutants that remain harmful at low concentration. As earlier described, at low concentrations, contaminants may provide insufficient energy for degrading populations to be identified from among a diverse background of microbial taxa. In one of the first studies to capitalize on the use of SIP in bioremediation studies, Manefield *et al.* (2002) sought to identify bacterial taxa responsible for the degradation of phenol in an aerobic industrial reactor. An industrial facility responsible for the treatment of 1800 l wastewater per minute was heavily contaminated with phenolic compounds (>200 µg/ml). Floc-based microbial community reactors with a combined volume of 1700 m$^3$ were used to reduce the concentration of phenolics in the outflow to < 10 µg/ml. A possible candidate responsible for phenol degradation at the site was *Pseudomonas putida* BS564 which was previously isolated from the reactor and shown to use phenol as a sole source of carbon. A pulse of $^{13}C_6$-phenol was used to label the nucleic acids of phenol degraders and bacterial RNA labelled with phenol-derived $^{13}C$ was separated from the RNA of the remaining community by equilibrium density centrifugation and fractionation. Sequence analysis of the $^{13}C$-labelled RNA revealed that most dominant member of the phenol-degrading bacterial community was a member of the genus *Thauera*. While little attention has previously been paid to the ability of this genus to degrade aromatic compounds, the use of SIP provides evidence that such organisms were an important degrader of phenol within the bioreactor system. Their approach demonstrates how molecular methods may be used to identify organisms dominating biodegradation processes and inform on the best choice or organisms to maintain and improve bioreactor performance. Importantly, the use of such technologies allow the issue of 'unculturable taxa' to be conveniently bypassed. In addition to identifying active pollutant degrading taxa within bioreactor communities, SIP has proven to be a useful tool to detect novel pollutant degraders *in situ* within contaminated soil and water environments (Lear *et al.*, 2007; Regonne *et al.*, 2013).

Nucleic acid stable-isotope probing approaches have also been used to inform on more complex mechanisms of pollutant co-metabolism. For example, van der Zaan *et al.* (2012) explored the degradative potential of anaerobic microbial communities towards benzene within a nitrate enriched chemostat. After eight years of operation, a diversity of microbial taxa remained abundant within the system, and yet despite multiple attempts, no benzene degraders could be isolated as pure cultures. Since multiple bacterial taxa were present in the reactor, this led to the hypothesis that the anaerobic degradation of benzene is via microbial consortia coexisting in a syntrophic, or cross-feeding interaction. After adding $^{13}C$-benzene to benzene-degrading cultures, labelled microbial DNA was separated by centrifugation. Uncultured

*Peptococcaceae*, *Rhodocyclaceae* and *Burkholderiaceae* were represented in the $^{13}C$-labelled DNA, providing evidence of their role in benzene degradation. The authors suggest a syntrophic interaction of *Peptococcaceae* with hydrogen consuming species, supported by observations that these species are inhibited at high hydrogen concentration and that biofilm was essential for benzene degradation, a mode-of-life that supports close physical cooperation. Thus, the use of SIP is proving useful to identify microbial strains, including unculturable taxa, that may be involved in the biofilm-mediated degradation of harmful substances as well as putative microbial interactions that may support contaminant degradation. Such knowledge is being used to inform on and enhance the rate and extent of pollutant degradation by microbial biofilms in industrial settings.

Modern molecular methods are also providing needed insight by allowing the identification of contaminant degrading bacteria within complex natural communities. In particular, stable isotope probing has allowed researchers to target functional subsets of microbial communities, reducing the amount of sequence analysis required to characterize pollutant-degrading taxa. At the same time, significant recent advances in sequencing methodologies are allowing more and more genetic data to be acquired per sample, providing new opportunities to investigate the microbial community diversity and functional potential of natural biofilm communities. Recently, the high-throughput shotgun sequencing analysis of microbial DNA was used to inform on the natural hydrocarbon degradation pathways used by marine microorganisms following the 2010 Deepwater Horizon oil spill. During this event ≈4 million barrels of oil were accidentally released into the Gulf of Mexico (Zukunft, 2010). Shortly after the incident, a number of studies tracked fate of the major contaminant plume within the deep-sea water, as well as populations of active hydrocarbon degrading bacteria, revealing that *Oceanospirallales* dominated, later to be replaced by *Colwellia* and *Cycloclasticus* (Valentine *et al.*, 2010; Redmond and Valentine, 2012; Mason *et al.*, 2014). During the degradation process, it was suggested that the production of substantial extracellular polymeric substances by hydrocarbon degrading bacteria led to the settling of oil aggregates in a large marine snow or 'dirty blizzard' event (Schrope, 2013). This fate of hydrocarbons deposited on the sea floor was investigated by Mason *et al.* (2014). Sediment samples (0–1 cm depth) were collected from 64 locations at a depth of ≈1500 m below the water's surface. The microbial communities present with in each sediment core was assessed using targeted sequencing of 16S rRNA genes, and shotgun metagenomic methods to further characterize the genetic attributes of 14 of these samples using an Illumina DNA sequencing approach. This metagenomic analysis revealed that the deposition of hydrocarbons appeared to increase the functional diversity of surface sediment-associated microorganisms. In particular, an increased relative abundance of genes associated with the degradation of various alkanes and aromatic hydrocarbons were noted in sites that exceeded EPA sediment benchmark concentrations of polycyclic aromatic hydrocarbons (PAHs). The divergence of data derived from both the sediment and water column provides evidence that their respective hydrocarbon degradation pathways differ. Both aliphatic and aromatic processes were observed in the surface sediment communities, while aliphatic degradation appears to dominate in the plume (Mason *et al.*, 2012). However, few transcripts for PAH degradation were identified in either the plume or the sediment. Indeed, the slow rate of PAH degradation in these sediments suggests that the communities heavily impacted by the oil spill were not degrading PAHs. These findings suggest that surface-sediment associated microbial communities contributed to the rapid depletion of oil that escaped during the major Deep Water Horizon event in 2010 and may help to inform on the most appropriate response to further spills. However, this approach also indicates why recalcitrant PAHs may persist in the deep water sediment, also impacting meiofauna diversity (Montagna *et al.*, 2013) such that additional procedures to hasten the loss of recalcitrant PAHs may be desirable.

A major advance in determining the identity and functional attributes of contaminant degraders within mixed biofilm communities is provided by the powerful combination of both SIP and metagenomic technologies. While the 'per base' costs of DNA sequencing continue to decrease, one of the drawbacks of environmental metagenomics remains the difficulty associated with detecting genes of interest among an enormous background of genetic diversity. However, the use of metagenomics to characterize the diversity of SIP-derived genetic material is one approach that can be adopted to overcome this major current obstacle. The selective enrichment of targeted populations can allow the functional attributes of low abundant taxa to be explored in detail, an approach which has now been incorporated into a number of recent scientific studies [refer to Uhlik *et al.* (2013) for a review of the recent literature].

Dumont *et al.* (2006) conducted a pioneering study which demonstrated how the genomes of ecologically relevant functional subpopulations can be isolated and reassembled from SIP-labelled DNA. In their study $^{13}C$-labelled methane was incubated with forest soil before SIP was used to separate the genes of bacterial taxa capable of methylotrophy (i.e. organisms with the ability to use a reduced carbon compound such as methane, or methanol, for growth). The $^{13}C$-labelled DNA was then used to construct a metagenomic library, resulting in the discovery of the *pmoCAB* operon, which encodes for a particulate methane monoxygenase enzyme. The operon was fully sequenced using a 'shotgun' approach. This approach may not have succeeded if the DNA of methylotrophic bacteria had not first been separated using a SIP-based approach to reduce the amount of DNA sequencing required to identify the target gene. A similar approach was used by Sul *et al.* (2009) to explore the genetics of PCB degradation within contaminated sediments in a Michigan State River, USA. Following incubation of the soil with $^{13}C$-labelled biphenyl, analysis of the $^{13}C$-labelled DNA confirmed the presence of a novel biphenyl dioxygenase (*bphA*) gene and was used to inform on the gene organization of *bph* genes in comparison with other biphenyl-degrading taxa. NA-SIP has also been used in collaboration with a number of other molecular-based methods, including microarrays targeting the presence and abundance of key functional genes (e.g. the Geochip array). For example, in a study also using $^{13}C$-labelled biphenyl, Leigh *et al.* (2007) examined the utilization of the substrate in soil surrounding an Austrian pine. The functional attributes of the $^{13}C$-labelled DNA were interrogated using a Geochip microarray, revealing an abundance of biphenyl-degrading bacteria which also possessed genes associated with the degradation of a far wider range of aromatic substrates. Studies such as these highlight the opportunities provided by the combination of NA-SIP, environmental metagenomics and other advanced methodologies, which may be particularly useful to identify the genetic attributes of low abundance taxa. Such investigations are suitable to identify if organisms capable of degrading a pollutant are already present at a site, and could be enhanced using a biostimulation approach to remediation, or if bioaugmentation of a site with novel taxa is likely to be required.

## Conclusions

Recent advances in molecular methods, as well as microscopy, have revolutionized our understanding of the role of microbial biofilms for the remediation of harmful contaminants. Far from being a disorganized mass of microbial slime, biofilms are confirmed to form highly organized structures in which individual taxa are provided protection from contaminants while also maximizing their exposure to vital resources. It is clear that biofilm communities adapt in a somewhat predictable manner to the presence of environmental contaminants. Increases in the abundance of pollutant tolerant taxa or contaminant degradation/resistance genes may indeed be used to indicate the presence of environmental pollutants in the environment, or the success of bioremediation strategies. However, until recently it has remained hard to determine the role and identity of many rare microbial taxa for the degradation of contaminants, particularly for contaminants that are toxic in low concentration. Improvements in DNA sequencing methodologies, combined with the use of novel methods to isolate the genetic material of substrate-degrading microorganisms are being used to identify novel biofilm-dwelling taxa, or consortia, from among diverse assemblages of non-degrading taxa. Our improved knowledge of the genetic diversity of microbial biofilms and their interactions with harmful pollutants is now providing vital opportunities for the improved removal of pollutants using a broad range of both *in situ* and *ex situ* treatment strategies.

## References

Ancion, P.-Y., Lear, G., Neale, M., Roberts, K., and Lewis, G.D. (2014). Using biofilms as a novel approach to assess stormwater treatment efficacy. Water Res. *49*, 406–415.

Andrews, J.S., Mason, V.P., Thompson, I.P., Stephens, G.M., and Markx, G.H. (2006). Construction of artificially structured microbial consortia (ASMC) uding dielectrophoresis: Examining bacterial interactions via metabolic intermediates within environmental biofilms. J. Microb. Methods *64*, 96–106.

Applegate, B.M., Kehrmeyer, S.R., and Sayler, G.S. (1998). A chromosomally based *tod-luxCDABE* whole-cell reporter for benzene, toluene, ethylbenzene and xylene (BTEX) sensing. Appl. Environ. Microbiol. *64*, 2730–2735.

Bar-Zeev, E., Berman-Frank, I., Girshevitz, O., and Berman, T. (2012). Revised paradigm of aquatic biofilm formation facilitated by microgel transparent exopolymer particles. Proc. Natl. Acad. Sci. U.S.A. *109*, 9119–9124.

Bassler, B.L., and Losick, R. (2006). Bacterially speaking. Cell *125*, 237–246.

Bayles, K.W. (2007). The biological role of death and lysis in biofilm development. Nat. Rev. Biotechnol. *5*, 721–726.

Bell, T.H., Yergeau, E., Martineau, C., Juck, D., Whyte, L.G., and Greer, C.W. (2011). Identification of nitrogen-incorporating bacteria in petroleum-contaminated Arctic soils by using $^{15}$N DNA-based stable isotope probing and pyrosequencing. Appl. Environ. Microbiol. *77*, 4163–4171.

Beller, H.R., Kane, S.R., Legler, T.C., and Alvarez, P.J. (2002). A real-time polymerase chain reaction method for monitoring anaerobic, hydrocarbon-degrading bacteria based on a catabolic gene. Environ. Sci. Technol. *36*, 3977–3984.

Bent, S.J., Pierson, J.D., and Forney, L.J. (2007). Measuring species richness based on microbial community fingerprints: the Emperor has no clothes. Appl. Environ. Microbiol. *73*, 2399–2401.

Bentley, D.R., Balasubramanian, S., Swerdlow, H.P., Smith, G.P., Milton, J., Brown, C.G., Hall, K.P., Evers, D.J., Barnes, C.L., Bignell, H.R., *et al.* (2008). Accurate whole human genome sequencing using reversible terminator chemistry. Nature *456*, 53–59.

Bordenave, S., Goni-Urriza, M.S., Caumette, P., and Duran, R. (2007). Effects of heavy fuel oil on the bacterial community structure of a pristine mat. Appl. Environ. Microbiol. *73*, 6089–6097.

Boschker, H.T.S., Nold, S.C., Wellsbury, P., Bos, D., de Graaf, W., Pel, R., Parkes, R.J., and Cappenberg, T.E. (1998). Direct linking of microbial populations to specific biogeochemical processes by 13C-labelling of biomarkers. Nature *392*, 801–805.

Branco, R., Cristovao, A., and Moriais, P.V. (2013). Highly sensitive, highly specific whole-cell bioreporters for the detection of chromate in environmental samples. PLoS One DOI: 10.1371/journal.pone.0054005.

Brazma, A., Hingamp, P., Quackenbush, J., Sherlock, G., Spellman, P., Stoeckert, C., Aach, J., Ansorge, W., Ball, C.A., Causton, H.C., *et al.* (2001). Minimal information about a microarray experiment (MIAME) – towards standards for microarray data. Nat. Genet. *29*, 365–371.

Brodie, E.L., DeSantis, T.Z., Joyner, D.C., Baek, S.M., Larsen, J.T., Andersen, G.L., Hazen, T.C., Richardson, P.M., Harman, D.J., Tokunaga, T.K., *et al.* (2006). Application of a high-density oligonucleotide microarray approach to study bacterial population dynamics during uranium reduction and reoxidation. Appl. Environ. Microbiol. *72*, 6288–6298.

Caporaso, J.G., Lauber, C.L., Walters, W.A., Berg-Lyons, D., Huntley, J., Fierer, N., Owens, S.M., Betley, J., Fraser, L., Bauer, M., *et al.* (2012). Ultra-high-throughput microbial community analysis on the Illumina HiSeq and MiSeq platforms. ISME J. *6*, 1621–1624.

Cattony, E.B.M., Chinalia, F.A., Ribeiro, R., Zaiat, M., Foresti, E., and Varesche, M.B.A. (2005). Ethanol and toluene removal in a horizontal-flow anaerobic immobilized biomass reactor in the presence of sulfate. Biotechnol. Bioeng. *91*, 244–253.

Chiellini, C., Iannelli, R., and Petroni, G. (2013). Temporal characterization of bacterial communities in an phytoremediation pilot plant aimed at decontaminating polluted sediments dredged from Leghorn harbor, Italy. N. Biotechnol. *30*, 772–779.

Chinalia, F.A., Paton, G.I., and Killham, K.S. (2008). Physiological and toxicological characterization of an engineered whole-cell biosensor. Bioresource Technol. *99*, 714–721.

Ciesielski, S., Bulkowska, K., Dabrowska, D., Kaczmarczyk, D., Kowal, P., and Mozejko, J. (2013). Ribosomal intergenic spacer analysis as a tool for monitoring methanogenic Archaea changes in an anaerobic digester. Curr. Microbiol. *67*, 240–248.

Dalahmeh, S.S., Jonsson, H., Hylander, L.D., Hui, N., Yu, D., and Pell, M. (2014). Dynamics and functions of bacterial communities in bark, charcoal and sand filters treating greywater. Water Res. *54*: 21–32.

De la Iglesia, R., Valenzuela-Heredia, D., Pavissich, J.P., Freyhoffer, S., Andrade, S., Correa, J.A., and Gonzalez, B. (2010). Novel polymerase chain reaction primers for the specific detection of bacterial copper P-type ATPases gene sequences in environmental isolates and metagenomic DNA. Lett. Appl. Microbiol. *50*, 552–562.

Dopheide, A., Lear, G., Stott, R., and Lewis, G.D. (2009). Relative diversity and community structure of ciliates in stream biofilms according to molecular and microscopy methods. Appl. Environ. Microbiol. *75*, 5261–5272.

Dopheide, A., Lear, G., He, Z., Zhou, J., and Lewis, G.D. (2015). Functional gene composition and redundancy in diverse stream biofilm communities. PLoS One, 10, e0123179.

Dorn, J.G., Brusseau, M.L., and Maier, R.M. (2005). Real-time, *in situ* monitoring of bioactive zone dynamics in heterogeneous systems. Environ. Sci. Technol. *39*, 8898–8905.

Dumont, M.G., Radajewski, S.M., Miguez, C.B., McDonald I.R., and Murrell, J.C. (2006). Identification of a complete methane monooxygenase operon from soil by combining stable isotope probing and metagenomic analysis. Environ. Microbiol. Rep. *8*: 1240–1250.

Gomez-Alvarez, V., Schrantz, K.A., Pressman, J.G., Speitel, G.E., and Wahman, D.G. (2013). Pyrosequencing Analysis of Bench-Scale Nitrifying Biofilters Removing Trihalomethanes. Environ. Eng. Sci. *30*, 582–588.

Gomez-Alvarez, V., Schrantz, K.A., Pressman, J.G., and Wahman, D.G. (2014). Biofilm community dynamics in bench-scale annular reactors simulating arrestmatn of chloraminated drinking water nitrification. Environ. Sci. Technol. *48*, 5448–5457.

Handelsman, J., Rondon, M.R., Brady, S.F., Clardy, J., and Goodman, R.M. (1998). Molecular biological access to the chemistry of unknown soil microbes: a new frontier for natural products. Chem. Biol. *5*, R245–249.

Handley, K.M., Wrighton, K.C., Piceno, Y.M., Anderson, G.L., deSantis, T.Z., Williams, K.H., Wilkins, M.J., Lucie N'Guessen, A., Peacock, A., Bargar, J., *et al.* (2012). High-density Phylochip profiling of stimulated aquifer microbial communities reveals a complex response to acetate amendment. FEMS Microbiol. Ecol. *81*, 188–204.

Hay, A.G., Rice, J.F., Applegate, B.M., Bright, N.G., and Sayler, G.S. (2000). A bioluminescent whole-cell reporter for detection of 2,4-dichlorophenoxyacetic acid and 2,4-dichlorophenol in soil. Appl. Environ. Microbiol. *66*, 4589–4594.

He, Z., Deng, Y., Van Nostrand, J.D., Tu, Q., Xu, M., Hemme, C.L., Li, X., Wu, L., Gentry, T.J., Yin, Y., *et al.* (2010). GeoChip 3.0 as a high-throughput tool for analyzing microbial community composition, structure and functional activity. ISME J. *4*, 1167–1179.

He, Z.L., Gentry, T.J., Schadt, C.W., Wu, L.Y., Liebach, J., Chong, S.C., Huang, Z.J., Wu, W.M., Gu, B.H., Jardine, P., *et al.* (2007). GeoChip: a comprehensive microarray for investigating biogeochemical, ecological and environmental processes. ISME J. *1*, 67–77.

Hery, M., Gault, A.G., Rowland, H.A.L., Lear, G., Polya, D.A., and Lloyd, J.R. (2008). Molecular and cultivation-dependent analysis of metal-reducing bacteria implicated in arsenic mobilisation in south-east Asian aquifers. Appl. Geochem. *23*, 3215–3223.

Hogardt, K., Trebesius, A.M., Geiger, M., Hornef, J., Rosemnecker, J., and Heesemann, J. (2000). Specific and rapid detection by fluorescent *in situ* hybridization of bacteria in clinical samples obtained from cystic fibrosis patients. J. Clin. Microbiol. *38*, 818–825.

Huang, L.N., Tang, F.Z., Song, Y.S., Wan, C.Y., Wang, S.L., Liu, W.Q., and Shu, W.S. (2011). Biodiversity, abundance, and activity of nitrogen-fixing bacteria during primary succession on a copper mine tailings. FEMS Microbiol. Ecol. *78*, 439–450.

Jechalke, S., Vogt, C., Reiche, N., Franchini, A.G., Borsdorf, H., Neu, T.R., and Richnow, H.H. (2010). Aerated treatment pond technology with biofilm promoting mats for the bioremediation of benzene, MTBE and ammonium contaminated groundwater. Water Res. *44*, 1785–1796.

Jouanneau, S., Durand, M.J., and Thouand, G. (2012). Oline detection of metals in environmental samples: Comparing two concepts of bioluminescent bacterial biosensors. Environ. Sci. Technol. *46*, 11979–11987.

Krohn-Molt, I., Wemheuer, B., Alawi, M., Poehlein, A., Güllert, S., Schmeisser, C., Pommerening-Röser, A., Grundhoff, A., Daniel, R., Hanelt, D., *et al.* (2013). Metagenome survey of a multispecies and alga-associated biofilm revealed key elements of bacterial-algal interactions in photobioreactors. Appl. Environ. Microbiol. *79*, 6196–6206.

Lawrence, J.R., Chenier, M.R., Roy, R., Beaumier, D., Fortin, N., Swerhone, G.D.W., Neu, T.R., and Greer, C.W. (2004). Microscale and molecular assessment of impacts of nickel, nutrients, and oxygen level on structure and function of river biofilm communities. Appl. Environ. Microbiol. *70*, 4326–4339.

Lear, G., and Lewis, G.D. (2009). Impact of catchment land use on bacterial communities within stream biofilms. Ecol. Indicators *9*, 848–855.

Lear, G., Harbottle, M.J., van der Gast, C.J., Jackman, S.A., Knowles, C.J., Sills, G., and Thompson, I.P. (2004). The effect of electrokinetics on soil microbial communities. Soil Biol. Biochem. *36*, 1751–1760.

Lear, G., Polya, D.A., Song, B., Gault, A., and Lloyd, J. (2007). Molecular analysis of arsenate-reducing bacteria within Cambodian sediments following amendment with acetate. Appl. Environ. Microbiol. *73*, 1041–1048.

Lear, G., Anderson, M.J., Smith, J.P., Boxen, K., and Lewis, G.D. (2008). Spatial and temporal heterogeneity of the bacterial communities in stream epilithic biofilms. FEMS Microbiol. Ecol. *65*, 463–473.

Lear, G., Niyogi, D., Harding, J., Dong, Y., and Lewis, G.D. (2009a). Biofilm bacterial community structure in streams affected by acid mine drainage. Appl. Environ. Microbiol. *75*, 3455–3460.

Lear, G., Turner, S.J., and Lewis, G.D. (2009b). Effect of light regimes on the utilisation of an exogenous carbon source by freshwater biofilm bacterial communities. Aquatic Ecol. *43*, 207–220.

Lear, G., McBeth, J.M., Boothman, C., Gunning, D.J., Ellis, B.L., Lawson, R.S., Morris, K., Burke, I.T., Bryan, N.D., Brown, N.D., *et al.* (2010). Probing the biogeochemical behaviour of technetium using a novel nuclear imaging approach. Environ. Sci. Technol. *44*, 156–162.

Lear, G., Ancion, P.-Y., Harding, J., and Lewis, G.D. (2012a). Using stream biofilm microbial communities as indicators of freshwater ecosystem health. N. Zealand J. Marine Freshwater Res. *43*, 291–301.

Lear, G., Dopheide, A., Ancion, P.-Y., Roberts, K., Washington, V., Smith, J., and Lewis, G.D. (2012b). Biofilms in freshwater. In Microbial Biofilms, Current Research and Applications, G. Lear, and G.D. Lewis, eds. (Caister Academic Press, Norfolk, UK), pp. 129–151.

Lear, G., Washington, V., Neale, M.W., Case, B., Buckley, H.L., and Lewis, G.D. (2013). The biogeography of stream bacteria. Global Ecol. Biogeography *22*, 544–554.

Leigh, M.B., Pellizari, V.H., Uhlik, O., Sutka, R., Rodrigues, J., Ostrom, N.E., Zhou, J., and Tiedje, J.M. (2007). Biphenyl-utilizing bacteria and their functional genes in a pine root zone contaminated with polychlorinated biphenyls (PCBs). ISME J. *1*, 134–148.

Liang, B., Cheng, H., Van Nostrand, J.D., Ma, J., Yu, H., Kong, D., Liu, W., Ren, N., Wu, L., Wang, A., *et al.* (2014). Microbial community structure and function of Nitrobenzene reduction biocathode in response to carbon source switchover. Water Res. *54*, 137–148.

Manefield, M., Whiteley, A.S., Griffiths, R.I., and Bailey, M.J. (2002). RNA stable isotope probing, a novel means of linking microbial community function to phylogeny. Appl. Environ. Microbiol. *68*, 5367–5373.

Mariner, R., Johnson, D.B., and Hallberg, K.B. (2008). Characterisation of an attenuation system for the remediation of MN(II) contaminated waters. Hydrometallurgy *94*, 100–104.

Mason, O.U., Hazen, T.C., Borglin, S., Chain, P.S., Dubinsky, E.A., Fortney, J.L., Han, J., Holman, H.-Y.N., Hultman, J., and Lamendella, R. (2012). Metagenome, metatranscriptome and single-cell sequencing reveal microbial response to Deepwater Horizon oil spill. ISME J. *6*, 1715–1727.

Mason, O.U., Scott, N.M., Gonzalez, A., Robbins-Pianka, A., Baelum, J., Kimbrel, J., Bouskill, N.J., Prestat, E., Borglin, S.E., Joyner, D.C., *et al.* (2014). Metagenomics reveals sediment microbial community response to Deepwater Horizon oil spill. ISME J. *8*, 1463–1475.

Monard, C., Martin-Laurent, F., Lima, O., Devers-Lamrani, M., and Binet, F. (2012). Estimating the biodegradation of pesticide in soils by monitoring pesticide-degrading gene expression. Biodegradation *24*, 203–213.

Montagna, P.A., Baguley, J.G., Cooksey, C., Hartwell, I., Hyde, L.J., Kalke, R.D., Kracker, L.M., Reuscher, M., and Rhodes, A.C.E. (2013). Deep-sea benthic footprint of the Deepwater Horizon blowout. PLoS One *8*, e70540.

Naik, D., Smith, E., and Coumming, J.R. (2009). Rhizosphere carbon deposition, oxidative stress and nutritional changes in two poplar species exposed to aluminium. Tree Physiol. *29*, 423–436.

Neufeld, J.D., Wagner, M., and Murrell, J.C. (2007). Who eats what, where and when? Isotope-labelling experimetns are coming of age. ISME J. *1*, 103–110.

Nielsen, A.T., Tolker-Nielsen, T., Barken, K.B., and Molin, S. (2000). Role of commensal relationships on the spatial structure of a surface-attached microbial consortium. Environ. Microbiol. *2*, 59–68.

Radajewski, S., Ineson, P., Parekh, N.R., and Murrell, J.C. (2000). Stable isotope probing as a tool in microbial ecology. Nature *403*, 646–649.

Ranjard, L., Poly, F., Lata, J.C., Mougal, C., Thioulouse, J., and Nazaret, S. (2001). Characterization of bacterial and fungal soil communities by automated ribosomal intergenic spacer analysis fingerprints: biological and methodological variability. Appl. Environ. Microbiol. *67*, 4479–4487.

Ranjard, L., Dequiedt, S., Prevost-Boure, N.C., Thioulouse, J., Saby, N.P.A., Lelievre, M., Maron, P.A., Morin, F.E.R., Bispo, A., Jolivet, C., *et al.* (2013). Turnover of soil bacterial diversity driven by wide-sclae environmental heterogeneity. Nat. Commun. *4*, 1434.

Redmond, M.C., and Valentine, D.L. (2012). Natural gas and temperature structured a microbial community response to the Deepwater Horizon oil spill. Proc. Natl. Acad. Sci. U.S.A. *109*, 20292–20297.

Regonne, R.K., Martoin, F., Mbawala, A., Ngassoum, M.B., and Jouanneau, Y. (2013). Identification of soil bacteria able to degrade phenanthrene bound to a hydrophobic sorbent *in situ*. Environ. Pollut. *180*, 145–151.

Rieder, S.R., and Frey, B. (2013). Methyl-mercury affects microbial activity and biomass, bacterial community structure but rarely the fungal community structure. Soil Biol. Biochem. *64*, 164–173.

Saha, R., Donofrio, R.S., Goeres, D.M., and Bagley, S.T. (2012). Rapid detection of rRNA group I pseudomponads in contaminated metalworking fluids and biofilm formation by fluorescent *in situ* hybridization. Appl. Microbiol. Biotechnol. *94*, 799–808.

Schrope, M. (2013). Dirty blizzard buried Deepwater Horizon oil. Nat. News. doi:10.1038/nature.2013.12304.

Shah, D., Zhang, Z., Khodursky, A.B., Kaldalu, N., Kurg, K., and Lewis, K. (2006). Persisters: a distinct physiological state of *E. coli*. BMC Microbiol. *6*, 53.

Shendure, J., and Lieberman Aiden, E. (2012). The expanding scope of DNA sequencing. Nature *30*, 1084–1094.

Shimomura, O., Johnson, F.H., and Saiga, Y. (1962). Extraction, purification and properties of aequorin, a bioluminescent protein from the luminous hydromedusan, *Aequorea*. J. Cell. Comparative Physiol. *59*, 223–239.

Shingler, V., and Moore, T. (1994). Sensing of aromatic compounds by the DmpR transcriptional activator of phenol-catabolizing *Pseudomonas* sp. strain CF600. J. Bacteriol. *176*, 1555–1560.

Steward, G.F., Jenkins, B.D., Ward, B.B., and Zehr, J.P. (2004). Development and testing of a DNA microarray to assess nitrogenase (*nifH*) gene diversity. Appl. Environ. Microbiol. *70*, 1455–1465.

Sul, W.J., Park, J., Quensen, J.F. III, Rodrigues, J.L., Seliger, L., Tsio, T.V., Zylstra, G.J., and Tiedje, J.M. (2009). DNA-stable isotope probing integrated with metagenomics for retrieval of biphenyl dioxygenase genes from polychjlorinated biphenyl-contaminated river sediment. Appl. Environ. Microbiol. *75*, 5501–5506.

Sun, Y., Polishchuk, E.A., Radoja, U., and Cullen, W.R. (2004). Identification and quantification of arsC genes in environmental samples by using real-time PCR. J. Microbiol. Methods *58*, 335–349.

Tecon, R., Wells, M., and van der Meer, J.R. (2006). A new green fluorescent protein-based bacterial biosensor for analysing phenanthrene fluxes. Environ. Microbiol. *8*, 697–708.

Tischer, K., Kleinsteuber, S., Schleinitz, K.M., Fetzer, I., Soptt, O., Strange, F., Lohse, U., Franz, J., Neumann, F., Gerling, S., *et al.* (2013). Microbial communities along biogeochemical gradients in a hydrocarbon-contaminated aquifer. Environ. Microbiol. *15*, 2603–2615.

Tu, Q., Yu, H., He, Z., Deng, Y., Wu, L., Van Nostrand, J.D., Zhou, A., Voordeckers, J., Lee, Y.-J., Qin, Y., *et al.* (2014). GeoChip 4: a functional gene-array-based high-throughput environmental technology for microbial community analysis. Mol. Ecol. Res. *14*, 914–928.

Turpeinen, R., Kairesalo, T., and Haggblom, M.M. (2004). Microbial community structure and activity in arsenic-, chromium- and copper-contaminated soils. FEMS Microbiol. Ecol. *47*, 39–50.

Tyson, G.W., Chapman, J., Hugenholtz, P., Allen, E.E., Ram, R.J., Richardson, P.M., Solovyev, V.V., Rubin, E.M., Rokhsar, D.S., and Banfield, J.F. (2004). Community structure and metabolism through reconstruction of microbial genomes from the environment. Nature *428*, 37–43.

Uhlik, O., Leewis, M.-C., Strejcek, M., Musilova, L., Mackova, M., Leigh, M.B., and Macek, T. (2013). Stable isotope probing in the metagenomics era: a bridge towards improved bioremediation. Biotechnol. Adv. *31*, 154–165.

Valentine, D.L., Kessler, J.D., Redmond, M.C., Mendes, S.D., Heintz, M.B., Farwell, C., Hu, L., Kinnaman, F.S., Yvon-Lewis S., *et al.* (2010). Propane respiration jump-starts microbial response to a deep oil spill. Science *330*, 208–211.

van der Zaan, B.M., Saia, F.T., Stams, A.J.M., Plugge, C.M., de Vos, W.M., Smidt, H., Langenhoff, A.A.M., and Gerritse,

J. (2012). Anaerobic benzene degradation under denitrifying conditions: *Peptococcaceae* as dominant benzene degraders and evidence for a syntrophic process. Environ. Microbiol. *14*, 1171–1181.

Washington, V., Lear, G., Neale, M., and Lewis, G.D. (2013). Environmental effects on biofilm bacterial communities: a comparison of natural and anthropogenic factors in New Zealand streams. Freshwater Biol. *58*, 2277–2286.

Whiteley, A.S., Griffiths, R.I., and Bailey, M.J. (2003). Analysis of the microbial functional diversity within water-stressed soil communities by flow cytometric analysis and CTC+ cell sorting. J. Microbiol. Methods *54*, 257–267.

Wilmes, P., Remis, J.P., Hwang, M., Auer, M.W., Thelen, M.P., and Banfield, J.F. (2009). Natural acidic biofilm communities reflect organismal and functional organisation. ISME J. *3*, 266–270.

Woods, A., Watwood, M., and Schwartz, E. (2011). Identification of a tolune-degrading bacterium from a soil sample through $H_2{}^{18}O$ stable isotope probing. Appl. Environ. Microbiol. *77*, 5995–5999.

Wouters, K., Maes, E., Spitz, J.-A., Roeffaers, M.B.J., Wattiau, P., Hofkens, J., and Springael, D. (2010). A non-invasive fluorescent staining procedure allows Confocal Laser Scanning Microscopy based imaging of *Mycobacterium* in multispecies biofilms colonizing and degrading polycyclic aromatic hydrocarbons. J. Microbiol. Methods *83*, 317–325.

Yamamura, T., Schwendenmann, L., and Lear, G. (2013). Tree species identity has little impact on the structure of soil bacterial communities in a 10-yr old tropical tree plantation. Biol. Fertility Soils *49*, 819–828.

Yao, H., Gao, Y., Nicol, G.W., Campbell, C.D., Prosser, J.I., Zhang, L., Han, W., and Singh, B.K. (2011). Links between ammonia oxidizer community structure, abundance, and nitrification potential in acidic soils. Appl. Environ. Microbiol. *77*, 4618–4625.

Yarwood, J.M., Bartels, D.J., Volper, E.M., and Greenberg, E.P. (2004). Quorum sensing in *Staphylococcus aureus* biofilms. J. Bacteriol. *186*, 1838–1850.

Yergeau, E., Lawrence, J.R., Waiser, M.J., Korber, D.R., and Greer, C.W. (2010). Metatranscriptomic analysis of the reponse of river biofilms to pharmaceutical products using anonymous DNA microarrays. Appl. Environ. Microbiol. *76*, 5432–5439.

Yergeau, E., Lawrence, J.R., Sanschagrin, S., Waiser, M.J., Korber, D.R., and Greer, C.W. (2012). Next-generation sequencing of microbial communties in the Athabasca River and its tributaries in relation to oil sands mining activities. Appl. Environ. Microbiol. *78*, 7626–7637.

Yergeau, E., Lawrence, J.R., Sanschagrin, S., Roy, J.L., Swerhone, G.D.W., Korber, D.R., and Greer, C.W. (2013). Aerobic biofilms grown from Athabasca watershed sediments are inhibited by increasing concentrations of bituminous compounds. Appl. Environ. Microbiol. *79*, 7398–7412.

Yergeau, E., Sanschagrin, A., Maynard, C., St-Arnaud, M., and Greer, C.W. (2014). Microbial expression profiles in the rhizosphere of willows depend on soil contamination. ISME J. *8*, 344–358.

Zukunft, P. (2010). Summary report for sub-sea and sub-surface oil and disperion detection: sampling and monitoring. Available online at: http://www.restorethegulf.gov/sites/default/files/documents/pdf/OSAT_Report_FINAL_17DEC.pdf.

# Part III

# Case Studies

# Biofilm-mediated Degradation of Polycyclic Aromatic Hydrocarbons and Pesticides

8

Marta Pazos, Laura Ferreira, Emilio Rosales and
Maria Ángeles Sanromán

## Abstract

Environmental pollution caused by emerging organic compounds, such as polycyclic aromatic hydrocarbons (PAHs) and pesticides has attracted considerable attention in recent years. PAHs constitute a class of organic substances consisting of two or more fused benzene rings and exhibiting recalcitrance and strong mutagenic/carcinogenic properties. On the other hand, pesticides are used for controlling, repelling, preventing or eradicating pests, and include not only a wide range of chemical compounds but also antimicrobial or disinfectant agents. Their composition often poses a threat for humans and the environment. As a result of the complex structure of both types of pollutants and their recalcitrant and persistent properties, the remediation of these compounds by conventional methods is often difficult. To solve this problem alternative treatment processes have been studied. Among them, biological treatment by biofilms has been found to be suitable for their remediation because of their high microbial biomass and ability to immobilize pollutants.

In this chapter a condensed overview on the role of biofilms for the remediation of PAHs and pesticides in the environment is presented and several studies in which different biofilms or immobilized microorganisms have been used for PAH biodegradation are summarized.

## Introduction

In the few last years, it has been observed that more and more pollutants, especially PAHs and pesticides, are released into the environment. The remediation of sites contaminated with these pollutants is an important environmental issue, particularly where various ecological factors prolong their presence in the environment. Critically, properties such as low solubility and slow desorption rates make pollutants including PAHs and various pesticides difficult to remove. Therefore, it is necessary to know the characteristics of these pollutants in order to design the most appropriate strategies for their remediation.

### Polycyclic aromatic hydrocarbons (PAHs)

Polycyclic aromatic hydrocarbons (PAHs) are ubiquitous environmental contaminants formed mainly during the incomplete combustion or pyrolysis of fossil fuels and biofuels. Total combustion of hydrocarbons in the presence of oxygen results in the complete oxidation of carbon and hydrogen present to carbon dioxide and water. However, the reaction may not always be this simple depending on the prevailing environmental conditions (Connell *et al.*, 1997). At temperatures above 500°C, some of the C–C and C–H bonds are broken giving rise to free radicals. Depending on the amount of oxygen present, many of these radicals can react with oxygen to form carbon dioxide and water vapour. However, sometimes the presence of oxygen is limited and therefore the formation of carbon dioxide and water vapour is reduced because there is insufficient oxygen to react with all free radicals arising. These radicals can then react together to form new molecules or larger radicals. In turn, these new molecules and radicals may react with each other, resulting in more complex structures, which may result in the formation of a variety of PAHs. The type and amount of PAHs formed depend on the conditions existing at the time of its formation and the temperature and type of burned organic matter. So, for example, the decomposition of coal, cellulose, snuff, polyethylene and polyvinyl chloride at 1000°C may produce PAHs, but the proportion with alkyl substituents decreases with increasing combustion temperature (Björseth and Ramdahl, 1985). The first study of this mechanism of PAH production was provided by Badger (1962). In Fig. 8.1, a possible mechanism of PAH formation is shown during incomplete combustion.

PAHs may also be created during the geological formation of fossil fuels such as oil. This mechanism is not as significant as the previous one, but it is important enough to be worthy of discussion. In this case, the temperatures are much lower than those of combustion (less than 200°C) and the breakdown of the molecules is favoured by the high pressures to which they are subjected. Under these conditions, PAHs may be created by mechanisms similar to those that occur during combustion, but the process is slower and the PAHs produced are of different chemical form. A higher proportion of PAHs with alkyl substituents, favoured by the relatively low temperatures at which the process occurs, is obtained. A third way of forming small minority PAHs, via microbial synthesis, remains highly debated (Harvey, 1997).

The US Environmental Protection Agency (EPA) created a list of PAHs most frequently found in environmental samples (Table 8.1), they are known as the 16 EPA–PAHs (US EPA, 1984) Eight of these PAHs are known to give rise to serious health concerns (Zedeck, 1980) and are often used as a marker for PAH contamination in general, although contributing only 1–20% of the total carcinogenicity found in samples from the environment (Šimko, 2002).

PAHs are widely distributed in the environment. PAHs emitted into the atmosphere by natural processes or industrial sources are largely adsorbed on particulate matter in the atmosphere. These adsorbed PAHs can be deposited into aquatic and soil environments by fallout, subsequently leading to water and soil pollution. In a soil environment, the behaviour of PAHs is closely related to their distribution and transportation between soil components such as clay minerals and organic matter (Harms, 2011; Pazos *et al.*, 2010). Toxic PAHs are persistent since they are strongly adsorbed on soils, sludges or sediments because of their strong hydrophobicity which makes them less bioavailable, while simultaneously limiting conventional remediation measures. These PAHs can be removed from the environment through adsorption, volatilization, photolysis, and chemical degradation (Alcántara *et al.*, 2008, 2010, 2012; Gómez *et al.*, 2010; Pazos *et al.*, 2013). However, microbial degradation is considered as one of the major degradation pathways (Cobas *et al.*, 2013;

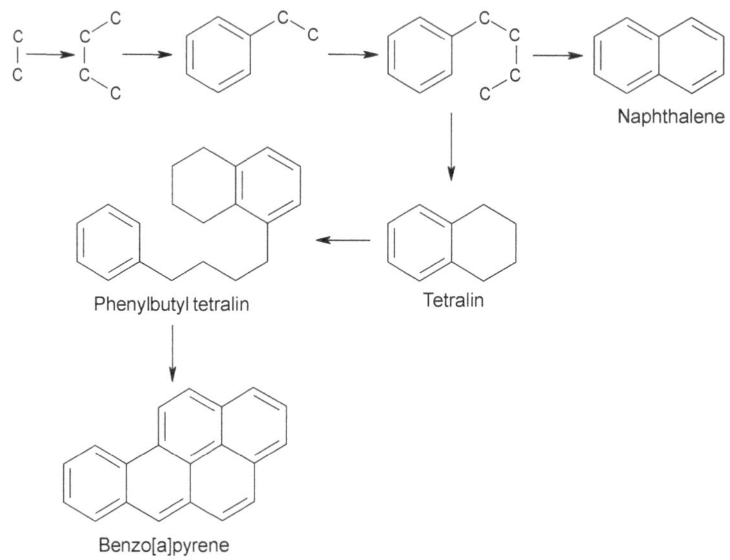


**Figure 8.1** Proposed mechanism of PAH formation of during incomplete combustion.

Ferreira *et al.*, 2013; Harayama, 1997; Haritash and Kaushik, 2009; Rosales *et al.*, 2012; Wilson and Jones, 1993; Yuan *et al.*, 2001).

PAHs are biodegraded/biotransformed into less complex metabolites, and through mineralization into inorganic minerals. The rate of biodegradation depends on pH, temperature, oxygen, microbial population, degree of acclimation, accessibility of nutrients, chemical structure of the compound, cellular transport properties, and chemical partitioning in the growth medium (Haritash and Kaushik, 2009). A number of microbial species are known to degrade PAHs and most of them are isolated from contaminated soil or sediments (Table 8.1). Among them, PAH-degrading bacteria such as *Pseudomonas aeruginosa, Pseudomonas fluorescens, Mycobacterium* spp., *Haemophilus* spp., *Rhodococcus* spp., *Paenibacillus* spp., *Trichoderma longibrachiatum* and *Arthrobacter viscosus*, are commonly studied. In addition, these bacteria can produce biosurfactants and light oils that may increase the bioavailability of PAHs and the metabolic potential of the bacterial community (Cébron *et al.*, 2008; Cobas *et al.*, 2013; Ferreira *et al.*, 2013; Haderlein *et al.*, 2006; Lu *et al.*, 2014; Rosales *et al.*, 2012). On the other hand, lignolytic fungi (e.g. *Phanerochaete chrysosporium, Bjerkandera adusta, Pleurotus ostreatus, Coriolopsis rigida*) can degrade PAHs due to the activity of the lignolytic enzymes such as lignin peroxidase, laccase, and manganese peroxidase. These enzymes catalyse radical formation by oxidation to destabilize bonds in a molecule (Gómez *et al.*, 2006; Qasemian *et al.*, 2011; Valentín *et al.*, 2010; Winquist *et al.*, 2014). In addition, PAH-degrading bacteria are fully capable of utilizing the infrastructure of the fungi moving across their surface in a process known as 'fungal braiding' to travel and propagate (Banitz *et al.*, 2013; Furuno *et al.*, 2012). The coexistence of bacteria and fungi in media and their known catabolic cooperation suggest that physical interactions between them may be important for contaminant degradation (Furuno *et al.*, 2012; Lladó *et al.*, 2012; Šnajdr *et al.*, 2011).

In this chapter, the ability of the fungus *Trametes versicolor* and the bacterium *Pseudomonas stutzeri* to degrade PAHs is later reported.

## Pesticides

The first pesticide synthesized was dichlorodiphenyl trichloroethane (DDT) in 1874. Since 1939, when Paul Müller discovered its insecticides properties, thousands of similar products have been developed, commercialized and used all over the world (Hellou *et al.*, 2013). Today, over 900 active chemical pesticides are used to manufacture more than 40,000 commercial preparations and pesticide production and application have progressively increased worldwide during recent decades. Indeed, the US-EPA estimates that pesticide use doubled between 1960 and 1980 (Frazar, 2000).

A degree of persistence is required for pesticides in order for them to remain effective in controlling, repelling, preventing or eradicating pests. To obtain these effects, pesticides frequently include not only a wide range of chemical compounds but also antimicrobial or disinfectant agents. The term pesticide is often misunderstood to refer only to insecticides; however, it also applies to herbicides, fungicides, and various other antibiotic substances. A large amount of applied pesticides often do not reach their intended target due to volatilization and leaching, which may result in serious environmental problems.

The human and environmental effects of pesticides depend on the type of pesticide in question. For example, DeLorenzo *et al.* (1999) determined that chlorpyrifos concentrations between 1 and 10 mg/l depress the growth of the blue-green algae *Anabaena* sp. and reduce numbers of ciliated protozoa in natural microbial communities. Human health may be threatened by pesticides in water at low-to-medium concentration (< 500 mg/l) (Malato *et al.*, 2002). Grouped according to their active chemical groups, the main categories of pesticides are as follows: organochlorines, organophosphates, organosulfurs, carbamates, formamidines, dinitrophenols, pyrethroids, nicotinoids, spiosinas, phenylpyrazoles, pyrroles and pyrazoles, among others. Some of them, such as the organophosphates and carbamates, affect the nervous system. Others may irritate the skin or eyes. Some pesticides may be carcinogens and others affect the human endocrine system (US EPA, 2014). Recently, public health concerns have been heightened by increased knowledge of the high toxicity and the associated environmental hazards caused by the occurrence of pesticides in water and wastewater (Eriksson *et al.*, 2007). Even at low concentration, these compounds may be a major health concern, for example due to their extremely

**Table 8.1** Properties and chemical structures of 16 priority PAHs and selection of research citations investigating their biodegradation

| Name | CAS Number | Chemical structure | Molecular weight (g/mol) | Water solubility (mg/l) | Microorganism | Degradation (%) | Reference |
|---|---|---|---|---|---|---|---|
| Acenaphthene | 82-32-9 | 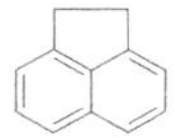 | 154.21 | 1.93 | *Penicillium frequentans* | 100 | Amezcua-Allieri *et al*. (2012) |
| | | | | | *Acinetobacter* sp. strain AGAT-W | 99 | Ghosal *et al*. (2013) |
| Acenaphthylene | 208-96-8 | 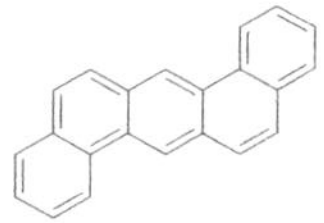 | 152.2 | 3.93 | *Stenotrophomonas* sp. RMSK | 80 | Nayak *et al*. (2009) |
| | | | | | *Rhizobium* sp. strain CU-A1 | 100 | Poonthrigpun *et al*. (2006) |
| Anthracene | 120-12-7 | 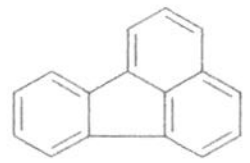 | 178.2 | 0.0076 | *Coriolopsis rigida* | 89 | Gómez *et al*. (2006) |
| | | | | | *Janibacter anophelis* strain JY11 | 80 | Zhang *et al*. (2009) |
| | | | | | *Sphingomonas* sp. VKM B-2434 | 90 | Baboshin *et al*. (2008) |
| Benz(a)anthracene | 56-55-3 | 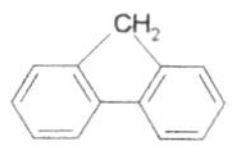 | 228.29 | 0.010 | *Trichoderma longibrachiatum* | 97.34 | Rosales *et al*. (2012) |
| | | | | | *Byssochlamys spectabilis* | 96.33 | Rosales *et al*. (2012) |
| | | | | | *Arthorbacter viscosus* | 100 | Ferreira *et al*. (2013) |
| Benzo(a)pyrene | 50-32-8 | 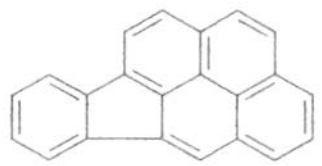 | 252.3 | 0.0023 | *Coriolopsis rigida* | 89 | Gómez *et al*. (2006) |
| | | | | | *Sphingomonas* sp. VKM B-2434 | 55 | Baboshin *et al*. (2008) |
| | | | | | *Arthrobacter oxydans* B4 | 68 | Peng *et al*. (2012) |
| Benzo(b) fluoranthene | 205-99-2 | 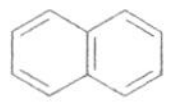 | 252.3 | 0.0012 | *Sphingomonas paucimobilis* EPA505 | 15 | Ye *et al*. (1996) |
| Benzo(ghi)perylene | 191-24-2 | 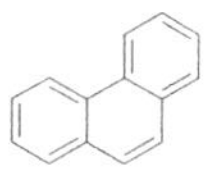 | 276.34 | 0.00026 | PAH degrading bacteria | 81 | Xia *et al*. (2006) |
| Benzo(k) fluoranthene[a] | 207-08-9 | 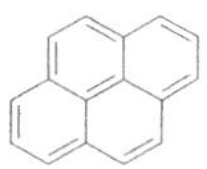 | 252.31 | 0.00055 | *Ochrobactrum* sp. VA1 | 57 | Arulazhagan and Vasudevan (2011) |
| Chrysene | 218-01-9 | 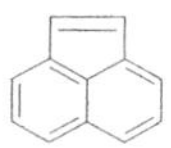 | 228.3 | 0.0028 | *Penicillium frequentans* | 85 | Amezcua-Allieri *et al*. (2012) |
| | | | | | *Sphingomonas* sp. VKM B-2434 | 45 | Baboshin *et al*. (2008) |
| | | | | | *Sphingomonas paucimobilis* EPA505 | 30 | Ye *et al*. (1996) |

| | | | | | | | |
|---|---|---|---|---|---|---|---|
| Dibenz(a,h) anthracene | 53-70-3 | 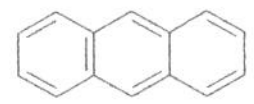 | 278.35 | 0.0005 | *Sphingomonas paucimobilis* EPA505 | 5 | Ye *et al*. (1996) |
| | | | | | *Stenotrophomonas maltophilia* VUN 10,003 | 30 | Juhasz *et al*. (2002) |
| | | | | | *Stenotrophomonas maltophilia* VUN 10,010 | 26.7 | Boonchan *et al*. (2000) |
| Fluoranthene | 206-44-0 | 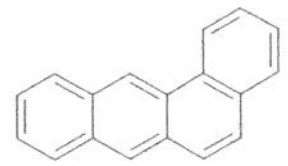 | 202.26 | 0.20–0.26 | *Penicillium frequentans* | 85 | Amezcua-Allieri *et al*. (2012) |
| | | | | | *Sphingomonas* sp. VKM B-2434 | 90 | Baboshin *et al*. (2008) |
| | | | | | *Pleurotus ostreatus* HP-1 | 54.09 | Patel *et al* 2009 |
| Fluorene | 86-73-7 | 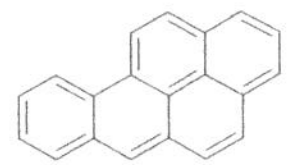 | 166.2 | 1.68–1.98 | *Penicillium frequentans* | 95 | Amezcua-Allieri *et al*. (2012) |
| | | | | | *Pseudomonas* sp. strain F274 | 100 | Grifoll *et al*. (1994) |
| Indeno(1,2,3-cd) pyrene | 193-39-5 | 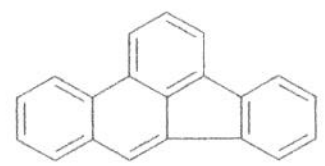 | 276.3 | 0.062 | *Pandoraea* sp. | 60 | Du *et al*. (2010) |
| | | | | | Isolated bacteria | 50.71 | Chen *et al*. (2010) |
| Naphthalene | 91-20-3 | 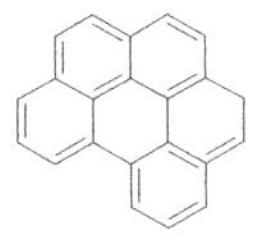 | 128.17 | 25.6341 | *Penicillium frequentans* | 92 | Amezcua-Allieri *et al*. (2012) |
| | | | | | *Marinobacter* NCE 312 | 100 | Hedlund *et al*. (2001) |
| Phenanthrene | 85-01-8 | 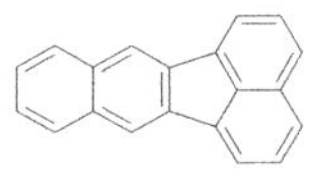 | 178.2 | 1.20 | *Trametes versicolor* | 80 | Rosales *et al*. (2013) |
| | | | | | *Pleurotus ostreatus* | 91 | Rosales *et al*. (2013) |
| | | | | | *Trichoderma longibrachiatum* | 63 | Cobas *et al*. (2013) |
| | | | | | *Arthrobacter viscosus* | 100 | Ferreira *et al*. (2013) |
| Pyrene | 129-00-0 | 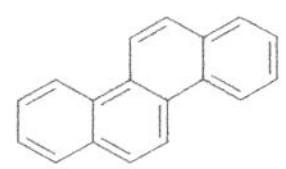 | 202.3 | 0.077 | Fusant bacterial strain F14 | 46.5 | Lu *et al*. (2014) |
| | | | | | *Trichoderma longibrachiatum* | 97.34 | Rosales *et al*. (2012) |
| | | | | | *Byssochlamys spectabilis* | 96.33 | Rosales *et al*. (2012) |
| | | | | | *Trametes versicolor* | 75 | Rosales *et al*. (2013) |
| | | | | | *Pleutorus ostreatus* | 95 | Rosales *et al*. (2013) |

high endocrine disruption potency and genotoxicity (Arques *et al.*, 2007).

At present, pesticide contamination originating from agricultural industries are contributing relatively high quantities of toxic pesticides into the environment. The most commonly used pesticides are the ones most likely to cause water pollution and this more effort is required to meet regulatory wastewater discharge standards for these compounds (Ginebreda *et al.*, 2014; López-Roldán *et al.*, 2013).

In general, pesticides are highly stable organic pollutants that may not be easily biodegradable in nature (Fulekar, 2010). According to the Stockholm Convention on Persistent Organic Pollutants, 9 of the 12 persistent organic chemicals are pesticides. Regulation No. 850/2004 given by the European Community (2004) declares: 'The persistent organic pollutant content in waste is to be destroyed or irreversibly transformed into substances that do not exhibit similar characteristics, unless other operations are environmentally preferable'. For this reason, treatment techniques which destroy the pollutants should always be the preferred option.

In several cases, intermediate degradation compounds are more toxic than the initial parent compounds. For example, Moscoso *et al.* (2013) investigated novel microbial agents such as *Pseudomonas stutzeri* and the previously obtained consortium LB2, with the ability to degrade chlorpyrifos from polluted effluents. They detected that cultures of consortium LB2 showed low degradation levels in parallel with obtaining very low biomass. This fact could be explained by an inhibitory effect due to the increased toxicity of an intermediate degradation compound such as TCP (3,5,6-trichloro-2-pyridinol) which was detected in the culture medium. These results are in agreement with the findings of Cáceres *et al.* (2007) and Pino and Peñuela (2011), who demonstrated an accumulation of TCP as being responsible for the high toxicity.

Pesticides can be decomposed in wastewaters and drinking waters by various treatment processes, such as physical treatments (lined evaporative beds and activated carbon adsorption), chemical treatments (photolysis, hydrolysis and chemical oxidation) and biological treatment (Bourgin *et al.*, 2013; Iglesias *et al.*, 2014; Ikehata and El Din, 2006; Moscoso *et al.*, 2013). To prevent water and soil pollution and to protect human health, appropriate soil and wastewater treatment systems that are easy to operate and suitable for use in diverse environmental conditions, are needed. Among several treatment technologies, microbial bioremediation technologies have been found to be very effective and economical in the treatment of hazardous waste (Jilani, 2013).

As mentioned by Singh and Walker (2006), several microorganisms have been isolated which are able to utilize pesticides as a source of energy. In Table 8.2, a selection of research on pesticide biodegradation by microorganisms is cited. There are some examples of fungi including *Trametes hirsutus*, *Phanerochaete chrysosporium*, *Phanerochaete sordida*, *Fusarium verticillioides*, *Ganoderma australe*, and *Cyathus bulleri* that are able to degrade lindane and other pesticides (Chen *et al.*, 2012; Dritsa *et al.*, 2009; Guillén-Jiménez *et al.*, 2012; Sagar and Singh, 2011; Singh *et al.*, 1999; Singh and Kuhad, 1999; Xiao, 2013). However, most evidence suggests that soil bacteria are the principal components responsible for enhanced biodegradation (Walker and Roberts, 1993). Several pure bacterial isolates such as *Pseudomonas aeruginosa*, *Bacillus cereus*, *Klebsiella* sp., *Serratia marscescens*, *Flavobacterium* sp., *Arthrobacter* sp., *Burkholderia* sp., *Mycobacterium* sp., with the ability to use specific pesticides as a sole source of carbon, nitrogen or phosphorus have been isolated and also mixed bacterial cultures (Ghanem *et al.*, 2007; Rauh *et al.*, 2012; Seo *et al.*, 2007; Singh *et al.*, 2003; Singh and Walker, 2006; Yang *et al.*, 2005). Therefore, bioremediation can offer an efficient and cheap option for the decontamination of polluted ecosystems and destruction of pesticide agents. For this reason, in this chapter, the biodegradation potential of pyrimethanil, by the bacterium *Pseudomonas stutzeri* is evaluated.

## PAH degradation by the immobilized white-rot fungus *Trametes versicolor*

Much research effort has been expended to find suitable methods for the remediation of terrestrial and aquatic environments contaminated with

PAHs. As earlier described, ligninolytic fungi are particularly suitable for the development of such processes, since they produce extracellular lignin-degrading enzymes such as lignin peroxidase, laccase, and manganese peroxidase which degrade a wide range of organic pollutants. Among them, laccase is of great interest due to its ability to catalyse the oxidation of both phenolic and non-phenolic lignin-related compounds (Bourbonnais and Paice, 1990) as well as highly recalcitrant environmental pollutants (Johannes *et al.*, 1996; Moldes *et al.*, 2003; Pointing, 2001; Rodríguez Couto *et al.*, 2006).

An efficient application of the PAH biodegradation requires the continuous performance of a system during which the aromatic ring is oxidized and then systematically broken down into more simple PAH metabolites and/or carbon dioxide (Bamforth and Singleton, 2005). Unfortunately, most reports that use white rot fungi are restricted to batch or semi-continuous operation. To overcome this disadvantage, fungal biofilms obtained from immobilization techniques are required.

The use of immobilized cells has potential utility in several bioprocesses and could provide additional advantages over freely suspended cells such as ease of regeneration, reuse of the biomass, easier liquid-solid separation and minimal clogging in continuous-flow systems (Domínguez *et al.*, 2007; Rodríguez Couto *et al.*, 2004a,b, 2006; Rosales *et al.*, 2007, 2012). There are several techniques in cell immobilization; however, in this study entrapment and attachment techniques were selected. In the entrapment, the organisms are caught in the interstices of porous material or are physically restrained within or on a solid or porous matrix such as a stabilized gel or membrane. During attachment, the microorganisms adhere to the surfaces or other organisms by self-adhesion or chemical bonding. Materials commonly employed for the attachment technique include synthetic materials like polyurethane foam (López *et al.*, 2010; Sanroman *et al.*, 1994, 1996) and nylon sponge (Cobas *et al.*, 2013; Moldes *et al.*, 2003; Rosales *et al.*, 2012).

The aim of the present study is to find a suitable technique for the immobilization of the white-rot fungus *T. versicolor* in order to develop an efficient laccase production system for the degradation of PAHs in wastewater.

## Materials and methods

### Microorganism

*T. versicolor* (CECT. 20148) was maintained at 4°C on Potato Dextrose Agar (PDA) and sub-cultured every 3 months.

### Immobilization methods

**Immobilization in alginate beads**

*T. versicolor* was entrapped in Ca-alginate polymer beads employing a concentration of alginate of 2.5%. The fungus sodium alginate solution was dropped through a syringe into a hardening solution composed of 0.25 M $CaCl_2$ to create spherical alginate beads. These particles were cured at 4°C for 2 h in the gelling solution, then filtered and washed repeatedly with distilled water. Finally, they were stored at 4°C in $CaCl_2$ solution. All solutions were previously sterilized at 121°C.

**Immobilization in polyurethane foam**

Cubes of polyurethane foam (characteristics: density, 20 kg/m$^3$; surface area, 414 ± 10 m$^2$/m$^3$) of size 0.5 cm$^3$ were used. They were pre-treated by washing once in methanol and twice in distilled water. Finally, the cubes were dried overnight at room temperature. The cubes of polyurethane foam were inoculated with three agar plugs (diameter, 3 mm), from an actively growing fungus cultured on PDA.

### Growth of the immobilized fungus in liquid media

Erlenmeyer flasks (250 ml) containing 100 ml of culture medium and 20 g of alginate beads or 1.8 g of polyurethane cubes were employed. The composition of the basal culture medium per litre was: 10 g glucose, 15 g yeast extract, 0.75 g $NH_4Cl$, 2 g $KH_2PO_4$, 0.5 g $MgSO_4{\cdot}7H_2O$, 0.1 g $CaCl_2{\cdot}2H_2O$, 0.5 g KCl, 0.2 g thiamine and 20 mM acetate buffer (pH 4.5). Basal medium was sterilized at 121°C for 20 min. After cooling, the vitamin thiamine, previously sterilized by microfiltration (0.22 μM), was added to the basal medium.

The Erlenmeyer flasks were capped with cotton stoppers, which permitted a passive aeration, and incubated in an orbital shaker at 150 rpm and 30°C. Samples were taken every 2 or 3 days, centrifuged (8000 g, 10 min) and analysed for laccase activity,

**Table 8.2** Properties and chemical structures of key pesticides and selection of research citations on their biodegradation

| Pesticide | CAS Number | Chemical Structure | Molecular weight | Water solubility (g/l) | Chemical class | Microorganism | Degradation (%) | Reference |
|---|---|---|---|---|---|---|---|---|
| Malathion | 121-75-5 | 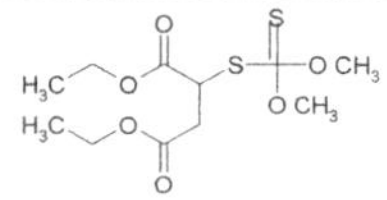 | 330.36 | 0.148 | Organophosphorus | *Brevibacillus* sp. KB2 | 36.22 | Singh *et al.* (2012) |
| | | | | | | *Bacillus cereus* PU | 49.31 | |
| | | | | | | *Pseudomonas xanthomarina* | 70.5 | Goyat *et al.* (2013) |
| Dimethoate | 60-51-5 | 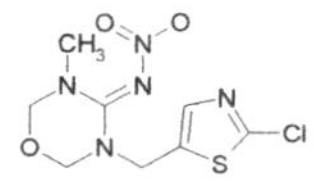 | 229.26 | 39.8 | Organophosphorus | *Paracoccus* sp. | 100 | Li *et al.* (2010) |
| | | | | | | *Fusarium* sp. | 100 | Zhao *et al.* (2009) |
| | | | | | | *Flavobacterium* sp. ATCC 27551 | 72.0 | Ortiz-Hernández *et al.* (2003) |
| Monocrotophos | 6923-22-4 | 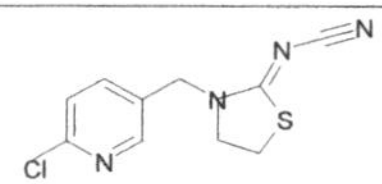 | 223.16 | 818 | Organophosphorus | *Aspergillus oryzae* | 100 | Bhalerao and Puranik (2009) |
| | | | | | | *Paracoccus* sp. | 92.47 | Jia *et al.* (2007) |
| Ethoprophos | 13194-48-4 | 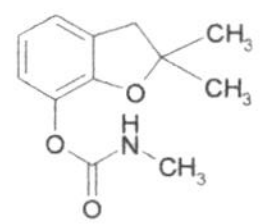 | 242.34 | 1.3 | Organophosphorus | *Flavobacterium* sp. | 100 | Karpouzas *et al.* (2005) |
| | | | | | | *Sphingomonas paucimobilis* | | |
| | | | | | | *Flavobacterium* sp. ATCC 27551 | 83.0 | Ortiz-Hernández *et al.* (2003) |
| | | | | | | *Cunninghamella echinulata* | 100 | Tillmanns *et al.* (1978) |
| Linuron | 330-55-2 | 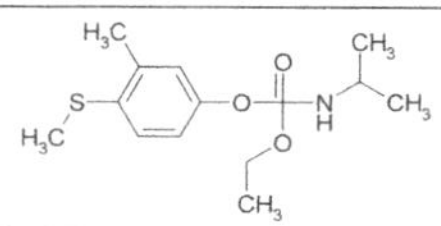 | 249.09 | 0.0638 | Phenylurea | *Cunninghamella echinulata* | 100 | Tillmanns *et al.* (1978) |
| | | | | | | *Variovorax* sp. | 100 | Breugelmans *et al.* (2010) |
| Carbendazim | 10605-21-7 | 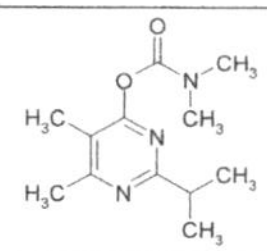 | 191.19 | 0.008 | Benzimidazole | *Rhodococcus* sp. | 83.8 | Xiao *et al.* (2013) |
| | | | | | | *Pseudomonas* sp. | | |
| | | | | | | *Sphingomonas* sp. | | |
| Cypermethrin | 52315-07-8 | 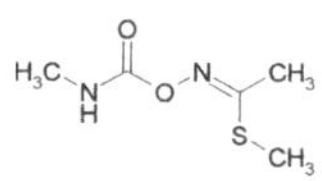 | 416.30 | 0.4 $10^{-5}$ | Pyrethroid | *Pseudomonas* IES-Ps-1 | 88 | Jilani (2013) |
| | | | | | | *Pseudomonas fluorescens* | 90 | Grant *et al.* (2002) |
| | | | | | | *Serratia plymuthica* | | |
| Prometryn | 7287-19-6 | 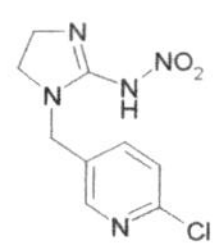 | 241.36 | 0.033 | Triazine | *Microbacterium* sp. | 100 | Pérez-Bárcena *et al.* (2013) |
| | | | | | | *Enterobacter* sp. | | |
| | | | | | | *Acinetobacter* sp. | | |
| | | | | | | *Flavobacterium* sp. | | |

| | | | | | | | | |
|---|---|---|---|---|---|---|---|---|
| Simazine | 122-34-9 | 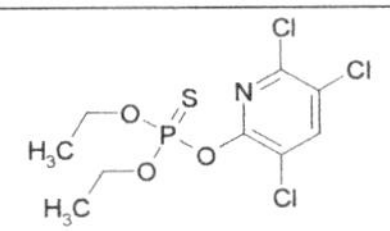 | 201.66 | 0.005 | Triazine | *Arthrobacter urefaciens* | 100 | Błaszak *et al*. (2011) |
| Thiamethoxam | 153719-23-4 | 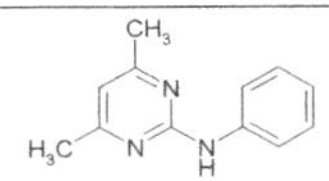 | 291.71 | 4.1 | Neonicotinoid | *Pseudomonas* sp. 1G. | 70 | Pandey *et al*. (2009) |
| | | | | | | *Ensifer adhaerens* TMX-23 | 27.2 | Zhou *et al*. (2013) |
| Thiacloprid | 111988-49-9 | 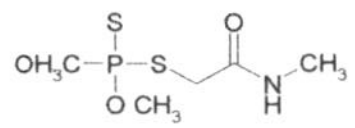 | 252.72 | 0.184 | Neonicotinoid | *Rhodotorula mucilaginosa* | 59.9 | Dai *et al*. (2010) |
| | | | | | | Soil microbial | 98.8 | Liu *et al*. (2011) |
| Carbofuran | 1563-66-2 | 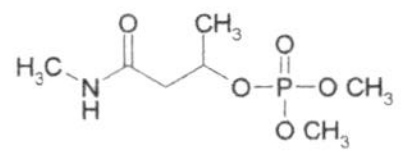 | 221.25 | 0.322 | Carbamate | *Burkholderia capacia* PCL3 | 100 | Plangklang and Reungsang (2013) |
| | | | | | | *Enterobacter* sp. | 71 | Mohanta *et al*. (2012) |
| | | | | | | *Pichia anomala* HQ-C-01 | 95.2 | Yang *et al*. (2011) |
| Fenamiphos | 22224-92-6 | 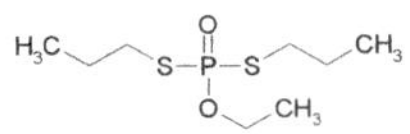 | 303.36 | 0.345 | Organophosphorus | *Pseudomonas putida*<br>*Acinetobacter rhiszospherae* | 100 | Chanika *et al*. (2011) |
| | | | | | | *Microbacterium esteraromacticum* MM1 | 100 | Cáceres *et al*. (2009) |
| Pirimicarb | 23103-98-2 | 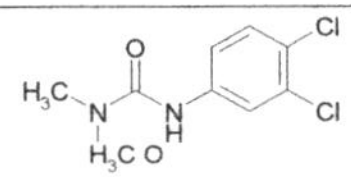 | 238.29 | 3.1 | Carbamate | *Aspergillus niger* | n.a. | Chen *et al*. (2009) |
| Methomyl | 16752-77-5 | 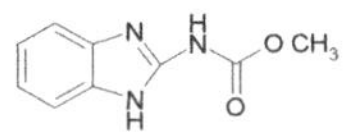 | 162.21 | 55 | Carbamate | *Paracoccus* sp. | 100 | Xu *et al*. (2009) |
| | | | | | | Mixture of activated sludge. | 100 | Farré *et al*. (2002) |
| Imidacloprid | 138261-41-3 | 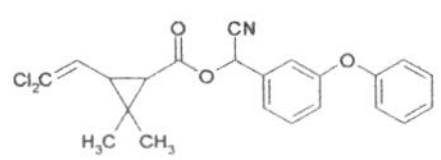 | 255.66 | 0.61 | Neonicotinoide | *Klebsiellla pseumoniae* BCH1 | 78 | Phugare *et al*. (2013) |
| | | | | | | Soil microbial | 22.5 | Liu *et al*. (2011) |
| | | | | | | *Penicillium oxalicum* | 14.1 | Ji *et al*. (2010) |
| Chlorpyrifos | 2921-88-2 | 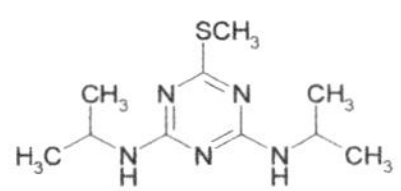 | 350.59 | 0.002 | Organophosphorus | *Sphingobacterium* sp. JAS3 | 100 | Abraham and Silambarasan (2013) |
| | | | | | | *Serratia marcescens* DI101 | 91 | Abo-Amer (2011) |
| | | | | | | *Pseudomonas stutzeri* | 99.72 | Moscoso *et al*. (2013) |
| Pyrimethanil | 53112-28-0 | 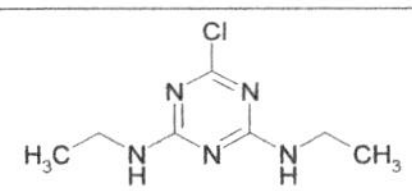 | 199.25 | 0.121 | Pyrimidine | *Scenedesmus obliquus*<br>*Scenedesmus quadricauda* | 10 | Dosnon-Olette *et al*. (2010) |

n.a., not available.

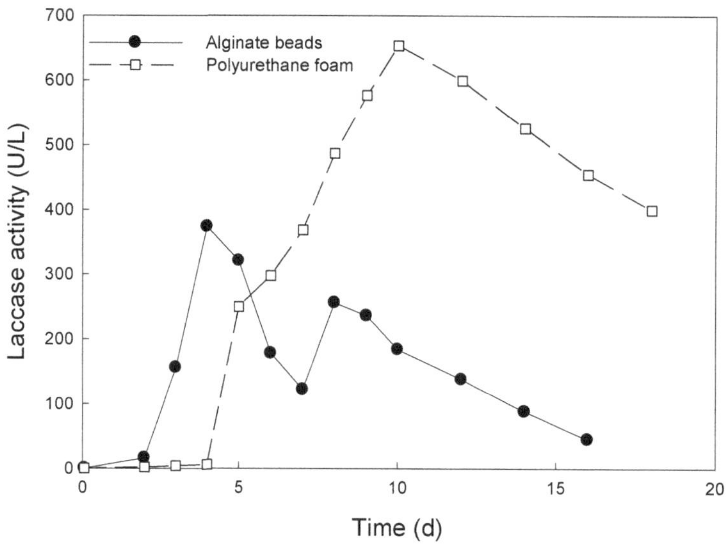


**Figure 8.2** Laccase production by *T. versicolor* immobilized on alginate beads and polyurethane foam.

glucose and PAH concentrations. Experiments were done in duplicate and samples analysed in triplicate, the standard deviation being less than 15%.

### Analytical methods

Laccase activity was determined spectrophotometrically as described by Niku-Paavola *et al.* (1991) with 2,2′-azino-di-[3-ethyl-benzo-thiazolin-sulphonate], or ABTS, as a substrate. One activity unit was defined as the amount of enzyme that oxidized 1 µmol of ABTS per min. Activities are expressed in U/l. Reducing sugars were measured by the dinitrosalicylic acid method using D-glucose as a standard (Ghose, 1978). PAH concentrations were recorded using an Agilent 1100 HPLC with a XDB-C8 reverse-phase column. The injection volume was set at 5 µl and the isocratic eluent (acetonitrile and water) was pumped at a rate of 1 ml/min. PAH detection was performed using a diode array detector at 200–400 nm (Alcántara *et al.*, 2009).

## Comparison of immobilization methods

A comparative study was carried out on the production of laccase by immobilized *T. versicolor* comparing entrapment and attachment techniques in alginate beads and polyurethane foam, respectively (Fig. 8.2).

When the fungus was immobilized via the entrapment technique with Ca-alginate beads, only two peaks with relatively high laccase activities were attained on the fourth (374 U/l) and eighth day (256 U/l). In addition, the fungus was not retained in the alginate beads. Thus, after only a few culture days considerable cell leakage from the polymeric matrix into the medium was observed. This caused several operational problems in the bioreactor hampering its proper performance.

As for the attachment technique, when the fungus was immobilized on cubes of polyurethane foam, laccase production started on the fifth day (250 U/l) and increased up to the tenth day (654 U/l). This laccase activity was around 75% higher than that obtained in cultures with the fungus immobilized in Ca-alginate beads.

Immobilization via attachment can be considered as a natural state for fungi, since most of them tend to attach firmly on surfaces in nature. Therefore, it is not surprising that artificially immobilized fungi can produce extracellular secondary metabolites. In addition, this technique provides additional advantages over Ca-alginate beads such as easier liquid–solid separation and minimal clogging. For this reason, it can be asserted that the alginate entrapment technique is poorly suited for the immobilization of *T. versicolor*. On the contrary, the attachment technique using cuttings of polyurethane foam gave very good results; this system was further employed to evaluate the ability of *T. versicolor* to degrade PAHs.

Several studies have shown that diverse fungi are capable of PAH mineralization and rates of degradation can be correlated with the ligninolytic enzymes produced (Alcalde *et al.*, 2002; Dodor *et al.*, 2004; Gómez *et al.*, 2006; Pickard *et al.*, 1999). Therefore, in this study the biodegradation of several high molecular weight compounds, anthracene and phenanthrene by cultures of *T. versicolor* immobilized on polyurethane foam was studied. The PAHs were added to 5-day-old cultures, as an aqueous solution (1% Tween 20 and 5 mM acetone), at final concentration of 100 mM. During the metabolism of anthracene by immobilized *T. versicolor*, accumulation of anthraquinone in the culture medium was identified. The maximum removal rate was observed after 5 days of the addition. This removal time coincides with increased enzymatic activity observed in these cultures, since the concentration of laccase peaked after 10 days. When phenanthrene was added, similar behaviour was detected. All cultures exhibited anthracene and phenanthrene removals in the range 83–87%, after only 5 days of incubation. Therefore, this shows the high potential of *T. versicolor* immobilized on cubes of polyurethane foam to degrade these pollutants.

## PAH degradation by *T. versicolor* grown in solid state fermentation

In the last few years, there has been an increasing trend towards the utilization of solid-state fermentation (SSF) techniques to produce several bulk chemicals and enzymes from agroindustrial wastes. SSF is defined as any fermentation process carried out on a solid material in absence of free-flowing liquid (Pandey *et al.*, 2001). The low moisture content means that fermentation can only be carried out by a limited number of microorganisms, mainly yeasts and fungi, although some bacteria have also been used. This methodology has been reported to produce a more stable product, with lower energy requirements, and conducted in smaller fermenters to treat smaller volumes of polluting effluents, than submerged fermentation (SmF) systems (Rodríguez Couto and Sanromán, 2006).

There is an increasing trend towards the efficient utilization and value-addition of agroindustrial wastes in SSF. Not only does the utilization of these wastes provide alternative substrates for chemical and/or enzyme production, but the degradation of the waste material simultaneously helps in solving pollution problems. Typically, the microorganisms in SSF grow under conditions close to their natural habitats to produce certain enzymes and metabolites, which usually will not be produced or be produced only at low yield in submerged cultures (Longo and Sanroman, 2009). Various combinations of agricultural substrates/byproducts and white rot fungi have been used successfully in solid state fermentation for ligninolytic enzyme production (Rodríguez Couto *et al.*, 2002).

The nature of the employed material is the most important parameter affecting SSF processes and its selection depends upon several factors mainly related with cost and availability. In the SSF process the material used as a solid substrate not only supplies nutrients to the culture but it also serves as an anchorage for the microbial cells to permit the development of fungal biofilm on the surface of these solid substrates. SSFs present significant potential advantages for filamentous fungi which are capable of penetrating into solid substrates using turgor pressure generated at the tip of the mycelium (Ramachandran *et al.*, 2004). Among the several factors which are important for encouraging microbial growth and activity associated with a particular substrate, particle size and moisture level/water activity are the most critical (Auria *et al.*, 1992; Barrios-González *et al.*, 1993; Echevarria *et al.*, 1991; Liu and Tzeng, 1999; Pastrana *et al.*, 1995). It is known that smaller substrate particles provide a larger surface area for microbial invasion, but if they are too small it may result in substrate agglomeration as well as poor growth. On the other hand, larger particles provide better aeration but a limited surface for microbial growth. For this reason, it is necessary to select a compromised particle size for each particular process and to determine the most suitable substrate for SSF to provide the highest enzyme production. The present study deals with the objective of enhancing the production of laccase from *T. versicolor* using SSF biofilms generated during PAH degradation.

## Materials and methods

In this study, the fungus, medium and analytical methods were the same as previously described. However, the glucose concentration was reduced to 4 g/l since agroindustrial wastes provided an additional substrate.

### Agroindustrial wastes

#### Barley bran

The particle length was about 3–7 mm. The chemical composition of barley bran is 21.4% lignin, 23% cellulose and 32.7% hemicellulose (Cruz *et al.*, 2000). The material was autoclaved at 121°C for 20 minutes and then stored under sterile conditions until use.

#### Ground orange peelings

Orange peelings are composed of both soluble and insoluble carbohydrates. The soluble sugars in orange peel are glucose, fructose and sucrose. The insoluble polysaccharides in cell walls of orange peel are composed of pectin, cellulose and hemicellulose. They also contain minor amounts of organic acids, mainly citric, malic, malonic and oxalic, proteins, mineral ions, phenolic compounds and polyols (Grohmann *et al.*, 1995). Prior to use, the peelings were cut to a mean size of 5 mm and pretreated as follows: they were first soaked for an hour in 30 ml of KOH 83.17 mM (10 g of fresh peelings) to neutralize organic acids (Stredansky and Conti, 1999). Then, they were thoroughly washed with distilled water and dried at moderate temperature. The peelings were autoclaved at 121°C for 20 minutes and then stored under sterile conditions until use.

#### Kiwi fruit waste

Unless otherwise stated, this substrate consisted of a mixture of peelings and pulp at a ratio 1:9, since this is the proportion present in the kiwi fruit. For all experiments this waste was pretreated and autoclaved in similar way to the orange peelings, as earlier described.

### Solid state fermentation (SSF)

The cultures were incubated in cotton-plugged Erlenmeyer flasks (250 ml) containing 2.5 g agroindustrial waste and 15 ml sterile culture medium. Inoculation was carried out directly in the Erlenmeyer flasks. Three agar plugs (diameter, 3 mm), from an actively growing fungus on PDA, per Erlenmeyer flask were used as inoculum. The Erlenmeyer flasks were incubated statically in a Sanyo MLR-350H incubator, in complete darkness. The medium was maintained at 30°C and 90% humidity, avoiding evaporation.

## Laccase production and PAH degradation

Recent papers on SSF highlight the enormous potential of this fermentation technique for the development of several bioprocesses (Longo and Sanromán, 2006; Rodríguez Couto *et al.*, 2002; Rodríguez Couto and Sanromán, 2005, 2006; Rosales *et al.*, 2007). The selection of suitable substrates for SSF processes depends on several factors mainly related with cost and availability and thus may involve the screening of several agroindustrial wastes. Moreover, the utilization of these types of support-substrates helps in solving pollution problems associated with their disposal.

In this section, the potential of barley bran, orange peelings or kiwi waste as a support-substrate for laccase production by *T. versicolor* under SSF conditions was investigated. As can be seen in Fig. 8.3, when *T. versicolor* was cultivated on agroindustrial waste cultures, the fungus grew forming a biofilm on the wastes.

Therefore, the fungus can use the nutrients of the wastes for its growth. In all cultures, glucose,

**Figure 8.3** Biofilm of *T. versicolor* on agroindustrial wastes.

measured as reducing sugars, diminished from the initial value to 1.5 g/l on the third day, and from there onwards it remained stable. These profiles, in reducing sugars, could be due to the release of some hydrolysis products from the wastes. Thus, these substrates could be employed without the addition of glucose in the culture medium, which would imply an important advantage from an economical point of view.

When *T. versicolor* was cultivated on barley bran cultures, laccase production began on the fourth day (150 U/l) (Fig. 8.4). Then, it increased until peaking on the eighth day (1186 U/l) and from there onwards it decreased. This peak was almost 3-fold higher than that found operating with inert support. As for kiwi waste cultures, laccase production started on the second day (172 U/l) and then, it rapidly increased reaching a maximum value of 2133 U/l on the sixth day of cultivation. This value was about 50% higher than that obtained using barley bran as a support.

Operating with orange peelings as the support-substrate, laccase production began on the fourth day (184 U/l) and then sharply increased up to a maximum value of 3152 U/l on the seventh day. This value was about sevenfold higher than that attained employing inert support and much higher than that found in the other incubations.

The obtained results clearly show the enormous potential of food wastes to promote laccase activity by *T. versicolor* under SSF conditions, since they led to much higher activities than inert supports. In addition, by employing these agroindustrial wastes the initial amount of the added glucose was reduced to 60% of that used with the inert supports. Therefore, the utilization of such wastes on the one hand promoted high enzymatic levels and on the other hand provided a considerable reduction in potential production costs. In view of the obtained results, it can be concluded that orange peelings are an excellent support-substrate for laccase production by *T. versicolor* under SSF conditions due to their high content in sugars, cellulose and pectin.

Therefore, taking into account the high laccase production from orange peelings-SSF, PAH degradation studies were performed. Here, the degradation of phenanthrene and pyrene by whole fungal cultures during SSF was studied. PAHs were added to 5-day-old cultures, to achieve a final concentration of 100 μM. The maximum phenanthrene removal rate observed in the first days coincided with a period of increased enzymatic activity in these cultures, where laccase peaks occurred after 7 days. When pyrene was added, the degradation rate was slower; however the initial concentration of pyrene was reduced to around 73% of the original concentration after 7 days. In order to determine the reliability of this fungus to degrade these pollutants an additional batch test was carried out. In this second batch, the PAHs were added on the seventh day at the same initial concentration. Similar behaviour was detected, although in all cases the

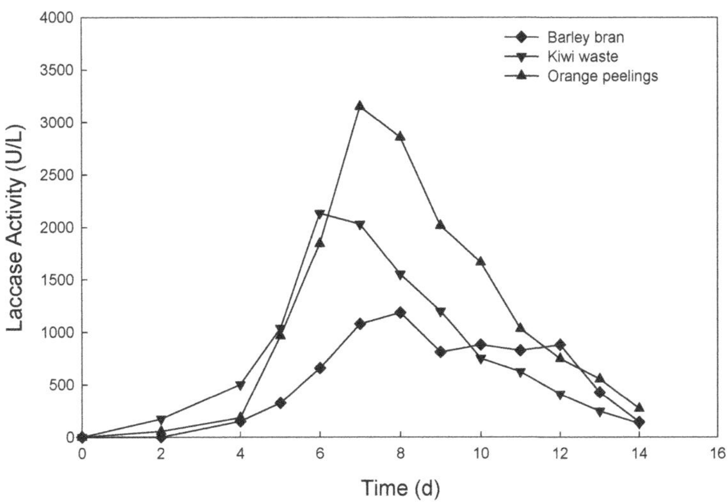


**Figure 8.4** Laccase production using *T. versicolor* under SSF conditions.

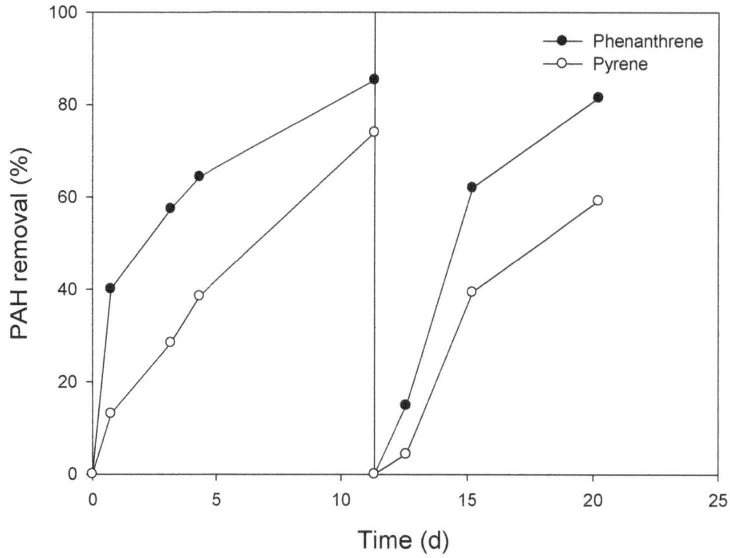


**Figure 8.5** Degradation of phenanthrene and pyrene by *T. versicolor* during SSF employing orange peelings as a support-substrate.

degradation rates were rather low, being reduced to concentrations of 87% and 60% of the original in 14 days (Fig. 8.5) for degradation of phenanthrene and pyrene, respectively. To determine the adsorption of PAHs onto the fungus, the fungus and orange peelings from the culture were treated with acetonitrile to extract and measure the possible residual PAHs. No residual PAHs were found, indicating that the PAHs were not adsorbed onto the SSF material or fungus. This means that PAH degradation was only due to enzymes produced by the fungus during cultivation and highlights the high potential of *T. versicolor* degrade these pollutants under SSF.

## PAH degradation by immobilized *Pseudomonas stutzeri*

A feasible bioremediation technology requires microorganisms being capable of quick adaptation to and efficient use of pollutants of interest. In the last two decades, different strains classified in the genus *Pseudomonas* have been identified as promising microorganisms for use in bioremediation (Liu *et al.*, 2011, 2014; Ma *et al.*, 2013; Moscoso *et al.*, 2013; Santos *et al.*, 2008). These bacteria are widely distributed in the environment and are often found to survive in nutrient deficient ecosystems perhaps owing to their versatile metabolism (Nanda *et al.*, 2010). For example, *Pseudomonas* spp. have shown a great capacity to degrade a wide range of aromatic compounds from benzene to benzopyrene (Cao *et al.*, 2009). Some of these strains have demonstrated their ability to degrade PAHs under aerobic conditions, using oxygen as an electron acceptor, or under anaerobic conditions, using nitrate as electron acceptor (Grishchenkov *et al.*, 2000).

It is well known that low PAH solubility can negatively impact bioremediation because pollutants need to be accessible to microorganisms for metabolism to occur. One way to overcome this drawback is to add surfactant compounds, which increase PAH solubility by decreasing the interfacial surface tension between the PAH and the soil/water interface. Among the existing surfactants, non-ionic surfactants are highly preferred and the most frequently applied in PAH biodegradation since they are less toxic to bacteria than anionic and cationic surfactants (Fernando Bautista *et al.*, 2009). There is a wide variety of non-ionic surfactants available for use; however, Triton X-100 and Tween 80 are the most common non-ionic surfactant investigated for enhancing bacterial PAH degradation (Lu *et al.*, 2011).

The scientific community has widely reported that bacterial immobilization offers many advantages over the use of free bacteria for the treatment of environmental pollutants (Mollaei *et al.*, 2010; Wasi *et al.*, 2013). Bacterial immobilization is a proven approach that increases the stability of bacteria under unfavourable conditions by providing an appropriate microenvironment that can improve tolerance against toxic compounds (Rezaee *et al.*, 2008). Synthetic and natural substrates have been shown as attractive and promising for use in bacterial immobilization. In recent times, the scientific community has focused attention on bio-hydrogels such as alginate, carrageenan, agar or pectate (Sevillano *et al.*, 2012; Mulla *et al.*, 2013). Bio-hydrogels are particularly useful due to their high water content, flexibility, bio-compatibility, and easily controlled diffusion characteristics (Pires *et al.*, 2011). Therefore, immobilization in bio-hydrogels offers the retention of bacterial biofilm in a homogeneous and solid medium that can be placed in the environment for contaminant removal purposes.

In this study, we monitored the degradation of PAHs in the presence of Tween 80 by *Pseudomonas stutzeri* CECT 930 immobilized in agar. After that, the degradation ability of the immobilized bacteria was further examined in fluidized and fixed bed reactors.

## Materials and methods

### Microorganism

*P. stutzeri* CECT 930 (ATCC 17588) was maintained at 4°C on solid agar petri plates composed of glucose 10 g/l, peptone 5 g/l, malt extract 3 g/l, yeast extract 3 g/l and agar 15 g/l and subcultured every 3 months.

### Bacterial growth and immobilization

Bacteria were grown in 250 ml Erlenmeyer flasks with 50 ml of MH growth medium (described by Hou *et al.*, 2005) containing either anthracene or phenanthrene as a carbon source, at a concentration of 100 μM. Tween 80 (1%) and acetone (2%) were included for PAH solubilization. The pH was initially adjusted to 7.5 with 0.5 M NaOH, and the MH autoclaved at 121°C for 20 min. The flasks were inoculated, plugged with sterile cellulose stoppers for passive aeration and incubated in an orbital shaker at 30°C and 100 rpm. Cells were harvested during the growth phase; the bacterial cell suspension was centrifuged for 10 min at 8000 rpm and resuspended in 25 ml of sterile 0.9% NaCl.

The agar entrapment was carried out according to Mulla *et al.* (2013). Agar was dissolved in 75 ml distilled water and autoclaved at 121°C for 20 min. Agar and bacterial cell suspensions were mixed by a magnetic stirrer obtaining a final agar concentration of 2% (w/v) and poured into glass petri dishes. After that, the solidified agar was cut with a puncher (diameter, 6 mm) and stored in sterile 0.9% NaCl at 4°C.

### Batch assays

Batch experiments were carried out in 250 ml Erlenmeyer flasks with 50 ml of the MH medium using immobilized bacteria exposed to different PAH concentrations. The flasks were inoculated with approximately 7 g agar with immobilized bacteria (approx. 1 mg of bacteria). Flasks were agitated in an orbital shaker at 100 rpm and 30°C; passive aeration was permitted by cellulose stoppers. Liquid samples were taken periodically to be analysed (Alcántara *et al.*, 2009). Experiments were done in duplicate and samples were analysed in triplicate, the standard deviation being less than 15%.

### Reactor configuration

Two different reactors were used to evaluate the degradation ability of the immobilized bacteria.

*Fluidized bed reactors* consisted of a glass column (16 cm height and 2 cm internal diameter) with a working volume of 40 ml. The reactor was autoclaved at 121°C for 20 minutes and then filled with *P. stutzeri* immobilized in agar (approx. 8 g). To maintain a constant flow in the column, a double head peristaltic pump was used to circulate the MH media which filled the column. For the experiments, two different concentrations of phenanthrene, 100 μM and 200 μM, were used. The temperature was regulated at 30°C using a thermostatic chamber, and constant aeration of 80 ml/min provided.

Our *fixed bed reactor* consisted of a jacketed glass column (20 cm height and 4.5 cm in internal diameter) with a working volume of 150 ml. The reactor was autoclaved at 121°C for 20 min to maintain sterile conditions. The column was then packed with seven layers of nylon sponge (4.3 cm diameter) interspersed with six agar layers (approx. 11 g of agar with *P. stutzeri* immobilized). The column was filled with MH medium and a double head peristaltic pump was used to circulate the liquid media. The temperature was again regulated at 30°C, this time by the circulation of temperature-controlled water through the jacketed glass column and constant aeration of 20 ml/min.

### Scanning electron microscopy (SEM)

A series of scanning electron microscopy (SEM) images were taken to provide a visual characterization of the immobilized bacteria. Briefly, the samples were dehydrated; critical point dried, submerged in liquid $N_2$, cut and coated with gold. Images were collected on a JEOL JSM6700F field emission scanning electron microscope using an accelerating voltage of 3.0–30 kV.

## PAH degradation in batch mode by immobilized *P. stutzeri*

Experiments were carried out using bacteria immobilized on agar at different initial concentrations of phenanthrene (350 to 1100 μM) or anthracene (15 to 60 μM). Because of the low water solubility of these pollutants, 0.0076 mg/l for anthracene and 1.2 mg/l for phenanthrene, the experiments were performed in presence of Tween 80 at 1%. The PAH concentration and the pH of the culture medium was monitored for 6 days and abiotic assays performed as control treatments. For all experiments, regardless of the pollutant, the pH of the culture medium increased constantly from pH 6.6 at 0 days to pH 8.7 at 6 days. These changes in pH were attributed to microbial activity because no changes were detected in control incubations.

Phenanthrene removal was perceptible after the first day, and around 10% of removal was achieved regardless of the initial pollutant concentration (Fig. 8.6). Afterwards, the removal profiles were different reaching greatest removal rate for the highest initial PAH concentrations. This is in accordance with pollutant concentration dependent degradation kinetics (Hussain *et al.*, 2009). At the end of the incubation period, the removal efficiency for all tested initial concentration treatments was higher than 90%. This confirms that microbial immobilization did not prevent the ability of the bacteria *P. stutzeri* to use phenanthrene as a carbon source. These data are consistent with previous work reported by Moscoso *et al.* (2012).

Next, the microbial bioconversion of anthracene was investigated. The biodegradation started after two days and at this time, different removal efficiencies were reached depending on the initial concentration, ranging from 5% to 77%, for the lowest and highest concentrations studied, respectively (Fig. 8.7). Similar to previous experiments the high removal rates were detected for the highest concentrations. During the analysis of the liquid samples, anthraquinone, an indicative product of anthracene degradation, was identified by HPLC (Fig. 8.8). Therefore, the oxidative degradation of the PAHs was demonstrated.

The biodegradation of PAHs using immobilized microorganisms was described using a logistic

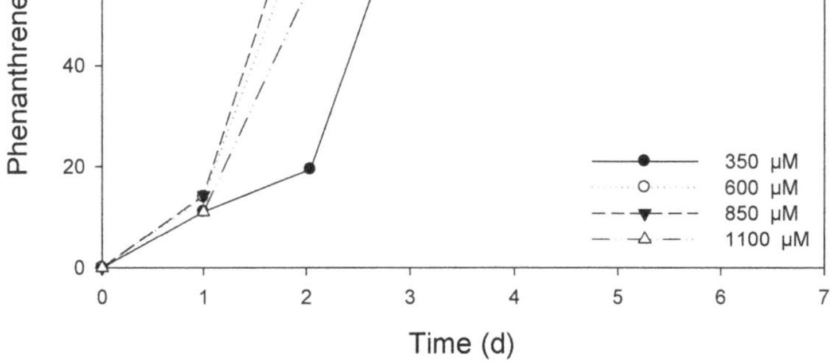


**Figure 8.6** Degradation profiles of phenanthrene by immobilized *P. stutzeri*.

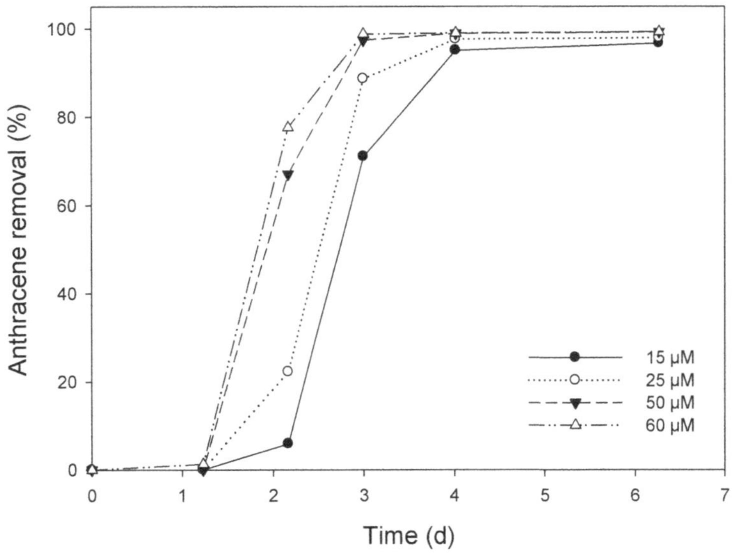


**Figure 8.7** Degradation profiles of anthracene by immobilized *P. stutzeri* exposed to different initial concentrations of the substrate.

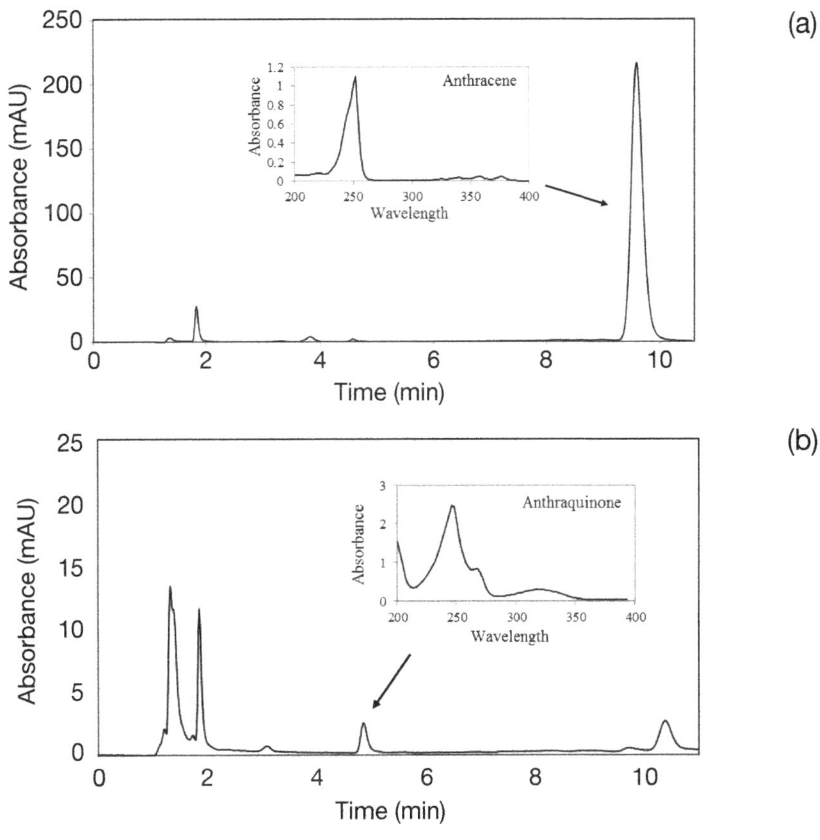


**Figure 8.8** HPLC chromatogram of liquid sample using immobilized bacteria to study the degradation of anthracene. Chromatograms show (a) the abundance of anthracene in the initial incubation and (b) the production of anthraquinone during incubation.

equation in which $D$ is percentage PAH removal (%) at a specific moment of incubation ($t$ days), $D_0$ and $D_{max}$ are the initial and maximum percentage removal (%), respectively, and $\mu_D$ is the maximum specific degradation rate ($d^{-1}$).

$$D = \frac{D_{max}}{1 + e^{\left[\ln\left(\frac{D_{max}}{D_0} - 1\right) - \mu_D t\right]}}$$

Sigma Plot 8.0 software was utilized to adjust the experimental data to the model using an iterative procedure, based on the Marquardt–Levenberg algorithm. For all concentrations tested the biodegradation of phenanthrene using immobilized *P. stutzeri* fitted the model well ($R^2 > 0.99$); the values of the constants are presented in the Table 8.3.

For all phenanthrene concentrations studied, the maximum specific degradation rate (μd) was higher than obtained by Moscoso *et al.* (2012) using free cells of the same bacterial strain. Therefore, it can be presumed that bacterial immobilization enhances the degradation ability of these microorganisms. It is postulated that a polymer matrix can confer protection from high levels of these toxic compounds due to alterations in the membrane composition of microorganisms growing in the matrix (Wasi *et al.*, 2011).

To evaluate the presence and distribution of bacteria, scanning electron microscopy (SEM) images were acquired after the assays (Fig. 8.9). Bacterial cells were uniformly distributed over the surface and embedded in the agar structure. The surface was porous enough so that PAH molecules can freely diffuse into the porous structure and reach the embedded bacteria. The microscopy images suggest that the bio-hydrogel used in this study possesses excellent physical characteristics as a support for *P. stutzeri* immobilization.

## PAH degradation in continuous operation mode by immobilized *P. stutzeri*

The microbial bioconversion of PAHs by immobilized *P. stutzeri* was investigated with the intent of developing reliable bioremediation technologies to treat polluted water streams. Treatment techniques for water and effluents contaminated with organic compounds often require the use of systems that can operate continuously. Therefore, the next step was the design a system in which the performance of the immobilized bacteria could be assayed in continuous mode. For this purpose, the design of two different reactors, both fixed and fluidized bed reactors, were carried out and the degradation ability of the immobilized bacteria was evaluated in each configuration.

First, the design of a reactor containing the bio-hydrogel with the immobilized bacteria, in a fixed bed configuration, was evaluated. In this experiment a continuous synthetic effluent containing 100 μM phenanthrene was provided. The system operated at two different hydraulic retention times (HRTs), 2 and 6 days, and the phenanthrene removal was followed over time (Fig. 8.10).

**Table 8.3** Parameters obtained by fitting a logistic model for phenanthrene and anthracene removal

**(a) Phenanthrene**

| Initial concentration | 1100 μM | 850 μM | 600 μM | 350 μM |
|---|---|---|---|---|
| $D_{max}$ | 97.96 | 96.71 | 96.16 | 93.36 |
| $d_0$ | 0.983 | 0.719 | 1.080 | 0.017 |
| $\mu_D$ | 2.405 | 3.136 | 2.690 | 3.584 |
| $R^2$ | 0.999 | 0.999 | 0.999 | 0.990 |

**(b) Anthracene**

| Initial concentration | 15 μM | 25 μM | 50 μM | 60 μM |
|---|---|---|---|---|
| $D_{max}$ | 96.06 | 97.79 | 98.63 | 98.95 |
| $d_0$ | $3.4 \times 10^{-4}$ | $3.2 \times 10^{-3}$ | $3.9 \times 10^{-4}$ | $7.2 \times 10^{-4}$ |
| $\mu_D$ | 4.529 | 4.203 | 6.078 | 6.057 |
| $R^2$ | 0.999 | 0.999 | 0.999 | 0.990 |

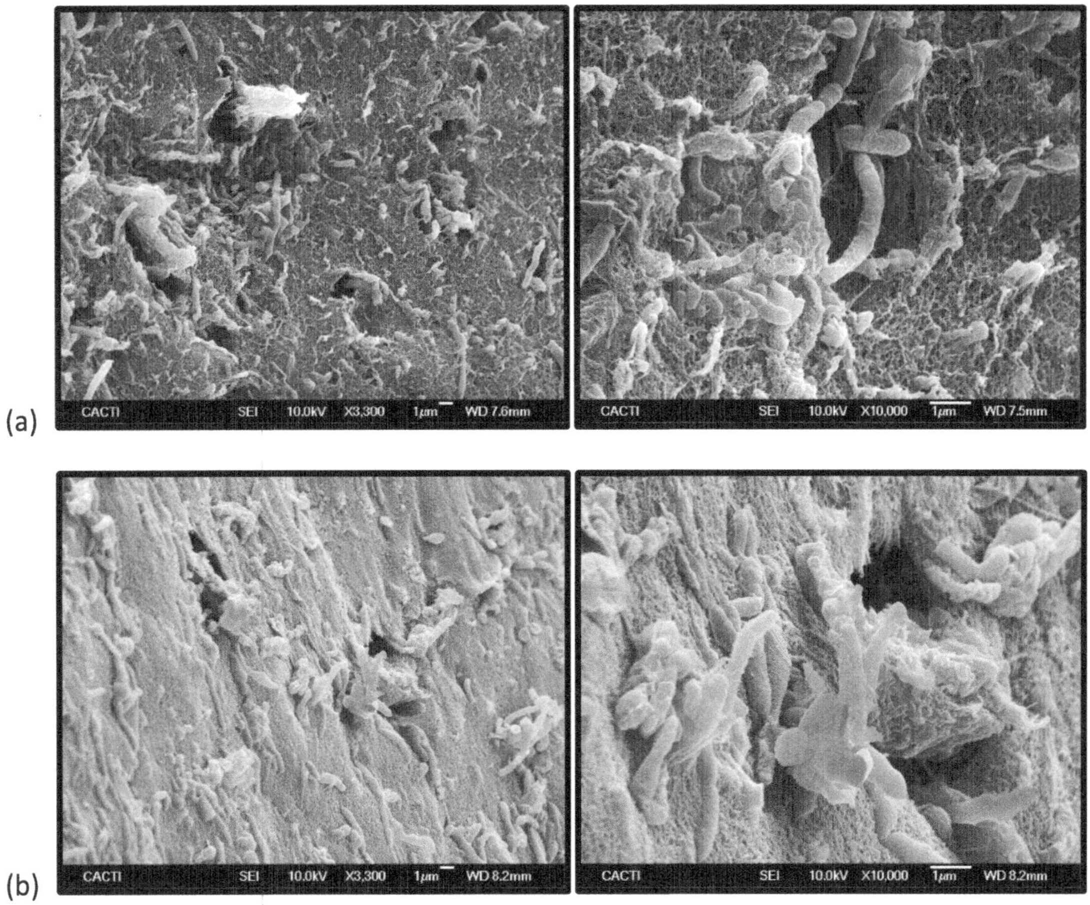


**Figure 8.9** SEM images of immobilized *P. stutzeri* (a) on the agar surface and (b) inside the agar.

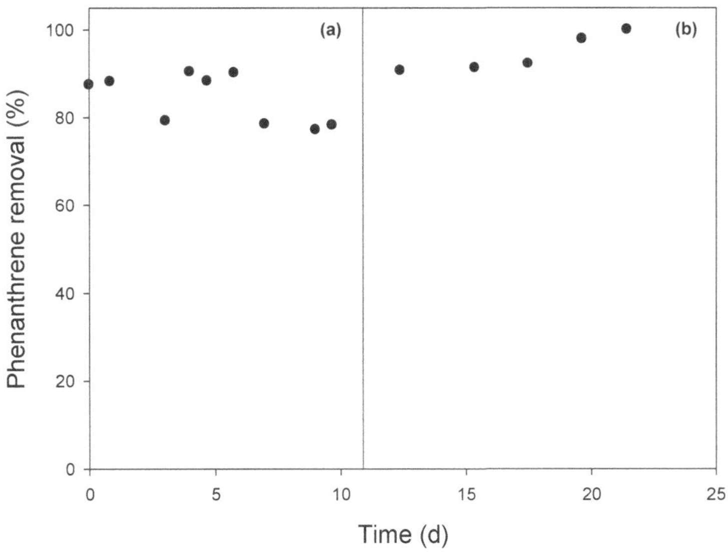


**Figure 8.10** Phenanthrene removal by the fixed bed reactor operating at a HRT of (a) 2 days and (b) 6 days, each with an initial phenanthrene concentration of 100 µM.

At HRT of 2 days, high fluctuations in the pollutant removal reached by the reactor were detected. These fluctuations were related with the high flow provided to the reactor. Nevertheless, the level of phenanthrene removal was always higher than 75%. As was expected, an increase in HRT reduced fluctuations in the removal degree; removals higher than 90% were consistently obtained.

To evaluate the effect of the reactor design on degradation levels, bio-hydrogels containing *P. stutzeri* were used to fill a fluidized bed reactor working in continuous mode. In this type of reactor, soft aeration moves the bio-hydrogels while a continuous wastewater stream is passed through it. The phenanthrene degradation levels operating at two different pollutant concentrations, 100 and 200 μM, with a HRT of 4 days, are shown in Fig. 8.11.

At initial phenanthrene concentrations of 100 μM, the reactor quickly reached a steady state removal efficiency of around 99%. A slight reduction in removal efficiency was obtained when the pollutant concentration was increased at 200 μM but the removal efficiency reached at the outlet of the reactor nevertheless exceeded 97%. The obtained results suggest that this system is suitable for the treatment of continuous wastewater streams.

## Pyrimethanil degradation

Pyrimethanil, N-(4,6-dimethylpyrimidin-2-yl) aniline, is a synthetic anilinopyrimidine chemical that acts as a contact fungicide, which gives protective and curative properties to crops and seeds. Pyrimethanil can be applied as a seed treatment or as a foliar spray to a number of crops and acts by inhibiting the biosynthesis of methionine, thereby preventing protein formation and cell division. Pyrimethanil prevents disease caused by a broad spectrum of fungi, including *Alternaria* spp., *Botrytis cinerea*, *Cercospora* spp., *Cladosporium* spp., *Colletotrichum* spp., *Monilia* spp., *Mycosphaerella* spp., *Penicillium* spp. and *Venturia* spp.

Pyrimethanil has moderate water solubility (121 mg/l) and recent research confirms that the pesticides pyrimethanil and tebuconazole are frequently detected in water samples in the wine-growing regions of Spain (Herrero-Hernández *et al.*, 2013). However, little research has yet focused on the removal of this pesticide. At the present time, the proposed methodologies for pyrimethanil removal are adsorption methods (Marín-Benito *et al.*, 2012), electrochemical treatment (Errami *et al.*, 2012), photocatalytic treatment (Navarro *et al.*, 2011) or phytoremediation (Dosnon-Olette *et al.*, 2010). However, little research has evaluated microbial biodegradation as a treatment for the removal of this pesticide. Bizaj *et al.* (2011)

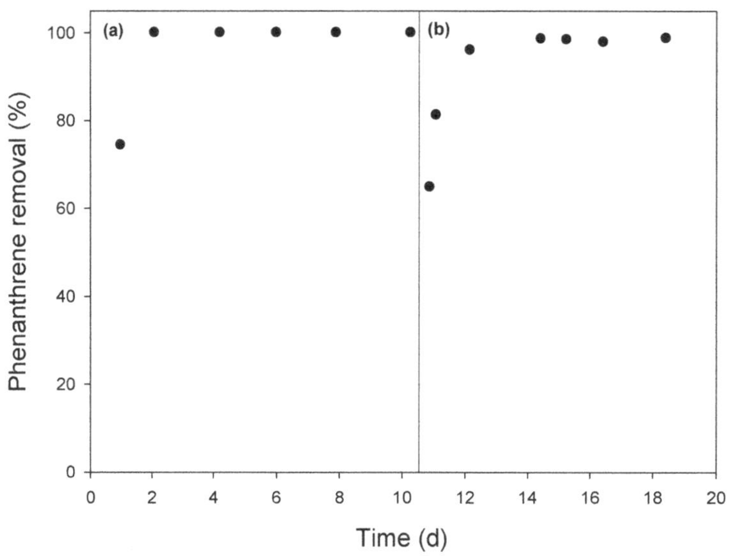


**Figure 8.11** Phenanthrene removal by the fluidized bed reactor working at a HRT of 4 days with an initial phenanthrene concentration of (a) 100 μM and (b) 200 μM.

found that *Saccharomyces cerevisiae* was able to eliminate pyrimethanil in liquid cultures and their results strongly suggest that the removal was a consequence of adsorption only, by the metabolic actions of these microorganisms. Similar results were reported by Ballesteros Martín *et al.* (2008). They determined that pyrimethanil was not toxic to *Pseudomonas putida* CECT 324 at wide range of concentrations; however, they also found that the pesticide was not biodegradable by this bacterium. Therefore, the search for microorganisms able to metabolize this pollutant is required. Based on the previous promising results using the bacteria *P. stutzeri* for the removal of aromatic compounds, the aim of this study was to evaluate the ability of this bacterium for pyrimethanil removal in liquid culture.

### Materials and methods

Here, the bacteria were grown in agar as described previously.

#### Batch assays

Batch experiments were carried out in 250 ml Erlenmeyer flasks with 50 ml of a mineral medium reported by Moscoso *et al.* (2013) containing pyrimethanil as a carbon source at a concentration of 10 mg/l. Each flask was inoculated with approximately 2 mg of active bacteria obtained directly from growth on solid agar plates. Flasks were agitated in an orbital shaker at 150 rpm and 37.5°C; passive aeration was permitted by the use of cellulose flask stoppers. Liquid samples were taken periodically to be analysed. Experiments were done in duplicate and samples analysed in triplicate, the standard deviation being less than 15%.

#### Pyrimethanil analysis

Pesticide concentrations in liquid were determined using an Agilent 1100 HPLC equipped with an XDB-C8 reverse-phase column (150 × 4.6 mm i.d., 5 mm). Prior to injection, samples were centrifuged (10,000 rpm, 5 min) and filtered through a 0.45 mm Teflon filter. The injection volume was set at 20 ml, and the isocratic eluent (acetonitrile–water) was pumped at a rate of 1 ml/min for 8 min. Detection was performed with a diode array detector from 200 to 400 nm; the column temperature was maintained at 20°C.

#### Biomass determination

Bacterial growth was determined by spectrophotometery using a Helios Beta Thermo device set at a wavelength of 600 nm. Samples were centrifuged (10,000 rpm, 5 min), and the pellet was resuspended in distilled water prior to biomass determination.

### Pyrimethanil degradation by *P. stutzeri*

The microorganism was inoculated into the medium containing pyrimethanil as a sole carbon source. In this study, no surfactants were added to the culture because the pyrimethanil is soluble at the concentration studied (10 mg/l). The growth of bacteria was followed during incubation; after the first day active growth was observed (Fig. 8.12).

However, the presence of the pollutant caused stress to the bacteria and bacterial aggregates appeared after the second day. The removal of pyrimethanil by the bacteria was followed by the reduction of bacterial concentrations in the liquid medium during 12 days. As is shown in Fig. 8.12, the reduction of the pesticide concentration was progressive during the 12 days of culture, reaching a maximum removal around 18%. This value is higher than previously reported in research developed by Dosnon-Olette *et al.* (2010). They found a maximum of 10% for pyrimethanil removal using the microalgae *Scenedesmus quadricauda*. Although, during the assays, no degradation products were detected, extraction of pyrimethanil from biomass using acetonitrile showed negligible pesticide concentration. Therefore, these results offer a promising approach for the future bioremediation of this pesticide using the bacteria *P. stutzeri.*

## Conclusions

A brief survey on the role of biofilms for the remediation of PAHs and pesticides from the environment has been presented. Although several processes such as adsorption, volatilization, photolysis, or chemical degradation are still the prevailing remediation strategies, the relevance and potential of bioremediation has been demonstrated. A considerable amount of recent research has focused on the utilization of microbial cultures for the remediation of these emerging pollutants and should continue to do so.

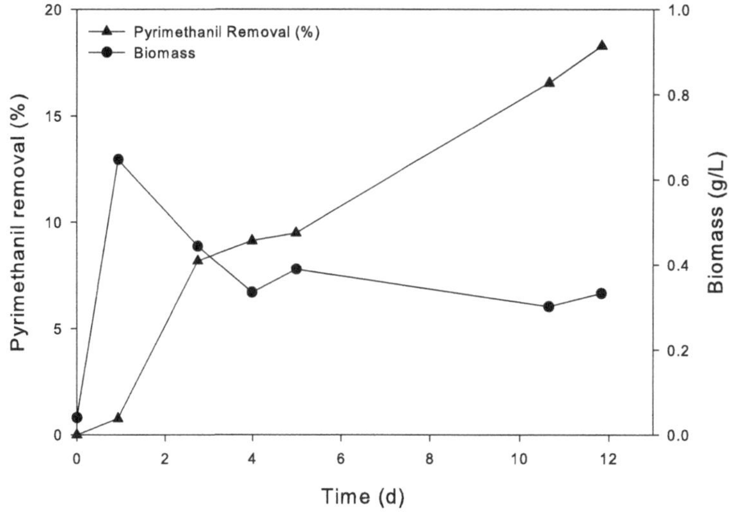


**Figure 8.12** Profiles of *P. stutzeri* growth and removal percentage of pyrimethanil.

In this chapter, it was demonstrated that bioremediation using biofilms of bacteria or fungi, is an attractive approach for the treatment of emerging pollutants such as PAHs or pesticides in wastewater streams. Immobilization of the microorganisms *P. stutzeri* and *T. versicolor* was feasible, allowing active biofilms to be continuously used for the effective treatment of polluted effluents. Moreover, it is demonstrated that the fungus *T. versicolor* and the bacterium *P. stutzeri* are promising candidates for degrading organic compounds including PAHs and pesticides.

In summary, the application of immobilization technology to improve the removal of emerging pollutants constitutes a very interesting and challenging topic, holding great promise both for academic research and engineering development.

## Acknowledgements

This work was supported by the Spanish Ministry of Economy and Competitiveness and ERDF funds (CTM2011-25389). The authors are thankful to the Ramón y Cajal programme for Marta Pazos financial support and Xunta de Galicia for financial support of researcher Emilio Rosales via a postdoctoral grant.

## References

Abo-Amer, A.E. (2011). Biodegradation of diazinon by *Serratia marcescens* DI101 and its use in bioremediation of contaminated environment. J. Microbiol. Biotechnol. *21*, 71–80.

Abraham, J., and Silambarasan, S. (2013). Biodegradation of chlorpyrifos and its hydrolyzing metabolite 3,5,6-trichloro-2-pyridinol by *Sphingobacterium* sp. JAS3. Process Biochem. *48*, 1559–1564.

Alcalde, M., Bulter, T., and Arnold, F.H. (2002). Colorimetric assays for biodegradation of polycyclic aromatic hydrocarbons by fungal laccases. J. Biomol. Screen. *7*, 547–553.

Alcántara, M.T., Gómez, J., Pazos, M., and Sanromán, M.A. (2008). Combined treatment of PAHs contaminated soils using the sequence extraction with surfactant-electrochemical degradation. Chemosphere *70*, 1438–1444.

Alcántara, M.T., Gómez, J., Pazos, M., and Sanromán, M.A. (2009). PAHs soil decontamination in two steps: Desorption and electrochemical treatment. J. Hazard. Mater. *166*, 462–468.

Alcántara, M.T., Gómez, J., Pazos, M., and Sanromán, M.A. (2010). Electrokinetic remediation of PAH mixtures from kaolin. J. Hazard. Mater. *179*, 1156–1160.

Alcántara, M.T., Gómez, J., Pazos, M., and Sanromán, M.A. (2012). Electrokinetic remediation of lead and phenanthrene polluted soils. Geoderma *173–174*, 128–133.

Amezcua-Allieri, M.A., Ávila-Chávez, M.A., Trejo, A., and Meléndez-Estrada, J. (2012). Removal of polycyclic aromatic hydrocarbons from soil: A comparison between bioremoval and supercritical fluids extraction. Chemosphere *86*, 985–993.

Arques, A., Amat, A.M., García-Ripoll, A., and Vicente, R. (2007). Detoxification and/or increase of the biodegradability of aqueous solutions of dimethoate by means of solar photocatalysis. J. Hazard. Mater. *146*, 447–452.

Arulazhagan, P., and Vasudevan, N. (2011). Biodegradation of polycyclic aromatic hydrocarbons by a halotolerant bacterial strain *Ochrobactrum* sp. VA1. Mar. Pollut. Bull. *62*, 388–394.

Auria, R., Palacios, J., and Revah, S. (1992). Determination of the interparticular effective diffusion coefficient for

$CO_2$ and $O_2$ in solid state fermentation. Biotechnol. Bioeng. *39*, 898–902.

Baboshin, M., Akimov, V., Baskunov, B., Born, T.L., Khan, S.U., and Golovleva, L. (2008). Conversion of polycyclic aromatic hydrocarbons by *Sphingomonas* sp. VKM B-2434. Biodegradation *19*, 567–576.

Badger, G.M. (1962). Mode of formation of carcinogens in human environment. Natl. Cancer Inst. Monogr. *9*, 1–16.

Ballesteros Martín, M.M., Sánchez Pérez, J.A., García Sánchez, J.L., Montes de Oca, L., Casas López, J.L., Oller, I., and Malato Rodríguez, S. (2008). Degradation of alachlor and pyrimethanil by combined photo-Fenton and biological oxidation. J. Hazard. Mater. *155*, 342–349.

Bamforth, S.M., and Singleton, I. (2005). Bioremediation of polycyclic aromatic hydrocarbons: Current knowledge and future directions. J. Chem. Technol. Biotechnol. *80*, 723–736.

Banitz, T., Johst, K., Wick, L.Y., Schamfuß, S., Harms, H., and Frank, K. (2013). Highways versus pipelines: Contributions of two fungal transport mechanisms to efficient bioremediation. Env. Microbiol. Rep. *5*, 211–218.

Barrios-González, J., González, H., and Mejía, A. (1993). Effect of particle size, packing density and agitation on penicillin production in solid state fermentation. Biotechnol. Adv. *11*, 539–547.

Bhalerao, T.S., and Puranik, P.R. (2009). Microbial degradation of monocrotophos by *Aspergillus oryzae*. Int. Biodeterior. Biodegrad. *63*, 503–508.

Bizaj, E., Cuš, F., and Raspor, P. (2011). Removal of pyrimethanil and fenhexamid from *Saccharomyces cerevisiae* liquid cultures. Food Technol. Biotechnol. *49*, 474–480.

Björseth, A., and Ramdahl, T. (1985). Handbook of Polycyclic Aromatic Hydrocarbons. Volume 2 Emission, Sources and Recent Progress in Analytical Chemistry (Marcel Dekker, Inc., New York).

Błaszak, M., Pełech, R., and Graczyk, P. (2011). Screening of microorganisms for biodegradation of simazine pollution (obsolete pesticide azotop 50 WP). Water Air Soil Pollut. *220*, 373–385.

Boonchan, S., Britz, M.L., and Stanley, G.A. (2000). Degradation and mineralization of high-molecular-weight polycyclic aromatic hydrocarbons by defined fungal-bacterial cocultures. Appl. Environ. Microbiol. *66*, 1007–1019.

Bourbonnais, R., and Paice, M.G. (1990). Oxidation of non-phenolic substrates. An expended role for laccase in lignin biodegradation. FEBS Lett. *267*, 99–102.

Bourgin, M., Albet, J., and Violleau, F. (2013). Study of the degradation of pesticides on loaded seeds by ozonation. J. Environ. Chem. Eng. *1*, 1004–1012.

Breugelmans, P., Leroy, B., Bers, K., Dejonghe, W., Wattiez, R., De Mot, R., and Springael, D. (2010). Proteomic study of linuron and 3,4-dichloroaniline degradation by *Variovorax* sp. WDL1: Evidence for the involvement of an aniline dioxygenase-related multicomponent protein. Res. Microbiol. *161*, 208–218.

Cáceres, T., He, W., Naidu, R., and Megharaj, M. (2007). Toxicity of chlorpyrifos and TCP alone and in combination to *Daphnia carinata*: the influence of microbial degradation in natural water. Water Res. *41*, 4497–4503.

Cáceres, T.P., Megharaj, M., Malik, S., Beer, M., and Naidu, R. (2009). Hydrolysis of fenamiphos and its toxic oxidation products by *Microbacterium* sp. in pure culture and groundwater. Bioresour. Technol. *100*, 2732–2736.

Cao, B., Nagarajan, K., and Loh, K.C. (2009). Biodegradation of aromatic compounds: Current status and opportunities for biomolecular approaches. Appl. Microbiol. Biotechnol. *85*, 207–228.

Cébron, A., Norini, M., Beguiristain, T., and Leyval, C. (2008). Real-time PCR quantification of PAH-ring hydroxylating dioxygenase (PAH-RHDα) genes from Gram positive and Gram negative bacteria in soil and sediment samples. J. Microbiol. Methods 73, 148–159.

Chanika, E., Georgiadou, D., Soueref, E., Karas, P., Karanasios, E., Tsiropoulos, N.G., Tzortzakakis, E.A., and Karpouzas, D.G. (2011). Isolation of soil bacteria able to hydrolyze both organophosphate and carbamate pesticides. Bioresour. Technol. *102*, 3184–3192.

Chen, H., Ding, A., Dou, J., Cheng, L., Fan, F., Du, Y., and Liu, X. Optimal conditions for biodegradation of indeno(1,2,3-cd)pyrene in soil slurry reactions. (18 June 2010 through 20 June 2010) Paper presented at 4th International Conference on Bioinformatics and Biomedical Engineering, iCBBE 2010. (Chengdu).

Chen, S., Luo, J., Hu, M., Lai, K., Geng, P., and Huang, H. (2012). Enhancement of cypermethrin degradation by a coculture of *Bacillus cereus* ZH-3 and *Streptomyces aureus* HP-S-01. Bioresour. Technol. *110*, 97–104.

Chen, T., Fu, F., Chen, Z., Li, D., Zhang, L., and Chen, G. (2009). Study on the photodegradation and microbiological degradation of pirimicarb insecticide by using liquid chromatography coupled with ion-trap mass spectrometry. J. Chromatogr. A *1216*, 3217–3222.

Cobas, M., Ferreira, L., Tavares, T., Sanromán, M.A., and Pazos, M. (2013). Development of permeable reactive biobarrier for the removal of PAHs by *Trichoderma longibrachiatum*. Chemosphere *91*, 711–716.

Connell, D., Hawker, D., Warne, M., and Vowles, P. (1997). Basic Concepts of Environmental Chemistry (Lewis Publishers, Boca Raton, FL, USA).

Cruz, J.M., Dominguez, J.M., Dominguez, H., and Parajo, J.C. (2000). Preparation of fermentation media from agricultural wastes and their bioconversion into xylitol. Food Biotechnol. *14*, 79–97.

Dai, Y.J., Ji, W.W., Chen, T., Zhang, W.J., Liu, Z.H., Ge, F., and Sheng, Y. (2010). Metabolism of the neonicotinoid insecticides acetamiprid and thiacloprid by the yeast *Rhodotorula mucilaginosa* strain IM-2. J. Agric. Food Chem. *58*, 2419–2425.

DeLorenzo, M.E., Scott, G.I., and Ross, P.E. (1999). Effects of the agricultural pesticides atrazine, deethylatrazine, endosulfan, and chlorpyrifos on an estuarine microbial food web. Environ. Toxicol. Chem. *18*, 2824–2835.

Dodor, D.E., Hwang, H.-M., and Ekunwe, S.I.N. (2004). Oxidation of anthracene and benzo[a]pyrene by immobilized laccase from *Trametes versicolor*. Enzyme Microb. Technol. 35, 210–217.

Domínguez, A., Gómez, J., Lorenzo, M., and Sanromán, Á. (2007). Enhanced production of laccase activity by *Trametes versicolor* immobilized into alginate beads by the addition of different inducers. World J. Microbiol. Biotechnol. 23, 367–373.

Dosnon-Olette, R., Trotel-Aziz, P., Couderchet, M., and Eullaffroy, P. (2010). Fungicides and herbicide removal in *Scenedesmus* cell suspensions. Chemosphere *79*, 117–123.

Dritsa, V., Rigas, F., Doulia, D., Avramides, E.J., and Hatzianestis, I. (2009). Optimization of culture conditions for the biodegradation of lindane by the polypore fungus Ganoderma Australe. Water Air Soil Pollut. *204*, 19–27.

Du, Y., Dou, J., Cheng, L., Ding, A., Fan, F., and Chen, H. Biodegradation of indeno(1,2,3-cd)pyrene by a pure bacterial culture of *Pandoraea* sp. (18 June 2010 through 20 June 2010) Paper presented at 4th International Conference on Bioinformatics and Biomedical Engineering, iCBBE 2010. (Chengdu).

Echevarria, J., Rodriguez Leon, J.A., Espinosa, M.E., and Delgado, G. (1991). Optimization of solid state fermentation of sugar cane by *Aspergillus niger* considering particles size effect. Acta Biotechnol. *11*, 15–22.

Eriksson, E., Baun, A., Mikkelsen, P.S., and Ledin, A. (2007). Risk assessment of xenobiotics in stormwater discharged to Harrestrup Å, Denmark. Desalination *215*, 187–197.

Errami, M., Salghi, R., Zarrouk, A., Assouag, M., Zarrok, H., Benali, O., Bazzi, E., Hammouti, B., and Al-Deyab, S.S. (2012). Electrochemical degradation of imazalil and pyrimethanil by anodic oxidation on boron-doped diamond. J. Chem. Pharm. Res. *4*, 3518–3525.

European Community. (2004). Regulation No 850/2004 of the European Parliament and of the Council of 29 April 2004 on Persistent Organic Pollutants and Amending Directive 79/117/EEC. Available online: http://eur-lex.europa.eu/legal-content/EN/TXT/PDF/?uri=CELEX:32004R0850&qid=1440759592390&from=EN (accessed 28 August 2015).

Farré, M., Fernandez, J., Paez, M., Granada, L., Barba, L., Gutierrez, H.M., Pulgarin, C., and Barceló, D. (2002). Analysis and toxicity of methomyl and ametryn after biodegradation. Anal. Bioanal. Chem. *373*, 704–709.

Fernando Bautista, L., Sanz, R., Carmen Molina, M., González, N., and Sánchez, D. (2009). Effect of different non-ionic surfactants on the biodegradation of PAHs by diverse aerobic bacteria. Int. Biodeterior. Biodegrad. *63*, 913–922.

Ferreira, L., Cobas, M., Tavares, T., Sanromán, M.A., and Pazos, M. (2013). Assessment of *Arthrobacter viscosus* as reactive medium for forming permeable reactive biobarrier applied to PAHs remediation. Environ. Sci. Pollut. Res. *20*, 7348–7354.

Frazar, C. (2000). The Bioremediation and Phytoremediation of Pesticide-contaminated Sites (U.S. Environmental Protection Agency, Washington, DC, USA).

Fulekar, M.H. (2010). Environmental Biotechnology (CRS Press, Enfield, UK).

Furuno, S., Foss, S., Wild, E., Jones, K.C., Semple, K.T., Harms, H., and Wick, L.Y. (2012). Mycelia promote active transport and spatial dispersion of polycyclic aromatic hydrocarbons. Environ. Sci. Technol. *46*, 5463–5470.

Ghanem, I., Orfi, M., and Shamma, M. (2007). Biodegradation of chlorpyrifos by *Klebsiella* sp. isolated from an activated sludge sample of waste water treatment plant in damascus. Folia Microbiol. *52*, 423–427.

Ghosal, D., Dutta, A., Chakraborty, J., Basu, S., and Dutta, T.K. (2013). Characterization of the metabolic pathway involved in assimilation of acenaphthene in *Acinetobacter* sp. strain AGAT-W. Res. Microbiol. *164*, 155–163.

Ghose, T.K. (1978). Measurement of cellulase activities. Pure Appl. Chem. 257–268.

Ginebreda, A., Kuzmanovic, M., Guasch, H., de Alda, M.L., López-Doval, J.C., Muñoz, I., Ricart, M., Romaní, A.M., Sabater, S., and Barceló, D. (2014). Assessment of multi-chemical pollution in aquatic ecosystems using toxic units: Compound prioritization, mixture characterization and relationships with biological descriptors. Sci. Total Environ. *468–469*, 715–723.

Gómez, J., Alcántara, M.T., Pazos, M., and Sanromán, M.A. (2010). Soil washing using cyclodextrins and their recovery by application of electrochemical technology. Chem. Eng. J. *159*, 53–57.

Gómez, J., Rodríguez Solar, D., Pazos, M., and Sanromán, M.Á. (2006). Applicability of *Coriolopsis rigida* for biodegradation of polycyclic aromatic hydrocarbons. Biotechnol. Lett. *28*, 1013–1017.

Goyat, P., Bhatia, D., and Malik, D.K. (2013). Isolation of Malathion degrading *Pseudomonas xanthomarina* with plant growth promoting activity. Res. J. Chem. Environ. *17*, 59–66.

Grant, R.J., Daniell, T.J., and Betts, W.B. (2002). Isolation and identification of synthetic pyrethroid-degrading bacteria. J. Appl. Microbiol. *92*, 534–540.

Grifoll, M., Selifonov, S.A., and Chapman, P.J. (1994). Evidence for a novel pathway in the degradation of fluorene by *Pseudomonas* sp. strain F274. Appl. Environ. Microbiol. *60*, 2438–2449.

Grishchenkov, V.G., Townsend, R.T., McDonald, T.J., Autenrieth, R.L., Bonner, J.S., and Boronin, A.M. (2000). Degradation of petroleum hydrocarbons by facultative anaerobic bacteria under aerobic and anaerobic conditions. Process Biochem. *35*, 889–896.

Grohmann, K., Cameron, R.G., and Buslig, B.S. (1995). Fractionation and pretreatment of orange peel by dilute acid hydrolysis. Bioresour. Technol. *54*, 129–141.

Guillén-Jiménez, F.D.M., Cristiani-Urbina, E., Cancino-Díaz, J.C., Flores-Moreno, J.L., and Barragán-Huerta, B.E. (2012). Lindane biodegradation by the *Fusarium verticillioides* AT-100 strain, isolated from Agave tequilana leaves: Kinetic study and identification of metabolites. Int. Biodeterior. Biodegrad. *74*, 36–47.

Haderlein, A., Legros, R., and Ramsay, B.A. (2006). Pyrene mineralization capacity increases with compost maturity. Biodegradation *17*, 293–302.

Harayama, S. (1997). Polycyclic aromatic hydrocarbon bioremediation design. Curr. Opin. Biotechnol. *8*, 268–273.

Haritash, A.K., and Kaushik, C.P. (2009). Biodegradation aspects of Polycyclic Aromatic Hydrocarbons (PAHs): A review. J. Hazard. Mater. *169*, 1–15.

Harms, H. (2011). Bioavailability and bioaccessibility as key factors in bioremediation. In Comprehensive Biotechnology, Moo-Young, M., and Agathos, S. eds., (Elsevier, Spain), pp. 83–94.

Harvey, R.G. (1997). Polycyclic aromatic hydrocarbons (Wiley-VCH, New York, NY, USA).

Hedlund, B.P., Geiselbrecht, A.D., and Staley, J.T. (2001). *Marinobacter* strain NCE312 has a *Pseudomonas*-like naphthalene dioxygenase. FEMS Microbiol. Lett. *201*, 47–51.

Hellou, J., Lebeuf, M., and Rudi, M. (2013). Review on DDT and metabolites in birds and mammals of aquatic ecosystems. Environ. Rev. *21*, 53–69.

Herrero-Hernández, E., Andrades, M.S., Álvarez-Martín, A., Pose-Juan, E., Rodríguez-Cruz, M.S., and Sánchez-Martín, M.J. (2013). Occurrence of pesticides and some of their degradation products in waters in a Spanish wine region. J. Hydrol. *486*, 234–245.

Hou, Y., Kong, Y., Yang, J., Zhang, J., Shi, D., and Xin, W. (2005). Biodesulfurization of dibenzothiophene by immobilized cells of *Pseudomonas stutzeri* UP-1. Fuel *84*, 1975–1979.

Hussain, S., Arshad, M., Shaharoona, B., Saleem, M., and Khalid, A. (2009). Concentration dependent growth/non-growth linked kinetics of endosulfan biodegradation by *Pseudomonas aeruginosa*. World J. Microbiol. Biotechnol. *25*, 853–858.

Iglesias, O., Gómez, J., Pazos, M., and Sanromán, M.A. (2014). Electro-Fenton oxidation of imidacloprid by Fe alginate gel beads. Appl. Catal. B Environ. *144*, 416–424.

Ikehata, K., and El Din, M.G. (2006). Aqueous pesticide degradation by hydrogen peroxide/ultraviolet irradiation and Fenton-type advanced oxidation processes: A review. J. Environ. Eng. Sci. *5*, 81–135.

Ji, W.W., Chen, T., Sang, Q., Dai, Y.J., Ge, F., and Yuan, S. (2010). Metabolism of chloronicotinyl neonicotinoid insecticides by fungi *Penicillium oxalicum* IM-3. J. Ecol. Rural Environ. *26*, 246–250.

Jia, K.Z., Li, X.H., He, J., Gu, L.F., Ma, J.P., and Li, S.P. (2007). Isolation of a monocrotophos-degrading bacterial strain and characterization of enzymatic degradation. Huanjing Kexue *28*, 908–912.

Jilani, S. (2013). Comparative assessment of growth and biodegradation potential of soil isolate in the presence of pesticides. Saudi J. Biol. Sci. *20*, 257–264.

Johannes, C., Majcherczyk, A., and Hüttermann, A. (1996). Degradation of anthracene by laccase of *Trametes versicolor* in the presence of different mediator compounds. Appl. Microbiol. Biotechnol. *46*, 313–317.

Juhasz, A., Stanley, G., and Britz, M. (2002). Metabolite repression inhibits degradation of benzo[a]pyrene and dibenz[a,h]anthracene by *Stenotrophomonas maltophilia* VUN 10,003. J. Ind. Microbiol. Biotechnol. *28*, 88–96.

Karpouzas, D.G., Fotopoulou, A., Menkissoglu-Spiroudi, U., and Singh, B.K. (2005). Non-specific biodegradation of the organophosphorus pesticides, cadusafos and ethoprophos, by two bacterial isolates. FEMS Microbiol. Ecol. *53*, 369–378.

Li, R., Zheng, J., Wang, R., Song, Y., Chen, Q., Yang, X., Li, S., and Jiang, J. (2010). Biochemical degradation pathway of dimethoate by *Paracoccus* sp. Lgjj-3 isolated from treatment wastewater. Int. Biodeterior. Biodegrad. *64*, 51–57.

Liu, B., and Tzeng, Y. (1999). Water content and water activity for the production of cyclodepsipeptides in solid-state fermentation by *Metarhizium anisopliae*. Biotechnol. Lett. *21*, 657–661.

Liu, Z., Dai, Y., Huang, G., Gu, Y., Ni, J., Wei, H., and Yuan, S. (2011). Soil microbial degradation of neonicotinoid insecticides imidacloprid, acetamiprid, thiacloprid and imidaclothiz and its effect on the persistence of bioefficacy against horsebean aphid *Aphis craccivora* Koch after soil application. Pest Manage. Sci. *67*, 1245–1252.

Lladó, S., Solanas, A.M., de Lapuente, J., Borràs, M., and Viñas, M. (2012). A diversified approach to evaluate biostimulation and bioaugmentation strategies for heavy-oil-contaminated soil. Sci. Total Environ. *435–436*, 262–269.

Longo, M.A., and Sanromán, M.A. (2006). Production of food aroma compounds: Microbial and enzymatic methodologies. Food Technol. Biotechnol. *44*, 335–353.

Longo, M.A., and Sanroman, M.A. (2009). Application of solid-state fermentation to food industry. In Innovation in Food Engineering: New Techniques and Products Vol. 1, Passos, M.L., and Ribeiro, C.P., eds. (CRC Press, Boca Raton, FL, USA), pp. 107–135.

López, E., Deive, F.J., Longo, M.A., and Sanromán, M.A. (2010). Culture conditions and investigation of bioreactor configurations for lipase production by *Rhizopus oryzae*. Chem. Eng. Technol. 33, 1023–1028.

López-Roldán, R., Jubany, I., Martí, V., González, S., and Cortina, J.L. (2013). Ecological screening indicators of stress and risk for the Llobregat river water. J. Hazard. Mater. *263*, 239–247.

Lu, J., Guo, C., Zhang, M., Lu, G., and Dang, Z. (2014). Biodegradation of single pyrene and mixtures of pyrene by a fusant bacterial strain F14. Int. Biodeterior. Biodegrad. *87*, 75–80.

Lu, X.Y., Zhang, T., and Fang, H.H.P. (2011). Bacteria-mediated PAH degradation in soil and sediment. Appl. Microbiol. Biotechnol. *89*, 1357–1371.

Ma, J., Xu, L., and Jia, L. (2013). Characterization of pyrene degradation by *Pseudomonas* sp. strain Jpyr-1 isolated from active sewage sludge. Bioresour. Technol. *140*, 15–21.

Malato, S., Blanco, J., Cáceres, J., Fernández-Alba, A.R., Agüera, A., and Rodríguez, A. (2002). Photocatalytic treatment of water-soluble pesticides by photo-Fenton and TiO2 using solar energy. Catal. Today *76*, 209–220.

Marín-Benito, J.M., Andrades, M.S., Sánchez-Martín, M.J., and Rodríguez-Cruz, M.S. (2012). Dissipation of fungicides in a vineyard soil amended with different spent mushroom substrates. J. Agric. Food Chem. *60*, 6936–6945.

Mohanta, M.K., Saha, A.K., Zamman, M.T., Ekram, A.E., Khan, A.S., Mannan, S.B., and Fakruddin, M. (2012). Isolation and characterization of carbofuran degrading bacteria from cultivated soil. Biochem. Cell. Arch. *12*, 313–320.

Moldes, D., Rodríguez Couto, S., Cameselle, C., and Sanromán, M.A. (2003). Study of the degradation of dyes by MnP of *Phanerochaete chrysosporium* produced in a fixed-bed bioreactor. Chemosphere *51*, 295–303.

Mollaei, M., Abdollahpour, S., Atashgahi, S., Abbasi, H., Masoomi, F., Rad, I., Lotfi, A.S., Zahiri, H.S., Vali, H., and Noghabi, K.A. (2010). Enhanced phenol degradation by *Pseudomonas* sp. SA01: Gaining insight into the novel single and hybrid immobilizations. J. Hazard. Mater. *175*, 284–292.

Moscoso, F., Deive, F.J., Longo, M.A., and Sanromán, M.A. (2012). Technoeconomic assessment of phenanthrene degradation by *Pseudomonas stutzeri* CECT 930 in a batch bioreactor. Bioresour. Technol. *104*, 81–89.

Moscoso, F., Teijiz, I., Deive, F.J., and Sanromán, M.A. (2013). Approaching chlorpyrifos bioelimination at bench scale bioreactor. Bioprocess Biosyst. Eng. *36*, 1303–1309.

Mulla, S.I., Talwar, M.P., Bagewadi, Z.K., Hoskeri, R.S., and Ninnekar, H.Z. (2013). Enhanced degradation of 2-nitrotoluene by immobilized cells of *Micrococcus* sp. strain SMN-1. Chemosphere *90*, 1920–1924.

Nanda, S., Sahu, S.S., and Abraham, J. (2010). Syudies on the biodegradation of natural and synthetic polyethylene by *Pseudomonas* (spp). J. Appl. Sci. Environ. Manage. *14*, 57–60.

Navarro, S., Fenoll, J., Vela, N., Ruiz, E., and Navarro, G. (2011). Removal of ten pesticides from leaching water at pilot plant scale by photo-Fenton treatment. Chem. Eng. J. *167*, 42–49.

Nayak, A.S., Veeranagouda, Y., Lee, K., and Karegoudar, T.B. (2009). Metabolism of acenaphthylene via 1,2-dihydroxynaphthalene and catechol by *Stenotrophomonas* sp. RMSK. Biodegradation *20*, 837–843.

Niku-Paavola, M.L., Raaska, L., and Itävaara, M. (1991). Detection of white-rot fungi by a non-toxic stain. Mycol. Res. *94*, 27–31.

Ortiz-Hernández, M.L., Quintero-Ramírez, R., Nava-Ocampo, A.A., and Bello-Ramírez, A.M. (2003). Study of the mechanism of *Flavobacterium* sp. for hydrolyzing organophosphate pesticides. Fundam. Clin. Pharmacol. *17*, 717–723.

Pandey, A., Szakacs, G., Soccol, C.R., Rodriguez-Leon, J.A., and Soccol, V.T. (2001). Production, purification and properties of microbial phytases. Bioresour. Technol. *77*, 203–214.

Pandey, G., Dorrian, S.J., Russell, R.J., and Oakeshott, J.G. (2009). Biotransformation of the neonicotinoid insecticides imidacloprid and thiamethoxam by *Pseudomonas* sp. 1G. Biochem. Biophys. Res. Commun. *380*, 710–714.

Pastrana, L.M., Gonzalez, M.P., Pintado, J., and Murado, M.A. (1995). Interactions affecting gibberellic acid production in solid-state culture: A factorial study. Enzyme Microb. Technol. *17*, 784–790.

Pazos, M., Rosales, E., Alcántara, T., Gómez, J., and Sanromán, M.A. (2010). Decontamination of soils containing PAHs by electroremediation: A review. J. Hazard. Mater. *177*, 1–11.

Pazos, M., Iglesias, O., Gómez, J., Rosales, E., and Sanromán, M.A. (2013). Remediation of contaminated marine sediment using electrokinetic-Fenton technology. J. Ind. Eng. Chem. *19*, 932–937.

Peng, H., Yin, H., Deng, J., Ye, J., Chen, S., He, B., and Zhang, N. (2012). Biodegradation of Benzo[a]pyrene by *Arthrobacter oxydans* B4. Pedosphere *22*, 554–561.

Pérez-Bárcena, J.F., Ahuatzi-Chacón, D., Castillo-Martínez, K.L., Ruiz-Ordaz, N., Galíndez-Mayer, J., Juárez-Ramírez, C., and Ramos-Monroy, O. (2013). Effect of herbicide adjuvants on the biodegradation rate of the methylthiotriazine herbicide prometryn. Biodegradation 1–11.

Phugare, S.S., Kalyani, D.C., Gaikwad, Y.B., and Jadhav, J.P. (2013). Microbial degradation of imidacloprid and toxicological analysis of its biodegradation metabolites in silkworm (*Bombyx mori*). Chem. Eng. J. *230*, 27–35.

Pickard, M.A., Roman, R., Tinoco, R., and Vazquez-Duhalt, R. (1999). Polycyclic aromatic hydrocarbon metabolism by white rot fungi and oxidation by *Coriolopsis gallica* UAMH 8260 laccase. Appl. Environ. Microbiol. *65*, 3805–3809.

Pino, N., and Peñuela, G. (2011). Simultaneous degradation of the pesticides methyl parathion and chlorpyrifos by an isolated bacterial consortium from a contaminated site. Int. Biodeterior. Biodegrad. *65*, 827–831.

Pires, C., Marques, A.P.G.C., Guerreiro, A., Magan, N., and Castro, P.M.L. (2011). Removal of heavy metals using different polymer matrixes as support for bacterial immobilisation. J. Hazard. Mater. *191*, 277–286.

Plangklang, P., and Reungsang, A. (2013). Biodegradation of carbofuran in sequencing batch reactor augmented with immobilised *Burkholderia cepacia* PCL3 on corncob. Chem. Ecol. *29*, 44–57.

Pointing, S.B. (2001). Feasibility of bioremediation by white-rot fungi. Appl. Microbiol. Biotechnol. *57*, 20–33.

Poonthrigpun, S., Pattaragulwanit, K., Paengthai, S., Kriangkripipat, T., Juntongjin, K., Thaniyavarn, S., Petsom, A., and Pinphanichakarn, P. (2006). Novel intermediates of acenaphthylene degradation by *Rhizobium* sp. strain CU-A1: Evidence for naphthalene-1,8-dicarboxylic acid metabolism. Appl. Environ. Microbiol. *72*, 6034–6039.

Qasemian, L., Guiral, D., Belghazi, M., Ferré, E., Gros, R., and Farnet, A.-. (2011). Identification of various laccases induced by anthracene and contribution to its degradation in a Mediterranean coastal pine litter. Chemosphere *84*, 1321–1328.

Ramachandran, S., Patel, A.K., Nampoothiri, K.M., Francis, F., Nagy, V., Szakacs, G., and Pandey, A. (2004). Coconut oil cake – a potential raw material for the production of a-amylase. Bioresour. Technol. *93*, 169–174.

Rauh, V.A., Perera, F.P., Horton, M.K., Whyatt, R.M., Bansal, R., Hao, X., Liu, J., Barr, D.B., Slotkin, T.A., and Peterson, B.S. (2012). Brain anomalies in children exposed prenatally to a common organophosphate pesticide. Proc. Natl. Acad. Sci. U.S.A. *109*, 7871–7876.

Rezaee, A., Godini, H., and Bakhtou, H. (2008). Microbial cellulose as support material for the immobilization of denitrifying bacteria. Environ. Eng. Manage. J. *7*, 589–594.

Rodríguez Couto, S., and Sanromán, M.A. (2005). Application of solid-state fermentation to ligninolytic enzyme production. Biochem. Eng. J. *22*, 211–219.

Rodríguez Couto, S., and Sanromán, M.A. (2006). Application of solid-state fermentation to food industry – a review. J. Food Eng. *76*, 291–302.

Rodríguez Couto, S., Gundín, M., Lorenzo, M., and Sanromán, M.A. (2002). Screening of supports and inducers for laccase production by *Trametes versicolor* in semi-solid-state conditions. Process Biochem. *38*, 249–255.

Rodríguez Couto, S., Sanromán, M.A., Hofer, D., and Gübitz, G.M. (2004a). Production of laccase by *Trametes hirsuta* grown in an immersion bioreactor and its application in the decolorization of dyes from a leather factory. Eng. Life Sci. *4*, 233–238.

Rodríguez Couto, S., Sanromán, M.A., Hofer, D., and Gübitz, G.M. (2004b). Stainless steel sponge: A novel carrier for the immobilisation of the white-rot fungus *Trametes hirsuta* for decolourization of textile dyes. Bioresour. Technol. *95*, 67–72.

Rodríguez Couto, S., Rosales, E., and Sanromán, M.A. (2006). Decolourization of synthetic dyes by *Trametes hirsuta* in expanded-bed reactors. Chemosphere *62*, 1558–1563.

Rosales, E., Rodríguez Couto, S., and Sanromán, M.A. (2007). Increased laccase production by *Trametes hirsuta* grown on ground orange peelings. Enzyme Microb. Technol. *40*, 1286–1290.

Rosales, E., Pérez-Paz, A., Vázquez, X., Pazos, M., and Sanromán, M.A. (2012). Isolation of novel benzo[a] anthracene-degrading microorganisms and continuous bioremediation in an expanded-bed bioreactor. Bioprocess Biosyst. Eng. *35*, 851–855.

Rosales, E., Pazos, M., and Ángeles Sanromán, M. (2013). Feasibility of Solid-State Fermentation Using Spent Fungi-Substrate in the Biodegradation of PAHs. Clean Soil Air Water *41*, 610–615.

Sagar, V., and Singh, D.P. (2011). Biodegradation of lindane pesticide by non white- rots soil fungus *Fusarium* sp. World J. Microbiol. Biotechnol. *27*, 1747–1754.

Sanromán, A., Pintado, J., and Lema, J.M. (1994). A comparison of two techniques (adsorption and entrapment) for the immobilization of *Aspergillus niger* in polyurethane foam. Biotechnol. Tech. *8*, 389–394.

Sanromán, A., Feijoo, G., and Lema, J.M. (1996). Immobilization of *Aspergillus niger* and *Phanerochaete chrysosporium* on polyurethane foam. Prog. Biotechnol. *11*, 132–135.

Santos, E.C., Jacques, R.J.S., Bento, F.M., Peralba, M.d.C.R., Selbach, P.A., Sá, E.L.S., and Camargo, F.A.O. (2008). Anthracene biodegradation and surface activity by an iron-stimulated *Pseudomonas* sp. Bioresour. Technol. *99*, 2644–2649.

Seo, J., Keum, Y., Harada, R.M., and Li, Q.X. (2007). Isolation and characterization of bacteria capable of degrading Polycyclic Aromatic Hydrocarbons (PAHs) and organophosphorus pesticides from PAH-contaminated soil in Hilo, Hawaii. J. Agric. Food Chem. *55*, 5383–5389.

Sevillano, X., Isasi, J.R., and Peñas, F.J. (2012). Performance of a fluidized-bed bioreactor with hydrogel biomass carrier under extremely low-nitrogen availability and effect of nitrogen amendments. J. Chem. Technol. Biotechnol. *87*, 402–409.

Šimko, P. (2002). Determination of polycyclic aromatic hydrocarbons in smoked meat products and smoke flavouring food additives. J. Chromatogr. B Biomed. Anal. Technol. Biomed. Life Sci. *770*, 3–18.

Singh, B., Kaur, J., and Singh, K. (2012). Biodegradation of malathion by *Brevibacillus* sp. strain KB2 and *Bacillus cereus* strain PU. World J. Microbiol. Biotechnol. *28*, 1133–1141.

Singh, B.K., and Kuhad, R.C. (1999). Biodegradation of lindane (γ-hexachlorocyclohexane) by the white-rot fungus *Trametes hirsutus*. Lett. Appl. Microbiol. *28*, 238–241.

Singh, B.K., and Walker, A. (2006). Microbial degradation of organophosphorus compounds. FEMS Microbiol. Rev. *30*, 428–471.

Singh, B.K., Kuhad, R.C., Singh, A., Lal, R., and Tripathi, K.K. (1999). Biochemical and molecular basis of pesticide degradation by microorganisms. Crit. Rev. Biotechnol. *19*, 197–225.

Singh, B.K., Walker, A., Morgan, J.A.W., and Wright, D.J. (2003). Role of soil pH in the development of enhanced biodegradation of fenamiphos. Appl. Environ. Microbiol. *69*, 7035–7043.

Šnajdr, J., Dobiášová, P., Vetrovský, T., Valášková, V., Alawi, A., Boddy, L., and Baldrian, P. (2011). Saprotrophic basidiomycete mycelia and their interspecific interactions affect the spatial distribution of extracellular enzymes in soil. FEMS Microbiol. Ecol. *78*, 80–90.

Stredansky, M., and Conti, E. (1999). Xanthan production by solid state fermentation. Process Biochem. *34*, 581–587.

Tillmanns, G.M., Wallnöfer, P.R., Engelhardt, G., Olie, K., and Hutzinger, O. (1978). Oxidative dealkylation of five phenylurea herbicides by the fungus *Cunninghamella echinulata* Thaxter. Chemosphere *7*, 59–64.

US EPA. (1984). Health effects assessment of polycyclic aromatic hydrocarbons (PAHs). EPA Report No. 540/1-86-013 (Environmental Protection Agency, Cincinnati, OH, USA).

US EPA. (2014). Human health issues. Available online: http://www.epa.gov/pesticides/health/human.htm (last accessed January 2014).

Valentín, L., Kluczek-Turpeinen, B., Willför, S., Hemming, J., Hatakka, A., Steffen, K., and Tuomela, M. (2010). Scots pine (*Pinus sylvestris*) bark composition and degradation by fungi: Potential substrate for bioremediation. Bioresour. Technol. *101*, 2203–2209.

Walker, A., and Roberts, S.J. Degradation, biodegradation and enhanced biodegradation (11 October 2013 through 13 October 2013). Paper presented at Degradation, Biodegradation and Enhanced Biodegradation. Proc. 9th Symp. Pesticide Chemistry: The chemistry, mobility and degradation of xenobiotics. Piacenza, Italy.

Wasi, S., Tabrez, S., and Ahmad, M. (2011). Suitability of immobilized *Pseudomonas fluorescens* SM1 strain for remediation of phenols, heavy metals, and pesticides from water. Water Air Soil Pollut. *220*, 89–99.

Wasi, S., Tabrez, S., and Ahmad, M. (2013). Use of *Pseudomonas* spp. for the bioremediation of environmental pollutants: A review. Environ. Monit. Assess. *185*, 8147–8155.

Wilson, S.C., and Jones, K.C. (1993). Bioremediation of soil contaminated with polynuclear aromatic hydrocarbons (PAHs): a review. Environ. Pollut. *81*, 229–249.

Winquist, E., Björklöf, K., Schultz, E., Räsänen, M., Salonen, K., Anasonye, F., Cajthaml, T., Steffen, K.T., Jørgensen, K.S., and Tuomela, M. (2014). Bioremediation of PAH-contaminated soil with fungi – From laboratory to field scale. Int. Biodeterior. Biodegrad. *86*, 238–247.

Xia, X.H., Yu, H., Yang, Z.F., and Huang, G.H. (2006). Biodegradation of polycyclic aromatic hydrocarbons in the natural waters of the Yellow River: Effects of high sediment content on biodegradation. Chemosphere *65*, 457–466.

Xiao, P.F. (2013). Biodegradation and biotransformation of lindane by DDT-degrading white rot fungi. Adv. Mater. Res. *726–731*, 39–42.

Xiao, W., Wang, H., Li, T., Zhu, Z., Zhang, J., He, Z., and Yang, X. (2013). Bioremediation of Cd and carbendazim co-contaminated soil by Cd-hyperaccumulator *Sedum alfredii* associated with carbendazim-degrading bacterial strains. Environ. Sci. Pollut. Res. *20*, 380–389.

Xu, J.L., Wu, J., Wang, Z.C., Wang, K., Li, M.Y., Jiang, J.D., He, J., and Li, S.P. (2009). Isolation and Characterization of a Methomyl-Degrading *Paracoccus* sp. mdw-1. Pedosphere *19*, 238–243.

Yang, L., Zhao, Y., Zhang, B., Yang, C., and Zhang, X. (2005). Isolation and characterization of a chlorpyrifos and 3,5,6-trichloro-2- pyridinol degrading bacterium. FEMS Microbiol. Lett. *251*, 67–73.

Yang, L., Chen, S., Hu, M., Hao, W., Geng, P., and Zhang, Y. (2011). Biodegradation of carbofuran by *Pichia anomala* strain HQ-C-01 and its application for bioremediation of contaminated soils. Biol. Fertil. Soils 47, 917–923.

Ye, D., Akmal Siddiqi, M., Maccubbin, A.E., Kumar, S., and Sikka, H.C. (1996). Degradation of polynuclear aromatic hydrocarbons by *Sphingomonas paucimobilis*. Environ. Sci. Technol. *30*, 136–142.

Yuan, S.Y., Chang, J.S., Yen, J.H., and Chang, B. (2001). Biodegradation of phenanthrene in river sediment. Chemosphere *43*, 273–278.

Zedeck, M.S. (1980). Polycyclic aromatic hydrocarbons: a review. J. Environ. Pathol. Toxicol. *3*, 537–567.

Zhang, G., Ling, J., Sun, H., Luo, J., Fan, Y., and Cui, Z. (2009). Isolation and characterization of a newly isolated polycyclic aromatic hydrocarbons-degrading *Janibacter anophelis* strain JY11. J. Hazard. Mater. *172*, 580–586.

Zhao, J., Zhao, D., and Han, J. Isolation and characterization of dimethoate degrading phytopathogen fungus from soil (11 June 2009 through 13 June 2009) Paper presented at 3rd International Conference on Bioinformatics and Biomedical Engineering, iCBBE 2009. Beijing.

Zhou, G.C., Wang, Y., Zhai, S., Ge, F., Liu, Z.H., Dai, Y.J., Yuan, S., and Hou, J.Y. (2013). Biodegradation of the neonicotinoid insecticide thiamethoxam by the nitrogen-fixing and plant-growth-promoting rhizobacterium *Ensifer adhaerens* strain TMX-23. Appl. Microbiol. Biotechnol. *97*, 4065–4074.

# Detoxification of Hexavalent Chromium from Industrial Wastewater using a Bacterial Biofilm System

9

Zainul Akmar Zakaria, Wan Azlina Ahmad, Wan Haslinda Wan Ahmad and Sindhu Mathew

## Abstract

The incomplete removal of chromium from industrial wastewaters, particularly Cr(VI) species is of importance due to its persistence, stability and high solubility. Chemical reduction followed by precipitation is the most common Cr(VI) treatment technique used in the industry. Such an approach has disadvantages as large volumes of sludge are generated, obnoxious gases may be released and the process can be associated with high operating costs. Microorganisms such as bacteria have long been reported to have the ability to reduce Cr(VI). The use of Cr(VI) resistant-reducing bacterial biofilm offers an interesting solution to treat Cr(VI) containing industrial wastewater as well as the bioremediation of Cr(VI)-polluted areas. Nevertheless, promising results obtained from laboratory scale experiments do not necessarily equate to success under large-scale operating conditions. This is because important operational parameters are normally controlled in the laboratory and huge natural fluctuations in the wastewater are avoided. Hence it is necessary to assess the robustness of the system at a larger scale where it can be subjected to fluctuations naturally occurring in wastewater, and as typically encountered in factory compounds or contaminated areas.

## Introduction

The name 'chromium' is derived from the Greek word 'chroma', meaning colour, as this metal forms a wide variety of compounds with different colours. $Na_2Cr_2O_7$ is bright orange, $Cr_2O_3$ is green and the Zn and Pb salts of $CrO_4^{2-}$ are bright yellow. Chromium (Cr) is an essential micronutrient in the diet of animals and humans and is required for the normal sugar, protein and lipid metabolism in mammals (Mordenti *et al.*, 1997); it is involved in a complex named glucose tolerance factor (GFC) which functions as an insulin cofactor. However Cr is a non-essential metal for microorganisms and plants (Cervantes *et al.*, 2001; Dhal *et al.*, 2013) and elevated levels of Cr are toxic to the environment and living organisms. Toxicity depends on the oxidation state of Cr. It can exist in oxidation states from −2 to +6, with 0, +3 and +6 being most common (Windholz *et al.*, 1976).

From total chrome ore production, 90% is used in metallurgical industries for steel, alloy and nonferrous alloy production, 5% in refractory (iron and steel, cement, glass, ceramics and machinery) and the remaining 5% in chemical industries such as leather tanning, electroplating, wood preservation, in pigment production and as an oxidizing agent (Dhal *et al.*, 2013). $CrO_4^{2-}$ is effective in 'passivating' steel, by reacting to form a protective oxide layer on the surface. Wastes from production operations are a prime source of Cr pollution. Cr is resistant to ordinary corrosive agents at room temperature which facilitates its application in the electroplating industry. Cr compounds have been used for many years as pigments in paints and other coatings. Cr is also important in converting animal hides into leather. In the tanning process, the hides are treated with basic solutions of Cr(III) salts, which cause the crosslinking of collagen proteins (Olmsted and Williams, 1994). The Cr ion in liquid tanning wastes occurs mainly in its trivalent form, which gets further oxidized to hexavalent forms (Garg *et al.*, 2007). Nearly 80–90% of leather is tanned with Cr chemicals (Papp, 2004), with 40% of Cr being discharged as mixtures of Cr(VI) and Cr(III) effluent (Saha and Orvig, 2010).

## Cr(VI)-containing industrial waste waters

Hexavalent chromium contamination of the environment is the result of extensive anthropogenic use of chromate and dichromate in industries such as stainless steel production, metal finishing, electroplating, leather tanneries, ink manufacture, pigment fabrication, automobile manufacturing, glass and ceramics, cement, corrosion inhibitors in cooling water, wood preservation, power plants and nuclear facilities (Dhal *et al.*, 2013; Viamajala *et al.*, 2003). About 35% of Cr released from all anthropogenic sources is Cr(VI). It is a hazardous contaminant as it readily spreads beyond the site of initial contamination through aquatic systems and ground water (Quintelas and Tavares, 2001).

## Trivalent and hexavalent chromium

In the environment, Cr exists primarily as trivalent, Cr(III), and hexavalent chromium, Cr(VI), which are the stable forms occurring in nature. Cr(VI) is rarely naturally occurring, is highly toxic, relatively soluble in aqueous systems and is readily transformed in groundwater (Hossain *et al.*, 2005). In contrast, Cr(III) exists naturally in the environment, is less toxic and is even required for human glucidic metabolism (Zayed and Terry, 2003). Cr(III) is prevalent in municipal waste water rich in organics (Fukai, 1967) and displays a high affinity for negatively charged ions and colloids resulting in the formation of complexes that precipitate as amorphous hydroxide (Sawyer *et al.*, 1978). Most of the Cr(III) present in water is adsorbed onto particulate matter, whereas Cr(VI) remains in solution.

The existence of Cr(VI) and Cr(III) in nature at different redox potentials (Eh) and pH levels have been reported (Reeder *et al.*, 2006). The change in Cr valence correlates with the pH and redox potential characteristics of the surrounding environment. Cr(III) is the most thermodynamically stable oxidation state under reducing conditions. Cr(III) predominates at pH < 3.0, while at pH > 3.5, hydrolysis of aqueous Cr(III) yields trivalent chromium hydroxyl species such as $CrOH^{2+}$, $Cr(OH)_2^{+}$, $Cr(OH)_3^{0}$ and $Cr(OH)_4^{-}$, where $Cr(OH)_3^{0}$ is the only solid species, existing as an amorphous precipitate. Cr(VI) exists primarily as salts of chromic acid ($H_2CrO_4$), hydrogen chromate ion ($HCrO_4^{-}$) and chromate ion ($CrO_4^{2-}$), depending on the pH. $H_2CrO_4$ predominates at pH < 1.0, $HCrO_4^{-}$ at pH between 1.0 and 6.0 and $CrO_4^{2-}$ at pH > 6.0. The dichromate ion ($Cr_2O_7^{2-}$), a dimer of $HCrO_4^{-}$, minus a water molecule, forms when the concentration of Cr exceeds approximately 1 g/l (Mohan and Pittman Jr., 2006).

## Toxicity of hexavalent chromium

Cr(VI) is extremely toxic and causes eczema, allergies and respiratory tract disorders. In humans, exposure to Cr(VI) has resulted in nasal irritation, skin irritation, ear drum perforation (Ahluwalia and Goyal, 2007), kidney, liver and gastric damage (Acar and Malkoc, 2004) and lung carcinoma (Rene'N *et al.*, 2006) It has been reported to possess carcinogenic (Lindsay and von Holy, 1997) and mutagenic effects (Quintelas *et al.*, 2008) on biological systems owing to its strong oxidizing nature. Cr(VI) has been classified as Group 'A' human carcinogen and one of the main pollutants by the US EPA (Dhal *et al.*, 2013). Cr(VI) can accumulate in the placenta impairing fetal development in mammals (Ahmad *et al.*, 2010) and also cause decreases in reproductive health (Kanojia *et al.*, 1998). The National Institute of Occupational Health and Safety (NIOHS) recommends an exposure limit of 1 mg/l for Cr(VI) and 500 μg/l for Cr(III) for a 10 h work day, 40 h week.

Cr(VI) is easily taken up through transport channels in the cell membrane of both prokaryotic and eukaryotic organisms (Liu *et al.*, 1995; Zouboulis *et al.*, 2004). Once inside the cell, Cr(VI) reacts spontaneously with intracellular reductants such as ascorbate or glutathione to generate intermediates such as Cr(V) and/or Cr(IV) and the end product Cr(III) (Kulczycki *et al.*, 2002). Cr(V) can undergo redox cycling to regenerate Cr(VI) and in the process it donates an electron to oxygen, generating reactive oxygen species (ROS) that can combine with DNA–protein complexes and modify them. The $LD_{50}$ (dose causing death of 50% of the population) for oral toxicity in rats is 50–100 mg/kg for Cr(VI) and 1900–3000 mg/kg for Cr(III) (De Flora *et al.*, 1990). The deleterious effects of Cr(VI) is due to the ROS generated during the one electron reduction of Cr(VI) to Cr(V) by the intracellular Cr(VI) reductants (Cheung and Gu, 2007). Cr(VI) can easily penetrate cellular membranes and undergo reduction reactions, giving rise to various reactive intermediates (Cervantes *et al.*,

2001). The biological effects of chromate appear to be related to its uptake, since Cr(VI) passes easily across cellular membranes, but the trivalent form is incapable of crossing cellular membranes (Vaiopoulou and Gikas, 2012).

When fish are exposed to hexavalent chromium dissolved in water, the gills function as the main route of its uptake (Knoll and Fromm, 1960) and this is usually associated with structural damage to the gill and impaired respiration and osmoregulatory function. Cr(VI) can pass readily through the gill membrane and accumulate rapidly in the various tissues and organs such as bone, spleen, kidney, gall bladder, gastrointestinal tract and brain (Buhler *et al.*, 1977). Generally trivalent chromium appears to be more acutely toxic to fish than hexavalent chromium. The $LC_{50}$ for Cr(III) pooling data for all species and water conditions was 22 mg/l, significantly lower than 100.7 mg/l for Cr(VI) (Holdway, 1988).

Cr(VI) contamination in the environment also alters the structure of soil microbial communities (Kang *et al.*, 2006). Short term Cr(VI) exposure inhibits soil biological properties such as phosphatase and sulfatase activities and may decrease microbial biomass (Viti and Giovannetti, 2007). Cr(VI) at 10 to 12 mg/l was observed to inhibit most bacterial isolates growing in semisynthetic media. Gram-negative bacteria were more sensitive to Cr than Gram-positives (Ross *et al.*, 1981). It was found to be mutagenic to *E. coli* (Lindsay and von Holy, 1997) and *S. typhimurium* (Petrilli and De Flora, 1977).

## Conventional removal techniques for Cr(VI) from wastewater

A wide array of technologies is available for the removal of Cr from wastewater. Some of these are well established methods that have been in practice for decades, others are more recent innovations. The discharge of Cr (VI) to the aqueous environment is regulated to below 0.05 mg/l by the US EPA (Baral and Engelken, 2002) and the European Union. The maximum allowable level of Cr (VI) in drinking water is 50–100 μg/l (LaGrega *et al.*, 2001). According to the Environmental Quality Regulation (EQR) (Industrial Effluent) 2009, amendment on Environmental Quality Act (EQA) 1974, the permissible discharge limit for Cr (VI) is 0.05 mg/l for both Standard A and Standard B (Department of Environment, 2009), 0.20 mg/l for Standard A and 1.0 mg/l for Standard B for Cr(III).

In practical applications, the choice of treatment depends on effluent characteristics such as concentration of Cr(VI), pH, temperature, flow volume, biological oxygen demand (BOD), the economics involved and social factors including the regulatory standards set by government agencies (Acharya *et al.*, 2009).

### Physico-chemical techniques for the removal of Cr(VI)

The most commonly used technique for the removal of Cr in wastewater is chemical reduction followed by precipitation, using reagents such as $SO_2$ gas or sodium bisulphite in an acid solution. Both of these reactants form the same active reducing agent, sulphurous acid, which reduces Cr(VI) to Cr(III) under acidic conditions and subsequent adjustment of the solution pH to near-neutral precipitates of Cr(III) as hydroxides, as shown in Equation 9.4.

$$SO_2 + H_2O \text{ ® } H_2SO_3 \tag{9.1}$$

$$NaHSO_3 + H^+ \text{ ® } H_2SO_3 + Na^+ \tag{9.2}$$

The theoretical requirement is 3 kg of sodium bisulphite plus 2–3 kg of sulphuric acid, to reduce 1 kg of hexavalent Cr. Equation 9.3 takes place almost instantaneously at a pH of 2.5.

$$4CrO_4^{2-} + 6NaHSO_3 + 3H_2SO_4 + 8H^+ \text{ ® } 2Cr_2(SO4)_3 + 3Na_2SO_4 + 10H_2O \tag{9.3}$$

Further treatment with NaOH precipitates the Cr(III) as indicated in Equation 9.4.

$$Cr_2(SO4)_3 + 6NaOH \text{ ® } 2Cr(OH)_3 + 3Na_2SO_4 \tag{9.4}$$

Since $Cr(OH)_3$ is precipitated, the resulting effluent will contain little or no residual Cr. The $SO_2$ reacts as follows (Equation 9.5):

$$2CrO4^{2-} + 3SO_2 + 4H^+ \text{ ® } Cr_2(SO4)_3 + 2H_2O \tag{9.5}$$

Following reduction, the effluent is treated with NaOH solution or $Ca(OH)_3$ slurry to neutralize the acidity and precipitate the Cr. The use of NaOH solution is advised where sludge formation must be reduced to an absolute minimum. However in general, the disadvantage of this process is the generation of secondary sludge due to the chemical reagents used and the high cost (Higgins *et al.*, 1997). The sludge presents difficulties in process management, sludge transportation and final disposal issues.

Other processes employed are ultrafiltration (Ghosh and Bhattacharya, 2006) adsorption, membrane technologies (Hsu *et al.*, 2011), electrocoagulation (Akbal and Camci, 2011), ion exchange (Rengaraj *et al.*, 2003), electro-dialysis, reverse osmosis (Ahluwalia and Goyal, 2007) and solvent extraction (Agrawal *et al.*, 2008) but the equipment, maintenance and operational costs of all these processes are exceedingly high for large-scale treatment (Beleza *et al.*, 2001). One of the major limitations of some of these processes is that it is economically viable only at high or moderate Cr concentrations and not at low concentrations (1–100 mg/l). Other disadvantages include incomplete metal removal, high energy requirements and contamination of ground water due to the generation of secondary waste (Cossich *et al.*, 2004). At low concentrations of Cr (10–100 mg/l), ion exchange is considered a better alternative but it is not economically appealing. Moreover they generate large volumes of sludge and also release obnoxious gases such as $H_2S$ (Ahmad *et al.*, 2010). Therefore, it is important to develop an innovative, low cost and ecofriendly method for removal of Cr from the wastewater (Zahoor and Rehman, 2009).

## Biological methods for the removal of Cr(VI)

Biological systems, such as biosorption, bioreduction or bioaccumulation using living cells have been examined for their Cr removal abilities (Dermou *et al.*, 2007; Fude *et al.*, 1994; Sharma and Forster, 1993) where these systems offer potential ecofriendly alternatives to existing technologies to remove Cr(VI) from wastewater (Chen and Gu, 2005). A wide variety of microorganisms such as bacteria, yeast, algae, protozoa and fungi have developed the capabilities to protect themselves from heavy metal toxicity by various mechanisms such as adsorption, uptake, methylation, oxidation and reduction (Zahoor and Rehman, 2009).

### Biosorption

Biosorbents were demonstrated to be more suitable for Cr removal at relatively low concentrations (10–100 mg/l) compared to other methods (Mona *et al.*, 2011). The Cr(VI) biosorption properties of algae, bacteria, cyanobacteria, yeast, fungi and agricultural products like wood bark, sawdust, bagasse and coir pith have been reviewed in detail (Saha and Orvig, 2010). Biosorption of hexavalent Cr by live and heat killed biomass of *Acinetobacter junnii*, an isolate from a chromite mine, showed the reduction of Cr(VI) to Cr(V) on the biomass surface upon Cr(VI) adsorption (Paul *et al.*, 2012). Several other researchers have also studied the adsorption of Cr(VI) from aqueous solution using cellulosic, lignocellulosic and agricultural waste materials (Ahmad *et al.*, 2010; Khan *et al.*, 2004).

The biggest barrier in the application of biosorption process by industry is the high cost of adsorbents presently available for commercial use. The cost of adsorption technology can be reduced if the adsorbent is inexpensive. So, there is a need to develop low cost and easily available adsorbent for the removal of Cr ions from the aqueous environment such as from agricultural waste. Cr(VI) biosorption still today is mainly confined to lab studies as the mechanism of biosorption is not fully understood and due to shortcomings in biosorption technology and limited research on chemically modified biosorbent mixtures. However, a biosorption system consisting of a bacterial biofilm of *Arthrobacter viscosus* supported on synthetic zeolites in which the biofilm reduces Cr(VI) to Cr(III) and trivalent chromium is captured in the zeolite by ion exchange has been reported (Tavares *et al.*, 2007).

### Bioaccumulation

Bioaccumulation is a metabolic-dependent process, whereby the pollutant or cation is accumulated intracellularly across the cell membrane or is removed through the cell metabolic cycle (Malik, 2004). Cr present in the form of $CrO_4^{2-}$ in groundwater has been demonstrated to be removed by the unicellular yeast *Saccharomyces cerevisiae* under both aerobic and anaerobic conditions with the rate being slightly higher under anaerobic conditions.

Freshly hydrated cells showed a rapid initial removal (55–67 mg/h/g biomass dry wt) of Cr(VI) followed by a slower uptake (0.6–1.1 mg/h/g biomass dry wt) which reduced with time. The main advantage of using growing cultures in bio-removal is avoiding the need for a separate biomass production process, for instance cultivation, harvesting, drying, processing and storage prior to use (Krauter *et al.*, 1996). However there are significant practical limitations with systems employing living cells, including the fact that bioremoval efficiency is limited by the potential toxic effects of the heavy metals in solution. For example, the growth of *A. niger* was greatly affected and inhibited by an initial Cr(VI) concentration of 75 mg/l (Dursun *et al.*, 2003).

### Bioreduction

The most commonly used approach for the detoxification of wastewaters containing Cr(VI) is its reduction to the much less toxic and immobile form of Cr(III). Microbes capable of reducing Cr(VI) have been discovered beginning from the 1970s (Romanenko and Koren'kov, 1977) and the pursuit of strains capable of efficient transformation has resulted in the isolation of several aerobic and anaerobic microbes. Since several microorganisms possess the capability to reduce Cr(VI) to Cr(III) (Camargo *et al.*, 2004; He *et al.*, 2011; Zakaria *et al.*, 2007), it gives immense opportunities for the development of technologies to detoxify Cr(VI)-contaminated wastewater as an alternative to the existing physico-chemical technologies (Dhal *et al.*, 2013).

The cell membrane is impermeable to Cr(III) and therefore its toxicity is only 1/1000 of the toxicity of Cr(VI) (Czako-Ver *et al.*, 1999). The insolubility of Cr(III) facilitates the removal of Cr, if Cr(VI) can be biotransformed to Cr(III). Both Gram-negative and Gram-positive bacteria are capable of converting Cr(VI) into Cr(III) (Pal and Paul, 2004). Enzymatic reduction of chromate by Cr(VI)-resistant bacteria proceeds aerobically, anaerobically or both (Thacker *et al.*, 2006). The physiological mechanisms involved in Cr(VI) reduction vary widely among species. In some cases, the enzyme catalysed reaction is membrane associated, whereas in others the enzyme is in the soluble fraction. A soluble Cr(VI) reductase was partially purified from *Pseudomonas putida* PRS2000 (Ishibashi *et al.*, 1990). Cr(VI) reductase from *Pseudomonas ambigua* G-1 was purified to 38 fold (Suzuki *et al.*, 1992) while 600 fold purity was obtained for ChrR, the reductase from *Pseudomonas putida* MK1 (Park *et al.*, 2000). However, the availability of effective Cr(VI) reducing organisms is an essential prerequisite for the bioreduction of Cr contaminated wastewater or soil (Pattanapipitpaisal *et al.*, 2001). Application of Cr(VI)-resistant bacteria for detoxification of Cr(VI) has been considered as an economical, effective and safe procedure over conventional physico-chemical methods (Pinon-Castillo *et al.*, 2010).

## Bacterial resistance and reduction of Cr(VI)

The Cr resistance of bacteria and yeast has been studied (Luli *et al.*, 1983; Trevors *et al.*, 1985). In pseudomonads the Cr resistance is plasmid encoded. Plasmids in *P. aeruginosa* belonging to the Inc P-2 group provided resistance to Cr. The plasmid encoded resistance mechanism enabled the organism to tolerate a two to four fold increase in metal concentrations. *P. fluorescens* LB300 harboured plasmid pLHB1 which conferred chromate resistance on the organism. As a result they were able to tolerate 40- to 200-fold higher levels of chromate (Bopp *et al.*, 1983). *P. ambigua* G-1 was resistant to 4000 mg/l Cr(VI) (Horitsu *et al.*, 1983). Prokaryotes are more resistant to Cr(VI) than eukaryotes (Kamaludeen *et al.*, 2003) and Gram-positive bacteria have been reported to be more chrome tolerant than Gram-negative bacteria.

## Application of a bacterial biofilm system in Cr(VI) detoxification

Over 99% of microorganisms on earth live in the form of biofilms (Vu *et al.*, 2009), which are populations of cells surrounded by a matrix of extracellular polymeric substances (EPS) (Denkhaus *et al.*, 2007). Bacteria have a high surface area to volume ratio and therefore can provide a large contact surface for interaction with metals from the surrounding environment (Zouboulis *et al.*, 2004). The structural polymers in the bacterial cell wall provide acidic functional groups like carboxyl, phosphoryl and hydroxyl groups that are responsible for the reactivity with metal ions (Kulczycki *et al.*, 2002).

The binding of Cr ions by the supported biofilms has been studied by several authors. Quintelas and Tavares (2001) used a biofilm of *Arthrobacter viscosus* to remove Cr(VI), while Kang *et al.* (2006) applied a biofilm of *Pseudomonas aeruginosa* for the removal of Cr(III). Ahmad *et al.* (2010) investigated the removal of Cr from industrial waste water using the ChromeBac™ system in which *Acinetobacter haemolyticus* was immobilized onto the carrier material of rubber wood sawdust. Complete reduction of Cr(VI) to Cr(III) was observed for neutralized electroplating wastewater with a Cr(VI) concentration of 17–81 mg/l. Quintelas *et al.* (2008) investigated the ability of three different bacterial species *Streptococcus equisimilis, Bacillus coagulans* and *Escherichia coli* supported on granular activated carbon (GAC) to remove Cr(VI) from low concentration liquid solutions in batch and column studies. Metal uptake values of 5.82, 5.35 and 4.12 mg/g biosolids, respectively was obtained for *S. equisimilis, B. coagulans* and *E. coli*, for an initial metal concentration of 100 mg/l. Studies with industrial effluent showed Cr uptake values of 0.083, 0.090 and 0.110 mg/g biosolids, respectively for *S. equisimilis, B. coagulans* and *E. coli* at an initial concentration of 4.2 mg/l. The reduction of negatively charged Cr(VI) to Cr(III), a cation, enhances the sorption ability on the Gram-negative *E. coli.* The lower values of Cr removal in industrial effluent are due to inhibition in the presence of other metal ions. The Gram-positive bacteria (*B. coagulans* and *S. equisimilis*) presented better metal removal percentages in batch studies than in column studies due to the higher contact time (Quintelas *et al.*, 2008).

Activated carbon might be able to retain Cr from liquid solutions in certain conditions but the fact that the hexavalent ion is negatively charged as chromate ($CrO_4^-$) or dichromate ($Cr_2O_7^{2-}$) and strongly hydrated drastically reduces the uptake of this metal by GAC. On the other hand, the cationic form $Cr^{3+}$ is easily retained by the adsorbent. The novelty of this study is the synergetic effect of the combination of the bacterial biofilms ability to reduce $Cr^{6+}$ to $Cr^{3+}$ and GAC, and ability to retain $Cr^{3+}$ on its surface. The biofilm by itself may not be able to retain appreciable amounts of Cr, but the carbon matrix will allow Cr accumulation for downstream processing. Without the biofilm, GAC would not adsorb the chromate or dichromate ions due to ionic repulsions and steric limitations (Quintelas *et al.*, 2008).

## Case study: Bioremediation of Cr(VI) wastewater by a biofilm system comprising the locally isolated Cr(VI) resistant-reducing *Acinetobacter haemolyticus*

### Introduction

Numerous reports have demonstrated the feasibility of using bacterial processes for the treatment of Cr (VI) containing industrial wastewaters by using either pure cultures or bacterial consortia. The use of pure cultures such as *Bacterium dechromaticans* (Romanenko *et al.*, 1976) and *Pseudomonas fluorescens* LB300 (Bopp and Ehrlich, 1988) is useful but not suitable for *in situ* applications as it is prone to contamination with other organisms, leading to declines in Cr(VI) reduction ability over time. The use of selected Cr(VI)-resistant and Cr(VI)-reducing indigenous microbes from the Cr(VI) contaminated wastewater is therefore advantageous.

In industrial applications, nutrient rich by-products of industrial processes such as molasses and maize steep liquor are preferred due to its abundance and cheap pricing (Eccles, 1999). In this case study, liquid pineapple waste was chosen as a nutrient for bacteria, as it is found in abundance in the Johor state of Malaysia and due to its extremely cheap pricing at RM 0.167/tonne (Lee Pineapple Sdn. Bhd., personal communication). One example of a bioremediation process that was short-lived, due to high nutrient pricing, is the bacterium-based metal removal system developed by Advanced Mineral Technologies, Inc. in Colorado that uses *Bacillus* sp. as biosorbent (Volesky, 1990).

Supports used for microbial attachment can be divided into two main categories, namely organic and inorganic, and it is widely accepted that organic carriers (polysaccharides, proteins and synthetic polymers) are better than the inorganic carriers (alumina, zirconia, magnesia, silica) due to the variety of reactive groups present on its surfaces such as carboxyl, amino and hydroxyl groups (Kolot, 1981a). Cellulose was chosen as the support material for bacterial immobilization in the column. Cellulose contains three easily substitutable hydroxyl groups on each anhydro glucose unit. Substitution with functional groups such as diethylaminoethyl and carboxymethyl provide weakly basic and acidic ion exchange conditions respectively (Kolot, 1981b) which enhance bacterial attachment. Regular cellulose has a fibrous shape, is porous and has high surface area (Kolot, 1981a). Supports having high porosity and surface area has been suggested by Macaskie and Dean (1990) as a prerequisite for bacterial attachment. The low cost and stability of cellulose prompted its selection as the support material for use in this study.

Even though *Acinetobacter haemolyticus* (*A. haemolyticus*) was isolated from Cr(VI)-containing batek wastewater, i.e. textile-related effluent, Cr(VI) concentrations in the batek wastewater was within the stipulated discharge limit according to Standard B (Department of Environment, 2004), i.e. 0.05 mg/l, and hence not very suitable for use in Cr(VI) reduction study. This prompted us to look for other Cr(VI)-containing industrial wastewater. In this study, electroplating wastewater was selected as the effluent to be treated in the developed bioremediation system. Even though most of these electroplating premises have proper waste treatment facilities, possibilities of untreated heavy metals such as Cr to make their way into the environment should not be taken lightly.

This case study assesses the potential application of a locally developed bioremediation system for treating Cr(VI) containing wastewater. The system was demonstrated using a 2-l flow-through column packed with solid agricultural waste to develop a biofilm primarily consisting of the Cr(VI)-reducing-resistant *Acinetobacter haemolyticus* EF369508.

## Materials

### Bacteria

In this work, *Acinetobacter haemolyticus* was isolated from the effluent of a batek (textile-related) manufacturing premise in Kota Bharu, Kelantan, Malaysia. It was grown in nutrient broth (NB) for 24 hours, and shaken at 150 rpm at 30°C before being immobilized onto the column support material. The batek effluent contains a mixture of heavy metals which was determined (ICP-MS, Toshiba UP6100) to be slightly higher than the permissible discharge limit according to Standard B in Malaysia (Department of Environment, 2004). *A. haemolyticus* was isolated using the method suggested by Greenberg *et al.* (1985) and identified using the Microlog® Biolog Identification System and via 16S rRNA gene analysis which was carried out by First BASE Laboratories Sdn. Bhd., Malaysia. Results from the 16S rRNA gene sequence analysis for *A. haemolyticus* are shown in Table 9.1. To date, *A. haemolyticus* has not been reported as either a chromium-resistant or a chromium-reducing bacterium.

### Nutrient

Liquid pineapple waste (LPW) was obtained from Lee Pineapple Sdn. Bhd., a pineapple processing premise in Tampoi, Johor Bahru, Malaysia. The LPW was sampled either from the sedimentation tanks or after the solid waste grinding section. At the solid waste grinding section, solid components from the pineapple fruit, such as the skin, were separated from the liquid part via grinding. Prior to use in the experiment, effluent from the waste grinding section was filtered several times through muslin cloth to remove suspended materials. Effluent from the sedimentation tanks requires less filtration because of less suspended materials present. Sediments containing small but coarse pineapple skin and husks were discarded. Some of the characteristics of LPW are as shown in Table 9.2.

### Cr(VI)-containing wastewater

The Cr(VI)-containing electroplating wastewater was obtained from one of the electroplating premises located in the Pasir Gudang Industrial Area, Johor, Malaysia. Some of the characteristics of

**Table 9.1** Identification of *A. haemolyticus* by 16S rRNA gene sequence analysis

| Species as close relatives | Accession No. | Similarity (%) |
|---|---|---|
| *Acinetobacter haemolyticus* | X81662 | 99.5 |
| *Acinetobacter haemolyticus* | AY586400 | 99.5 |
| *Acinetobacter* sp. | AY639376 | 99.5 |
| *Acinetobacter* sp. | AY902243 | 98.6 |
| Uncultured *Acinetobacter* sp. | AF467306 | 98.6 |

**Table 9.2** Some characteristics of liquid pineapple waste

| Parameter | Characteristic |
|---|---|
| pH | 3.19 ± 0.28 |
| Colour | Yellowish green |
| Temperature (°C) | 33.9 ± 1.61 |
| Bacterial count | 4 colonies isolated |
| *$BOD_5$ (mg/l) | 16830 ± 593.97 |
| *COD (mg/l) | 55725 ± 1449.57 |
| $SO_4^{2-}$ (mg/l) | 58.2 ± 1.41 |

*Values shown are mean values of duplicate samples from two sampling trips made between February and October 2005; values for pH and temperature are average of $n = 5$.

**Table 9.3** Some characteristics of Cr (VI) containing electroplating wastewater

| Parameter | Profile |
|---|---|
| Colour | Yellow to dark orange |
| Turbidity (NTU) | Clear |
| pH | 2.30–2.70 ± 0.05 |
| Temperature (°C) | 30.7 ± 3.39 |
| Bacterial count | No colony isolated |
| $SO4_2^-$ (mg/l) | 4.45 ± 3.32 |
| Pb (mg/l) | 1.97 ± 0.43 |
| As (mg/l) | 0.29 ± 0.13 |
| Hg (mg/l) | 0.02 ± 0.01 |
| Cu (mg/l) | 0.63 ± 0.11 |
| Fe (mg/l) | 1.49 ± 0.18 |
| Ni (mg/l) | 4.76 ± 1.67 |
| Cd (mg/l) | 0.08 ± 0.01 |
| Cr (mg/l) | 328.77 ± 82.38 |
| Pb (mg/l) | 1.97 ± 0.43 |

*Values shown are mean of three samples.

Cr(VI) wastewater collected are as shown in Table 9.3. The choice of the parameters characterized was based on their potential influence on bacterial Cr(VI) reduction (Wang, 2000). Low concentrations of sulphate ions at 4.45 mg/l indicate that it may not compete with Cr(VI) uptake by *A. haemolyticus*. This was based on an earlier finding that Cr(VI) reduction by *A. haemolyticus* was unaltered even at 10 mM $SO_4^{2-}$, or 2101.6 mg/l. Colour of the effluent depends on types of cleaning and electroplating baths. For example, Cr(VI) would yield a yellow to dark orange effluent (Cushnie Jr, 1985) while Cu would give a blue solution. The clear effluent is a result of the extensive use of water to rinse excess metallic compound on the plated surface. Some electroplating premises have a number of electroplating baths that run concurrently depending on production.

The use of different metals such as Zn, Cu, Cd, Ni and Cr in each electroplating bath results in wastewater containing a mixed metal solution. The value of 328.77 mg/l for Cr was much higher than other metals with the closest being Ni and Pb (4.76 and 1.97 mg/l respectively). Of the 328.77 mg/l Cr, Cr(VI) was determined as 256 mg/l via a diphenylcarbazide (DPC) method. Higher concentrations of Cr(VI) were expected due to the acidic pH (2.30–2.70) of electroplating wastewater. However, the difference between Cr value determined by ICP-MS and DPC was due to the difference in sensitivity of the two methods where the ability of ICP-MS to detect concentrations as low as 0.05 µg/l far outweighs the DPC method with a sensitivity limit of 5 µg/l. Pb, Cr and Hg were present at concentrations 4–10 times higher than permissible discharge limit according to Standard B (Department of Environment, 2004), while the rest of the metals were discharged close or below the permissible concentrations. High concentrations determined for Pb and Cr may be the result of their extensive use in production activities. Another important parameter, total organic carbon (TOC) was not analysed due to high concentrations of Hg as its presence may disrupt analysis by the TOC analyser (TOC V-E, Shimadzu) which only permits Hg concentrations of up to 0.005 mg/l in the solution. TOC is important as it can be used to indicate whether the effluent contains enough carbon to sustain bacterial growth.

## 2-I Cr(VI) bioremediation setup

The experimental setup for the Cr(VI) reduction system consists of a holding tank, peristaltic pump, column (bioreactor), precipitation tank and disinfection tank and is shown in Fig. 9.1.

As a support material, a commercial cellulose bath sponge was sliced into 1 $cm^3$ cubes. Additionally, inert stone (not-uniform size, 2 mm to 5 mm) was packed at the inlet and outlet sections of the column to ensure even liquid flow distribution as well as maintaining the column content (Fig. 9.2). Cellulose was chosen as the support material to immobilize *A. haemolyticus* in the column based on several factors mentioned earlier. Before *A. haemolyticus* cells were attached to the cellulose support material to form biofilm, the column was first flushed with 1 L of deionized water to prevent clogging by large particulate substances on the support material and the column. A modified procedure from von Canstein *et al.* (1999) was used during the immobilization of *A. haemolyticus* cells onto cellulose matrix packed in the column which consists of three important steps that is flushing of the column content (cellulose) with deionized water, attachment of *A. haemolyticus* cells on the cellulose followed by initial formation of biofilm by the attached bacteria.

Field emission scanning electron microscopy (Hitachi S-4500 FESEM) was carried out at the Material Science Laboratory, Universiti Teknologi Malaysia, Johor Bahru, to determine the development of biofilm on cellulose in the column. After 7 and 14 days of column operation, 3–5 cellulose cubes at the top section of the column were sampled using ethanol sterilized plastic forceps before viewing under FESEM. After 42 days, the same number of cellulose cubes were removed from three different sections of the column namely the bottom, middle and top, for viewing under the FESEM.

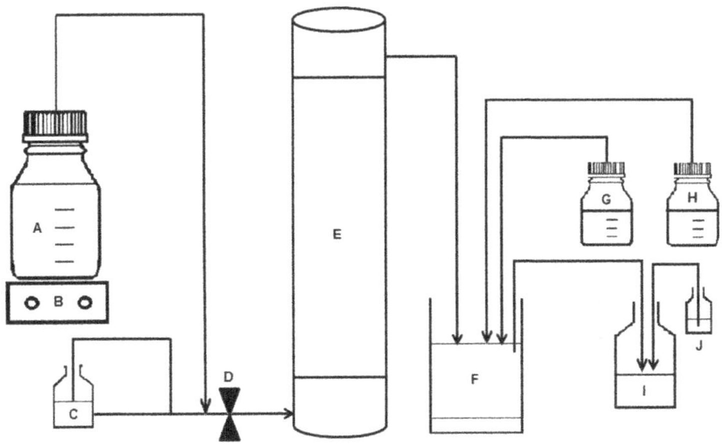


**Figure 9.1** Schematic representation of the experimental setup for Cr (VI) reduction system: (A) holding tank, (B) magnetic stirrer, (C) culture of *A. haemolyticus*, (D) peristaltic pump, (E) column (bioreactor), (F) precipitation tank, (G) NaOH solution, (H) coagulant solution, (I) disinfection tank, (J) disinfectant.

**Figure 9.2** (left) Commercial cellulose as support material (right) Inert stone.

## Immobilization of *A. haemolyticus*

In this study, 6 h was required for attachment of *A. haemolyticus* cells to the cellulose support material. During the 6 h period, 1 L of *A. haemolyticus* cells was recirculated approximately nine times (based on column retention time ($R_t$) of 0.67 h or 40 min) through the column. This step was carried out to give *A. haemolyticus* ample time to attach itself to the cellulose support material. Annachhatre and Bhamidimarri (1992) suggested that continuous supply of cells is a pre-requisite for microbial attachment onto a support. The 6 h period used for bacterial attachment was deemed sufficient compared to other bacteria such as *Bacillus* sp. and *P. fluorescens* LB 300 that requires only 10 minutes (without circulation; Table 9.4).

In both cases, however, this short bacterial attachment period was complemented by the longer time used for initial biofilm formation, that is 6 days compared to 1 day, as used in this study. The following approaches are suggested for biofilm formation on solid support material. Either a short period of initial bacterial attachment, followed by a longer period of nutrient supplementation to allow biofilm formation (Chirwa and Wang, 1997a,b) or a long period of initial bacterial attachment, followed by longer period of nutrient supply (Dermou *et al.*, 2005). The latter approach was used in this study. It is noteworthy that even though a long initial bacterial attachment period of 6 h was used, the total time utilized to complete both bacterial attachment and biofilm formation is still shorter than reported by Chirwa and Wang (1997a,b). This can be attributed to the known ability of *A. haemolyticus* to form biofilm on solid surfaces (Holt *et al.*, 1994). Hence, it can be said that the bacterial immobilization period for this study is highly time efficient which may prove beneficial in industrial applications.

The initial microbial attachment to solid surfaces and eventual colonization of the surface is thought to be due to several transport and biological processes, namely adsorption of dissolved organics on the wetted surface, transport of microorganisms to the wetted surface and microbial adhesion to the surface (Veenstra *et al.*, 1999). Microbial transfer could be a result of diffusion, convection or Brownian motion. The biofilm then develops through cellular reproduction and microbial production of extracellular polymers. For example, as the cells penetrate the viscous boundary layer of a biofilm, resistance to microbial transport increases where diffusion might be the controlling mechanism. Once a mobile bacterium is in the vicinity of a charged surface, it may chemotactically respond to any concentration gradient that exists in the interfacial region (Speitel Jr and DiGiano, 1987). The surface characteristics of the support material such as unit surface area, porosity and surface roughness have been found to influence biofilm formation during start-up. Surface roughness and specific surface area were important characteristics as it enhances biofilm formation. It was reported that initial bacterial attachment occurs in the crevices of support material where attached biomass is well protected from fluid shear that allows the colonization (formation of biofilm) of support material surfaces. This well protected site is provided by support materials with high surface roughness such as

**Table 9.4** Duration used for attachment of bacteria on support material in column

| Bacteria | Metal | Support material | Time allocated for bacterial attachment | Time allocated for initial formation of biofilm | Cells used as | Reference |
|---|---|---|---|---|---|---|
| *Bacillus* sp. | Cr(VI) | Glass beads | 10 min | 6 days | Pure culture | Shakoori *et al.* (2000) |
| *Acinetobacter* sp. | Cr(VI) | Gravel | 30 days | 20 days | Activated sludge | Dermou *et al.* (2005) |
| *P. fluorescens* LB300 | Cr(VI) | Siran beads | 10 min | 6 days | Pure culture | Holt *et al.* (1994) |
| *P. putida* | Hg(II) | Siran beads | 1 day | 10 days | Pure culture | Von Canstein *et al.* (1999) |
| *A. haemolyticus* | Cr(VI) | Cellulose | 6 h | 1 day | Pure culture | This study |

granular activated carbon and cellulose (Speitel Jr and DiGiano, 1987).

In this study, the column comprised of cellulose and attached cells of *A. haemolyticus*, was not pre-adapted to Cr(VI) due to its innate Cr(VI) resistant ability. This is advantageous in terms of industrial application as it can save time needed to start the Cr(VI) reduction process. Wang and Xiao (1995) reported that 14 days was needed to precondition a Cr(VI) reduction column where the column was flushed with tap water containing Cr(III) at 0.001 to 0.002 mg/l. Chirwa and Wang (1997b) used a higher Cr(VI) concentration of 10 mg/l Cr(VI) to condition the *P. fluorescens* LB300 attached column for 2 days before real wastewater was treated. The upwards influent flow, together with the presence of inert stones, ensures even distribution of the effluent and maximizes contact time between Cr(VI) and the column content. The influent was also circulated to obtain a mixed flow pattern in the bioreactor and to ensure maximum utilization of the bioreactor for Cr(VI) reduction (Krishna and Philip, 2005). Other inert materials such as polyethylene and gravel may be used for the same purpose as reported by Ekenberg *et al.* (2005) and Dermou *et al.* (2005) respectively. Krishna and Philip (2005), however, pumped the influent downwards through a packed corrugated PVC material (Fujino media) via a fixed nozzle placed at the top of reactor to ensure even distribution of influent to the whole bioreactor area.

## Cr(VI) bioremediation performance

Overall, the bioremediation of Cr(VI) from the electroplating wastewater was carried out for 42 days and the profile obtained is represented in Fig. 9.3. In general, the study comprised four main steps: optimization of influent Cr(VI) concentration, column regeneration, variation in influent flow rate and sustainability of the column system in Cr(VI) reduction in continuous operation. The regeneration step is necessary to reactivate bacterial populations in the column that were expected to dwindle resulting from Cr(VI) toxicity during the optimization to influent Cr(VI) concentrations. Segregation of column study into different parts (I–IV, Fig. 9.3) as used in this study was also reported by other researchers such as Chirwa and Wang (1997a), Krishna and Philip (2005) and Shen and Wang (1995).

Most studies on biological reduction of Cr(VI) were conducted in batch reactors using pure cultures. For example, Wang and Xiao (1995) studied several factors affecting Cr(VI) reduction by pure cultures of bacteria in shake flasks. The same technique was also used by Shakoori *et al.* (2000) and Megharaj *et al.* (2003) who worked with pure bacterial cultures isolated from tannery effluent. Fujie *et al.* (1994) used a 1-l bioreactor in studying various factors affecting Cr(VI) reduction by *E. cloacae* strain HO1. Bopp and Ehrlich (1988) used 250 ml Erlenmeyer flasks to assess the Cr(VI) reduction ability of *P. fluorescens* LB 300. It is only recently that continuous flow and fixed film bioreactors were used for biological reduction of Cr(VI). Shen and Wang (1993) demonstrated Cr(VI) reduction in a two-stage continuous flow suspended growth bioreactor system. Cells of *E. coli* were grown in a completely mixed reactor before pumping into the second stage plug flow reactor where Cr(VI)

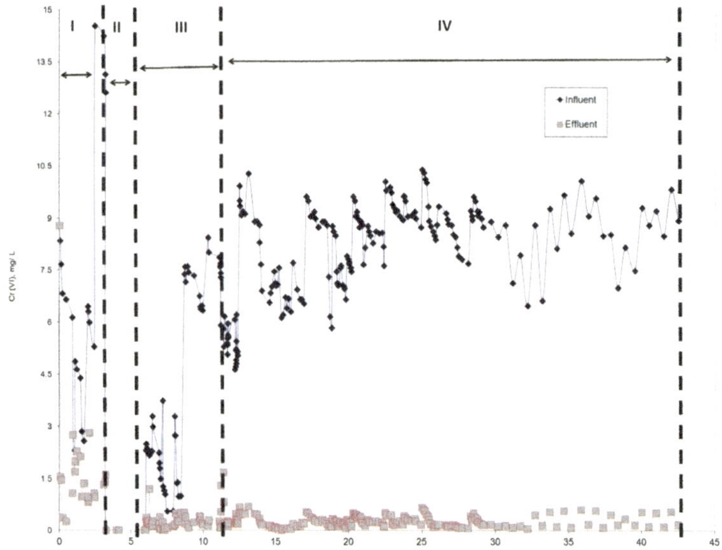


**Figure 9.3** Profile for bioremediation of Cr (VI) from electroplating wastewater by biofilm–immobilized column system for 42 days. I, optimization of influent Cr(VI) concentration; II, column regeneration; III, –variation in influent flow rate; and IV, sustainability of the column system.

reduction takes place. The first report on the use of fixed film bioreactors for Cr(VI) reduction was by Chirwa and Wang (1997a), where a continuous flow laboratory-scale biofilm reactor was demonstrated to reduce Cr (VI) without the need to constantly resupply fresh Cr(VI) reducing cells. Ekenberg *et al.* (2005) reported the use of biofilm formed on plastic carriers to treat Cr(VI) containing leachate originating from ferrochromium slag where an anaerobic column system was operated in upwards flow. Another reported use of anaerobic condition in microbial reductions of Cr(VI) was by Krishna and Philip (2005), where biofilm formed on Fujino media consisting of Cr(VI)-resistant microbial consortia was used for the treatment of leachate from Cr(VI) contaminated soil.

Since, *A. haemolyticus* is a strictly aerobic bacteria, the anaerobic condition might be beneficial for invading anaerobic bacteria from the influent. The decrease in pH from anaerobic fermentation can be visualized by the colour change from slightly brownish (influent) to pale yellow (effluent). Ekenberg *et al.* (2005) reported gradual colour change in effluent from yellow to grey. The brownish effect observed in the influent can be attributed to a reaction known as Maillard browning (Ivy, 2005). The effect can be observed at pH 7.0 and is a result of a series of reactions producing melanoidins, which have distinct flavours and brown pigments (Martins, 2003). However, this browning effect is gradually inhibited at pH lower than 6.5 resulting in a pale yellow effluent indicating acidic conditions.

The ability of *A. haemolyticus* to use pineapple waste as a carbon source has been reported by Ivy (2005). It is important to note that the same batch of pineapple waste was used in this study as that reported by Ivy (2005). Even though, *A. haemolyticus* was reported to grow best in a mixture of 60% v/v pineapple waste (pH 7.0) in nutrient broth (NB) medium, the lower pineapple waste concentration of 15% v/v used in this study should satisfy the energy requirement for bacterial growth and the subsequent Cr(VI) reduction. Glucose was identified as the main sugar component utilized by the bacteria compared to sucrose and fructose. Around 80% of 2.97 g glucose/l was consumed by *A. haemolyticus* that corresponded to an $OD_{600}$ value of 1.15, which was typical of the optimum value obtained when bacteria was grown in nutrient broth only (Ivy, 2005). In this study, glucose was considered to be the main carbon source utilized by *A. haemolyticus* since its concentration was highest in pineapple wastewater and because substantial utilization of glucose solution during growth was earlier reported. The effective role of glucose as a carbon source in Cr (VI) reduction can be described by the following equation (Chirwa and Wang, 1997a):

$$C_6H_{12}O_6 + 8CrO_4^{2-} + 34H^+ \rightarrow 8Cr^{3+} + 6HCO_3^- + 2OH^- \quad (9.6)$$

From Equation 9.6, complete breakdown of 1 mole glucose would yield sufficient energy to reduce 8 moles Cr(VI). For example, complete breakdown of 5 g/l glucose (28 mM) is expected to yield sufficient energy to reduce 500 mg/l Cr(VI) (192 mM). Based on the molar equilibrium ratio between glucose to Cr(VI) of 1:8, breakdown of 28 mM glucose would yield enough energy to reduce approximately eight times the amount of glucose that is 204 mM Cr(VI), which is higher than the 192 mM Cr(VI) targeted for reduction (Shakoori *et al.*, 2000; Wagner-Dobler *et al.*, 2000). The raw glucose concentration was determined as 4.95 g/l. However, the column influent mixture contains 15% v/v of pineapple that corresponds to 0.74 g/l or 4.16 mM glucose. Influent Cr(VI) concentrations ranged from 10 to 15 mg/l or 3.84 to 5.76 mM. From the molar equilibrium ratio, complete breakdown of 4.16 mM glucose would produce enough energy to reduce 33.26 mM Cr(VI). Taking into consideration the range of influent Cr(VI) concentrations of 3.84 to 5.76 mM, it can be said that excess carbon source was supplied for the Cr(VI) reduction system throughout the study. Besides providing sufficient amounts of energy for Cr(VI) reduction, another important reason to maintain high influent substrate concentration is the possibility to quickly recover biomass loss in the column that result from shearing forces created from the operational condition. High substrate concentrations would enhance cell division on the support material. This knowledge is important in the design of column operating conditions especially at low substrate concentrations. This is because biofilm shearing may be more closely related to microbial growth rate than the amount of attached microbial biomass (Speitel Jr and DiGiano, 1987).

## Development of biofilm

The development of bacterial biofilm in the column during the Cr(VI) reduction process was investigated using SEM on day 0, day 7, day 14 and day 42 which was the final day of column operation in this study (Fig. 9.4). Fig. 9.4 shows that on day 0, the surface of the cellulose support material appeared empty, without the presence of bacteria. Different sizes of pores were also observed. After 7 days of column operation, bacterial colonization of the cellulose surface begins to appear. The presence of cocci-shaped cells may represent stationary phase cells of *A. haemolyticus* clumped to one another forming a possible microcolony (Fig. 9.5B, boxed area). This condition was also reported by von Canstein *et al.* (2002) in *P. putida* immobilized columns for mercury reduction. However, most of the cellulose surface area and pores remained uncolonized. Such a condition may be caused from the removal of large sections of biofilm, a condition known as 'biofilm sloughing' (Annachhatre and Bhamidimarri, 1992). During biofilm development, portions of biofilm peel away from the surface and are entrained in the fluid flow. Detachment is the process of continuous biofilm removal and is highly dependent on hydrodynamic conditions such as flow rate and toxicity. The presence of toxic substances or limitations in byproduct diffusion from the biofilm, further aggravate the situation and would lead to death or lysis of the bacteria (Annachhatre and Bhamidimarri, 1992). In this study, 'biofilm sloughing' might be encountered during the early phase of study especially during the initial 3.5 days of column operation. Even though sufficient concentration of pineapple waste was present in the influent, the development of biofilm may be impacted by the toxicity of influent Cr(VI), which was reflected by the low Cr(VI) reduction capacity of the column system. However, this effect gradually diminished over time, as depicted by the high Cr(VI) reduction capacity obtained from day 6 onwards. The condition can be explained by adjustment of the bacterial cells to the environmental conditions (Dermou *et al.*, 2005) such as nutrients, temperature, toxic substances and the presence of other resistant microorganisms that can compete for substrates in the influent (Krishna and Philip, 2005).

After 14 days, biofilm thickness increased and microorganisms of various shapes and sizes begin to show (Fig. 9.4C). Since the column operation was carried out under non-sterile conditions, the possible mono species biofilm established (day 0 to day 7) may therefore be subjected to colonization by ubiquitous Cr(VI) resistant bacteria. At this time, it can be stated that the invading microorganisms were beginning to have a role in the biofilm as

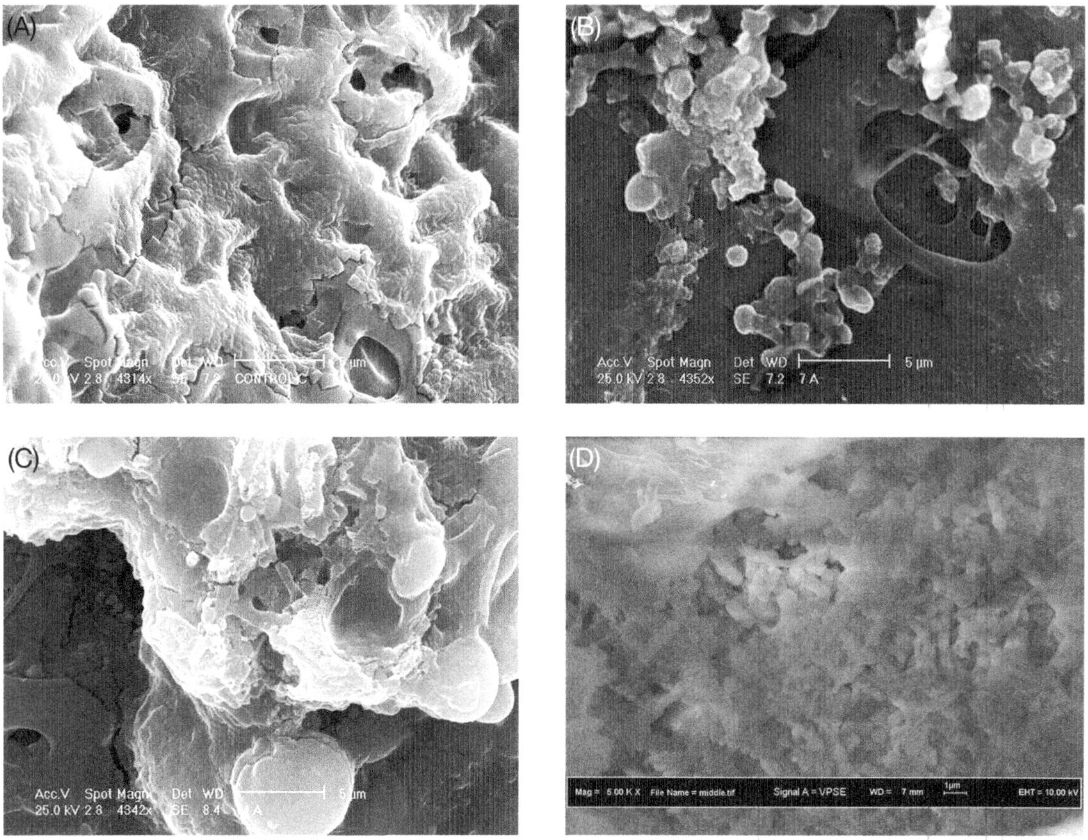


**Figure 9.4** Development of biofilm on cellulose support material. (A) day 0, (B) day 7, (C) day 14 (arrow indicate growth of thick biomass), (D) day 42 (magnification 5000).

evident from its good survival ability. The invading bacteria colonized the biofilm during the operation of the column which had been washed thoroughly with 90% ethanol prior to packing with cellulose material. After 42 days which marked the end of column operation, the cellulose surface was filled with bacterial mass (Fig. 9.4D). Viable cell count increased tremendously compared with days 7 and 14. This could explain the high Cr(VI) reduction capacity of cells. However, there is also the potential for decrease levels of reductase activity per unit cell resulting from Cr(VI) toxicity as, suggested by Krishna and Philip (2005). In their report, high Cr(VI) reduction capacity was maintained by increases in the number of active Cr(VI)-reducing cells even though per-cell levels of reductase activity decreased due to Cr(VI) toxicity. Even though development of slime layers in biofilm may include steps such as the creation of slime capsule layers, growth of globular masses, webs, thin sheets and growth of thick pore-filling masses of slime (Idris *et al.*, 2007), none of these features were clearly distinguished (as shown in Fig. 9.5). This could be due to the use of raw electroplating and pineapple wastewaters during column operation. Variation of important parameters in the wastewaters such as water composition, pH and nutrients may influence bacterial growth, morphology and accompanying exopolysaccharides.

Coccus- and rod-shaped cells predominated the cellulose surface, possibly representing *A. haemolyticus* cells at stationary and log phase respectively. Formation of a layer of slime was also noticed on the cellulose surface (Fig. 9.5A) and may be attributed to the presence of suspended materials in the pineapple waste, which is known to form fine layers on the surface of solid support material (Wagner-Dobler *et al.*, 2000) or the presence of slime forming water borne bacteria such as *Streptococcus* and *Alcaligenes* sp. (Madigan *et al.*, 2001).

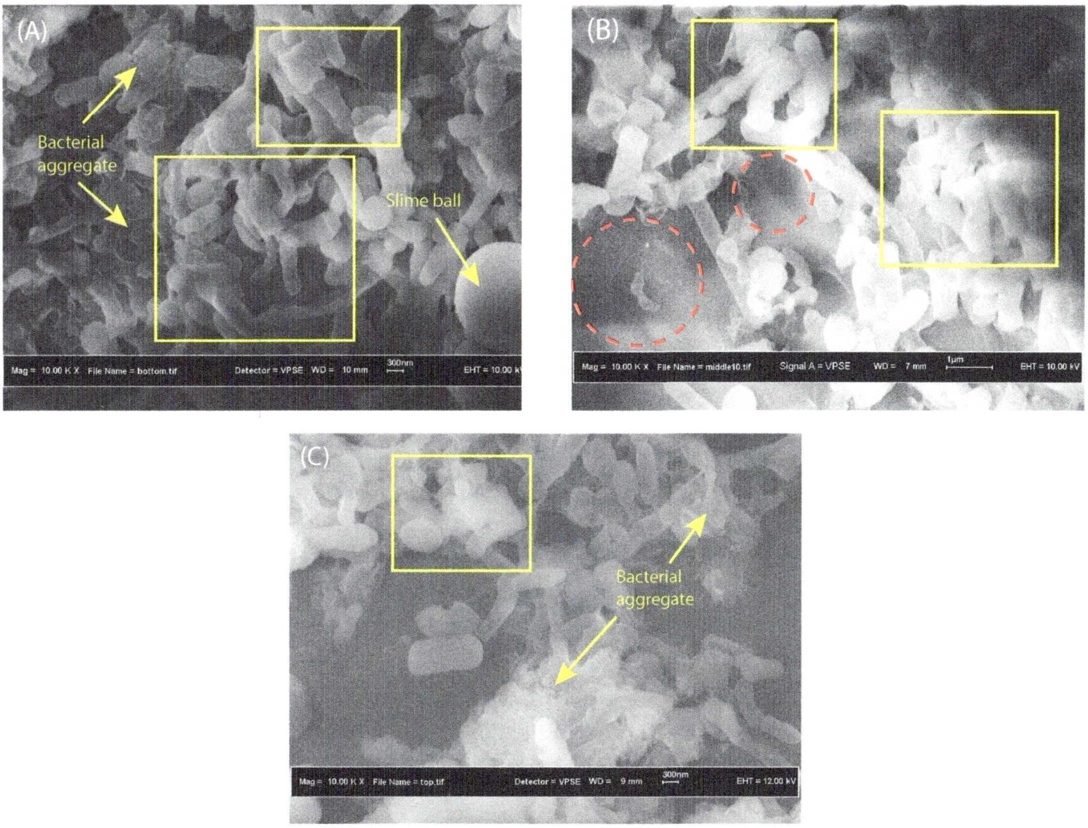


**Figure 9.5** FESEM electron micrographs of biofilm on cellulose support material after 42 days of column operation (A) Bottom section (B) Middle section (C) Top section; Magnification=10,000×; boxes show microcolonies of bacteria; a slime-like layer is represented by red-dotted circles.

It is generally agreed that microbial communities work better than pure cultures in bioremediation (Loreau and Hector, 2001; von Canstein *et al.*, 2002). However, in the case of Cr(VI) reduction, the reaction carried out by the bacteria is enzymatic detoxification that is necessary for survival but not for sustaining growth (von Canstein *et al.*, 2002). In a completely mixed system, the principle of competitive exclusion would predict that strains with the highest detoxification rate would have the largest selective advantage. They should therefore out-compete strains with lower detoxification rates in a mixed culture community and perform best as pure cultures (Holt *et al.*, 1994). Loreau and Hector (2001) reported an increase in community stability and efficiency via microbial diversity. Wagner-Dobler *et al.* (2000) reported that in a bioreactor filled with metal resistant/reducing bacteria, each strain would be responsible for different ecological niches within the bioreactor. This was based on the difference between growth and metal reducing characteristics where their co-existence was mediated by differential growth and death rates at fluctuating wastewater compositions. In addition, direct mutualistic interactions might occur between different strains resulting in different mixed microcolonies, where the activity of one organism would be beneficial for another (Moller *et al.*, 1998; Nielsen *et al.*, 2000). Therefore, it is suggested that the invasion of ubiquitous Cr(VI) resistant bacteria onto the biofilm of the inoculum strain that is *A. haemolyticus*, did not interfere with the high Cr(VI) reduction capacity. Instead, these bacteria increase biofilm diversity and contribute to the observed robustness and stability of the process under a wide range of physical and chemical conditions. The use of Cr(VI) resistant strains would also enable fast reactor start-up without the need to adapt the inoculum strain to the wastewater conditions.

From the energy-dispersive X-ray (EDAX) profiles obtained (profiles not shown) similar intensity ratios for C:O were found on the biofilm containing cellulose for all sections, that is 1.5639: 1.0189 (top), 1.6311: 1.2345 (middle) and 1.7601: 1.0631 (bottom) respectively. For Cr, the intensity increased in the order top (0.7432), middle (0.2823) and bottom (0.3423) corresponding to per cent weight of 0.63, 4.31, 15.21 respectively. This high value of precipitated Cr especially 15.21%, may account for the undetected 16.23% Cr in the effluent. This indicates a higher Cr deposition, most probably as chromium hydroxide (Ekenberg *et al.*, 2005) at the bottom section of the column followed by middle and the top sections. A similar finding was also reported by Ekenberg *et al.* (2005) where analysis of the biofilm from Cr(VI) reduction columns revealed that approximately 18% of the dry weight consisted of chromium. One plausible explanation for this condition is higher cell concentrations present at the bottom section of the column due to the up-flowing nature of influent. It is estimated that most of the Cr(VI) reduction, nutrients and DO consumption occurred in this section. This is in agreement with Krishna and Philip (2005) who stated that most of the available DO supplied in influent would be consumed within 10% of the column bed height by bacteria growing on the biofilm.

## Biofilm microbial diversity

The microbial diversity in the biofilm of this Cr(VI) reducing environment was studied using a similar experimental setup as described but with sugarcane bagasse as the support material for *A. haemolyticus*. Microbial identification was carried out using a culture independent approach where the total genomic DNA was first extracted from the packed bed reactor and then used as template for the 16S rRNA gene fragment amplification by a polymerase chain reaction (PCR) technique. The 16S rRNA gene library was constructed for each sample and the microbial diversity determined by performing restriction fragment length polymorphism (RFLP) analysis. The finding obtained in this study not only provide new insight into the bacterial community formed in the Cr(VI) reducing environment of *A. haemolyticus* immobilized sugarcane bagasse (SCB) treatment system, but generated useful information that might assist in future studies and further enhancement of metal–microbe interactions in the presence of SCB as a support medium. Furthermore, the results obtained may be useful in searching for novel microorganisms that could be useful in Cr(VI) reduction. The total genomic DNA from treated sugarcane bagasse of a control column (C1) and a column packed with immobilized *A. haemolyticus* (i.e. test T1) were successfully extracted and used as a template for amplification of 16S rRNA genes. Purified 16S rRNA gene fragments were then cloned into cloning vector yT&A (Yeastern Biotech Co. Ltd., Taiwan) and transformed into *E. coli* DH5α competent cells. A total of 693 white positive colonies of control sample (C5) and 422 of a sample (T5) were screened respectively. Out of these, 145 and 186 positive clones were determined to carry the desired insert of approximately 1500 bp and were selected for subsequent analysis.

To differentiate the identity of cloned DNA fragments, RFLP analysis was carried out. The analysis revealed that the control and test samples consisted of 81 and 114 RFLP patterns, respectively. Clones carrying an identical RFLP pattern were presumptively the same bacteria. Nevertheless, none of these patterns resemble RFLP patterns similar to *A. haemolyticus*. This might be because the portion of the sample collected did not cover the whole reactor content. As mentioned earlier, the total genomic DNA of treated SCB was extracted from the middle section of reactors for both control and test samples. Ideally, the bacterial population is expected to be the highest at a point closest to the inlet and gradually decreasing with increased reactor length. This is because the bacterial population nearer to the inlet section would receive the highest concentrations of dissolved oxygen and nutrients which is crucial for bacterial growth and maintenance. In this study, bacterial populations at the top section were expected to give the highest Cr(VI) reduction. However, it should also be noted that it received the highest incoming Cr(VI) concentration, and thus greater chance for Cr(VI) toxicity (Ahmad *et al.*, 2009). This was the basis for the selection of samples from the middle section of the column.

Sequencing analysis of these unique RFLP patterns revealed a diverse bacterial community in the sample studied. For control sample, out of the 81 patterns, five dominant patterns represent 26.2%

(38 clones) of the total library clones. While for the test sample, eight dominant patterns out of 114 patterns represent 21.0% (39 clones) of the total library clones. To determine the identity of these clones, the cloned DNA sequences were sent for analysis and were compared to other bacteria in the NCBI GenBank database. Sequencing analysis of the nearly complete 16S rRNA gene sequences of the dominant taxa (22 clones, 6.6%) showed highest sequence homology to the *Chitinophaga terrae* strain KP01 (GenBank Accession No. NR041540), originally isolated from soil in South Korea (Kim and Jung, 2007). The second most abundant pattern as represented by 11 clones (3.3% abundance) was most closely related (94% sequence homology) to *Laribacter hongkongensis* strain HKU1 (GenBank Accession No. NR025167). The biochemical profile of *L. hongkongensis* strain HKU1 shows it can produce catalase, cytochrome oxidase, urease, arginine dihydrolase and reduces nitrate. The bacterium was initially isolated from the blood and empyema of a cirrhotic patient (Yuen *et al.*, 2001). The third (10 clones, 3.0%) and fourth (nine clones, 2.7%) most abundant occurrences were related to uncultured bacterium clone WW2_146 (GenBank Accession No. GQ264320) and *Ottowia thiooxydans* strain K11 (GenBank Accession No. NR029001) with 94% and 97% sequence similarities, respectively. *O. thiooxydans* strain K11 was reported as a facultatively anaerobic bacteria which reduces nitrate and nitrite and produces $N_2O$ as the sole end product of nitrate respiration. It has also been reported to utilize carboxylic acids and amino acids as substrates. The type strain, K11 was originally isolated from activated sludge of a high load basin of a municipal wastewater treatment plant (Spring *et al.*, 2004). Other patterns were matched with 96% sequence similarity (nine clones, 2.7%) to the uncultured bacterium clone WW3_39 (GenBank Accession No. GQ264495) followed by 93% sequence similarity (four clones, 1.2% abundance) to *Rhizobium cellulosilyticum* strain ALA10B2 (GenBank Accession No. NR043985), Candidate division OP10 bacterium Gsoil 348, 96% (GenBank Accession No. GQ339893), *Pedobacter* sp. oral clone AV100, 94% (GenBank Accession No. AF385531) and uncultured bacterial clone GI5-005-F01, 95% (GenBank Accession No. FJ192729). *R. cellulosilyticum* strain ALA10B2 was reported to actively produce cellulases on media containing CM–cellulose as a substrate and utilizes glucose, L-arabinose, mannose, mannitol, N-acetylglucosamine and malate as carbon sources and also grows on glutamate and methionine as carbon and nitrogen sources. The type strain ALA10B2 was isolated from sawdust of *Populus alba* in Spain (Garcia-Fraile *et al.*, 2007). Meanwhile, 16S rRNA gene sequence analysis of Candidate division OP10 bacterium Gsoil 348 indicated that the strain is related to members of the phylum *Armatimonadetes*. 16S rRNA gene sequences representing the phylum *Armatimonadetes* have been identified by culture-independent methods from various environments, including aerobic and anaerobic wastewater treatment processes, the rhizosphere, hypersaline microbial mats and subsurface geothermal water streams. Strain Gsoil 348 could not grow in high strength nutrients (higher than 1.5 g/l of total organic carbon) such as nutrient agar and showed slow assimilation of limited carbon substrates (Im *et al.*, 2012).

Among the selected dominant patterns for both control and test samples, 54 (16.3%) were affiliated to cultivated counterparts whereas 23 (6.9%) were affiliated to yet-uncultivated or uncultured counterparts in the GenBank databases. It is interesting to observe that phylum of *Bacteroidetes*, *Proteobacteria* and *Armatimonadetes* were the most abundant group of bacteria observed in the SCB biofilm with 7.9% (26 clones), 7.3% (24 clones) and 1.2% (four clones) occurrences, respectively. In addition, *Bacteroidetes* could also contribute significantly to lignocellulose constituents, including cellulose, hemicelluloses and pectin in the soil and some of the order *Sphingobacteriales* have been reported as cellulolytic taxa (Rattanachomsri *et al.*, 2011; Coughlan and Mayer, 1992; Field *et al.*, 2010). A phylogenetic tree was constructed based on the nearly complete 16S rRNA gene sequences of predominant clones (Fig. 9.6).

Besides the well-known cultivated bacteria, the analysis has revealed a rather large community of uncultured bacteria present in the SCB biofilm. It was observed that almost 7% of the library is constituted by still unknown or yet uncultivated bacterial members. These clones are mainly associated with uncultured bacterial clones from environmental samples of simulated low level radioactive waste sites, including cellulosic waste in the form of paper towels, cardboard boxes or wood contaminated with heavy metals and radionuclides

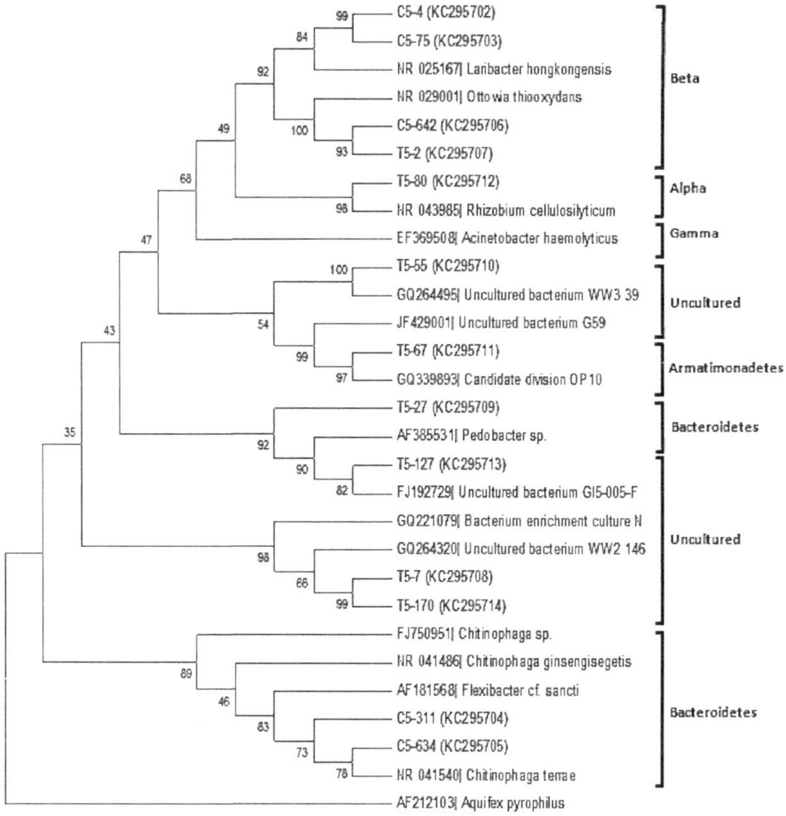


**Figure 9.6** Phylogram showing the phylogenetic relationships among 13 dominant taxa as determined by RFLP profiles of 16S rRNA gene fragments.

such as chromium and uranium (Field *et al.*, 2010) and a spacecraft assembly clean room (La Duc *et al.*, 2009). The predominating *Bacteroidetes* and *Proteobacteria* in the treatment reactors are in good agreement with Rattanamchomsri *et al.* (2011). The research group conducted culture independent phylogenetic analysis of the microbial community in industrial sugarcane bagasse feedstock piles and revealed that the microbial community was composed of nine major bacterial lineages dominated by *Firmicutes, Acidobacteria, Proteobacteria* and *Bacteroidetes* with >85% abundance. Other minor phyla include *Actinobacteria, Planctomycetes, Sphirochaetes, Deinococcus–Thermus* and Candidate division TM7 (Rattanamchomsri *et al.*, 2011). *Actinobacteria* was reported to play important roles, especially in the degradation of macromolecules in waste material, such as hemicelluloses, lignin and cellulose, at the later stages of composting (Steger *et al.*, 2007; Woan Ying Liew *et al.*, 2009).

## Conclusion

Biofilm-assisted remediation of Cr(VI)-contaminated wastewater process showed encouraging results both at laboratory and pilot-scale levels. Use of locally isolated Cr(VI) reducing-resistant bacterium and cheap agricultural waste as nutrient may propel the industrial application of processes such as this. This process is envisaged to be useful for the remediation of contaminated groundwater or stagnant water, point-source treatments for running water and to be incorporated into existing in-house wastewater treatment systems of small- to medium-scale industries.

## References

Acar, F., and Malkoc, E. (2004). The removal of chromium (VI) from aqueous solutions by *Fagus orientalis* L. Bioresour. Technol. *94*, 13–15.

Acharya, J., Sahu, J., Sahoo, B., Mohanty, C., and Meikap, B. (2009). Removal of chromium (VI) from wastewater by activated carbon developed from Tamarind wood activated with zinc chloride. Chem. Eng. J. *150*, 25–39.

Agrawal, A., Pal, C., and Sahu, K. (2008). Extractive removal of chromium (VI) from industrial waste solution. J. Hazard. Mater. *159*, 458–464.

Ahluwalia, S.S., and Goyal, D. (2007). Microbial and plant derived biomass for removal of heavy metals from wastewater. Bioresour. Technol. *98*, 2243–2257.

Ahmad, W.A., Zakaria, Z.A., Razali, F., and Samin, J. (2009). Evaluation of the combined Cr (VI) removal capacity of sawdust and sawdust-immobilized *Acinetobacter haemolyticus* supplied with brown sugar. Water Air Soil Pollut. *204*, 195–203.

Ahmad, W.A., Zakaria, Z.A., Khasim, A.R., Alias, M.A., and Ismail, S.M.H.S. (2010). Pilot-scale removal of chromium from industrial wastewater using the ChromeBac™ system. Bioresour. Technol. *101*, 4371–4378.

Akbal, F., and Camcı, S. (2011). Copper, chromium and nickel removal from metal plating wastewater by electrocoagulation. Desalination *269*, 214–222.

Annachhatre, A., and Bhamidimarri, S. (1992). Microbial attachment and growth in fixed-film reactors: process startup considerations. Biotechnol. Adv. *10*, 69–91.

Baral, A., and Engelken, R.D. (2002). Chromium-based regulations and greening in metal finishing industries in the USA. Environ. Sci. Policy *5*, 121–133.

Beleza, V., Boaventura, R., and Almeida, M. (2001). Kinetics of chromium removal from spent tanning liquors using acetylene production sludge. Environ. Sci. Technol. *35*, 4379–4383.

Bharti, A., Pandey, G., and Saxena, R. (1979). Physiological imbalances due to hexavalent chromium in freshwater algae [India]. Indian J. Environ. Health *21*, 234–243.

Bopp, L., Chakrabarty, A., and Ehrlich, H. (1983). Chromate resistance plasmid in *Pseudomonas fluorescens*. J. Bacteriol. *155*, 1105–1109.

Bopp, L.H., and Ehrlich, H.L. (1988). Chromate resistance and reduction in *Pseudomonas fluorescens* strain LB300. Arch. Microbiol. *150*, 426–431.

Buhler, D.R., Stokes, R.M., and Caldwell, R.S. (1977). Tissue accumulation and enzymatic effects of hexavalent chromium in rainbow trout (*Salmo gairdneri*). J. Fish. Res. Board Can. *34*, 9–18.

Camargo, F., Okeke, B., Bento, F., and Frankenberger, W. (2004). Hexavalent chromium reduction by immobilized cells and the cell-free extract of *Bacillus* sp. ES 29. Bioremediation J. *8*, 23–30.

Cervantes, C., Campos-García, J., Devars, S., Gutiérrez-Corona, F., Loza-Tavera, H., Torres-Guzmán, J.C., and Moreno-Sánchez, R. (2001). Interactions of chromium with microorganisms and plants. FEMS Microbiol. Rev. *25*, 335–347.

Chen, Y., and Gu, G. (2005). Preliminary studies on continuous chromium (VI) biological removal from wastewater by anaerobic–aerobic activated sludge process. Bioresour. Technol. *96*, 1713–1721.

Cheung, K., and Gu, J.D. (2007). Mechanism of hexavalent chromium detoxification by microorganisms and bioremediation application potential: a review. Int. Biodeter. Biodegr. *59*, 8–15.

Chirwa, E.M., and Wang, Y.T. (1997a). Hexavalent chromium reduction by *Bacillus* sp. in a packed-bed bioreactor. Environ. Sci. Technol. *31*, 1446–1451.

Chirwa, E.M., and Wang, Y.T. (1997b). Chromium (VI) reduction by *Pseudomonas fluorescens* LB300 in fixed-film bioreactor. J. Environ. Eng. *123*, 760–766.

Chrysochoou, M., and Johnston, C.P. (2012). Reduction of chromium (VI) in saturated zone sediments by calcium polysulfide and nanoscale zerovalent iron derived from green tea extract. Paper presented at: GeoCongress 2012 at State of the Art and Practice in Geotechnical Engineering (ASCE), 25–29 March 2012, Oakland, CA, USA.

Cossich, E.S., da Silva, E.A., Tavares, C.R.G., Cardozo Filho, L., and Ravagnani, T.M.K. (2004). Biosorption of chromium (III) by biomass of seaweed *Sargassum* sp. in a fixed bed column. Adsorption *10*, 129–138.

Coughlan, M.P., and Meyer, F. (1992). The cellulose decomposing bacteria and their enzyme systems. In: A. Ballows, H.G. Truper; M. Dworkin, W. Harder and K.H. Schleifer, eds. The Prokaryotes, A Handbook on the Biology of Bacteria, Ecophysiology, Isolation, Identification, Applications. Vol. 1, 2nd edn (Springer, New York, NY, USA), pp. 460–516.

Cushnie Jr, G.C. (1985). Electroplating Wastewater Pollution Control Technology (Noyes Publications, Park Ridge, NJ, USA).

Czako-Ver, K., Batiè, M., Raspor, P., Sipiczki, M., and Pesti, M. (1999). Hexavalent chromium uptake by sensitive and tolerant mutants of *Schizosaccharomyces pombe*. FEMS Microbiol. Lett. *178*, 109–115.

De Flora, S., Bagnasco, M., Serra, D., and Zanacchi, P. (1990). Genotoxicity of chromium compounds. A review. Mutation Res./Rev. Genetic Toxicol. *238*, 99–172.

Denkhaus, E., Meisen, S., Telgheder, U., and Wingender, J. (2007). Chemical and physical methods for characterisation of biofilms. Microchim. Acta *158*, 1–27.

Department of Environment (2004). Environmental Quality Report 2003, Ministry of Science, Technology and Innovation, Malaysia.

Department of Environment (2009). Annual Report. Ministry of Natural Resources and Environment, Malaysia.

Dermou, E., Velissariou, A., Xenos, D., and Vayenas, D. (2005). Biological chromium (VI) reduction using a trickling filter. J. Hazard. Mater. *126*, 78–85.

Dermou, E., Velissariou, A., Xenos, D., and Vayenas, D. (2007). Biological removal of hexavalent chromium in trickling filters operating with different filter media types. Desalination *211*, 156–163.

Dhal, B., Thatoi, H., Das, N., and Pandey, B. (2013). Chemical and microbial remediation of hexavalent chromium from contaminated soil and mining/metallurgical solid waste: a review. J. Hazard. Mater. *250–251*, 272–291.

Dursun, A.Y., Uslu, G., Cuci, Y., and Aksu, Z. (2003). Bioaccumulation of copper(II), lead(II) and chromium(VI) by growing *Aspergillus niger*. Process Biochem. *38*, 1647–1651.

Eccles, H. (1999). Treatment of metal-contaminated wastes: why select a biological process? Trends Biotechnol. *17*, 462–465.

Ekenberg, M., Martander, H., and Welander, T. (2005). Biological reduction of hexavalent chromium – a field study. Water Environ. Res. *77*, 425–428.

Field, E.K., D'Imperio, S., Miller, A.R., VanEngelen, M.R., Gerlach, R., Lee, B.D., Apel, W.A., and Peyton, B.M. (2010). Application of molecular techniques to elucidate the influence of cellulosic waste on the bacterial community structure at a simulated low-level-radioactive-waste site. Appl. Environ. Microbiol. *76*, 3106–3115.

Fude, L., Harris, B., Urrutia, M.M., and Beveridge, T.J. (1994). Reduction of Cr (VI) by a consortium of sulfate-reducing bacteria (SRB III). Appl. Environ. Microbiol. *60*, 1525–1531.

Fujie, K., Tsuchida, T., Urano, K., and Ohtake, H. (1994). Development of a bioreactor system for the treatment of chromate wastewater using *Enterobacter cloacae* HO1. Water Sci. Technol. *30*, 235–243.

Fukai, R. (1967). Valency state of chromium in seawater. Nature *213*, 901.

Garcia-Fraile, P., Rivas, R., Willems, A., Peix, A., Martens, M., Martínez-Molina, E., Mateos, P.F., and Velázquez, E. (2007). *Rhizobium cellulosilyticum* sp. nov., isolated from sawdust of *Populus alba*. Int. J. Syst. Evol. Microbiol. *57*, 844–848.

Garg, U.K., Kaur, M., Garg, V., and Sud, D. (2007). Removal of hexavalent chromium from aqueous solution by agricultural waste biomass. J. Hazard. Mater. *140*, 60–68.

Ghosh, G., and Bhattacharya, P.K. (2006). Hexavalent chromium ion removal through micellar enhanced ultrafiltration. Chem. Eng. J. *119*, 45–53.

Greenberg, A.E., Trussell, R.R., and Clesceri, L.S. (1985). Standard Methods for the Examination of Water and Wastewater, 16th edn. (American Public Health Association, New York, NY, USA).

He, M., Li, X., Liu, H., Miller, S.J., Wang, G., and Rensing, C. (2011). Characterization and genomic analysis of a highly chromate resistant and reducing bacterial strain *Lysinibacillus fusiformis* ZC1. J. Hazard. Mater. *185*, 682–688.

Higgins, T.E., Halloran, A.R., and Petura, J.C. (1997). Traditional and innovative treatment methods for Cr (VI) in soil. Soil Sediment Contam. *6*, 767–797.

Holdway, D. (1988). The toxicity of chromium to fish. Advan. Environ. Sci. Technol. *20*, 369–397.

Holt, J.G., Krieg, N.R., Sneath, P.H., Staley, J.T., and Williams, S.T. (1994). Bergey's Manual of Determinative Bacteriology (Williams and Wilkins Co., Baltimore, MD, USA).

Horitsu, H., Futo, S., Ozawa, K., and Kawai, K. (1983). Comparison of characteristics of hexavalent chromium-tolerant bacterium, *Pseudomonas ambigua* G-1, and its hexavalent chromium-sensitive mutant. Agric. Biol. Chem. *47*, 2907–2908.

Hossain, M.A., Alam, M., and Yonge, D.R. (2005). Estimating the dual-enzyme kinetic parameters for Cr (VI) reduction by *Shewanella oneidensis* MR-1 from soil column experiments. Water Res. *39*, 3342–3348.

Hsu, H.-T., Chen, S.-S., and Chen, Y.-S. (2011). Removal of chromium (VI) and naphthalenesulfonate from textile wastewater by photocatalysis combining ionic exchange membrane processes. Sep. Purif. Technol. *80*, 663–669.

Idris, A., Hashim, R., Rahman, R.A., Ahmad, W.A., Ibrahim, Z., Abdul Razak, P.R., Mohd Zin, H., and Bakar, I. (2007). Application of bioremediation processes for textile wastewater treatment using pilot plant. Int. J. Eng. Technol., *4*, 228–234.

Im, W.T., Hu, Z.Y., Kim, K.H., Rhee, S.K., Meng, H., Lee, S.T., and Quan, Z.X. (2012). Description of *Fimbriimonas ginsengisoli* gen. nov., sp. nov. within the Fimbriimonadia class nov., of the phylum Armatimonadetes. Antonie van Leeuwenhoek *102*, 307–317.

Ishibashi, Y., Cervantes, C., and Silver, S. (1990). Chromium reduction in *Pseudomonas putida*. Appl. Environ. Microbiol. *56*, 2268–2270.

Ivy, A. (2005). Bioreduction of chromium (VI) from electroplating wastewater using *Acinetobacter haemolyticus* grown in pineapple waste. BSc thesis, Universiti Teknologi Malaysia.

Kamaludeen, S.P., Megharaj, M., Juhasz, A.L., Sethunathan, N., and Naidu, R. (2003). Chromium–microorganism interactions in soils: remediation implications. Rev. Environ. Contam. Toxicol. *178*, 93–164.

Kang, S.Y., Bremer, P.J., Kim, K.W., and McQuillan, A.J. (2006). Monitoring metal ion binding in single-layer *Pseudomonas aeruginosa* biofilms using ATR-IR spectroscopy. Langmuir *22*, 286–291.

Kanojia, R.K., Junaid, M., and Murthy, R.C. (1998). Embryo and fetotoxicity of hexavalent chromium: a long-term study. Toxicol. Lett. *95*, 165–172.

Khan, N.A., Ibrahim, S., and Subramaniam, P. (2004). Elimination of heavy metals from wastewater using agricultural wastes as adsorbents. Malaysian J. Sci. *23*, 43.

Kim, M.K., and Jung, H.-Y. (2007). *Chitinophaga terrae* sp. nov., isolated from soil. Int. J. Syst Evol. Microbiol. *57*, 1721–1724.

Knoll, J., and Fromm, P. (1960). Accumulation and elimination of hexavalent chromium in rainbow trout. Physiol. Zool. 33, 1–8.

Kolot, F. (1981a). Microbial carriers – strategy for selection. Process Biochem. *16*, 2–4.

Kolot, F. (1981b). Microbial carriers – strategy for selection, part 2. Process Biochem. *16*, 30.

Krauter, P., Martinelli, R., Williams, K., and Martins, S. (1996). Removal of Cr (VI) from ground water by *Saccharomyces cerevisiae*. Biodegradation 7, 277–286.

Krishna, K.R., and Philip, L. (2005). Bioremediation of Cr (VI) in contaminated soils. J. Hazard. Mater. *121*, 109–117.

Kulczycki, E., Ferris, F., and Fortin, D. (2002). Impact of cell wall structure on the behavior of bacterial cells as sorbents of cadmium and lead. Geomicrobiol. J. *19*, 553–565.

La Duc, M.T., Osman, S., Vaishampayan, P., Piceno, Y., Andersen, G., Spry, J., and Venkateswaran, K. (2009). Comprehensive census of bacteria in clean rooms by using DNA microarray and cloning methods. Appl. Environ. Microbiol. 75, 6559–6567.

LaGrega, M.D., Buckingham, P.L., and Evans, J.C. (2001). Hazardous Waste Management, 2nd edn (McGraw Hill, New York, NY, USA).

Lindsay, D., and von Holy, A. (1997). Evaluation of dislodging methods for laboratory-grown bacterial biofilms. Food Microbiol. *14*, 383–390.

Liu, K.J., Jiang, J., Shi, X., Gabrys, H., Walczak, T., and Swartz, H.M. (1995). Low-frequency EPR study of chromium (V) formation from chromium (VI) in living plants. Biochem. Biophys. Res. Co. *206*, 829–834.

Loreau, M., and Hector, A. (2001). Partitioning selection and complementarity in biodiversity experiments. Nature *412*, 72–76.

Luli, G., Talnagi, J., Strohl, W.R., and Pfister, R. (1983). Hexavalent chromium-resistant bacteria isolated from river sediments. Appl. Environ. Microbiol. *46*, 846–854.

Macaskie, L.E., and Dean, A.C. (1990). Metal-sequestering biochemicals. In Biosorption of Heavy Metals, B. Volesky, ed. (CRC Press, Boca Raton, FL, USA), pp. 199–248.

Madigan, M.T., Martinko, J.M., and Parker, J. (2001). Brock: Biology of Microorganisms, 9th edn. (Prentice-Hall, New York, NY, USA).

Malik, A. (2004). Metal bioremediation through growing cells. Environ. Int. *30*, 261–278.

Martins, S.I.F.S. (2003). Unravelling the maillard reaction network by multiresponse kinetic modeling. PhD thesis, Wageningen University.

Megharaj, M., Avudainayagam, S., and Naidu, R. (2003). Toxicity of hexavalent chromium and its reduction by bacteria isolated from soil contaminated with tannery waste. Current Microbiol. *47*, 51–54.

Mohan, D., and Pittman Jr., C.U. (2006). Activated carbons and low cost adsorbents for remediation of tri and hexavalent chromium from water. J. Hazard. Mater. *137*, 762–811.

Moller, S., Sternberg, C., Andersen, J.B., Christensen, B.B., Ramos, J.L., Givskov, M., and Molin, S. (1998). In situ gene expression in mixed-culture biofilms: evidence of metabolic interactions between community members. Appl. Environ. Microbiol. *64*, 721–732.

Mona, S., Kaushik, A., and Kaushik, C. (2011). Biosorption of chromium (VI) by spent cyanobacterial biomass from a hydrogen fermentor using Box-Behnken model. Int. Biodeter. Biodegr. *65*, 656–663.

Mordenti, A., Piva, G., and Piva, G. (1997). Chromium in animal nutrition and possible effects on human health. In Chromium Environmental Issues, Canali, S., Tittarelli, F., and Sequi, P., eds. (Franco-Angeli Editore, Bologna, Italy), pp. 131–151.

Nielsen, A.T., Tolker-Nielsen, T., Barken, K.B., and Molin, S. (2000). Role of commensal relationships on the spatial structure of a surface-attached microbial consortium. Environ. Microbiol. *2*, 59–68.

Olmsted, J., and Williams, G.M. (1994). Chemistry: The Molecular Science (Mosby, St. Louis, MO).

Pal, A., and Paul, A. (2004). Aerobic chromate reduction by chromium-resistant bacteria isolated from serpentine soil. Microbiol. Res. *159*, 347–354.

Papp, J.F. (2004). Chromium use by market in the United States. Paper presented at Proceedings: Tenth International Ferroalloys Congress, Cape Town, South Africa, The South African Institute of Mining and Metallurgy, pp. 770–778.

Park, C., Keyhan, M., Wielinga, B., Fendorf, S., and Matin, A. (2000). Purification to homogeneity and characterization of a novel *Pseudomonas putida* chromate reductase. Appl. Environ. Microbiol. *66*, 1788–1795.

Patrick, R. (1978). Effects of Trace Metals in the Aquatic Ecosystem: The diatom community, base of the aquatic food chain, undergoes significant changes in the presence of trace metals and other alterations in water chemistry. Am. Sci. *66*, 185–191.

Pattanapipitpaisal, P., Brown, N., and Macaskie, L. (2001). Chromate reduction and 16S rRNA identification of bacteria isolated from a Cr (VI) contaminated site. Appl. Microbiol. Biotechnol. *57*, 257–261.

Paul, M.L., Samuel, J., Chandrasekaran, N., and Mukherjee, A. (2012). Comparative kinetics, equilibrium, thermodynamic and mechanistic studies on biosorption of hexavalent chromium by live and heat killed biomass of *Acinetobacter junii* VITSUKMW2, an indigenous chromite mine isolate. Chem. Eng. J. *187*, 104–113.

Petrilli, F.L., and De Flora, S. (1977). Toxicity and mutagenicity of hexavalent chromium on *Salmonella typhimurium*. Appl. Environ. Microbiol. 33, 805–809.

Pinon-Castillo, H., Brito, E., Goñi-Urriza, M., Guyoneaud, R., Duran, R., Nevarez-Moorillon, G., Gutiérrez-Corona, J., Caretta, C., and Reyna-López, G. (2010). Hexavalent chromium reduction by bacterial consortia and pure strains from an alkaline industrial effluent. J. Appl. Microbiol. *109*, 2173–2182.

Quintelas, C., and Tavares, T. (2001). Removal of chromium (VI) and cadmium (II) from aqueous solution by a bacterial biofilm supported on granular activated carbon. Biotechnol. Lett. *23*, 1349–1353.

Quintelas, C., Fernandes, B., Castro, J., Figueiredo, H., and Tavares, T. (2008). Biosorption of Cr (VI) by three different bacterial species supported on granular activated carbon-a comparative study. J. Hazard. Mater. *153*, 799–809.

Rattanachomsri, U., Kanokratana, P., Eurwilaichitr, L., Igarashi, Y., and Champreda, V. (2011). Culture-independent phylogenetic analysis of the microbial community in industrial sugarcane bagasse feedstock piles. Biosci. Biotechnol. Biochem. *75*, 232–239.

Reeder, R.J., Schoonen, M.A., and Lanzirotti, A. (2006). Metal speciation and its role in bioaccessibility and bioavailability. Rev. Mineral. Geochem. *64*, 59–113.

Rene'N, H., Apel, W.A., Thompson, V.S., and Sheridan, P.P. (2006). Low temperature reduction of hexavalent chromium by a microbial enrichment consortium and a novel strain of *Arthrobacter aurescens*. BMC Microbiol. *6*, 5.

Rengaraj, S., Joo, C.K., Kim, Y., and Yi, J. (2003). Kinetics of removal of chromium from water and electronic process wastewater by ion exchange resins: 1200H, 1500H and IRN97H. J. Hazard. Mater. *102*, 257–275.

Romanenko, V.I., Kuznetsov, S.I., and Korenkov, V.N. (1976). Method of biological purification of industrial effluents from chromates and bichromates. US Patent 3941691.

Romanenko, V., and Koren'kov, V. (1977). Pure culture of bacteria using chromates and bichromates as hydrogen acceptors during development under anaerobic conditions. Mikrobiologiia *46*, 414.

Ross, D., Sjogren, R., and Bartlett, R. (1981). Behavior of chromium in soils: IV. Toxicity to microorganisms. J. Environ. Qual. *10*, 145–148.

Saha, B., and Orvig, C. (2010). Biosorbents for hexavalent chromium elimination from industrial and municipal effluents. Coordin. Chem. Rev. *254*, 2959–2972.

Sawyer, C., McCarty, P., and Parkin, G. (1978). Chemistry for Environmental Engineers (McGraw Hill Book Company, New York).

Shakoori, A., Makhdoom, M., and Haq, R. (2000). Hexavalent chromium reduction by a dichromate-resistant Gram-positive bacterium isolated from effluents of tanneries. Appl. Microbiol. Biotechnol. *53*, 348–351.

Sharma, D.C., and Forster, C.F. (1993). Removal of hexavalent chromium using sphagnum moss peat. Water Res. *27*, 1201–1208.

Shen, H., and Wang, Y. (1993). Characterization of enzymatic reduction of hexavalent chromium by *Escherichia coli* ATCC 33456. Appl. Environ. Microbiol. *59*, 3771–3777.

Speitel Jr, G.E., and DiGiano, F.A. (1987). Biofilm shearing under dynamic conditions. J. Environ. Eng. *113*, 464–475.

Spring, S., Jäckel, U., Wagner, M., and Kämpfer, P. (2004). *Ottowia thiooxydans* gen. nov., sp. nov., a novel facultatively anaerobic, $N_2O$-producing bacterium isolated from activated sludge, and transfer of *Aquaspirillum gracile* to *Hylemonella gracilis* gen. nov., comb. nov. Int. J. Syst. Evol. Microbiol. *54*, 99–106.

Steger, K., Jarvis, Å., Vasara, T., Romantschuk, M., and Sundh, I. (2007). Effects of differing temperature management on development of Actinobacteria populations during composting. Res. Microbiol. *158*, 617–624.

Suzuki, T., Miyata, N., Horitsu, H., Kawai, K., Takamizawa, K., Tai, Y., and Okazaki, M. (1992). NAD (P) H-dependent chromium (VI) reductase of *Pseudomonas ambigua* G-1: a Cr (V) intermediate is formed during the reduction of Cr (VI) to Cr (III). J. Bacteriol. *174*, 5340–5345.

Tavares, J.S.C., Maria, T., and Pontes Correia Neves, M. (2007). Biosorption system produced from biofilms suported on faujasite (FAU) zeolite, process obtaining it and its usage for removal of hexavalent chromium Cr (VI) (WO Patent 2,007,020,588).

Thacker, U., Parikh, R., Shouche, Y., and Madamwar, D. (2006). Hexavalent chromium reduction by *Providencia* sp. Process Biochem. *41*, 1332–1337.

Trevors, J., Oddie, K., and Belliveau, B. (1985). Metal resistance in bacteria. FEMS Microbiol. Lett. 32, 39–54.

Vaiopoulou, E., and Gikas, P. (2012). Effects of chromium on activated sludge and on the performance of wastewater treatment plants: a review. Water Res. *46*, 549–570.

Veenstra, J.N., Sanders, D., and Ahn, S. (1999). Impact of chromium and copper on fixed-film biological systems. J. Environ. Eng. *125*, 522–531.

Viamajala, S., Peyton, B.M., and Petersen, J.N. (2003). Modeling chromate reduction in *Shewanella oneidensis* MR-1: Development of a novel dual-enzyme kinetic model. Biotechnol. Bioeng. *83*, 790–797.

Viti, C., and Giovannetti, L. (2007). Bioremediation of soils polluted with hexavalent chromium using bacteria: a challenge. In Environmental Bioremediation Technologies, S.N. Singh, R.D. Tripathi, eds. (Springer, Berlin, Germany), pp. 57–76.

Volesky, B. (1990). Removal and Recovery of Heavy Metals by Biosorption (CRC Press, Boca Raton, FL, USA).

von Canstein, H., Li, Y., Timmis, K., Deckwer, W.-D., and Wagner-Dobler, I. (1999). Removal of mercury from chloralkali electrolysis wastewater by a mercury-resistant *Pseudomonas putida* strain. Appl. Environ. Microbiol. *65*, 5279–5284.

von Canstein, H., Kelly, S., Li, Y., and Wagner-Dobler, I. (2002). Species diversity improves the efficiency of mercury-reducing biofilms under changing environmental conditions. Appl. Environ. Microbiol. *68*, 2829–2837.

Vu, B., Chen, M., Crawford, R.J., and Ivanova, E.P. (2009). Bacterial extracellular polysaccharides involved in biofilm formation. Molecules *14*, 2535–2554.

Wagner-Dobler, I., Lünsdorf, H., Lübbehüsen, T., Von Canstein, H., and Li, Y. (2000). Structure and species composition of mercury-reducing biofilms. Appl. Environ. Microbiol. *66*, 4559–4563.

Wang, Y.T., and Xiao, C. (1995). Factors affecting hexavalent chromium reduction in pure cultures of bacteria. Water Res. *29*, 2467–2474.

Wang, Y. (2000). Microbial Reduction of Chromate. In Environmental Microbe–Metal Interactions, Lovley, D., ed. (ASM Press, Washington, DC, USA), pp. 225–235.

Windholz, M., Budavari, S., Stroumtsos, L.Y., and Fertig, M.N. (1976). The Merck Index. An Encyclopedia of Chemicals and Drugs (Merck and Co., Inc., Kenilworth, NJ, USA).

Woan Ying Liew, P., Chyan Jong, B., Meng Goh, C., and Ahmad, M. (2009). Bacterial diversity associated with empty oil palm fruit bunch compost as revealed by cultivation-independent analyses of PCR-amplified 16S rRNA genes. J. Gen. Appl. Microbiol. *55*, 233–240.

Yuen, K.-Y., Woo, P.C., Teng, J.L., Leung, K.-W., Wong, M.K., and Lau, S.K. (2001). *Laribacter hongkongensis* gen. nov., sp. nov., a novel Gram-negative bacterium isolated from a cirrhotic patient with bacteremia and empyema. J. Clin. Microbiol. *39*, 4227–4232.

Zahoor, A., and Rehman, A. (2009). Isolation of Cr (VI) reducing bacteria from industrial effluents and their potential use in bioremediation of chromium containing wastewater. J. Environ. Sci. *21*, 814–820.

Zakaria, Z.A., Zakaria, Z., Surif, S., and Ahmad, W.A. (2007). Hexavalent chromium reduction by *Acinetobacter haemolyticus* isolated from heavy-metal contaminated wastewater. J. Hazard. Mater. *146*, 30–38.

Zayed, A.M., and Terry, N. (2003). Chromium in the environment: factors affecting biological remediation. Plant Soil *249*, 139–156.

Zouboulis, A., Loukidou, M., and Matis, K. (2004). Biosorption of toxic metals from aqueous solutions by bacteria strains isolated from metal polluted soils. Process Biochem. *39*, 909–916.

# Hydrocarbonoclastic Biofilms

10

Dina M. Al-Mailem and Samir S. Radwan

**Abstract**

Biofilms harbouring hydrocarbonoclastic (hydrocarbon-degrading) bacteria occur naturally, and are widely distributed in aquatic, terrestrial and even atmospheric environments. They seem to play a role in the 'self-cleaning' of environments contaminated with oil, oil vapour and oil derivatives. Such biofilms may develop on inanimate substrates such as gravel particles, small stones and waste metal, wood and plastic pieces. They may also be associated with animate biotic substrates, e.g. biofouling materials, macroalgae, cyanobacterial mats, the roots (rhizospheres) and leaves (phyllospheres, phylloplanes) of higher plants, as well as with fish and other aquatic animals. Hydrocarbonoclastic bacteria also live naturally in association with phototrophic picoplankton in surface waters and mud flats.

Biofilms comprising hydrocarbonoclastic bacteria can be established artificially on inert substrates such as gravel particles, or on glass and plastic plates. These man-made biofilms are promising tools for the bioremediation of oil-contaminated materials in bioreactors.

In biofilms, hydrocarbonoclastic bacteria enjoy various benefits which facilitate and enhance their oil-biodegradation activity, especially when coexisting with phototrophic and diazotrophic microorganisms which enrich the biofilm with oxygen and nitrogenous compounds. This principally occurs because oxygen and nitrogen are limiting factors to the mineralization of hydrocarbons by bacteria.

## Introduction

Environmental pollution with crude oil and oil derivatives has become a global phenomenon, spreading and extending into all types of ecosystems. Owing to occasional accidents, legal activities, e.g. oil production and transport, as well as illegal activities, e.g. oily waste disposal, military conflicts and others, aquatic, terrestrial and atmospheric environments are already impacted with these pollutants, worldwide. Nevertheless, being a natural product, crude oil (and its derivatives), is readily biodegradable in nature. Soils, waters and air harbour a great variety of hydrocarbonoclastic prokaryotes and eukaryotes, which are capable of naturally removing such pollutants from these environments.

Crude oil is a mixture of thousands of organic compounds mainly composed of carbon and hydrogen, along with small amounts of oxygen, nitrogen and sulfur. Trace amounts of metals, e.g. nickel, vanadium and others are also usually present. Irrespective of the origin of crude oil, four major classes of compounds are routinely recognized therein: aliphatic hydrocarbons, aromatic hydrocarbons, asphaltenes and resins (Colwell and Walker, 1977). The first class, the aliphatics comprises straight chain and branched chain alkanes of varying length; they have the general formula $C_nH_{2n+2}$. Low molecular weight alkanes (with fewer than five carbon atoms) are gaseous at room temperature, alkanes with about 5–15 carbon atoms are liquid and higher molecular weight alkanes are viscous to solid waxy materials, their chain lengths may reach about 70 carbon atoms. Aromatic hydrocarbons are cyclic compounds based on benzene rings. They may be simple with only one benzene ring or polyaromatic with five or more condensed rings. The 'hetero-compounds' comprise the asphaltenes and resins. In addition to carbon and hydrogen their molecules contain other atoms like oxygen, sulfur and nitrogen. Chemically asphaltenes may be comprised of phenols, fatty acids, ketones, esters, porphyrins and others. Resins may contain pyridines, quinolines, carbazols, sulfoxides, amides and others.

As already mentioned, various environments naturally harbour hydrocarbonoclastic microorganisms. This subject has been repeatedly reviewed during the past five decades (Atlas and Bartha, 1972; Colwell, 1977; Harayam, 1999; 2004; Head and Swannell, 1999; Head *et al.*, 2006; Klug and Markovetz, 1971; Levi *et al.*, 1979; Radwan and Sorkhoh, 1993; Radwan, 2008, 2009). These reviews show that the list of known hydrocarbonoclastic microorganisms has continued to grow over time; this list will certainly keep growing. They comprise bacteria, actinomycetes, yeast, moulds and archaea. Most of the reports were based on organisms detected in nonextreme environments, but in recent years extreme environments (Margesin and Schinner, 2001; Le Borgne *et al.*, 2008) have become focal points of interest.

The characteristic physiological feature of hydrocarbonoclastic microorganisms is their possession of oxygenases, also called hydroxylases. These enzyme systems catalyse the splitting of oxygen molecules into oxygen atoms and introduce the latter into hydrocarbon molecules leading to alcohols. Monooxygenases are involved in the hydroxylation of aliphatics, whereas dioxygenases are involved in hydroxylation of benzene rings. The produced alcohols are further oxidized via aldehydes or ketones to the corresponding fatty acids (Rehm and Reiff, 1981; Boulton and Ratledge, 1984; Buhler and Shindler, 1984). Fatty acids are subsequently catabolized via a β-oxidation sequence to acetyl CoA, which is a key intermediate in cell material and energy metabolism. Through this mechanism, hydrocarbonoclastic microorganisms utilize hydrocarbon substrates as sole sources of carbon and energy.

Regarding the identities of the hydrocarbonoclastic microorganisms, these vary according to the environment and the prevailing environmental parameters, a subject which will be mentioned through this chapter. Here, we emphasize that those organisms do not occur alone in the various habitats, not even in those with a long history of hydrocarbon pollution. Other non-hydrocarbonoclastic organisms also occur and impact directly or indirectly the activity of hydrocarbonoclastic organisms. It is therefore important to understand the nature of such 'microbial associations' which range from 'just loosely coexisting organisms' to 'almost completely integrated microbial societies', as is the case in biofilms. The latter appear to be associations particularly suited for the remediation of pollutants 'because of their high microbial biomass and ability to immobilize pollutants' (Singh *et al.*, 2006; Nilanjana *et al.*, 2012). Substrate colonization to produce biofilms involves the initial production of extracellular polysaccharide substances forming glycocalyx, that mediates cellular adhesion. The glycocalyx traps nutrients, conserves and concentrates enzymes released by bacteria and constitutes a physical barrier against antibacterial agents (Costerton *et al.*, 1978, 1987). As reviewed by Morton and Suman (1994), microorganisms growing in the presence of hydrophobic substrates follow two types of interaction; the uptake of minute substrate droplets from the substrate bulk (pseudosolubilization) and the uptake of fractions arising from direct cell contact to the substrate droplets. Many hydrocarbonoclastic microorganisms are long known to produce oil-in-water emulsions (Erikson and Nakahar, 1975; Zajic and Panchal, 1976; Margaritis *et al.*, 1979; Cooper and Zajic, 1980), also via extracellular bioemulsifiers. Such organisms frequently form dense films at the oil–water interface, considered as biofilms by earlier authors (Prince and Morton, 1989).

The major objective of this chapter is to describe natural and man-made microbial associations in which biofilm-associated hydrocarbonoclastic microorganisms are active counterparts, highlighting their potential application for hydrocarbon bioremediation.

## Natural microbial associations

### Planktonic associations in water and air

Although the term plankton implies that microorganisms occur in aquatic environments as separate individuals, in reality, this does not appear to be the case. Aquatic 'planktonic' cells tend to adhere to animate and inanimate surfaces, to each other and to cells of other planktonic species. The resulting associations are to be looked at as 'floating' biofilms with varying dimensions (Radwan *et al.*, 2005a). Most of such associations comprise photosynthetic and heterotrophic counterparts (Sudha *et al.*, 2011). The phototrophs are either cyanobacterial or microalgal species. The heterotrophic partners are usually heterotrophic bacteria, but yeasts and moulds may also be present. That the phototrophic and heterotrophic counterparts intimately interact in those associations is quite apparent from the fact that the establishment of axenic (free from bacteria) phototrophic cultures is often a very difficult task (Rippka, 1972). This is normally done by trying to eliminate the heterotrophic bacteria using antibiotics. It has been reported frequently that with progressive axinity the phototrophic cells progressively lose their vitality (Rippka, 1972; Al-Hasan *et al.*, 1994; Radwan and Al-Hasan, 2000).

Planktonic associations comprising hydrocarbonoclastic microorganisms occur naturally in coastal mud flats (Al-Mailem *et al.*, 2010; Coulon *et al.*, 2012), in surface (Al-Hasan *et al.*, 2001; Radwan *et al.*, 2005a) and subsurface (Al-Bader *et al.*, 2013), seawater and even as air-dust-borne associations (Al-Bader *et al.*, 2012a). The microbial composition of associations from two remote mud flats, also called 'tidal biofilms', and their role in removing hydrocarbon pollutants have been reported (Table 10.1).

Although the two mud flat samples were from two geographically different regions, Kuwait and the UK, they showed a striking similarity in their microbial composition. Thus, the phototrophic counterparts in both associations consisted predominantly of cyanobacteria and diatoms. The hydrocarbon-utilizing microbial partners in both mud flats belonged predominantly to the rather newly discovered group of the obligate hydrocarbonoclastic bacteria (OHCB) (Yakimov *et al.*, 2007) reported to be globally responsible for mineralizing hydrocarbons in the marine ecosystem. Oil consumption measurements revealed that the microbial consortia in both mud flats were effective in removing considerable proportions of hydrocarbons artificially added to mud flat samples. It has even been reported that there was a 10-fold increase in the amount of cyanobacteria in the oil-contaminated mud samples (Coulon *et al.*, 2012). This result consolidates earlier reports on enhancing the growth of pure filamentous cyanobacteria from the Arabian Gulf by crude oil addition (Al-Hasan *et al.*, 1994; Radwan and Al-Hasan, 2000). A question arises regarding whether or not cyanobacteria may contribute directly to hydrocarbon mineralization; a subject which has been discussed in earlier reviews (Radwan and Al-Hasan, 2000, 2001a). There are reports in the literature that cyanobacteria can oxidize hydrocarbons (Cerniglia *et al.*, 1980a,b; 1981; Al-Hasan *et al.*, 1998). However, recent knowledge indicates that such hydrocarbonoclastic activity was obviously the responsibility of the hydrocarbonoclastic bacteria associated with the studied cyanobacteria. Nevertheless, the cyanobacteria still seem to play an indirect role in this process because they represent substrates where heterotrophic bacteria adhere (immobilization), and provide these bacteria with molecular oxygen and fixed nitrogen needed for hydrocarbon consumption (Klug and Markovetz, 1971; Colwell and Walker, 1977; Rehm and Reiff, 1981; Radwan and Sorkhoh, 1993). Furthermore, there is evidence that unicellular and colonial cyanobacteria accumulate hydrocarbons from the environment in their interthylakoid spaces (Al-Hasan *et al.*, 2001; see also Fig. 10.1), a result which may explain, at least in part, why hydrocarbonoclastic bacteria usually have a special affinity towards cyanobacteria.

Planktonic associations are quite common in surface waters of the seas and oceans around the world. Millions of picocyanobacterial cells per ml surface seawater (uppermost 5 cm) belonging

**Table 10.1** Composition of tidal biofilms from the Arabian Gulf (Kuwait) and the Clone Estuary (UK)

| Microbial counterparts | Arabian Gulf association (Al-Mailem *et al.*, 2010) | Clone Estuary association (Coulon *et al.*, 2012) |
|---|---|---|
| 1. Phototrophs | Picocyanobacteria, filamentous cyanobacteria, diatoms | Cyanobacteria, diatoms |
| 2. Hydrocarbonoclastic bacteria | *Halomonas aquamarina*, *Marinobacter hydrocarbonoclasticus*, *Marinobacter* sp., *Alcanivorax* sp., *Dietzia maris* | *Alcanivorax borkumensis*, *Oceanospirillum* spp., *Oleibacter* sp., *Cycloclasticus* sp. |

(A1)

(A2)

(B)

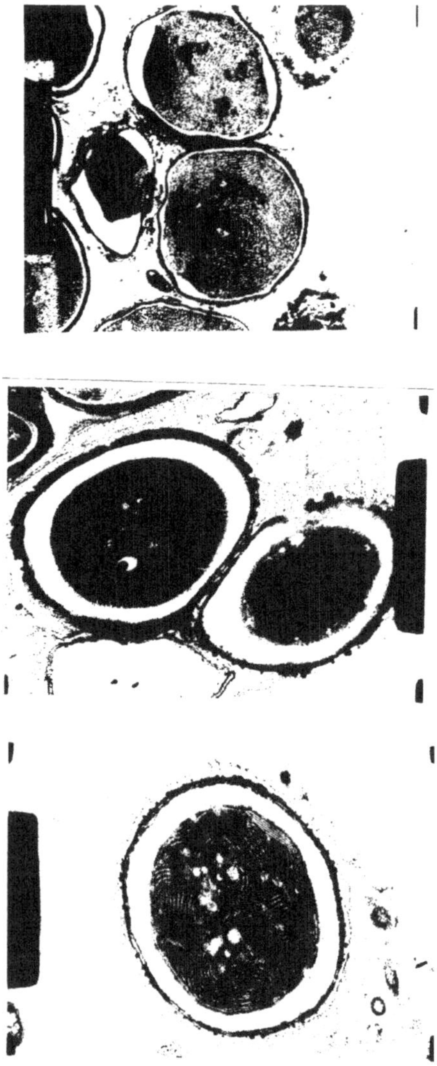

**Figure 10.1** Transmission electron micrographs showing (A) Accumulation of *n*-heptadecane by cells of the colonial cyanobacterium *Pleurocapsa* sp.. After 12 h (A1), the hydrocarbon inclusions (transparent areas) have appeared below the plasma membranes, and after 36 h (A2), they start to fill the interthylakoid spaces. (**B**) Accumulation of hydrocarbons below the plasma membrane and in interthylakoid spaces of the cyanobacteria *Synechococcus* sp.

predominantly to the genera *Synechococcus* and *Synechocystis* are usually found (Radwan *et al.*, 2004, 2005a). Like in mud flats, the cyanobacterial cells are often associated with hydrocarbonoclastic bacteria. Samples from the Arabian Gulf water surface revealed that bacteria associated with the picocyanobacteria belonged predominantly to the genus *Marinobacter* (Al-Wahaib *et al.*, unpublished results) which, as mentioned previously, is related to the OHCB group. Nonaxenic *Synechococcus* and *Synechocystis* cultures from water and air in the Gulf region, with their associated hydrocarbonoclastic bacteria consumed crude oil and pure aliphatic and aromatic hydrocarbons in batch cultures (Al-Bader *et al.*, 2012; Al-Wahaib *et al.*, unpublished results).

Subsurface water samples (2–3 m below the surface) from the Arabian Gulf also contained similar but not identical microbial associations. The phototrophic partner was related to the picocyanobacterium *Acaryochloris*, and the hydrocarbonoclastic bacteria belonged to the OHCB and others (Al-Bader *et al.*, 2013a).

Surprisingly, even air harbours on its dust particles, associations of phototrophic and hydrocarbonoclastic microorganisms (Al-Bader *et al.*, 2012). In this context, volatile hydrocarbons (Pearce *et al.*, 2009) and fatty acids (Warneck, 1988) have been detected in clouds and the upper atmosphere. Reportedly, the dust associations at a height of 15 m contained, as phototrophs, the cyanobacteria *Nostoc commune* and *Leptolynbya thermalis* and the chlorophyte *Glocotila* sp., whereas a seawater sample analysed for comparison contained *Leptolynbya* sp. and *Acaryochloris* sp.. The hydrocarbonoclastic bacteria in these air–dust associations were *Acinetobacter* sp., *Aeromonas caviae*, *Alcanivorax jedensis*, *Bacillus ashahii*, *B. pumilus*, *Marinobacter aquaeolei*, *Paenibacillus* sp. and *Stenotrophomonas maltophilia*. To recall, all these bacterial species are of marine origin and *Alcanivorax* and *Marinobacter* are related to the OHCB. When these associations were used as inocula in batch cultures, considerable proportions of crude oil in the surrounding medium were consumed.

## Coastal microbial mats

Although microbial mats, also as fossil stromotolites, are long known to be widely distributed in coastal areas around the globe, it is only in the last few decades that hydrocarbonoclastic bacteria were discovered as inhabitants of such natural biofilms. A pioneering approach to this subject was provided from our laboratory (Sorkhoh *et al.*, 1992) and published one year after the Second Gulf War (1991). The deliberate release of crude oil from the Al-Ahmadhi terminal by the Iraqi forces during the occupation of Kuwait (2 August 1990 to 26 February 1991) resulted in the Arabian Gulf environment suffering one of the greatest man made pollution catastrophes known. The $5 \times 10^5$ tons of crude oil that was released, drifted from the Kuwaiti coasts to pollute about 770 km of intertidal zone surrounding the Saudi Arabian coast. A Saudi research station, about 200 km south of the Kuwaiti borders, became covered with up to 20 kg $m^2$ oil sediments. The free rock surfaces received about 1.4 kg/$m^2$ oil. Regular inspection of these severely polluted areas showed that there was no visual change in crude oil volume or composition occurring over an 8-month period, except that the intensive growth of blue green microbial mats was observed over oil sediments. Similar mats were also recorded by our group in the much less polluted coasts of Kuwait (Fig. 10.2). The microbial mats were intimately associated with oil sediments, such that oil-free areas were also free of such mats. Mat samples from various localities were similar in their structure.

**Figure 10.2** Cyanobacterial mats on one of the small oil sediments trapped in the intertidal sand of the Kuwaiti coast in late 1991. A year later the oil sediment had disappeared via the hydrocarbonoclastic activities of the mat bacteria.

The predominant phototrophic counterparts were the filamentous cyanobacteria *Microcoleus* sp. and *Phormidium* sp.. Such organisms produce thick mucilage sheathes in which single cells and filaments are embedded. As mentioned before, there is no rigorous evidence that cyanobacteria and other phototrophs can directly biodegrade hydrocarbons, although there is evidence that they may accumulate these compounds in their interthylakoid spaces (see Fig. 10.1). Interestingly, up to $10^6$ cells of hydrocarbonoclastic bacteria per gram fresh mat have been found embedded in the mucilage of cyanobacteria.

Biofouling materials are biomasses that develop adhering to inanimate substrates submerged for prolonged time in seawater, e.g. navigation vehicles. The growth may be so intensive that the vehicles become heavier and thus, consume more fuel. The term biofouling is based on the notion that this growth is not recommended, vehicle paints usually contain substances that control it. Such bacteria could grow on crude oil as well as various pure aliphatic and aromatic hydrocarbons as sole sources of carbon and energy. The hydrocarbonoclastic bacteria that appear to perform this degradation may be immobilized in cyanobacterial mucilage, and thus are protected from being washed away. Furthermore, oxygen released as a byproduct of the cyanobacterial photosynthesis is suggested to enhance the bacterial hydrocarbonoclastic potential. As already mentioned, the initial step of bacterial degradation of the hydrocarbon molecule involves its hydroxylation using molecular oxygen. Diazotrophy (nitrogen fixation) is also known to be widely distributed among cyanobacteria. In microbial communities of oil-contaminated marine microcosms, cyanobacteria, were the dominant diazotrophs (Musat *et al.*, 2006; Sudha *et al.*, 2011). Environmental, light-incubated microcosm samples revealed dinitrogenase reductase genes (*nif*H) that affiliated with those of heterocystous cyanobacteria. Reportedly, upon a dark shift, *nif*H genes related to those of nonheterocystous cyanobacteria were expressed. To recall, nitrogen fertilization is a limiting factor to microbial hydrocarbon biodegradation. Therefore, the natural occurrence of diazotrophic cyanobacteria in these coastal mats makes it an effective biological material for hydrocarbon bioremediation in intertidal and coastal zones. Similar work has now been published on mats discovered in various regions of the globe.

Hydrocarbonoclastic bacteria immobilized in microbial mats from oily Arabian Gulf coasts were successfully inoculated and established in desert soil samples artificially polluted with crude oil (Sorkhoh *et al.*, 1995). Reportedly, the established bacteria removed 50% of the contaminating oil within 10–20 weeks. *Rhodococcus* and other nocardioforms prevailed in the first few weeks, but after 22 weeks, *Bacillus* and *Arthrobacter* dominated. These results highlight the potential application of natural biofilms for oil-bioremediation.

In this context, we have provided experimental evidence that the oil biodegradation potential at the Arabian Gulf coasts is higher than in the water body (Radwan *et al.*, 1999; Al-Awadhi *et al.*, 2002). Littoral materials from the intertidal zone were loaded with much more hydrocarbonoclastic microorganisms than inshore and offshore water samples. Animate materials (mats, epilithic growth) were much richer in such microorganisms than inanimate materials (sand, rocks, shells). Coast-immobilized bacterial strains of *Acinetobacter calcoaceticus* and *Micrococcus* sp. revealed a higher hydrocarbon removal potential than planktonic strains of both organisms in batch cultures.

## Coastal epilithic biofilms

We mentioned previously that coastal inert materials from Kuwait were found coated with biofilms rich in hydrocarbonoclastic bacteria. Such materials included rock pieces, iron bars, cloth pieces, shells, plastics, and sand (Radwan *et al.*, 1999, 2010) where bacterial numbers were in the magnitude of $10^6$ cells per $cm^2$ of the analysed materials. The numbers of hydrocarbonoclastic microorganisms associated with these materials were much higher in the autumn, winter and spring than in summer when the maximum temperatures in this arid region usually exceeds 60°C. Most dominant hydrocarbonoclastic bacteria in all samples were *Acinetobacter calcoaceticus* and *Micrococcus* sp., but fewer numbers of nocardioforms and other bacterial taxa were also identified. Evidence was provided that these bacterial species could effectively remove *n*-octadecane, an alkane, and phenanathrene, a polynuclear aromatic hydrocarbon, in batch cultures. In another study (Al-Awadhi *et al.*, 2002), evidence was offered that coastal materials viz. small stones and sand were more effective in hydrocarbon removal in batch cultures than water samples from the Gulf. Radwan and Al-Hasan (2001b) described hydrocarbon-degrading microbial consortia immobilized in biofilms coating gravel particles in the intertidal zone of the Arabian Gulf. Also here, the predominant hydrocarbonoclastic bacteria were *Acinetobacter calcoaceticus, Micrococcus* sp. and nocardioforms, which were embedded in blue green filamentous biomass. Each gravel particle, about 2 cm in diameter was coated with about 100 mg biomass (cyanobacteria and bacteria). The predominant cyanobacteria were identified as *Dermocarpelle* and *Lyngbya*. These gravel particles were used successfully in five successive cycles of crude oil removal when added into seawater. In contrast, control gravel particles from the same area, but lacking any visible biomass, brought about oil removal in the first two cycles only, after which they gradually lost their oil removal potential. These findings demonstrate the role played by the phototrophs in immobilization of the hydrocarbonoclastic bacteria in these natural biofilms. It will be shown later in this chapter we may establish such consortia as biofilms on inert surfaces.

## Biofilms on macroalgae

So far, little attention has been given to hydrocarbonoclastic microorganisms in biofilms coating macroalgae, although the latter are widely distributed in aquatic ecosystems. However, a single publication directly dealing with this subject (Radwan *et al.*, 2002) includes extensive and interesting information, which merits further consideration. Nine different macroalgal genera were collected from the intertidal and subtidal zones of different sites along the Kuwait City coast of the Arabian Gulf, where they represented the predominant macroalgae in the region (Al-Hasan and Jones, 1989). These macroalgae were *Enteromorpha* (Chlorophyta), *Colpomenia, Cystosiera, Hormophysa, Iyengaria, Padina* and *Sargassum* (Phaeophyta), *Gelidium* and *Gracillaria* (Rhodophyta). All freshly collected macroalgal samples were associated with hydrocarbonoclastic bacteria ranging in number between 2 and $30 \times 10^6$ cells/g. These bacteria belonged predominantly to the genus *Acinetobacter* and the group of the nocardioforms. Electron microscopic examination revealed that the macroalgal surfaces were coated with biofilms comprising dense

**Figure 10.3** Scanning electron micrograph showing rich bacterial communities in a biofilm coating a microalga from the Arabian Gulf.

populations of those bacteria, together with some phototrophic microalgae, particularly diatoms (Fig. 10.3). In batch cultures using 5 g fresh algal samples (with their bacterial biofilms) as inocula, between 38% and 56% of added phenanthrene and between 64% and 98% of added *n*-octadecane in the media were removed after incubation for 2 weeks at room temperature. Meanwhile, the numbers of the associated hydrocarbonoclastic bacteria increased by 32- to 490-fold, depending on the macroalgae and hydrocarbons studied. Only small numbers of such bacteria were released from the algal biofilms into the surrounding medium. This demonstrates how firmly the hydrocarbonoclastic bacteria were adhering to the macroalgal surfaces. It was concluded from this study that bacterial biofilms on macroalgae naturally contribute to the removal of hydrocarbon pollutants in the marine environment.

## Biofilms on biofouling materials

Our group collected biofouling materials from 30 offshore substrates in the Arabian Gulf in eight sites and analysed them for hydrocarbonoclastic microorganisms (Al-Awadhi *et al.*, 2012a). The different materials harboured between $7 \times 10^4$ and $4.1 \times 10^6$ bacterial cells/g. Those bacteria were affiliated to the genera *Alcanivorax*, *Dietzia*, *Gordonia*, *Hoyosella*, *Kokuria*, *Marinobacter*, *Micrococcus*, *Mycobacterium*, *Planococcus*, *Planomicrobium*, *Pseudoalteromonas*, *Salinivibrio* and *Vibrio*, two of which, *Alcanivorax* and *Marinobacter*, are related to the OHCB group. Most of these genera in addition to *Agrobacterium*, *Lobrenzia*, *Microbacterium*, *Nitratireductor*, *Pseudomonas* and *Rhodococcus* were diazotrophic. Agarolytic bacteria with hydrocarbonoclastic potential were also associated with certain biofouling materials (Sorkhoh *et al.*, 2010b). Pure cultures of all these bacterial isolates could grow on a wide range of *n*-alkanes with varying chain lengths, and the aromatics phenanthrene, naphthalene and biphenyl as sole sources of carbon and energy. Experimental evidence was offered indicating that individual isolates could remove substantial proportions of crude oil, pure *n*-octadecane and pure naphthalene in the surrounding media of batch cultures. Furthermore, microbial consortia in biofilms coating fresh environmental samples removed oil effectively in batch cultures in the presence and absence of $NaNO_3$. It was concluded that hydrocarbonoclastic bacteria associated with biofouling materials in the marine environment most probably contribute to the self-cleaning of this environment.

## Biofilms on plant roots

There is huge wealth of information in the literature about microorganisms associated with plant roots (see a review by Radwan, 2009). It is not within the objectives of this chapter to review this subject; this chapter is concerned with hydrocarbonoclastic microorganisms only. However, it seems useful to introduce this subject by recalling some basic facts related to the rhizosphere environment in general.

The rhizosphere, as the term indicates, is the unique layer of the soil lining, and consequently under the direct influence of plant roots. This microenvironment is more suitable for microbial life than the rest of the soil body. The rhizospheric area continuously receives organic nutrients, such as amino acids, monosaccharides, vitamins and other substrates, as root exudates and as dead root-tissue-debris arising from continuous root friction with rough soil particles. In view of the fact that atmospheric oxygen is continuously pumped down through the plant roots, these rhizosphere layers are always much better aerated than the bulk soil at depth. These and other factors lead to that the rhizosphere microflora is quantitatively and qualitatively different from the microflora in the bulk soil. Rhizosphere microorganisms are known to have more complex nutrient requirements, and consequently higher metabolic potential than most non-rhizospheric microorganisms. From a systematic viewpoint, rhizospheric bacterial communities are predominantly Gram-negative, i.e. they belong predominantly to the subdivision Proteobacteria. Prominent examples are species of the genus *Pseudomonas* and the nitrogen-fixing rhizobacteria.

One of the first approaches exploring the subject of hydrocarbonoclastic bacteria in rhizospheres of higher plants was through a communication from our laboratory (Radwan *et al.*, 1995). We reported in this publication on the observation that desert plants naturally growing in areas heavily polluted with dark crude oil always possessed white clean roots. Even their adhering rhizosphere soil particles were oil free. Microbiological analysis showed that the rhizospheres of such plants contained hydrocarbonoclastic bacteria in numbers of the magnitude of $10^8$ cells/g fresh root and hydrocarbonoclastic fungi in numbers of the magnitude of $10^5$ propagules per g. These bacteria consisted to >95% of *Arthrobacter* spp., whereas in the bulk soil *Arthrobacter* made up < 5% and the predominant genera were *Rhodococcus, Pseudomonas* and *Bacillus*. Rhizospheric, hydrocarbonoclastic fungi comprised mainly *Penicillium* and *Fusarium*, but in the non-rhizospheric soil *Trichoderma* predominated. This, and a few other related reports (e.g. Lee and Banks, 1993), focused researchers' interest on hydrocarbonoclastic microorganisms in the rhizospheres of higher plants and their application potential for the 'phytoremediation' of hydrocarbon polluted soil. Table 10.2 comprises some of these results and highlights that the numbers and identities of such hydrocarbonoclastic microorganisms vary according to the identity of the host plant. Phytoremediation is currently considered a powerful technology for the bioremediation of polluted soils (Frick *et al.*, 1999). This technology has basically been developed to accumulate and sometimes even to volatilize pollutants such as mercury. Its application for hydrocarbon contaminated sites relies on a different concept, however, namely pollutant mineralization by rhizospheric microorganisms. Earlier studies in this specific field include those by Qiu *et al.* (1997), Carman *et al.* (1998), Flathman and Lanza (1998), Pradhan *et al.* (1998), Radwan *et al.* (1998), Banks *et al.* (1999), Kulakov *et al.* (2000) and Tesar *et al.* (2002).

Our group contributed to studies on the phytoremediation of oily desert soils under the harsh arid conditions of the Arabian Gulf area. We offered experimental evidence that conventional cropping is an effective approach for the bioremediation of oil pollution in this region (Radwan *et al.*, 2000). The crops were legumes including broad beans which could still grow well in the presence of 10% crude oil. Obviously, cropping practices such as irrigation and turning the soil enhance conditions for growth and activities of soil microorganisms. Furthermore, legume roots are associated with nodules which harbour bacteria that fix molecular nitrogen. These nodules are also coated with biofilms containing hydrocarbonoclastic bacteria (Fig. 10.4). As already mentioned, fixed nitrogen is a limiting factor to microbial hydrocarbon mineralization.

The above facts and the well-known enhancing rhizosphere effect on the numbers of hydrocarbonoclastic microflora obviously lead to enhanced

**Table 10.2** Numbers and identities of hydrocarbonoclastic microorganisms in rhizospheres of higher plants

| Higher plants | Numbers per g root | Hydrocarbonoclastic genera | References |
|---|---|---|---|
| Maize, tomato, lupine, *Senccio glances* | $10^8$ bacterial cells, $10^5$ fungal propagules | *Arthrobacter*, *Fusarium* | Radwan *et al.* (1995) |
| *Cyprus* sp., *Pieris* sp., *Salsola* sp. | $10^8$ bacterial cells, few fungal cells | *Cellulomonas flavigena*, *Rhodococcus erythropolis*, *Arthrobacter* sp., | Radwan *et al.* (1998) |
| Black poplar, herbal plant | Not given | *Pseudomonas*, *Achromobacter*, *Xanthomonas*, *Zoogloea*, *Clavibacter*, *Acinetobacter*, *Variovorax*, *Alcaigenes* | Tesar *et al.* (2002) |
| Peas, beans, squash, peppers, cucumber, tomato | $1–10 \times 10^6$ bacterial cells | *Bacillus*, *Ochrobacterum*, *Enterobater*, *Rhodococcus*, *Arthrobacter*, *Pantoea*, *Nocardia*, *Pseudoxanthomonas*, *Micrococcus* | Al-Awadhi *et al.* (2009) |
| Peas, beans, peanuts, sunflower, tomato, mallow | $1–1.4 \times 10^6$ bacterial cells | *Microbacterium*, *Pseudomonas*, *Acidovorax*, *Rhodococcus*, *Nocardia*, *Mycobacterium* | Sorkhoh *et al.* (2010a) |

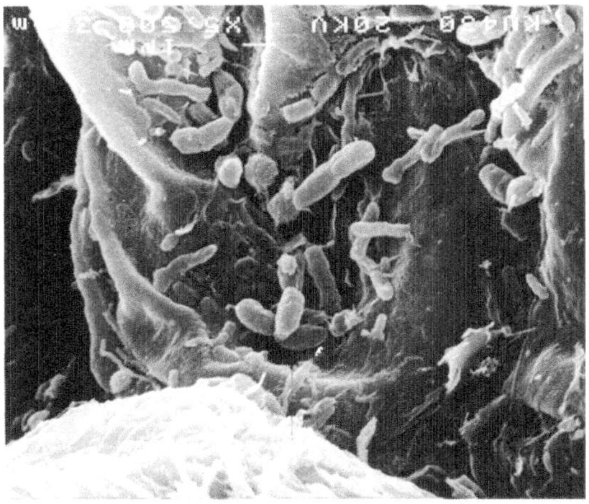

**Figure 10.4** Scanning electron micrograph showing bacterial consortia coating nodules of the legume broad bean (*Vicia faba*).

hydrocarbon bioremediation potential (Lee and Banks, 1993; Radwan *et al.*, 2000). Indeed, we and other authors found that the amounts of extractable hydrocarbons recovered from soils supporting crops, were usually dramatically less than those recovered from uncultivated soil controls. In this context, a group of rhizospheric bacteria, e.g. *Pseudomonas, Serratia* and *Flavobacterium*, have been reported to enhance plant growth, and were consequently called plant growth-promoting rhizobacteria (PGPR) (Polonenko *et al.*, 1987; Zhang *et al.*, 1996, 1997). Inoculating roots of broad beans with PGPR-suspensions enhanced plant growth and increased their phytoremediation potential (Radwan *et al.*, 2005b). Interestingly, nodule bacteria, viz. *Rhizobium* and *Bradyrhizobium* and the PGPR viz. *Pseudomonas, Serratia* and *Flavobacterium* are themselves hydrocarbonoclastic (Radwan *et al.*, 2007b). Legume root nodules are also coated with bacteria with the combined potential for nitrogen fixation and hydrocarbon degradation, which makes these periphytic consortia effective contributors to the bioremediation of oily, nitrogen poor desert soils (Dashti *et al.*, 2009). Bacteria with this combined activity seem to be common in the rhizosphere of crop plants; they comprise the genera *Bacillus, Ochrobactrum, Enterobacter, Rhodococcus, Arthrobacter, Pottola, Nocardia, Pseudoxanthomonas* and others (Al-Awadhi *et al.*, 2009). There is evidence that their numbers increase in the rhizospheres in response to soil contamination with crude oil (Sorkhoh *et al.*, 2010a). The above facts indicate that the rhizospheric biofilms naturally coating roots of higher plants accommodate mixtures of microbial communities whose combined physiological activities lead to the enhanced mineralization of hydrocarbon contaminants in soil.

## Biofilms on plant leaves

In contrast to the underground rhizospheric biofilms, biofilms associated with leaves and other aerial plant organs are exposed to the atmospheric environment. We have already referred to air-dust-borne associations of hydrocarbonoclastic bacteria with phototrophic microorganisms, and their potential role in removing atmospheric hydrocarbon pollutants (see 'Planktonic associations in water and air', above). Being associated with biofilms coating plant leaves, the phyllospheric, hydrocarbonoclastic bacteria are spatially close to soil, the origin of volatile hydrocarbons. These epiphytic microorganisms have an adequate supply of molecular oxygen evolving as byproduct of leaf photosynthesis and moisture transpired by the aerial plant organs. Furthermore, the microflora is provided with organic nutrients including hydrocarbons, even though the surrounding environment may not be hydrocarbon-contaminated. Higher plants are known to produce ethylene (as a plant hormone) and the cuticle coating leaves and shoots contains higher hydrocarbons (Swaminathan and Kochlar, 1989; Shlegel, 1995). This brief description highlights plant aerial organs as suitable sites for the colonization and degradation activity of hydrocarbonoclastic microorganisms.

The history of phyllospheric (this term will be used here to comprise all aerial plant organs, not only leaves) hydrocarbonoclastic microorganisms is more recent than that of their rhizospheric counterparts, and consequently the numbers of publications dealing with this subject are much fewer. Table 10.3 shows that plant leaves accommodate numbers of such microorganisms ranging between $10^3$ and $10^7$ cells per gram of leaf material. Variations of such numbers are due to variations of the host plants as well as the environmental conditions under which the plants have been raised. Unlike the rhizospheric, hydrocarbonoclastic bacteria which belong predominantly to the Gram-negative Proteobacteria, the phyllospheric counterparts comprise both Gram-negative

**Table 10.3** Number and identities of hydrocarbonoclastic microorganisms on the phylloplanes of higher plants

| Higher plants | Numbers per g leaves | Hydrocarbonoclastic genera | References |
|---|---|---|---|
| *Anthocleista*, *Sarcophyrynium*, *Canna*, *Colocassia*, *Musa*, *Colas*, *Citrus*, *Mangifera*, *Terminalia*, *Annona* | $10^6$ bacterial cells | *Acinetobacter*, *Flavobacterium*, *Pseudomonas*, *Bacillus* | Ilori *et al.* (2006) |
| American grass, broad beans | $10^4$–$10^5$ bacterial cells | *Rhodococcus*, *Pseudomonas* | Sorkhoh *et al.* (2011) |
| Peas, beans (legumes), tomato, sunflower (non legumes) | $9 \times 10^7$ bacterial cells | *Microbacterium*, *Rhodococcus*, *Citrobacter* | Ali *et al.* (2012) |
| Black-eyed peas, peanut, peas, treasure flower, tar vine, hoary stock, yellow trumpet brush, cucumber, sow thistle, ice plant, salola, sowbane, goosefoot | $10^3$–$10^4$ bacterial cells | *Rhodococcus*, *Pseudomonas*, *Microbacterium*, *Arthrobacter*, *Klebsiella*, *Achromobacter*, *Basea*, *Kocuria*, *Dyadobacter*, *Agromyces*, *Agrococcus*, *Paenibacillus*, *Bacillus*, *Brevibacilus*, *Cellulosimicrobium*, *Methylobacterium*, *Gordonia*, *Dietzia*, *Exiguobacterium*, *Acinetobacter*, *Planomicrobium*, *Chysebacterium*, *Flavobacterium*, *Afipia*, *Planococcus*, *Nesterenkonia*, *Halomonas*, *Salinococcus*, *Staphylococcus*, *Burkholdenia* | Al-Awadhi *et al.* (2012) |

and Gram-positive bacteria. Prominent Gram-negative partners are the genera *Pseudomonas, Flavobacterium, Acinetobacter, Klebsiella* and others. Gram-positive partners belonging to the tough-walled Actinobacteria and Firmicutes comprise *Rhodococcus, Arthrobacter, Microbacterium, Bacillus, Staphylococcus* and others. Ilori *et al.* (2006) were among the pioneer researchers who reported bacteria in the phyllospheres of tropical plants capable of growth on diesel as a substrate. Interestingly, the authors suggested these phyllospheric microflora as seeds for bioaugmentation to remediate polluted environments. Al-Awadhi *et al.* (2009) recorded a great variety of hydrocarbonoclastic bacteria on the leaves of four crop plants (Table 10.3). Many of these bacteria were also diazotrophic (nitrogen-fixers); this makes the whole phyllospheric biofilm self-sufficient as far as microbial nitrogen nutrition is concerned. Other authors who found hydrocarbonoclastic bacteria associated with aerial organs of higher plants were Olusoji *et al.* (2006), Sandhu *et al.* (2007, 2009), Yutthammo *et al.* (2010) and Al-Awadhi *et al.* (2012b).

Several authors tried to answer the question regarding whether phyllospheric bacteria were involved in removal of atmospheric hydrocarbon pollutants *in situ*. Thus, De Kempeneer *et al.* (2004) suggested the removal of toluene from indoor air via bioaugmentation of the phyllospheres of indoor plants with hydrocarbon degrading microorganisms. Van Aken *et al.* (2004) reported that phytosymbiotic bacteria associated with poplar tissues removed nitro-substituted explosives including trinitrotoluene. Sorkhoh *et al.* (2011) showed that incubating fresh leaves of American grass in closed microcosms containing volatile hydrocarbons resulted in up to more than 80% removal of these volatile contaminants. Experimental evidence suggests that the leaf tissues did not contribute to this removal.

## Biofilms on fish

The previous sections show that hydrocarbonoclastic microorganisms are frequently associated with phototrophic microorganisms as well as with macroflora in various environments. On the other hand, there is currently an information gap regarding the association of such microorganisms with members of the fauna of various habitats. Hence, research in this area is still needed. Our group in Kuwait made some contribution to knowledge in this field. We studied hydrocarbonoclastic bacteria associated with biofilms coating fish from the Arabian Gulf (Radwan *et al.*, 2007a). All sorts of fish studied in the Arabian Gulf and fish farms connected to this water body harboured rich consortia of oil-utilizing bacteria. More bacteria were found in association with the gills and inside the guts than on the fish surfaces. This was probably because the gills and guts contained more hydrocarbons than the fish surfaces, as chemical analysis has revealed. The numbers were in the range of about half a million to somewhat less than 20 million cells per square cm fish surface and per gram of fish gills or fish guts. As far as the bacterial identities are concerned, the bacterial species belonged to the genera *Psychrobacter, Vibrio, Planococcus, Pseudomonas, Acinetobacter,* i.e. typical hydrocarbonoclastic marine bacterial genera.

In another study, our group offered experimental evidence for the contribution of bacteria associated with live fish to crude oil removal in the surrounding seawater (Mahmoud *et al.*, 2009). Keeping live fish in seawater artificially polluted with crude oil or pure aliphatic or aromatic hydrocarbons for 3 weeks led to effective removal of these compounds. To rule out the probability that live fish may have been contributing to direct hydrocarbon consumption, we repeated the experiment using, instead of life fish, the bacterial consortia scraped off of fish surfaces and obtained similar results. Interestingly, similar results were also obtained irrespective of whether the surrounding seawater samples were fertilized with $NH_4NO_3$ or not. This was indicative of the nitrogen-fixation potential of the hydrocarbonoclastic bacteria. Confirming this assumption, parallel counts of hydrocarbonoclastic bacteria associated with fish on a nitrate-containing medium and a nitrogen-free medium gave similar bacterial numbers. Those bacteria had the potential for acetylene reduction, which confirms their possession of the nitrogenase system responsible for $N_2$ fixation.

## Man-made biofilms

Currently, studies on man-made microbial biofilms that comprise hydrocarbonoclastic bacteria are rather rare. This is true although there is extensive work on the application of biofilms in several other areas e.g. in human diseases and for the bioremediation of several environmental pollutants (for recent reviews, see Lear and Lewis, 2012, and other chapters of this volume). The first publication on hydrocarbonoclastic biofilms dates back about two decades, although the use of biofilms in water treatment dates back to over a century. The establishment of such biofilms relies on the concept of 'biobarriers' which can be directly placed *in situ* across the flow of contaminated plumes.

We have already described hydrocarbonoclastic biofilms naturally immobilized on gravel particles along the Arabian Gulf coasts (see 'Coastal epilithic biofilms', above). There, we described how we established hydrocarbon-degrading biofilms artificially using those epilithic microbial consortia (Al-Awadhi *et al.*, 2003). As inert substrates, we used clean gravel particles and glass plates. As source of microbial consortia, we collected fresh, coastal gravel particles, 2–3 cm in diameter, coated with blue green epilithic growth. As substrates, we collected biomass free (naked) gravel particles. To establish biofilms on those naked particles, every 25 such particles were distributed into a Petri dish, and two naturally coated particles were inserted among them as inocula. Portions, 150 ml, of seawater was added to every Petri dish. Some dishes were not further treated, others were supplied with 0.15%, w/v, $KNO_3$ as a nitrogen source, others with 5% crude oil, others with both crude oil and $KNO_3$ and others with all these materials plus 0.05% $KH_2PO_4$ as a source of phosphorous. The dishes were covered and incubated at about 22°C under continuous illumination, $14\,\mu E/m^2/s$ for 1 month. Monthly, the particles were transported into fresh aliquots of corresponding mixtures. In this way, five transport cycles were performed. The results showed that the naked gravel particles were successfully colonized by deep green-blue biomass. The most important factor here was the addition of $KNO_3$. It took two transfers for a faint biofilm to appear. However, after the fourth transfer, the biofilms reached their full development.

To establish biofilms on glass plates, a total of 3.2 g fresh biomass was scrapped off from coated gravel particles at the end of the sixth cycle and used as inoculum. This biomass was homogenized in seawater fertilized with $KNO_3$ in a glass jar. Glass plates, $10 \times 10$ cm, were vertically dipped in the suspension for 3 months under continuous light. Our analysis showed that the magnitude of man-made biofilms established on gravel particles and glass plates were higher than the magnitude of those naturally established on gravel particles. The microbial composition of man-made biofilms on gravel particles and glass plates were similar. Phototrophic counterparts which were pioneer colonizers comprised filamentous cyanobacteria, picocyanobacteria and diatoms. Hydrocarbonoclastic counterparts consisted of nocardioforms and *Acinetobacter calcoaceticus*. Microscopic examination revealed that many of these bacteria were packed vertically along cyanobacterial filaments. It was also found that those man-made biofilms removed crude oil in seawater effectively.

De Lorenzo *et al.* (2005) described the application of hydrocarbon-degrading biofilms in the treatment of contaminated groundwater. Such biofilms consisted of column-packed pumice granules colonized by a microbial consortium from a gasoline-contaminated ground water sample. This consortium was enriched by the addition of toluene to a mineral medium supporting the microbial consortium. After 12 months, 99% of toluene in the system was removed. Evidence was presented that exopolymeric materials surrounding microbial cells became steadily enriched over time on the pumice particles. The microorganisms predominating the biomass at the end of the bioremediation process were the nocardioform *Rhodococcus erythropolis*, and the Gammaproteobacterium *Pseudomonas marginalis*.

Andrews *et al.* (2006) reported on the construction of artificial biofilms using two hydrocarbonoclastic bacteria, *Acinetobacter* sp. and *Pseudomonas putida*. Bacterial cells were suspended in a low-conductivity medium, directed towards specific areas in a microelectrode array via dielectrophoresis, and finally immobilized by polyethylamine. The authors demonstrated the feasibility of this method for studying metabolic interactions in architectured biofilms, i.e. constructed biofilms in which bacterial taxa are spatially arranged in a specific pattern'. *Acinetobacter* sp. areas of the biofilm reportedly degraded benzyl alcohol supplied to the system, producing benzoate which was subsequently mineralized in *P. putida* immobilizing areas of the biofilm. Evidence was provided that *Acinetobacter* cells occupied the top layer and *P. putida* cells the bottom layer of the biofilm. Several earlier investigators also noted that distinct microscopic layers were frequently formed in natural biofilms by physical and chemical interactions among the cells (Costerton *et al.*, 1987; Christensen *et al.*, 1998; Davies *et al.*, 1998; De Souza *et al.*, 1998; Tolker-Nilsen and Molin, 2000; Nielsen *et al.*, 2000 Andersen *et al.*, 1998).

Gertier *et al.* (2009) studied the possibility of oil-sorbent applications for the bioremediation of marine oil spills. In the presence of the oil-sorbents, microbial communities in microcosms with effective oil removal were selected in favour of the obligate hydrocarbonoclastic marine bacterium *Alcanivorax borkumensis*, whereby fast removal of *n*-alkanes occurred for one week, after which slower consumption of branched and polyaromatic nuclear hydrocarbons was recorded. Chavan and Mukherji (2010) studied the effect of phenol on the bioremediation of hydrocarbon-contaminated wastewater in a lab-scale rotating biological contractor with biofilm consisting mainly of the hydrocarbonoclastic species *Burkholderia cepacia* and a freshwater algal culture. Reportedly, hydrocarbon removal by algal-bacterial systems was more effective than by heterotrophic systems alone. The effect of phenol loading from 0.11 to 0.69 g phenol/$m^2$/day on diesel removal was measured with a 21 h hydraulic retention time and total petroleum hydrocarbon loading of 27.33 g hydrocarbons/$m^2$/day. With increases in phenol loading, phenol was removed completely, but the petroleum hydrocarbon consumption decreased from 99% to 94%. Reportedly, the presence of algal biomass in the biofilm enabled the operation at a high organic loading; hydrocarbonoclastic bacteria were better immobilized and aerated.

Golby *et al.* (2012) described the isolation of environmental mixed species biofilms from an oil sand tailing pond; these biofilms were kept under laboratory conditions. The species mixture consisted of the genera *Pseudomonas*, *Thauera*, *Hydrogenophaga*, *Rhodoferax* and *Acidovorax*. At least some of these genera via *Pseudomonas* and *Acidovorax* are well known hydrocarbonoclastic bacteria.

Our group in Kuwait established biofilm simultaneously harbouring anoxygenic and oxygenic phototrophic bacteria, diazotrophic bacteria and hydrocarbonoclastic bacteria (Al-Bader *et al.*, 2013b). This physiological diversity was reflected in effective removal of hydrocarbon pollutants in seawater by these biofilms. In this context, it is only recently that more reports on hydrocarbonoclastic biofilms started to appear (Chandran and Das, 2011; Vaysse *et al.*, 2011; Golby *et al.*, 2012). The biofilms with the above mentioned multiple physiological activities were established in our laboratory as follows. Seawater samples were collected from the Arabian Gulf, 5 m offshore and 5 cm below the surface of the water. The water was dispensed in all-glass tanks. One tank was left without any further treatment, as a control. Another tank received 0.1% crude oil and the third received oil plus 0.1% $KNO_3$. In each tank, 40 sterile microscopic glass slides were dipped vertically just below the water surface. Each tank was provided with a magnetic stirrer to mix the tank contents. The tanks were kept at 25°C under daylight (12 $\mu E/m^2/s$, 14 h) dark night (10 h) cycles for 7 months. The biofilms started to become visible to the naked eye after only 1 week and continued thickening for about 1 month. The biomass fresh weight of the control slides was 9 $mg/cm^2$ and of the differently treated samples 17 to 18 $mg/cm^2$. These values remained more or less constant through the remaining 6 months. Oil in the tanks persisted for one month, after which it visually disappeared by microbial emulsification and biodegradation. Microbial development in the biofilms was studied by combined fingerprinting of 16S rRNA genes amplified from the total DNA extracted directly from biofilm samples, and DNA sequencing of the electrophoresis gel bands. The obtained sequences were compared with those in the GenBank database. The glass slides that were dipped in seawater for only 15 min already revealed bands corresponding to the plastid of the diatom *Cymbella elegans*, the hydrocarbonoclastic Gammaproteobacterium *Cycloclasticus* and the picocyanobacteria *Chrococcidiopsis* and *Synechococcus*. The biofilms also comprised anoxygenic phototrophs belonging to the Alphaproteobacteria, oxygenic phototrophic, filamentous cyanobacteria, two species of the diazotrophic *Azospirillum* and the two hydrocabonoclastic Gammaproteobacteria *Cycloclasticus* and *Oleibacter*. The last two genera belong to the group of the obligate hydrocarbonoclastic bacteria (OHCB) believed to be the predominant hydrocarbon biodegraders in the marine ecosystem (Yakimov *et al.*, 2007). The coexistence of all these microbial taxa with various physiological activities in the biofilm obviously made the whole biomass nutritionally self-sufficient and adequately aerated, a condition quite promotive to biodegradation of aquatic pollutant hydrocarbons. Further experimental evidence for this has been obtained in our laboratory and the results are currently being prepared for publication.

Biofilms comprising bacterial consortia from untreated domestic sewage effluent were established on glass surfaces, as described above. The predominant hydrocarbonoclastic bacteria in these biofilms, as studied using a culture-dependent method, comprised among others *Shinella granuli*, *Microvirga subterrania*, *Bosea thiooxidans*, *Zavarzinia compransoris*, *Ancylobacter polymorphus* and *Mycobacterium smegmatis*. Using a culture-independent molecular method, the microorganisms in the same biofilms were affiliated with the Gammaproteobacterial genera *Pseudomonas*, *Stenotrophomonas*, the Alphaproteobacterial genera *Chelatococcus*, *Zavarzinia*, *Ancylobacter*, *Bosea*, *Kaitia*, *Microvirga*, *Shinella* and *Sphingobium*, the Betaproteobacterial genera *Achromobacter* and *Acidovorax* and the Actinobacterial genera *Microbacterium*, *Gordonia* and *Mycobacterium*, all genera arranged in decreasing order of predominance. As should be expected, the hydrocarbonoclastic potential of such man-made biofilms depended on the environmental factors prevailing during bioremediation. Thus, light enhanced hydrocarbon removal by about 33% more than the dark incubation. Addition of thioglycollate to reduce the redox potential of the surrounding medium reduced the hydrocarbon consumption by up to 90%. Doubling and tripling the exposed biofilm surface area enhanced oil removal by 20% and 31%, respectively. Using the same biofilms in five successive operations of bioremediation showed that the crude oil removal during the first two or three operations was enhanced by up to 33% but became inhibited in the fourth and fifth operations by 50% or more. This inhibition was probably due to biofilm mechanical loss during successive transfers.

## The lifestyle of organisms within hydrocarbonoclastic biofilms

For the treatment of wastewater rich in waste oil and/or its derivatives, physico-chemical systems consisting of American Petroleum Institute separators, tilted plate interceptors and dissolved air floatation units are applied. Subsequently, biological processes similar to the activated sludge process are usually used. These latter processes face a major problem related to the hydrophobicity of the hydrocarbon pollutants. Other problems related to the coexistence of toxic antimicrobial substances in the system may also prevail, inhibiting the biological clean up. The use of special bioreactors may represent a good solution to such problems. In comparison to the conventional suspended growth systems, attached growth bioreactors simplify bioremediation operations, make the microorganisms more tolerant of shock loading conditions, and offer many other advantages. The fact that biofilms in such reactors usually comprise phototrophic counterparts (cyanobacteria or algae) is interesting and useful, because, being active in extracellular polysaccharide production, these phototrophs, lead to effective immobilization of hydrocarbonoclastic cells preventing their continuous loss from the biofilm. Oxygen released as a byproduct of oxygenic photosynthesis is not only essential for the initial step of hydrocarbon degradation, but it also helps the degradation of toxic and antimicrobial substances. That is why various bioreactor types comprising phototrophic microorganisms have been described by many earlier investigators. Diazotrophic cyanobacteria are of special interest here, because they make the bioreactor consortium self-sufficient regarding nitrogen nutrition. In this context, and as described before, most of the natural hydrocarbonoclastic consortia, including hydrocarbonoclastic bacteria themselves are also diazotrophic. In other words, their lifestyle in the biofilms is quite suitable for effective hydrocarbon mineralization. Recently, evidence was provided that hydrocarbonoclastic bacteria in biofilms adapt to the biofilm lifestyle and maintain a higher ability to reinitiate biofilm development at the hydrocarbon–water interface. Vaysse *et al.* (2011) studied cells dispersed from *Marinobacter hydrocarbonoclasticus* biofilm and found that two proteins in the cell envelope were related to the ability to initiate biofilm development as well as to a higher cell affinity for hexadecane, the hydrocarbon studied. Tribelli *et al.* (2012) presented experimental evidence that the biofilm lifestyle enhanced diesel bioremediation and biosurfactant production in the Antarctic polyhydroxyalkanate producer *Pseudomonas extremaustralis*. This strain was used to study the effect of culture conditions in biofilm versus shaken flask cultures on diesel biodegradation. Biofilms exhibited increased cell growth, biosurfactant production and diesel removal compared with shaken flask cultures.

## Conclusions

Natural biofilms comprising hydrocarbonoclastic microorganisms are frequent in the biosphere and seem to play an important role in the self-cleaning of soil, aquatic and atmospheric environments. These biofilms usually comprise diverse microorganisms with diverse physiological activity which cooperate in enhancing hydrocarbon removal. Thus, oxygenic phototrophic partners secure better immobilization and aeration of the whole consortia. Diazotrophic microorganisms enrich biofilms with nitrogenous compounds known to be limiting to hydrocarbon biodegradation. The biofilms are either naturally architectured, or could be artificially architectured in such a way to enhance the bioremediation process. On the other hand, hydrocarbonoclastic bacteria in the biofilms are of course susceptible to attacks by specific viruses, protozoa and other organisms. Man-made, hydrocarbonoclastic biofilms can also be constructed, taking into consideration information gained from the study of natural biofilms. The future will probably witness their intensive application in treating oily wastes in bioreactors.

## Acknowledgement

Thanks are due to Mr Mohammed Eliyas for his technical help in preparing the manuscript.

## References

Al-Awadhi, H., Al-Hasan, R.H., and Radwan, S.S. (2002). Comparison of the potential of coastal materials loaded with bacteria for bioremediating oily sea water in batch culture. Microbiol. Res. *157*, 331–336.

Al-Awadhi, H., Al-Hasan, R.H., Sorkhoh, N.A., Salamah, S., and Radwan, S.S. (2003). Establishing oil-degrading biofilms on gravel particles and glass plates. Int. Biodeter. Biodeg. *51*, 181–185.

Al-Awadhi, H., El-Nemr, I., Mahmoud, H., Sorkhoh, N.A., and Radwan, S.S., (2009). Plant associated bacteria as tools for phytoremediation of oily nitrogen poor soils. Int. J. Phytoremed. *11*, 11–27.

Al-Awadhi, H., Dashti, N., Kansour, M., Sorkhoh, N.A., and Radwan, S.S. (2012a). Hydrocarbon-utilizing bacteria associated with biofouling materials from offshore waters of the Arabian Gulf. Int. Biodeter. Biodegr. *69*, 10–16.

Al-Awadhi, H., Al-Mailem, D., Dashti, N., Hakam, L., Eliyas, M., and Radwan, S.S. (2012b). The abundant occurrence of hydrocarbon-utilizing bacteria in the phyllospheres of cultivated and wild plants in Kuwait. Int. Biodeter. Biodegr. 73, 73–79.

Al-Bader, D., Eliyas, M., Rayan, R., and Radwan, S.S. (2012). Air-dust-borne associations of phototrophic and hydrocarbon-utilizing microorganisms: promising consortia in volatile hydrocarbon bioremediation. Environ. Sci. Pollut. Res. *19*, 3997–4005.

Al-Bader, D., Eliyas, M., Rayan, R., and Radwan, S.S. (2013a). Subsurface association of *Acaryochloris* related picocynobacteria and oil-utilizing bacteria in the Arabian Gulf water body: promising consortia in oil sediment bioremediation. Microb. Ecol. *65*, 555–565.

Al-Bader, D., Kansour, M., Rayan, R., and Radwan, S.S. (2013b). Biofilm comprising phototrophic, diazotrophic, and hydrocarbon-utilizing bacteria: a promising consortium in the bioremediation of aquatic hydrocarbon pollutants. Environ. Sci. Pollut. Res. *20*, 3252–3262.

Al-Hasan, R.H., and Jones, W.E. (1989). Marine algal flora and sea grasses of the coast of Kuwait. J. Univ. Kuwait (Sci.) *16*, 289–341.

Al-Hasan, R.H., Sorkhoh, N.A., Al-Bader, D., and Radwan, S.S. (1994). Utilization of hydrocarbons by cyanobacteria from microbial mats on oily coasts of the Gulf. Appl. Microbiol. Biotechnol. *41*, 615–619.

Al-Hasan, R.H., Al-Bader, D.A., Sorkhoh, N.A., and Radwan, S.S. (1998). Evidence for *n*-alkane consumption and oxidation by filamentous cyanobacteria from oil-contaminated coasts of the Arabian Gulf. Mar. Biol. *130*, 521–527.

Al-Hasan, R.H., Khanafer, M., Eliyas, M., and Radwan, S.S. (2001). Hydrocarbon accumulation by picocyanobacteria from the Arabian Gulf. J. Appl. Microbiol. *91*, 533–540.

Ali, N., Sorkhoh, N.A., Salamah, S., Eliyas, M., and Radwan, S.S. (2012). The potential of epiphytic hydrocarbon-utilizing bacteria on legume leaves for attenuation of atmospheric hydrocarbon pollutants. J. Environ. Manag. *93*, 113–120.

Al-Mailem, D.M., Sorkhoh, N.A., Salamah, S., Eliyas, M., and Radwan, S.S. (2010). Oil-bioremediation potential of Arabain Gulf mud flats rich in diazotrophic hydrocarbon-utilizing bacteria. Int. Biodeter. Biodeg. *64*, 218–225.

Atlas, R.M., and Bartha, R. (1972). Degradation and mineralization of petroleum in seawater: limitation by nitrogen and phosphorous. Biotechnol. Bioeng. *14*, 309–317.

Andersen, J.B., Sternberg, C., Poulsen, L.K., Bjorn, S.P., Givskov, M., and Molin, S. (1998). New unstable variants of green fluorescent protein for studies of transient gene expression in bacteria. Appl. Environ. Microbiol. *64*, 2240–2246.

Andrews, J.S., Mason, V.P., Thompson, I.P., Stephens, G.M., and Markx, G.H. (2006). Construction of artificially structured microbial consortia (ASMC) using dielectrophoresis: Examining bacterial interactions via metabolic intermediates within environmental biofilms. J. Microbiol. Meth. *64*, 96–106.

Banks, M.K., Schwab, A.P., Govindaraju, R.S., and Kulakow, P. (1999). Field demonstration. In: Fiorenza, S., Oubre, C.L., Ward, C.H., eds., Phytoremediation of Hydrocarbon-Contaminated Soil (Lewis Publishers, Boca Raton, FL, USA), pp. 3–88.

Boulton, C.A., and Ratledge, C. (1984). The physiology of hydrocarbon utilizing microorganisms. In: A. Wiesman, ed., Topics in Fermentation and Enzyme Technology, Vol 9 (Ellis Horwood, Chichester, UK), pp. 11–77.

Bühler, M., and Schindler, J. (1984). Aliphatic hydrocarbons. In Biotechnology – A Comprehensive Treatise in 8 Volumes, H.-J. Rehm and G. Reed, eds, Vol 6a (Verlag Chemie, Weinheim, Germany), pp. 329–385.

Carman, E.P., Crossman, T.L., and Gatliff, E.G. (1998). Phytoremediation of no. 2 fuel oil-contaminated soil. J. Soil Cont. 7, 455–466.

Cerniglia, C.E., Van Baalen, C., and Gibson, D.T. (1980a). Oxidation of biphenyl by the cyanobacterium *Oscillatoria* sp. strain JCM. Arch. Microbiol, *125*, 203–207.

Cerniglia, C.E., Van Baalen, C., and Gibson, D.T. (1980b). Metabolism of naphthalene by the cyanobacterium *Oscillatoria* sp strain JCM. J. Gen. Microbiol. *116*, 485–494.

Cerniglia, C.E., Freeman, J.P., and Van Baalen, C. (1981). Biotransformation and toxicity of aniline and aniline derivatives in cyanobacteria. Arch. Microbiol. *130*, 272–275.

Chandran, P., and Das, N. (2011). Degradation of diesel oil by immobilized *Candida tropicalis* and biofilm formed on gravel. Biodegradation *22*, 1181–1189.

Chavan, A., and Mukherjii, S. (2010). Effect of co-contaminant phenol on performance of al laboratory-scale RBC with algal-bacterial biofilm treating petroleum hydrocarbon-rich wastewater. J. Chem. Technol. Biotechnol. *85*, 851–859.

Christensen, B.B., Sternberg, C., Andersen, J.B., and Molin, S. (1998). In situ detection of gene transfer in a model biofilm engaged in degradation of benzyl alcohol. Appl. Environ. Microbiol. *106*, 25–28.

Colwell, R. (1977). Ecological aspects of microbial degradation of petroleum in the marine environment. CRC Crit. Rev. Microbiol. *5*, 423–445.

Colwell, R.R., and Walker, J.D. (1977). Ecological aspects of microbial degradation of petroleum in the marine environment. Crit. Rev. Microbiol. *5*, 423–445.

Cooper, D.G., and Zajic, J.E. (1980). Surface-active compounds from microorganisms. Adv. Appl. Microbiol. *26*, 229–253.

Costerton, J.W., Geesey, G.G., and Cheng, K.J. (1978). How bacteria stick. Sci. Am. *238*, 86–95.

Costerton, J.W., Cheng, K.J., Geesey, G.G., Ladd, T.I., Nickel, J.C., Dasgupta, M., and Marrie, T.J. (1987). Bacterial biofilms in nature and disease. Annu. Rev. Microbiol, *41*, 435–464.

Coulon, F., Chronopoulou, P.M., Fahy, A., Paisse, S., Goni-Urriza, M., Peperzak, L., Alvarez, L.A., McKew, B.A., Brussard, C.P.D., Underwood, G.J.C., Timmis, K.N., Duran, R., and McGenity, T.J. (2012). Central role of dynamic tidal biofilms dominated by aerobic hydrocarbonclastic bacteria and diatoms in the biodegradation of hydrocarbon in coastal mudflats. Appl. Environ. Mocrobiol. *78*, 3638–3648.

Dashti, N., Khanafer, M., Ali, N., El-Nemr, I., Sorkhoh, N., and Radwan, S. (2009). The potential of oil-utilizing bacterial consortia associated with legume root nodules for cleaning oily soils. Chemosphere *74*, 1354–1359.

Davies, D.G., Parsek, M.R., Pearson, J.P., Iglewski, B.H., Costerton, J.W., and Greenberg, E.P. (1998). The involvement of cell-to-cell signals in the development of a bacterial biofilm. Science *280*, 295–298.

De Kempeneer, L., Sercu, B., Vanbrabant, W., Van Langenhove, H., and Verstraete, W. (2004).

Bioaugmentation of the phyllosphere for the removal of toluene from indoor air. Appl. Microbiol. Biotechnol. *64*, 284–288.

De Lorenzo, A.D., Varcamonti, M., Parascandola, P., Vignola, R., Bernardi, A., Sacceddu, P., Sisto, R., and Alteriis, E.D. (2005). Characterization and performance of a toluene-degrading biofilm developed on pumice stones. Microb. Cell Fact. *4*, 4.

De Souza, M.L., Newcombe, D., Alvey, S., Crowley, D.E., Hay, A., Sadowsky, M.J., and Wackett, L.P. (1998). Molecular basis of a bacterial consortium: interspecies catabolism of atrazine. Appl. Environ. Microbiol. *64*, 178–184.

Erikson, L.E., and Nakahara, T. (1975). Growth in cultures with two liquid phases: hydrocarbon uptake and transport. Proc. Biochem. *10*, 9–13.

Flathman, P.E., and Lanza, G.R. (1998). Phytoremediation: current views on an emerging green technology. J. Soil Cont. *7*, 415–432.

Frick, C.M., Farrell, R.E., and Germida, J.J. (1999). Assessment of Phytoremediation as an *In Situ* Technique for Cleaning Oil-contaminated Sites (Petroleum Technology Alliance of Canada, Calgary, AB, Canada).

Gertier, C., Gerdts, G., Timmins, K.N., Yakimov, M.M., and Golyshin, P.N. (2009). Populations of heavy fuel oil-degrading marine microbial community in presence of sorbent materials. J. Appl. Microbiol. *107*, 590–605.

Golby, S., Ceri, H., Gieg, L.M., Chatterjee, I., Margues, L.L., and Turner, R.J. (2012). Evaluation of microbial biofilm communities from an Alberta oil sands tailings pond. FEMS Microb. Ecol. *79*, 240–250.

Harayama, S., Kasai, Y., and Hara, A. (2004). Microbial communities in oil-contaminated seawater. Curr. Opin. Biotechnol. *15*, 205–214.

Harayama, S., Kishira, H., Kasai, Y., and Shutsubo, K. (1999). Petroleum biodegradation in marine environments. J. Mol. Microbiol. Biotechnol. *1*, 63–70.

Head, I.M., and Swannell, R.P. (1999). Bioremediation of petroleum hydrocarbon contaminants in marine habitats. Curr. Opin. Biotechnol. *10*, 234–239.

Head, I.M., Jones, D.M., and Roling, W.F. (2006). Marine microorganisms make a meal of oil. Nat. Rev. Microbiol. *4*, 173–182.

Ilori, M.O., Amund, O.O., Ezeani, C.J., Omoijiahina, S.D., and Sunday, A. (2006). Occurrence and growth potentials of hydrocarbon degrading bacteria on the phylloplane of some tropical plants. Afr. J. Biotechnol. *5*, 542–545.

Klug, M.J., and Markovetz, A.J. (1971). Utilization of aliphatic hydrocarbons by microorganisms. Adv. Microb. Physiol. *5*, 1–43.

Kulakow, P.A., Schwab, A.P., and Banks, M.K. (2000). Screening plant species M. Tesar *et al.*/Soil Biology & Biochemistry 34 (2002) 1883–1892 1891 for growth on weathered, petroleum hydrocarbon-contaminated sediments. Int. J. Phytoremed. *2*, 297–317.

Lear, G., and Lewis, G.D. (2012). Microbial Biofilms: Current Research and Applications (Caister Academic Press, Norwich, UK).

Le Borgene, S., Paniagua, D., and Vazquez-Duhalt, R. (2008). Biodegradation of organic pollutants by halophilic bacteria and archaea. J. Mol. Microbiol. Biotechnol. *15*, 74–92.

Lee, E., and Banks, M.K. (1993). Bioremediation of petroleum contaminated soils using vegetation: A microbial study. J. Environ. Sci. Health A28. *2*, 2187–2198.

Levi, J.D., Shenman, J.L., and Ebbon, G.P. (1979). Biomass from liquid n-alkanes. In Economic Microbiology, Vol. 4, A.H. Rose, ed. (Academic Press, London, UK).

Mahmoud, H.M., Al-Hasan, R.H., Sorkhoh, N.A., Eliyas, M., and Radwan, S.S. (2009). Attenuation of oil pollutants in the Arabian Gulf water by bacteria naturally associated with live fish. Int. Biodeter. Biodegr. *63*, 615–620.

Margaritis, A., Zajic, J.E., and Gerson, D.F. (1979). Production and surface properties of microbial surfactants. Biotech. Bioeng. *21*, 1151–1162.

Margesin, R., and Schinner, F. (2001). Biodegradation and bioremediation of hydrocarbons in extreme environments. Appl. Microbiol. Biotechnol. *56*, 650–663.

Morton, L.H.G., and Surman, S.M. (1994). Biofilms in biodeterioration: a review. Int. Bideterior. Biodeg. *37*, 203–221.

Musat, F., Harder, J., and Widdel, F. (2006). Study of nitrogen fixation in microbial communities of oil contaminated sediment. Environ. Microbiol. *8*, 1834–1843.

Nielsen, A.T., Tolker-Nielsen, T., Barken, K.B., and Molin, S. (2000). Role of commensal relationships on the spatial structure of a surface-attached microbial consortium. Environ. Microbiol. *2*, 59–68.

Nilanjana, D., Lakshmi, V., Geetanjali, B., Jaseetha, A.S., and Evy, A.A.M. (2012). Application of biofilms on remediaiton of pollutants – an overview. J. Microbiol. Biotech. Res. *2*, 783–790.

Olusoji, I.M., Oladipupo, A.O., John, E.C., Sarah, O., and Adekunle, A.S. (2006). Occurrence and growth potentials of hydrocarbon degrading bacteria on the phyllosphereof some tropical plants. Afr. J. Biotechnol. *5*, 542–545.

Pearce, D.A., Bridge, P.D., Hughes, K.A., Sattler, B., Psenner, R., and Russell, N.J. (2009). Microorganisms in the atmosphere over Antarctica. FEMS Microb. Ecol. *69*, 143–157.

Pradhan, S.P., Conrad, J.P., Paterek, J.R., and Srivastava, V.J. (1998). Potential of phytoremediation for treatment of PAHs in soil at MGP sites. J. Soil Cont. *7*, 467–480.

Prince, E.L., and Morton, L.H.G. (1989). Biofilms and bioemulcifiers in hydrocarbon degradation. Int. Biodeter. *25*, 385–391.

Poloneko, D.R., Scher, F.M., Kloepper, J.W., Singleton, C.A., Laliberete, M., and Zalesk, I. (1987). Effect of root colonizing bacteria of nodulation of soybean roots by *Bradyrhizobium japonicum*. Can. J. Microbiol. *33*, 498–503.

Qiu, X., Leland, T.W., Shah, S.I., Sorensen, D.L., and Kendall, E.W. (1997). Chapter 14 field study: grass remediation for clay soil contaminated with polycyclic aromatic hydrocarbons. In Phytoremediation of Soil and Water Contaminants, E.L. Kruger, T.A. Anderson, J.R. Coats, eds., ACS Symposion Series 664 (American Chemical Society, Washington, DC, USA), pp. 186–199.

Radwan, S.S. (2008). Microbiology of oil-contaminated desert soils and coastal areas in the Arabian Gulf. In Microbiology of Extreme Soils. Soil Biology Vol 13, P.

Dion, C.S. Nautiyal, eds. (Springer, Berlin, Germany), pp. 275–297.

Radwan, S.S. (2009). Phytoremediation for oily desert soils. In Advances in Applied Bioremediation, A. Singh, R.C. Kuhad, and O.P. Ward, eds. (Springer-Verlag, Heidelberg, Germany), pp. 279–298.

Radwan, S.S., and Al-Hasan, R.H. (2000). Oil pollution and cyanobacteria. In The Ecology of Cyanobacteria, B.A. Whitton and M. Potts, eds. (Kluwer Academic Publishers, Netherlands).

Radwan, S.S., and Al-Hasan, R.H. (2001a). Cyanobacteria and the self-cleaning of oily environments. In: Photosynthetic Microorganisms in Environmental Biotechnology, H. Kojima and Y.K. Lee, eds. (Springer, New York, NY, USA), pp. 181–193.

Radwan, S.S., and Al-Hasan, R.H. (2001b). Potential application of coastal biolfilm-coated gravel particles for treating oily waste. Aqu. Microb. Ecol. *23*, 113–117.

Radwan, S.S., and Sorkhoh, N.A. (1993). Lipids of *n*-alkane utilizing microorganisms and their application potential. Adv. Appl. Microbiol. *39*, 29–90.

Radwan, S.S., Sorkhoh, N.A., and El-Nemr, I.M. (1995). Oil-biodegradation around roots. Nature (London) *376*, 302.

Radwan, S.S., Al-Awadhi, H., Sorkhoh, N.A., and El-Nemr, I.M. (1998). Rhizospheric hydrocarbon-utilizing microorganisms as potential contributors to phytoremediation of the oily Kuwaiti desert. Microb. Res. *153*, 247–251.

Radwan, S.S., Al-Hasan, R.H., Al-Awadhi, H., Salamah, S., and Abdullah, H.M. (1999). Higher oil biodegradation potential at the Arabian Gulf coast than in the water body. Mar. Biol. *135*, 741–745.

Radwan, S.S., Al-Awadhi, H., and El-Nemr, I.M. (2000). Cropping as a phytoremediation practice for oily soil with reference to crop safety as food. Int. J. Phytoremed. *2*, 383–396.

Radwan, S.S., Al-Hasan, R.H., Salamah, S., and Al-Dabbous, S. (2002). Bioremediation of oily sea water by bacteria immobilized in biofilms coating macroalgae. Int. Biodeter. Biodeg. *50*, 55–59.

Radwan, S.S., Al-Hasan, R.H., Salamah, S., and Khanafer, M. (2004). Enhancing effects of picocyanobacteria on Growth and hydrocarbon consumption potential of the associated oil utilizing bacteria. Arctic and Marine Oil Spill Program (AMOP). Ottawa, Canada, 8–10 June 2004, vol II, pp. 967–976.

Radwan, S.S., Al-Hasan, R.H., Salamah, S., and Khanafer, M. (2005a). Oil-consuming microbial consortia floating in the Arabian Gulf. Int. Biodeter. Biodeg. *56*, 28–33.

Radwan, S.S., Dashti, N., and El-Nemr, I. (2005b). Enhancing the growth of *Vicia faba* plants by microbial inoculation to improve their phytoremediation potential for oily desert areas. Int. J. Phytoremed. *7*, 19–32.

Radwan, S.S., Al-Hasan, R.H., Mahmoud, H.M., and Eliyas, M. (2007a) Oil-utilizing bacteria associated with fish from the Arabian Gulf. J. Appl. Microbiol. *103*, 2160–2167.

Radwan, S.S., Dashti, N., El-Nemr, I., and Khanafer, M. (2007b). Hydrocarbon utilization by nodule bacteria and plant-growth promoting rhizobacteria. Int. J. Phytoremed. *9*, 1–11.

Radwan, S.S., Mahmoud, H., Khanafer, M., Al-Habib, A., and Al-Hasan, R. (2010). Bacterial consortia with combined potential for hydrocarbon-utilization and nitrogen-fixation from epilithic biofilms along the Arabian Gulf coasts. Microb. Ecol. *60*, 354–363.

Rehm, H.J., and Reiff, I. (1981). Mechanisms and occurrence of microbial oxidation of long-chain alkanes. Adv. Biochem. Eng. *19*, 175–216.

Rippka, R. (1972). Photoheterotrophy and chemoheterotrophy among unicellular blue-green algae. Arch. Mikrobiol. *87*, 93–98.

Sandhu, A., Halverson, L.J., and Beattie, G.A. (2007). Bacterial degradation of airbornephenol in the phyllosphere. Environ. Microbiol. *2*, 383–392.

Sandhu, A., Halverson, L.J., and Beattie, G.A. (2009). Identification and genetic characterization of phenol-degrading bacteria from leaf microbial communities. Microb. Ecol. *57*, 276–285.

Schlegel, H.G. (1995). Degradation of naturally occurring substances (hydrocarbons). In General Microbiology, 7th edn, Zaborosch, C., ed. (Cambridge University Press, Cambridge, UK), pp. 423–432.

Singh, R., Paul, D., and Jain, R.K. (2006). Biofilms: implications and bioremediation trends in microbiology. Trends Microbiol. *14*, 389–397.

Sorkhoh, N., Al-Hassan, R., Hopner, T., and Radwan, S. (1992). Self-cleaning of the Gulf. Nature *359*, 109.

Sorkhoh, N.A., Al-Hasan, R.H., Khanafer, M., and Radwan, S.S. (1995). Establishment of oil-degrading bacteria associated with cyanobacteria in oil polluted soil. J. Appl. Bacteriol. *78*, 194–199.

Sorkhoh, N.A., Ali, N., Al-Awadhi, H., Dashti, N., Al-Mailem, D.M., Eliyas, M., and Radwan, S.S. (2010a). Phytoremediation for mercury in pristine and crude oil contaminated soils: contributions of rhizobacteria and their host plants to mercury removal. Ecotoxicol. Environ. Saf. *73*, 1998–2003.

Sorkhoh, N.A., Al-Awadhi, H., Al-Mailem, D.M., Kansour, M.K., Khanafer, M., and Radwan, S.S. (2010b). Agarolytic bacteria with hydrocarbon-utilization potential in fouling material from the Arabian Gulf coast. Int. Biodeter. Biodegr. *64*, 554–559.

Sorkhoh, N.A., Al-Mailem, D., Ali, N., Al-Awadhi, H., Salamah, S., Eliyas, M., and Radwan, S.S. (2011). Bioremediation of volatile oil hydrocarbons by epiphytic bacteria associated with American grass (*Cynodon* sp.) and broad bean (*Vicia faba*) leaves. Int. Biodeter. Biodegr. *65*, 797–802.

Sudha, S.S., Karthic, R., Francis, M., Soumya, T.S., and Renga Ramanujam, J. (2011). Isolation and preliminary characterization of associated microorganisms from *Spirulina* products and their silver mediated nano-particle synthesis. J. Algal Biomass Utiliz. *2*, 1–8.

Swaminathan, M.S., and Kochlar, S.L. (1989). Hydrocarbon yielding plants. In Plants and Society, Swaminathan, M.S., and Kochar, S.L., eds. (Macmillan Publishers Limited, London, UK), pp. 577–589.

Tesar, M., Reichenauer, T.G., and Sessitsch, A. (2002). Bacterial rhizosphere populations of black poplar and herbal plants to be used for phytoremediation of diesel fuel. Soil Biol. Bioche. *34*, 1883–1892.

Tolker-Nielsen, T., and Molin, S. (2000). Spatial organization of microbial biofilm communities. Microb. Ecol. *40*, 75–84.

Tribelli, P.M., Di Martino, C., Lopez, N.I., and Raiger Lustman, L.J. (2012). Biofilm life style enhances diesel bioremediation and biosurfactant production in the Antartic polyhydroxyalkanoate producer *Pseudomonas extremaustralis*. Biodeg. *5*, 645–651.

Van Aken, B., Yoon, J.M., and Schnoor, J.L. (2004). Biodegradation of nitro-substituted explosives 2,4,6-trinitrotoluene, hexahydro-1,3,5-trinitro-1,3,5-triazine, an octahydro-1,3,5,7-tetranitro-1,3,5-tetrazocine by a phytosymbiotic *Methylobacterium* sp. associated with poplar tissues (*Populus deltoides* × nigra DN34). Appl. Environ. Microbiol. *70*, 508–517.

Vaysse, P.J., Sivadon, P., Goulas, P., and Grimaud, R. (2011). Cells dispersed from *Marinobacter hydrocarbonocasticus* SP17 biofilm exhibit a specific protein profile associated with a higher ability to reinitiate biofilm development at the hexadecane–water interface. Environ. Microbiol. *13*, 737–746.

Warneck, P. (1988). Chemistry of the Natural Atmosphere (Academic Press, New York, NY, USA).

Yakimov, M.M., Timmis, K.N., and Golyshin, P.N. (2007). Obligate oil-degrading marine bacteria. Curr. Opin. Biotechnol. *18*, 257–266.

Yutthammo, C., Thongthammachat, N., Pinphanichakarn, P., and Luepromchai, E. (2010). Diversity and activity of PAH-degrading bacteria in the phyllosphere of ornamental plants. Microb. Ecol. *59*, 357–368.

Zajic, J.E., and Panchal, C.J. (1976). Bioemulsifiers. Crit. Rev. Microbiol. *5*, 39–66.

Zhang, F., Dashti, N., Hynes, R., and Smith, D.L. (1996). Plant growth promoting rhizobacteria and soybean (*Glycine max* (L.) Merr) nodulation and nitrogen fixation at suboptimal root zone temperature. Ann. Bot. *77*, 453–459.

Zhang, F., Dashti, N., Hynes, R., and Smith, D.L. (1997). Plant growth-promoting rhizobacteria and soybean (*Glycine max* (L.) Merr). Growth and physiology at suboptimal root zone temperatures. Ann. Bot. *79*, 243–249.

# Use of Biofilm Permeable Reactive Barriers for the *in Situ* Remediation of Mobile Contaminants

11

Youngwoo (Young) Seo

## Abstract

Bioremediation has been used for the destruction of organic chemicals in soils, groundwater and industrial wastewaters and can be categorized into both *in situ* and *ex situ* remediation strategies. Among *in situ* bioremediation technologies, the use of engineered permeable reactive barriers (PRBs) within contaminated aquifers has grown as a way to prevent the migration of mobile pollutants from contaminated plumes with minimal maintenance costs. During the past decade, PRBs have become an important *in situ* remediation option for contaminated groundwater. However, the influence of biofilm formation on the performance of PRBs is not well understood yet. This chapter reviews the influence of biofilm formation on the performance of PRBs for the continued removal of mobile contaminants. First, the general application of biofilm-mediated processes for the remediation of soil and groundwater bioremediation is discussed. Second, the use of PRBs incorporating biofilm grown on different supporting media is reviewed. Third, the influence of biofilm formation on the long term performances of PRBs is evaluated. Finally, emerging research trends and needs of relevance to the use of PRBs in remediation are considered.

## Introduction

Soil and groundwater contamination by complex organic and inorganic mixtures (e.g. coal tar, creosote, fuels, polycyclic aromatic hydrocarbons, herbicides, pesticides and heavy metals) is a ubiquitous problem around the world due to the fast growth of chemical industries, improper disposal or accidental release of chemicals into the environment (Eberhardt and Grathwohl, 2002). According to a National Research Council report, there are between 300,000 and 400,000 contaminated sites in the USA alone, which require cleanup costs up to one trillion dollars (National Research Council, 1994). Depending on the type of contaminants and subsurface aquifer conditions, these polluted sites can contaminate large amounts of groundwater causing serious environmental problems and public health concerns (Singh *et al.*, 2006). To mitigate contaminant transport and subsequent environmental problems, various remediation techniques using physical, chemical and biological methods have been developed and employed at test sites. Among them, 'pump and treat' methods have most commonly been applied for the treatment of contaminated soil and groundwater. However, even after many years of operation at various sites, this physicochemical treatment method often falls short of expected remediation targets, primarily due to low treatment efficiencies and high operational costs (Scherer *et al.*, 2000; Henderson and Demond, 2007).

Currently, bioremediation is considered as a viable alternative for the treatment of contaminated soil and groundwater to support the removal of both organic and inorganic compounds (Alexander, 1999). Bioremediation implies the application of living microorganisms to degrade hazardous organic compounds into non-toxic substances such as carbon dioxide and water and may include both *ex situ* and *in situ* remediation strategies. As detailed in Chapter 1, *ex situ* remediation methods include land farming, the use of bioreactors, composting, excavated soil washing, and soil-solvent extraction. *In situ* remediation strategies include the use of bio-stimulants which may include the addition of degrading microorganisms (bioaugmentation), bioventing and biosparging (the addition of nutrients and air to enhance biodegradation by indigenous microorganisms), as well as the use of permeable reactive barriers which is later described in this chapter. Usually, *ex situ* treatment is more expensive than *in situ* treatment because the former requires storage facilities to hold the contaminated media, and there are additional costs associated with both extracting and moving the contaminated material, which include both power supply and construction costs (Alexander, 1999).

Among *in situ* bioremediation technologies, the use of engineered permeable reactive barriers (PRBs) within contaminated aquifers has grown as a way to prevent further migration of mobile pollutants from contaminated plumes with minimal maintenance costs (Powell *et al.*, 1998; Scherer *et al.*, 2000; Kao *et al.*, 2001). For this reason, during the past decade, subsurface PRBs have become an important *in situ* remediation option for contaminated groundwater (Scherer *et al.*, 2000). Several review papers are available which focus on the use of PRBs for the treatment of various mobile contaminants (Scherer *et al.*, 2000; Henderson and Demond, 2007; Thiruvenkatachari *et al.*, 2008; Obiri-Nyarko *et al.*, 2014). These review papers mostly cover PRB design, type of contaminants, and choice of reactive media, largely focusing on zero-valent iron (ZVI). To date, however, the influence and impact of biofilm formation, which is essential for the performance of PRBs, remains little discussed.

This chapter reviews the influence of biofilm formation on the performance of PRBs for the treatment of mobile contaminants. First, the general application of biofilm technologies for the treatment of contaminated soil and groundwater are discussed. Second, the use and efficacy of PRBs with biofilms developed on different supporting media is reviewed. Third, the impact of biofilm formation on the long term performances of PRBs is evaluated. Finally, I conclude by discussing emerging and future research areas in the use of microbial PRB biofilms for the treatment of contaminated soil and water.

## Biofilm applications in bioremediation

### Biofilm

Microbial cells attach to almost any surfaces in soil or in aquatic environments. The immobilized cells grow, reproduce, and produce extracellular polymeric substances, which provide structural and functional integrity to the biofilm (Characklis and Marshall, 1990). Fig. 11.1A shows successive steps in the development of biofilm on the substratum, which are also described in Chapter 2. These stages are composed of (1) initial cell attachment on surfaces through weak reversible adhesion created by van der Waals forces, (2) colonization and maturation with the help of extracellular polymeric substances (EPS), and (3) dispersion and redistribution on new surfaces (Characklis and Marshall, 1990). Mature microbial biofilm comprises both microbial cells and extracellular polymeric substances (EPS) which help biofilm organisms to adhere to surfaces and each other to construct complex structures with voids and pores (Fig. 11.1B).

Biofilm EPS is well known to impact the physical and chemical characteristics of biofilms and perform complex roles in the biofilm life cycle (Wolfaardt *et al.*, 1995; Seo and Bishop, 2007; Wang *et al.*, 2012; Xue *et al.*, 2012; Xue and Seo, 2013). One of the main and most important roles of EPS is to build up an adhesive matrix that can provide the mechanical stability of biofilms while also promoting cell attachment to inert surfaces or to other microbial cells. Through this function of EPS, biofilms can support high densities of microorganisms per unit volume. For instance, the micro-channel structure developed in mature biofilms with the aid of EPS permits the aqueous transport of nutrients and other materials to the internal structures of the biofilm. Another important role of EPS is to protect microbial cells from toxic chemicals, such as disinfectants and antibiotics via preferential reactions of biocides with EPS, retarding the transport of biocides in EPS matrices (Xue *et al.*, 2012, 2013, 2014; Xue and Seo, 2013). In this way, EPS can scavenge and accumulate nutrients such as organic and inorganic compounds from the environment (Hans-Cur, 1995). As an adaptive strategy for ecological success, biofilm communities are known to use EPS as a reservoir to accumulate growth limiting nutrients from the environment and utilize

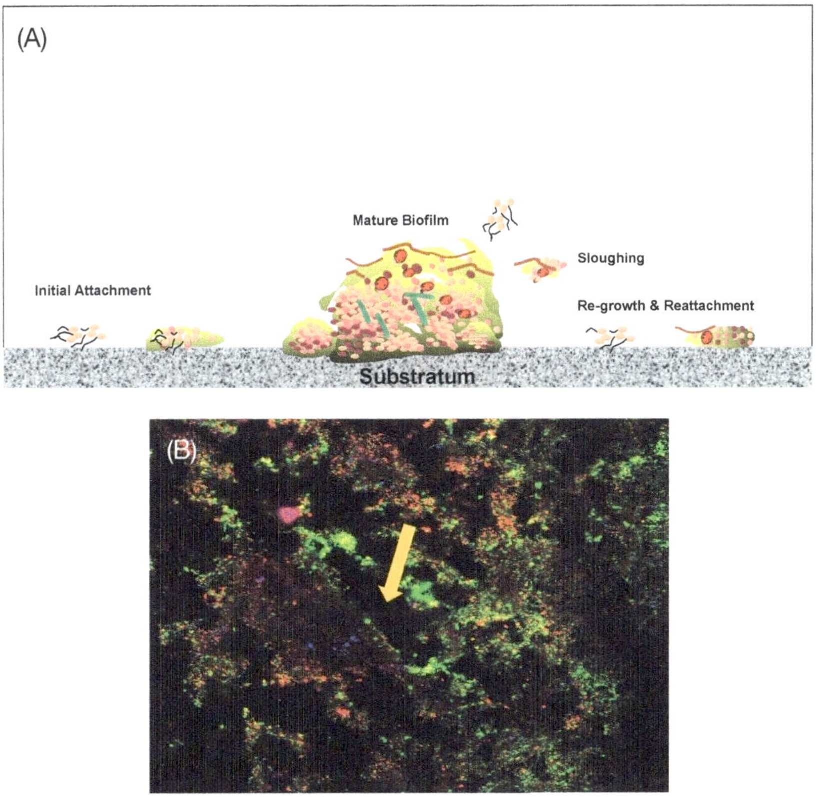


**Figure 11.1** (A) Basic steps involved in biofilm formation, (B) CLSM Images of multi species biofilm formation: arrows indicate void and channel structure of biofilm.

them for growth (Wolfaardt *et al.*, 1995). However, the relative importance of nutrient adsorption to EPS over direct adsorption to cells for bacterial growth and survival in low nutrient environment is still not well understood and the extent of adsorbed substrate utilization needs further study (Bellin and Rao, 1993; Wolfaardt *et al.*, 1995; Wang *et al.*, 2012; Bai *et al.*, 2014). Besides providing protective barriers and adsorption sites for nutrient accumulation, EPS can hold significant amounts of water and can prevent the desiccation of microorganisms under water-deficient conditions.

Previous studies have been conducted to characterize EPS from activated sludge flocs as well as attached biofilm from various sources. These studies reported that EPS are composed of different classes of organic macromolecules such as polysaccharides, proteins, humic substances, nucleic acids, phospholipids, and other polymeric compounds (Brown and Lester, 1982; Frølund *et al.*, 1996; Jahn and Nielsen, 1998; Nielsen and Jahn, 1999), as shown in Table 11.1. Currently, most research on EPS formation and its composition has focused on the study of polysaccharides and proteins, including enzymes (Wingender *et al.*, 1999).

## Application of biofilms in bioremediation

Biofilms offer numerous advantages for the treatment of contaminated water in comparison to the use of suspended bacteria (Characklis and Marshall, 1990; Lazarova and Manem, 1995; Wuertz *et al.*, 2003). First, the complex three-dimensional architecture of biofilms mean that they have the ability to support a variety of microbial populations at various spatial locations in the biofilm. The increased diversity of microorganisms in these biofilm communities can facilitate the degradation of different organic substances, which cannot be degraded by single bacterial species (e.g. allowing both aerobic and anaerobic degradation processes to operate side by-side). Second, biofilms support high microbial densities per unit volume. This enables faster substrate removal rates, making it possible to build

**Table 11.1** General composition of bacterial extracellular polymeric substances (adapted from Wingender *et al.*, 1999)

| Extracellular polymeric substance | Components | Main type of linkage between subunits | Structure of polymer backbone |
|---|---|---|---|
| Polysaccharides | Monosaccharides | Glycosidic bonds | Linear |
| | Uronic acid | | Branched |
| Proteins | Amino acids | Peptide bonds | Linear |
| Nucleic acids | Nucleotides | Phosphodiester bonds | Linear |
| Phospholipids | Fatty acids | Ester bonds | Side-chains |
| | Phosphate | | |
| | Ethanolamine | | |
| | Serine | | |
| | Choline | | |
| | Sugar | | |
| Humic substances | Phenolic compounds | Ether bonds, C–C bonds | Cross-linked |
| | Amino acids | Peptide bonds | |

effective small-scale reactors (Bishop, 1997; Muñoz *et al.*, 2009). Third, the biofilm mode-of-life confers resistance against various contaminants by protecting the internal community from shock loadings of high contaminant concentration (Singleton, 1994; Pesce *et al.*, 2011). This is because the substantial EPS component of biofilm can adsorb various contaminants in aquatic systems including natural organic matter (Wang *et al.*, 2012), heavy metals (Liu *et al.*, 2001; Gadd, 2010; Wei *et al.*, 2011), radionuclides (Barkay and Schaefer, 2001), and xenobiotic compounds (Schorer and Eisele, 1997; Wang *et al.*, 2002; Pan *et al.*, 2010; Jia *et al.*, 2011). Therefore, biofilm formation can enhance the removal of contaminants via biosorption, whereby these contaminants in biofilm may be further degraded or transformed.

Biofilm processes have been successfully applied for the treatment of wastewater, but also hold great promise for the remediation of contaminated soil and groundwater, including from highly recalcitrant compounds (Schorer and Eisele, 1997; Wang *et al.*, 2002; Johnsen *et al.*, 2005; Pan *et al.*, 2010; Jia *et al.*, 2011). Previous studies report that bacterial consortia in biofilm can metabolize or co-metabolize various xenobiotic compounds and immobilize organic compounds by biosorption of EPS. Additionally, the degradation of toxic compounds is aided by the role that biofilms play in sharing extracellular enzymes required for contaminant degradation (Wolfaardt *et al.*, 1995; Alexander, 1999). Chemotaxis which guides bacteria to swim towards substrate sources (e.g. hydrocarbons) plays an important role in direct bacterial attachment and subsequent biofilm formation on various contaminants (Parales *et al.*, 2000; Pandey and Jain, 2002). Having direct biofilm formation on the substrate (contaminants) with low solubility, biofilms can achieve an increased diffusion rate of substrate (following the Fick's laws of diffusion) into the cell and facilitate substrate degradation (Guerin and Jones, 1988; Guerin and Boyd, 1992; Eriksson *et al.*, 2002; Wick *et al.*, 2003; Rodrigues *et al.*, 2005; Seo and Bishop, 2007) (Fig. 11.2) . Indeed, studies report direct biofilm formation directly onto contaminant surfaces may help to overcome the limited bioavailability of hydrophobic organic compounds [e.g. polycyclic aromatic hydrocarbons (PAHs)], removing the need to solubilize these contaminants at low concentrations prior to degradation. Erikson at al. (2002) reported direct biofilm formation and growth on both phenanthrene and pyrene crystals as sole carbon substrates, indicating direct utilization of solid substrate. Other studies similarly report the influence of direct bacterial attachment and biofilm formation on contaminants [e.g. PAHs and non-aqueous phase liquids (NAPLs)] for biodegradation. Stelmack *et al.* (1999) reported inhibited adhesion of PAH degraders (a *Mycobacterium* strain and a *Pseudomonas* strain) to NAPLs under the presence of a surfactant, which affected biodegradation (Stelmack *et al.*, 1999). In their study for

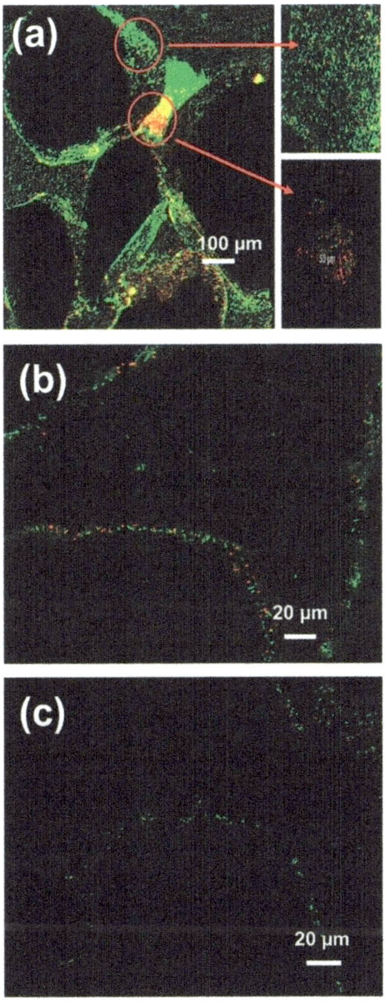


**Figure 11.2** In situ monitoring of biofilm formation on PAH spiked soil. (a) 3-D projection image of biofilm formation with no surfactant addition (with 10× objective): enlarged images on the right side were obtained with 40× objectives; (b) biofilm formation with no surfactant addition (with 40× objective); and (c) biofilm formation with 1000 mg/l surfactant addition; green represents live cells, and red represents dead cells [Reprinted with permission from Environmental Science & Technology, Copyright 2007 American Chemical Society].

anthracene biodegradation, Chen *et al.* (2000) reported that anthracene biodegradation was inhibited by the presence of a nonionic surfactant which inhibited direct biofilm formation on a solid anthracene substrate (Chen *et al.*, 2000). In contrast, however, other studies report the enhanced biodegradation of hydrophobic chemicals under the presence of synthetic surfactants, which may inhibit direct biofilm formation on chemical or soil surfaces, owing to the high bioavailability of solubilized hydrocarbons (Tiehm, 1994; Boonchan *et al.*, 1998). The secretion of biosurfactants by biofilm-dwelling bacteria may also enhance the biodegradation of hydrocarbons (Deziel *et al.*, 1996; Desai and Banat, 1997; Whang *et al.*, 2008). Bacterial biofilms appear to have the ability to adapt themselves to utilize hydrocarbons with their efficacy being dependent on biodegradation pathways used by the bacteria and environmental stimuli (Johnsen *et al.*, 2005; Seo and Bishop, 2007).

## Influence of biofilm formation on the performance of permeable reactive barriers (PRBs)

### Permeable reactive barrier (PRB)

The PRB is an *in situ* remediation process that uses a hydraulically permeable reaction zone located either at the edge of a contaminant plume or as a barrier to the horizontal flow of groundwater. Ideally, as contaminants in the groundwater migrate slowly through the reactive barrier, they are transformed into nontoxic products or immobilized in the barrier by biological and/or chemical processes (Scherer *et al.*, 2000; Henderson and Demond, 2007). Fig. 11.3 shows a schematic of a PRB which are also commonly termed 'microbial fences', *in situ* reactive zones, or 'biowalls' (Yerushalmi *et al.*, 1999; Leeson, 2001).

PRBs are usually implemented where the *ex situ* remediation of a pollutant is technically or economically infeasible. They have been extensively studied and tested for the *in situ* detention and removal of mobile contaminant plumes, which usually requires treating large volumes of groundwater containing low concentrations of contaminant. Conventional pump and treat methods are often not economically feasible to treat large volumes of groundwater due to the high energy costs associated with this approach. In contrast, PRBs can provide long-term treatment with limited maintenance costs since the approach frequently capitalizes on the natural flow of the groundwater, and hence the pollutant, and does not require the extensive pumping of contained liquids (Miller *et al.*, 2001; Henderson and Demond, 2007). Even though initial installation costs of PRBs with ground excavation can be higher than that of the pump and treat method, minimum

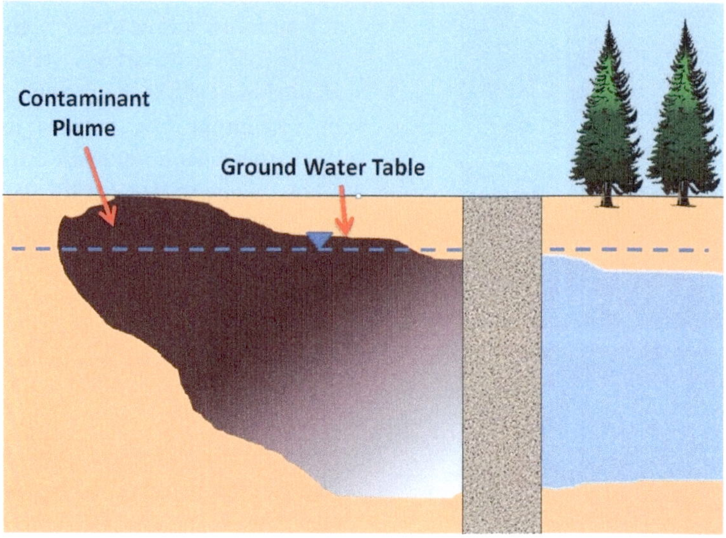


**Figure 11.3** Schematic of a permeable reactive barrier.

energy costs can offset high initial installation costs. In addition, PRBs do not require facilities for contaminant storage, transport and treatment other than monitoring wells (Thiruvenkatachari *et al.*, 2008).

PRBs have been tested for the removal of various organic compounds, heavy metals, radionuclides, and nutrients with either lab scale or full scale systems (Blowes *et al.*, 2000; Lamarche *et al.*, 2001; Liu *et al.*, 2001; Miller *et al.*, 2001; Salanitro *et al.*, 2001; Ludwig *et al.*, 2002; Gadd, 2010; Wei *et al.*, 2011). Among tested contaminants, PRBs have been extensively tested for the removal of chlorinated solvents (Powell *et al.*, 1998; Puls *et al.*, 1999; Vogan *et al.*, 1999; Shen and Wilson, 2007) and have also been applied to other hydrocarbons such as PAHs, methyl tert-butyl ether (MTBE), benzene, toluene, ethylbenzene, xylene (BTEX), herbicides and pesticides (Lamarche *et al.*, 2001; Miller *et al.*, 2001; Cole, 2012; Herrera-González *et al.*, 2013). Lamarche *et al.* (2001) used a 'funnel and gate system' (later described) filled with a coarse-medium silica sand containing 1% by volume of granular activated carbon (GAC) to test the removal efficacy of a naphthalene plume, achieving a 98% removal rate over 2 years. Miller *et al.* (2001) similarly reported that a full scale biobarrier system achieved a treatment efficiency of more than 99.9% for dissolved MTBE and BTEX (Miller *et al.*, 2001).

Despite its proven potential and many field installations of PRBs, PRB technologies are still considered in the experimental stage as few data on the long-term performance of PRBs is available. For this reason, we continue to collect long-term PRB operation data for zero-valent iron installed PRBs, recording changes in microbial community structure and corrosion products (Powell *et al.*, 1998; Scherer *et al.*, 2000; Henderson and Demond, 2007; Phillips *et al.*, 2010; Obiri-Nyarko *et al.*, 2014). However, most studies remain laboratory based and use column reactors (Fig. 11.4) to indirectly evaluate the long-term performances of PRBs exposed to high flow rates, which may not represent the ageing processes of reactive materials in the field (Seo and Bishop, 2008; Obiri-Nyarko *et al.*, 2014). More long-term studies with robust monitoring parameters are needed using both controlled lab scale reactors and field scale systems to better understand PRBs and to improve their performance.

## PRB configuration

Currently, both continuous and funnel-and-gate systems are common design configurations for the full scale field installation of PRBs, (Powell *et al.*, 1998; Thiruvenkatachari *et al.*, 2008) with the selection of optimum configuration conditions being dependent on groundwater flow characteristics, and the size and type of contaminant plume (Scherer *et al.*, 2000). Fig. 11.5A shows a continuous PRB

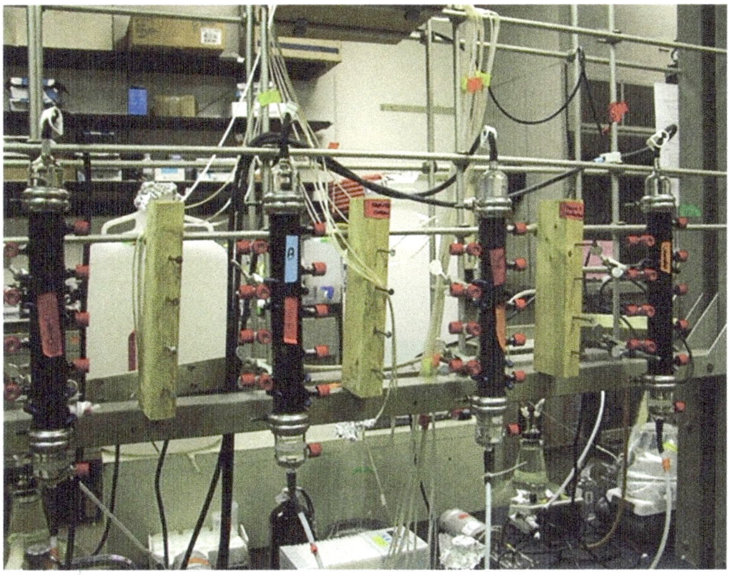

**Figure 11.4** Column reactors for lab scale PRB tests.

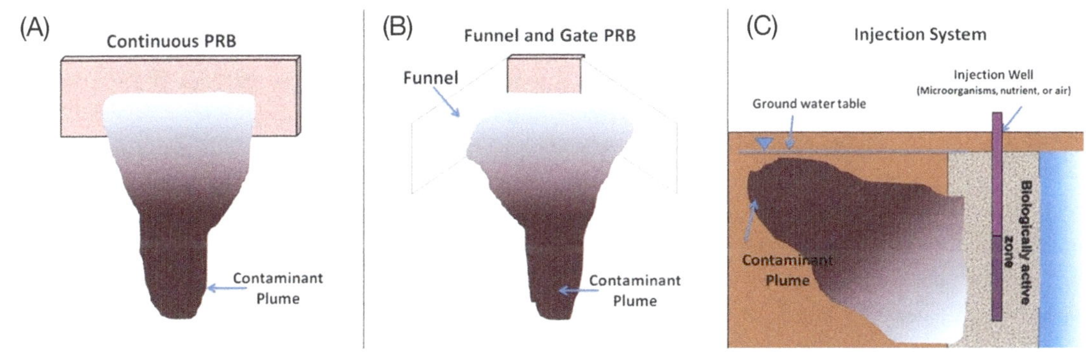


**Figure 11.5** Schematic of various PRB configurations: (A) continuous PRB, (B) funnel and gate PRB and (C) injection system.

system with a single reactive zone covering the vertical and horizontal dimensions of the contaminant plume. In the funnel-and-gate configuration, PRBs have funnels with low permeability that can divert groundwater flow towards a permeable reactive zone where chemical and/or biological reactions occur (Fig. 11.5B). The funnels are usually constructed with slurry walls or sheet piling and are set into bed rock or other impermeable layer. The continuous trench system usually has less impact on the groundwater flow while funnel-and-gate systems may significantly affect groundwater flow due to the reduced cross sectional area of the reactive gate. Thus, the maintained permeability of the reactive material in the gate is particularly important for successful operation of funnel-and-gate PRBs. In both design configurations, PRBs must completely encompass the contaminated groundwater flow. Backfilled reactive medium is commonly mixed with non-reactive media not only to minimize installation costs but also to maintain the hydraulic permeability of PRBs, especially for funnel-and-gate PRB, to help minimize flow restriction. When high cost reactive materials are used for PRBs, a funnel-and-gate configuration is preferred to reduce the initial installation cost of the PRB. Furthermore, considering the presence of contaminant mixtures, multiple reactive media can be installed in series and in this case, the funnel-and-gate PRB design may cost-effective (Powell *et al.*, 1998).

Instead of excavating the ground for PRB construction, injection systems (biosparging and bioaugmentation) are used to create reactive zones by adding suitable materials either to enhance the biological activities of indigenous microorganisms in soil (e.g. electron donors, acceptors, and nutrients) or, alternatively, to add the active microorganisms (Fig. 11.5C). By injecting substrates or microorganisms with the ability to degrade target contaminants close to contaminant plumes, metabolically active bacteria can form biofilms with EPS and prevent contaminants from migrating subsurface aquifer. This process is also called as bioaugmentation (Gentry *et al.*, 2004). PRBs with biofilm alone can be an attractive option over conventional PRBs as they do not necessarily require excavation, reactive support or even additional bacteria, if the activities of indigenous bacteria can be utilized (Hiebert *et al.*, 2001; Cunningham *et al.*, 2003; Komlos *et al.*, 2004).

Previous studies injecting active bacteria to enhance a specific biological activity have reported the limited transfer of bacterial cells within subsurface aquifers due to collision against the porous media. Studies also report premature clogging near to the injection well which may prevent bacteria from forming a suitable reactive zone (Richardson and Nicklow, 2002). To deliver bacteria to subsurface aquifer more effectively, various techniques have been considered which include reducing the size of bacteria by starvation (Lappin-Scott and Costerton, 1990; Camper *et al.*, 1993; Cunningham *et al.*, 2007) and surfactant mediated injection (Li and Logan, 1999). Another consideration needed for biofilm PRB is the biodegradability of contaminants. Some contaminants are not easily biodegradable by bacteria alone (Singh *et al.*, 2006) and additional reactive media or chemicals may be needed to catalyse the biological degradation of these contaminants.

## Supporting media for PRBs

In the early stage of PRB technology, most studies and field implementations of PRBs utilized zero-valent iron (ZVI) for the abiotic degradation of contaminants (Obiri-Nyarko *et al.*, 2014). In PRBs with ZVI, chemical reaction occurs between the surface of ZVI and contaminants, which are reduced by ZVI via electron transfer processes (Henderson and Demond, 2007). Previously, biological activities in ZVI PRBs were not considered and were more commonly regarded as one of inhibition factors for contaminant removal. However, a few studies have reported enhanced contaminant removal in the presence of biofilms (Van Nooten *et al.*, 2008; Wang and Tseng, 2009). The other supporting materials (or filling materials) such as granulated activated carbon, peat moss, saw dust, leaf compost, and organic mulch can enhance the efficiency of PRBs via enhanced biofilm formation (Scherer *et al.*, 2000; Kao *et al.*, 2001). These inexpensive organic materials have been considered in bioremediation and bioretention systems to enhance sorption capacity, as well as to provide alternative carbon sources (Kao and Borden, 1997; Schipper and Vojvodić-Vuković, 1998; Waybrant *et al.*, 1998). Fig. 11.6 shows types of supporting materials adapted for PRB applications from 124 projects (Scherer *et al.*, 2000). Among the inexpensive organic filling materials used for PBR, organic mulch has received attention for biofiltration and the bioremediation of hydrophobic organic pollutants (Morris, 2009; Wei and Seo, 2010; Öztürk *et al.*, 2012).

Organic mulches may include many biopolymers such as cellulose, lignin and proteins (Schwarzenbach *et al.*, 2005). Among these biopolymers, lignin is a complex polymer constructed from phenolic compounds. Fig. 11.7 shows monomeric phenylpropanoid units that polymerize to form lignin.

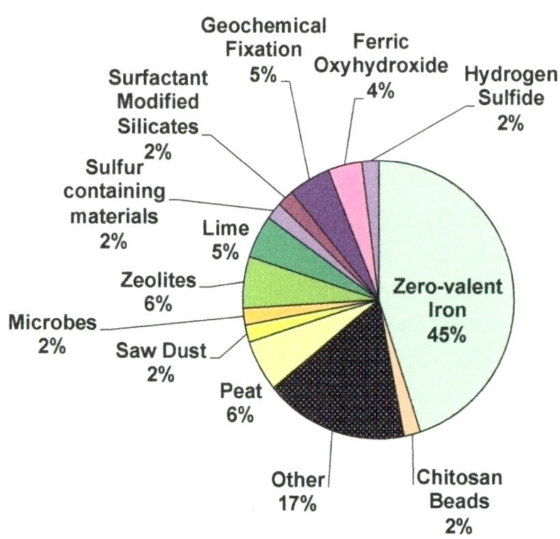


**Figure 11.6** Supporting materials used for PRB applications of 124 projects (adapted from Scherer *et al.*, 2000).

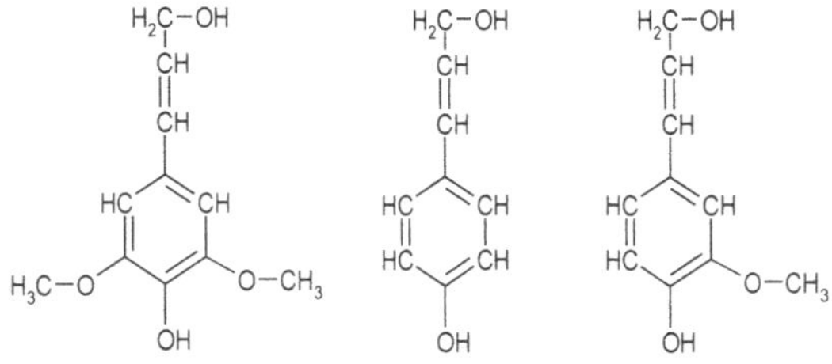


**Figure 11.7** Monomeric phenylpropanoid units that polymerize to form lignin (from Gadd, 2001).

Lignin is known to have a high affinity and sorption capacity for non-ionic organic compounds (Garbarini and Lion, 1986). In addition to its sorption capacity for these compounds, bark chips or shredded plant materials can also be used as complex additive fertilizers (Kästner and Mahro, 1996). Mulch and peat moss have been used as supporting materials for anaerobic degradation of chlorinated solvents in ground water. Results from previous studies reported that enhanced degradation of chlorinated solvents were observed after the formation of bioactive zones (Öztürk *et al.*, 2012).

Hydrophobic organic compound (HOC) sorption is the most common and important process governing the fate and transport of these chemicals in groundwater (Holmén and Gschwend, 1996; Scherer *et al.*, 2000). Therefore, enhanced sorption capacity to hydrophobic organic compounds can be a key factor for effective groundwater remediation. HOC sorption is also important for stable process operation, because biological treatment processes for HOCs must contend with the low growth rate of the degrading microorganisms and the fast migration of the organic pollutants in subsurface aquifer (Seo *et al.*, 2009). Therefore, a stable removal of HOC might be achieved in a biological treatment process, if the adsorbed organic compounds can be desorbed, diffused out from the adsorptive supporting material and then can be metabolized by attached microorganisms. In this aspect, chemical compound sorption can be helpful for stabilizing the biological degradation of toxic organic compounds by dampening concentration changes in the influent stream (Carvalho *et al.*, 2001). In the use of a PRB for hydrocarbon removal, sorption to the barrier's supporting materials can be helpful for preventing the migration of hydrocarbons and for supporting slow growing degraders. In a field application, Aziz *et al.* (2001) tested an organic mulch biowall for the remediation of chlorinated solvent-impacted groundwater (Aziz *et al.*, 2001). They found the mulch biowall was effective for the dechlorination of trichloroethylene (TCE) in the impacted groundwater, reporting that, for over 18 months, mulch could provide sufficient organic matter to stimulate biological reductive dechlorination, and TCE removal could reach 90%.

However, there is also a well-known adverse effect concerning hydrocarbon sorption associated with biodegradation. Many investigators have reported that hydrocarbons adsorbed to natural organic matter become less bioavailable to microorganisms and thus less susceptible to biodegradation. Hydrocarbons strongly adsorbed by geosorbents have also been shown to be quite persistent (Guerin and Boyd, 1992). Owing to sorption to a supporting material, adsorbed hydrocarbons might not be easily bioavailable to microorganisms, slowing rates of degradation. On the other hand, some researchers have also reported that sorption does not inhibit the degradation of hydrocarbons, since some hydrocarbon degraders can use the adsorbed hydrocarbons (Ortega-Calvo and Saiz-Jimenez, 1998; Puglisi *et al.*, 2007). Therefore, for long-term operation of a reactive permeable barrier, the bioavailability of adsorbed hydrocarbons on the new supporting materials needs further evaluation. In addition, compared to field scale data available for ZVI, very little information is available for the longevity of the organic support materials and their long term performances in PRBs (Geranio, 2007; Obiri-Nyarko *et al.*, 2014). Table 11.2 summarizes PRBs with biofilm for various contaminants.

## Influence of biofilm formation on the performance of permeable reactive barriers

### Hydraulic permeability of PRBs

In the case of the bioremediation of xenobiotic compounds in soils, both lab- and field-scale PRBs have been tested and examined to understand biofilm formation and its influence on contaminated soil and groundwater treatments (Gu *et al.*, 2002; Cunningham *et al.*, 2003; Vesela *et al.*, 2006). In biobarrier systems for xenobiotic compounds, biofilm EPS is known to play an important role in the sorption of organic pollutants (Wang *et al.*,

**Table 11.2** Biofilm biobarriers

| Type of contaminant | Type of biobarrier (supporting media for biofilm formation) | Operation type | Reference |
|---|---|---|---|
| Nitrate | Nutrient injection for biofilm formation | Field scale | Dutta *et al.* (2005), Michalsen *et al.* (2009), Cunningham *et al.* (2003) |
| | Anammox biofilm reactor | Lab scale | Meng *et al.* (2014) |
| | Granular activated carbon | | Liu *et al.* (2013) |
| | Mulch | | Cole (2012) |
| PAHs | Mulch | Lab scale | Seo and Bishop (2008), Seo *et al.* (2009) |
| | Pumice stone | Field scale | Di Lorenzo *et al.* (2005) |
| | Nylon sponge | | Cobas *et al.* (2013) |
| | Sepiolite | | Ferreira *et al.* (2013) |
| | Oxyhumolite (oxidized young lignite) | | Vesela *et al.* (2006) |
| Methyl tert-butyl ether (MTBE) | Soil | Lab scale | Saponaro *et al.* (2009) |
| | Oxyhumolite (oxidized young lignite) | Field scale | Vesela *et al.* (2006) |
| Plume | Electron acceptor (oxygen) | Lab scale | Kim *et al.* (2006) |
| TCE | ZVI | Lab scale | Teerakun *et al.* (2011) |
| | Shredded tires | Field scale | Lee *et al.* (2007) |
| | Mulch | | Morris (2009), Öztürk *et al.* (2012) |
| | Glass beads | | Komlos *et al.* (2004) |
| | Peat | | Kao and Lei (2000) |
| | Oxyhumolite | | Vesela *et al.* (2006) |
| Ametryn, 2,4-dichlorophenoxyacetic acid (2,4-D) | Fixed film reactor with recirculation of oxygenated liquid | Lab scale | Sandoval-Carrasco *et al.* (2013) |
| Hexavalent chromium | Zeolite or activated carbon | Lab scale | Quintelas *et al.* (2006), Fonseca *et al.* (2012) |
| | Quartz sand and HFO-coated sand | | Viamajala *et al.* (2008) |
| | Fractured-limestone and biostimulation with substrate | Lab scale | Mann (2012) |
| Heavy metals | Mulch (bioretention) | Lab scale | Lee *et al.* (2009) |
| | ZVI | | Lee *et al.* (2007) |
| | Shredded tires | | |
| Atrazine | Mulch | Lab scale | Cole (2012) |
| Heterocyclic nitramines | Mulch | Lab scale | Ahmad *et al.* (2007) |
| Propanil and its by-products | Soil | Lab scale | Herrera-González *et al.* (2013) |

2002; Aksu, 2005). As mentioned in the previous section, hydrocarbon sorption rates are of primary importance for the stable operation of engineered bioremediation systems (Headley *et al.*, 1998; Writer *et al.*, 2011) and biofilm-associated organisms and EPS may show a strong affinity to organic hydrocarbons (Aksu, 2005). With the help of the increased sorption capacity, if the biofilm EPS can trap organic compounds in the groundwater, the sequential biodegradation of the compounds can be achieved by the biochemically active microorganisms in the biofilm matrix, also increasing the number of degraders present. Through the entrapment and simultaneous degradation of hydrocarbons in the biofilm, the continued removal of hydrocarbons might be achieved. In this respect, the formation and the microbial community structure of the biofilm are crucial for the soil and groundwater remediation with PRBs. However, the biofilm formation in the subsurface aquifer is also known to cause adverse effects on contaminant removal. One of the commonly reported problems associated with biofilm is the reduced hydraulic conductivity caused by excessive biofilm formation (Taylor and Jaffé, 1990; Pintelon *et al.*, 2012). As the biofilm grows it produces EPS, and increased EPS biomass can block micro-pores in subsurface aquifers eventually decrease the permeability of the PRB. This will prevent the flow of contaminants into the reactive zone, reduce contaminant removal rates and may eventually divert the groundwater flow out of the barrier. Besides the excessive biofilm growth, previous studies report that the reduced permeability of the PRB can be caused by the significant precipitation and adsorption of minerals induced by biofilm formation and associated pH changes (Hans-Cur, 1995; Cooke *et al.*, 1999; Henderson and Demond, 2007). Additionally, even biogas (e.g. $N_2$) production by biofilms may impact the hydraulic permeability in PRBs (Liang *et al.*, 2000). However, the presence of biofilm may enhance permeability of PRBs in some cases. For example, previous studies report that hydrogen gas bubbles generated by the corrosion of ZVI under anaerobic conditions as being one of the major causes of permeability reduction in PRBs, thereby requiring the venting of hydrogen (Scherer *et al.*, 2000; Henderson and Demond, 2011). Since microorganisms in ZVI PRBs (e.g. methanogens) can utilize hydrogen gas as an electron donor for their growth under anaerobic conditions, consumption of produced hydrogen gas may attenuate permeability reductions and provide stable operations of PRBs (Scherer *et al.*, 2000).

### Reactivity of supporting media

Many laboratory and field studies report that sulfate- and iron-reducing bacteria play an important role in the long-term performance of the PRB with ZVI, affecting not only hydraulic permeability but also hydrocarbon biodegradation (Gu *et al.*, 1999; Scherer *et al.*, 2000; Van Nooten *et al.*, 2008; Wang and Tseng, 2009; Park *et al.*, 2010; Phillips *et al.*, 2010; Cullen *et al.*, 2011). Though the importance of microbial biofilm formation on the performances of the PRB is well accepted, there are still relatively small numbers of available studies that examined the role of biofilm on the performances of PRBs (Singh *et al.*, 2006; Obiri-Nyarko *et al.*, 2014).

The early stage implementation of the PRB extensively used ZVI to control both organic and inorganic contaminants via reductive degradation, where direct interactions between ZVI surfaces and contaminants in aqueous phase are required. Therefore, there have been great concerns over the reduction of the reactive sites on ZVI surface due to direct biofilm formation by indigenous microorganisms such as sulfate and iron reducing bacteria that are commonly found in subsurface aquifers (Scherer *et al.*, 2000). However, there are no systematic studies available and only a few studies report reduced reactivity of ZVI due to biological activities. Gu *et al.* (1999) reported the appearance of sulfide and the formation of insoluble ferrous sulfide precipitates due to microbial reduction of sulfate along with iron oxyhydroxide precipitates (Gu *et al.*, 1999). Though the study was conducted within lab scale column systems over a short experimental period, they observed potential pore plugging in column reactors packed with ZVI due to biofilm formation. The decreased degradation of chlorinated hydrocarbons due to the reduced abiotic reaction of ZVI was also reported by Gregory *et al.* (2000), where the addition of methogenic enrichment appeared to remove hydrogen and increase ZVI corrosion (Gregory *et al.*, 2000). Besides the ZVI, Lowry and Reinhart (2000) reported that sulfate-reducing bacteria biofilm formation deactivated palladium/aluminium oxide catalysis and decreased TCE dechlorination. However, they also reported that the catalyst could be regenerated by the addition of a disinfectant (sodium hypochlorite) (Lowry and Reinhard 2000).

In contrast to the potential adverse effect of biofilm formation on the reactivity of supporting media, other studies report the enhanced degradation of hydrocarbons by synergistic interactions between bacteria and ZVI (Lampron *et al.*, 2001; Wrenn, 2004; Wang and Tseng, 2009; Park *et al.*, 2010). Wang and Tseng (2009) report that the combination of physical and biological systems cause the enhanced dechlorination of TCE compared to the use of physicochemical or biological systems in isolation. The reactivity of ZVI was also maintained compared to that of abiotic columns (Wang and Tseng, 2009). Lampron *et al.* (2001) reported the increased degradation rate of TCE under the presence of anaerobic cultures, which directly immobilized and formed biofilm on ZVI surfaces. They observed the activities of methanogens utilizing hydrogen gases produced during the anaerobic corrosion process.

In addition to directly contributing to contaminant removal by microbial degradation, biofilm formation can indirectly improve contaminant removal by enhancing or maintaining the reactivity of ZVI (Gerlach *et al.*, 2000; Phillips *et al.*, 2010). For the PRBs with ZVI, bacterial biofilm can reduce concentrations of molecular oxygen and may increase the reactivity and longevity of ZVI. In addition, under anaerobic conditions, $H_2$ produced by ZVI can increase sulfate- and iron-reducing bacteria populations and subsequently enhance the biodegradation of redox sensitive hydrocarbons (Weathers *et al.*, 1997; Gu *et al.*, 2002). Other studies report the regeneration of reactive ZVI by iron reducing bacteria that utilize Fe(III) as an electron acceptor to reduce Fe(III) to Fe(II) (Gerlach *et al.*, 2000; Park *et al.*, 2010). Specifically, Gerlach *et al.* (2000) reported that biofilm-associated dissimilatory iron-reducing bacteria restored the reactivity of ZVI, thereby increasing the transformation rate of carbon tetrachloride to chloroform. They hypothesized that the restored reactivity is attributable to the reductive dissolution of Fe(III) corrosion products or the reduction of Fe(III) to Fe(II) by iron-reducing biofilms. To monitor degradation efficiencies of ZVI under the presence of different bacterial inocula, Van Nooten *et al.* (2008) conducted column tests using both a single-strain of *Geobacter sulfurreducens* culture and a bacterial consortium recovered from anaerobic soil samples (Van Nooten *et al.*, 2008). They observed the formation of highly reactive carbonate green rust as well as the high reactivity of the inoculated column linked to Fe(III)-reducing bacteria, while the reactivity of a non-inoculated column, including ZVI, was not changed.

While most studies have tried to understand the influence of biofilm formation on ZVI, the influence of biofilm formation on the reactivity of other supporting media is not well understood. As described in the previous section, information is available for field scale applications of PRBs supplemented with sustainable organic materials or activated carbons (Scherer *et al.*, 2000; Aziz *et al.*, 2001; Lamarche *et al.*, 2001). However, only a few laboratory scale studies have examined the reactivity changes of supporting materials by biofilm formation (Seo and Bishop, 2008; Seo *et al.*, 2009).

## Field-scale test results

Uncertainties in the long-term performance of PRBs are major concerns that hinder the broad use of PRBs for contaminated soil and groundwater remediation. Even though there are significantly increased numbers of PRB installations, we still do not have robust information that can be used to establish design principles for PRBs with different media (Henderson and Demond, 2007). Currently, most available information for field scale PRBs is for PRBs with ZVI, which were mostly examined for groundwater chemistry characteristics, generation of corrosion products and stable contaminant removal rates. However, most studies did not monitor changes in microbial populations and related impact on the performance of PRBs. Only a small amount of information is available from full scale PRB tests (O'Hannesin and Gillham, 1998; Powell *et al.*, 1998; Scherer *et al.*, 2000; Phillips *et al.*, 2010).

A field-scale PRB study conducted at a site in Borden, Ontario examined cores from a reactive ZVI wall after a two year test and reported no significant populations of iron-oxidizing or sulfate reducing bacteria (O'Hannesin and Gillham 1998; Powell *et al.*, 1998). During field PRB studies at a New York site (after 2 years of operation) and at the Lowry Air Force Base (after 18 months of operation), core samples were also collected and examined for microbial activities in PRBs with ZVI using both scanning electron microscopy (SEM) and conventional plate counting. Only small microbial numbers were recorded and no significant biofilm activities were identified (Powell *et al.*, 1998). Other studies conducted at the Sunnyvale and Moffett Fields (California) and Somersworth (New Hampshire) sites similarly reported low microbial activities (www.rtdf.org, Gu *et al.*, 1999). These field scale study results were somewhat expected as geochemical characteristics in the PRB are quite different from those in common subsurface aquifers. The ZVI creates a reducing environment and significantly increases pH in the reactive zone in a PRB, which discourages bacterial activities. In contrast, other field studies conducted at Newbury Park, CA, with ZVI and sand reactive barriers, and at a Portsmouth, OH, site with a filter column reported biofilm formation (www.rtdf.org, Liang *et al.*, 1997). Wilkin *et al.* (2003) also reported the highest microbial biomass accumulation at the interface between the contaminant and ZVI based PRBs, based on lipid biomarker analyses of viable microbial biomass (Wilkin *et al.*, 2003).

Recently, Phillips *et al* (2010) reported on the long-term performance of a field-scale ZVI PRB along with microbial community structure data. The performance results were obtained from Europe's oldest commercially installed PRB, Monkstown PRB in Northern Ireland after 10 years of operation. According to microbial community structure analyses using PCR-DGGE with specific group primers, authors reported that *Eubacteria*, *Archaea*, denitrifying bacteria, iron reducing bacteria, sulfate reducing bacteria, and methanogens were all detected from collected materials near the inlet zone of the PRB. They also reported that the high microbial activity at the inlet zone corresponded to constricted water flow due to fouling and drastic reductions in TCE concentration. They also reported a significant shift in microbial community structure comparing sites up- and down-gradient of the PRB. However, their function and influence on ZVI corrosion and the performance of the PRB were not discussed.

## Concluding remarks and future research needs

This review briefly summarizes our current understanding on PRBs with biofilm for their improved performance and long term stability for the removal of various organic contaminants. PRBs have shown great potential as a promising technology for the remediation of contaminated soil and groundwater. As discussed in previous sections, there are many lab scale studies available utilizing PRBs. There are also many reported field-scale PRBs using ZVI with the oldest PRB with ZVI being implemented two decades ago (Henderson and Demond, 2007; Obiri-Nyarko *et al.*, 2014); many PRBs have since been installed using sustainable organic materials. However despite the proven potential and multiple field installations of PRBs, PRB technologies are still considered in the experimental stage as we do not have unifying principles that can guide process design and long-term performance. More long-term studies, with robust and regular monitoring for biogeochemical changes, are needed using both controlled lab-scale reactors and field-scale systems.

There are several areas for future research needs to better understand and optimize PRB processes for them to be widely adopted for *in situ* soil and groundwater treatment. Future research needs are as follows:

- Though most PRB processes adapt ZVI as supporting materials, the biocorrosion of ZVI is not understood well. Since the biocorrosion of ZVI can both enhance or decrease the reactivity and longevity of ZVI, understanding the process of iron corrosion by biofilms will help engineers to maintain the reactivity of ZVI and better estimate the performance of PRB for long term contaminant removal. Specifically, the role of Fe(II) and iron oxide precipitates in reductive degradation processes needs to be further explored as they can serve as not only reductants but also oxidants for contaminant removal (Scherer *et al.*, 2000). The direct interaction between iron and biofilm, as well as the role of EPS on biocorrosion, need to be further elucidated to accurately predict the biocorrosion of ZVI PRBs (Beech and Sunner 2004).
- Recently, in order to improve the subsurface mobility and longevity of both microscale ZVI and nanoscale ZVI particles, there has been interest in the use of novel coating methods with new materials (e.g. starch, polystyrene, sulfonate, carboxyl methyl cellulose, cellulose acetate and polyelectrolyte co-polymers) (Baric *et al.*, 2012; Velimirovic *et al.*, 2012). Though new coating methods may enhance the subsurface mobility, their impact on microbial activities, direct biofilm formation, and biocorrosion processes are not yet understood. Therefore, bacteria interactions with coated ZVI and its influence on the reactivity of ZVI requires further assessment In addition, bimetallic ZVIs have received much attention to enhance contaminant removal (Liu *et al.*, 2014). However, the corrosion of bimetallic ZVI and its influence on catalytic degradation of recalcitrant pollutants remains poorly examined, however.
- Besides ZVI, there is growing interest in adapting sustainable organic materials as support media for biofilm formation in PRBs. Currently, most studies have focused on the

removal of hydrocarbons by PRBs, but mass transport mechanisms and structural forms of the biofilm exposed to the hydrocarbons are not well understood. Especially, the utilization of contaminants adsorbed to the biofilm matrix and enzymatic degradation of contaminants in biofilm needs further assessment. In addition, changes in biofilm community structure and beneficial or detrimental effects on the long term performance of PRBs need to be examined, using modern molecular methods.

- Beginning with their use by petroleum industries for oil recovery, surfactants have been applied for the remediation of soils contaminated with hydrophobic organic pollutants with low bioavailability (Volkering *et al.*, 1997). Surfactants can solubilize hydrophobic organic compounds which have low water solubility, incorporating them into the hydrocarbon-like core of the surfactant micelle. The increased solubility of organic compounds can provide more available substrate for microorganisms which can degrade and utilize the compounds leading to enhanced rates of removal. However, surfactant enhanced soil remediation has adverse effects. Though surfactant aided flushing has shown great potential, increasing the solubility of hydrocarbons, immobilized hydrocarbons can be transferred through groundwater and contaminate soil and groundwater without proper treatment (Moon *et al.*, 2003; Seo and Bishop, 2008). In addition, surfactants may impact the mobility of other substrates present in the environment and the fate of indigenous microorganisms (Volkering *et al.*, 1997; Seo and Bishop, 2007). It is crucial that we better understand the responses of microbial biofilms to surfactant addition in order to develop practical surfactant enhanced bioremediation strategies. Much more research is required to confirm the efficacy of surfactant additions for the efficiency of PBR-based bioremediation and biofiltration systems.

## Acknowledgements

The author appreciates financial supports from the National Science Foundation (CBET CBET-0852803) and the National Institute of Environmental Health Science (NIEHS) Superfund Basic Research Program. The author also appreciates Ms Christina Onyskiw at the University of Toledo and Dr Gavin Lear at the University of Auckland for the manuscript editing.

## References

Ahmad, F., Schnitker, S.P., and Newell, C.J. (2007). Remediation of RDX- and HMX-contaminated groundwater using organic mulch permeable reactive barriers. J. Contaminant Hydrol. *90*, 1–20.

Aksu, Z. (2005). Application of biosorption for the removal of organic pollutants: a review. Process Biochem. *40*, 997–1026.

Alexander, M. (1999). Biodegradation and Bioremediation (Academic Press, London, UK).

Aziz, C.E., Hampton, M., Schipper, M., and Haas, P. (2001). Organic mulch biowall treatment of chlorinated solvent-impacted groundwater. Sixth International In Situ and On Site Bioremediation Symposium (San Diego, CA, USA).

Bai, X., Casey, F.X., Hakk, H., DeSutter, T.M., Oduor, P.G., and Khan, E. (2014). Sorption and degradation of 17β-estradiol-17-sulfate in sterilized soil–water systems. Chemosphere, *119*, 1322–1328.

Baric, M., Majone, M., Beccari, M., and Papini, M.P. (2012). Coupling of polyhydroxybutyrate (PHB) and zero valent iron (ZVI) for enhanced treatment of chlorinated ethanes in permeable reactive barriers (PRBs). Chem. Eng. J. *195*, 22–30.

Barkay, T., and Schaefer, J. (2001). Metal and radionuclide bioremediation: issues, considerations and potentials. Curr. Opin. Microbiol. *4*, 318–323.

Beech, I.B., and Sunner, J. (2004). Biocorrosion: towards understanding interactions between biofilms and metals. Curr. Opin. Biotechnol. *15*, 181–186.

Bellin, C., and Rao, P. (1993). Impact of bacterial biomass on contaminant sorption and transport in a subsurface soil. Appl. Environ. Microbiol. *59*, 1813–1820.

Bishop, P.L. (1997). Biofilm structire and kinetics. Water Sci. Technol. *36*, 287–294.

Blowes, D.W., Ptacek, C.J., Benner, S.G., McRae, C.W., Bennett, T.A., and Puls, R.W. (2000). Treatment of inorganic contaminants using permeable reactive barriers. J. Contaminant Hydrol. *45*, 123–137.

Boonchan, S., Britz, M.L., and Stanley, G.A. (1998). Surfactant-enhanced biodegradation of high molecular weight polycyclic aromatic hydrocarbons by *Stenotrophomonas maltophilia*. Biotechnol. Bioeng. *59*, 482–494.

Brown, M.J., and Lester, J. (1982). Role of bacterial extracellular polymers in metal uptake in pure bacterial culture and activated sludge – I. Effects of metal concentration. Water Res. *16*, 1539–1548.

Camper, A.K., Hayes, J.T., Sturman, P.J., Jones, W.L., and Cunningham, A.B. (1993). Effects of motility and adsorption rate coefficient on transport of bacteria through saturated porous media. Appl. Environ. Microbiol. *59*, 3455–3462.

Carvalho, M., Vasconcelos, I., Bull, A., and Castro, P. (2001). A GAC biofilm reactor for the continuous degradation of 4-chlorophenol: treatment efficiency and microbial analysis. Appl. Microbiol. Biotechnol. *57*, 419–426.

Characklis, W.G., and Marshall, K.C. (1990). Biofilms (John Wiley & Sons Inc., New York, NY, USA).
Chen, P., Pickard, M.A., and Gray, M.R. (2000). Surfactant inhibition of bacterial growth on solid anthracene. Biodegradation *11*, 341–347.
Cobas, M., Ferreira, L., Tavares, T., Sanromán, M., and Pazos, M. (2013). Development of permeable reactive biobarrier for the removal of PAHs by *Trichoderma longibrachiatum*. Chemosphere *91*, 711–716.
Cole, A.L. (2012). Co-removal of Atrazine and nitrate from groundwater using a Mulch biofilm. MS Thesis, University of Nebraska.
Cooke, A.J., Rowe, R., Rittmann, B., and Fleming, I. (1999). Modeling biochemically driven mineral precipitation in anaerobic biofilms. Water Sci. Technol. *39*, 57–64.
Cullen, L.G., Tilston, E.L., Mitchell, G.R., Collins, C.D., and Shaw, L.J. (2011). Assessing the impact of nano-and micro-scale zerovalent iron particles on soil microbial activities: particle reactivity interferes with assay conditions and interpretation of genuine microbial effects. Chemosphere *82*, 1675–1682.
Cunningham, A.B., Sharp, R.R., Hiebert, R., and James, G. (2003). Subsurface biofilm barriers for the containment and remediation of contaminated groundwater. Bioremediation J. *7*, 151–164.
Cunningham, A.B., Sharp, R.R., Caccavo, F. Jr, and Gerlach, R. (2007). Effects of starvation on bacterial transport through porous media. Adv. Water Resources *30*, 1583–1592.
Desai, J.D., and Banat, I.M. (1997). Microbial production of surfactants and their commercial potential. Microbiol. Mol. Biol. Rev. *61*, 47–64.
Deziel, E., Paquette, G., Villemur, R., Lepine, F., and Bisaillon, J. (1996). Biosurfactant production by a soil pseudomonas strain growing on polycyclic aromatic hydrocarbons. Appl. Environ. Microbiol. *62*, 1908–1912.
Di Lorenzo, A., Varcamonti, M., Parascandola, P., Vignola, R., Bernardi, A., Sacceddu, P., Sisto, R., and de Alteriis, E. (2005). Characterization and performance of a toluene-degrading biofilm developed on pumice stones. Microbial Cell Factories *4*, 4.
Dutta, L., Nuttall, H.E., Cunningham, A., James, G., and Hiebert, R. (2005). In situ biofilm barriers: case study of a nitrate groundwater plume, Albuquerque, New Mexico. Remediation J. *15*, 101–111.
Eberhardt, C., and Grathwohl, P. (2002). Time scales of organic contaminant dissolution from complex source zones: coal tar pools vs. blobs. J. Contaminant Hydrol. *59*, 45–66.
Eriksson, M., Dalhammar, G., and Mohn, W.W. (2002). Bacterial growth and biofilm production on pyrene. FEMS Microbiol. Ecol. *40*, 21–27.
Ferreira, L., Cobas, M., Tavares, T., Sanromán, M., and Pazos, M. (2013). Assessment of Arthrobacter viscosus as reactive medium for forming permeable reactive biobarrier applied to PAHs remediation. Environ. Sci. Pollut. Res. *20*, 7348–7354.
Fonseca, B., Pazos, M., Tavares, T., and Sanromán, M. (2012). Removal of hexavalent chromium of contaminated soil by coupling electrokinetic remediation and permeable reactive biobarriers. Environ. Sci. Pollut. Res. *19*, 1800–1808.
Frølund, B., Palmgren, R., Keiding, K., and Nielsen, P.H. (1996). Extraction of extracellular polymers from activated sludge using a cation exchange resin. Water Res. *30*, 1749–1758.
Gadd, G.M. (2001). Fungi in Bioremediation (Cambridge University Press, Cambridge, UK).
Gadd, G.M. (2010). Metals, minerals and microbes: geomicrobiology and bioremediation. Microbiology *156*, 609–643.
Garbarini, D.R., and Lion, L.W. (1986). Influence of the nature of soil organics on the sorption of toluene and trichloroethylene. Environ. Sci. Technol. *20*, 1263–1269.
Gentry, T., Rensing, C., and Pepper, I. (2004). New approaches for bioaugmentation as a remediation technology. Crit. Rev. Environ. Sci. Technol. *34*, 447–494.
Gerlach, R., Cunningham, A.B., and Caccavo, F. (2000). Dissimilatory iron-reducing bacteria can influence the reduction of carbon tetrachloride by iron metal. Environ. Sci. Technol. *34*, 2461–2464.
Gregory, K.B., Mason, M.G., Picken, H.D., Weathers, L.J., and Parkin, G.F. (2000). Bioaugmentation of Fe (0) for the remediation of chlorinated aliphatic hydrocarbons. Environ. Eng. Sci. *17*, 169–181.
Gu, B., Phelps, T., Liang, L., Dickey, M., Roh, Y., Kinsall, B., Palumbo, A., and Jacobs, G. (1999). Biogeochemical dynamics in zero-valent iron columns: Implications for permeable reactive barriers. Environ. Sci. Technol. *33*, 2170–2177.
Gu, B., Watson, D.B., Wu, L., Phillips, D.H., White, D.C., and Zhou, J. (2002). Microbiological characteristics in a zero-valent iron reactive barrier. Environ. Monitor. Assess. *77*, 293–309.
Guerin, W.F., and Boyd, S.A. (1992). Differential bioavailability of soil-sorbed naphthalene to two bacterial species. Appl. Environ. Microbiol. *58*, 1142–1152.
Guerin, W.F., and Jones, G.E. (1988). Mineralization of phenanthrene by a *Mycobacterium* sp. Appl. Environ. Microbiol. *54*, 937–944.
Hans-Cur, F. (1995). Sorption sites in biofilms. Water Sci. Technol. *32*, 27–33.
Headley, J.V., Gandrass, J., Kuballa, J., Peru, K.M., and Gong, Y. (1998). Rates of sorption and partitioning of contaminants in river biofilm. Environ. Sci. Technol. *32*, 3968–3973.
Henderson, A.D., and Demond, A.H. (2007). Long-term performance of zero-valent iron permeable reactive barriers: a critical review. Environ. Eng. Sci. *24*, 401–423.
Henderson, A.D., and Demond, A.H. (2011). Impact of solids formation and gas production on the permeability of ZVI PRBs. J. Environ. Eng. *137*, 689–696.
Herrera-González, V.E., Ruiz-Ordaz, N., Galíndez-Mayer, J., Juárez-Ramírez, C., Santoyo-Tepole, F., and Montiel, E.M. (2013). Biodegradation of the herbicide propanil, and its 3, 4-dichloroaniline by-product in a continuously operated biofilm reactor. World J. Microbiol. Biotechnol. *29*, 467–474.
Hiebert, R., Sharp, R., Cunningham, A., and James, G. (2001). Development and demonstration of subsurface biofilm barriers using starved bacterial cultures. Contam. Soil Sed. Water, August 45–47.

Holmén, B.A., and Gschwend, P.M. (1996). Estimating sorption rates of hydrophobic organic compounds in iron oxide-and aluminosilicate clay-coated aquifer sands. Environ. Sci. Technol. *31*, 105–113.

Jahn, A., and Nielsen, P.H. (1998). Cell biomass and exopolymer composition in sewer biofilms. Water Sci. Technol. 37, 17–24.

Jia, C., Li, P., Li, X., Tai, P., Liu, W., and Gong, Z. (2011). Degradation of pyrene in soils by extracellular polymeric substances (EPS) extracted from liquid cultures. Process Biochem. *46*, 1627–1631.

Johnsen, A.R., Wick, L.Y., and Harms, H. (2005). Principles of microbial PAH-degradation in soil. Environ. Pollut. *133*, 71–84.

Kao, C.-M., and Borden, R.C. (1997). Enhanced TEX biodegradation in nutrient briquet–peat barrier system. J. Environ. Eng. *123*, 18–24.

Kao, C., and Lei, S. (2000). Using a peat biobarrier to remediate PCE/TCE contaminated aquifers. Water Res. *34*, 835–845.

Kao, C., Chen, S., and Liu, J. (2001). Development of a biobarrier for the remediation of PCE-contaminated aquifer. Chemosphere *43*, 1071–1078.

Kästner, M., and Mahro, B. (1996). Microbial degradation of polycyclic aromatic hydrocarbons in soils affected by the organic matrix of compost. Appl. Microbiol. Biotechnol *44*, 668–675.

Kim, G., Lee, S., and Kim, Y. (2006). Subsurface biobarrier formation by microorganism injection for contaminant plume control. J. Biosci. Bioeng. *101*, 142–148.

Komlos, J., Cunningham, A.B., Camper, A.K., and Sharp, R.R. (2004). Biofilm barriers to contain and degrade dissolved trichloroethylene. Environ. Progr. *23*, 69–77.

Lamarche, P., Lauzon, F., Tetreault, M., and Barker, J. (2001). Biodegradation of a naphthalene plume in a Funnel-and-Gate system. Sixth International In Situ and On Site Bioremediation Symposium (San Diego, CA, USA).

Lampron, K., Chiu, P., and Cha, D. (2001). Reductive dehalogenation of chlorinated ethenes with elemental iron: the role of microorganisms. Water Res. *35*, 3077–3084.

Lappin-Scott, H., and Costerton, J. (1990). Starvation and penetration of bacteria in soils and rocks. Experientia *46*, 807–812.

Lazarova, V., and Manem, J. (1995). Biofilm characterization and activity analysis in water and wastewater treatment. Water Res. *29*, 2227–2245.

Lee, J.-Y., Moon, C.-H., Kim, J.-H., and Oh, B.-T. (2007). Feasibility study of the bio-barrier with biologically-active tire rubbers for treating chlorinated hydrocarbons. Geosciences J. *11*, 131–136.

Lee, J.-Y., Youm, S.Y., Choi, S.-I., and Oh, B.-T. (2009). Removal of mixed contaminants by Fe0-based biobarrier in flow-through columns using recycled waste materials. J. Mater. Cycles Waste Manage. *11*, 214–221.

Leeson, A. (2001). Bioaugmentation, biobarriers, and biogeochemistry. In The Sixth International *In Situ* and On-site Bioremediation Symposium: San Diego, California, 4-7 June 2001 (Battelle Press, Columbus, OH, USA).

Li, Q., and Logan, B.E. (1999). Enhancing bacterial transport for bioaugmentation of aquifers using low ionic strength solutions and surfactants. Water Res. *33*, 1090–1100.

Liang, L., West, O., Korte, N., Goodlaxson, J., Pickering, D., Zutman, J., Anderson, F., Welch, C., Pelfrey, M., and Dickey, M. (1997). A field-scale test of trichloroethylene dechlorination using iron filings for the X-120/X-749 groundwater plume. ORNL/TM-13217.

Liang, L., Korte, N., Gu, B., Puls, R., and Reeter, C. (2000). Geochemical and microbial reactions affecting the long-term performance of *in situ* 'iron barriers'. Adv. Environ. Res. *4*, 273–286.

Liu, S.-J., Zhao, Z.-Y., Li, J., Wang, J., and Qi, Y. (2013). An anaerobic two-layer permeable reactive biobarrier for the remediation of nitrate-contaminated groundwater. Water Res. *47*, 5977–5985.

Liu, W.-J., Qian, T.-T., and Jiang, H. (2014). Bimetallic Fe nanoparticles: Recent advances in synthesis and application in catalytic elimination of environmental pollutants. Chem. Eng. J. *236*, 448–463.

Liu, Y., Lam, M., and Fang, H. (2001). Adsorption of heavy by EPS of activated sludge. Water Sci. Technol. *43*, 59–66.

Lowry, G.V., and Reinhard, M. (2000). Pd-catalyzed TCE dechlorination in groundwater: solute effects, biological control, and oxidative catalyst regeneration. Environ. Sci. Technol. *34*, 3217–3223.

Ludwig, R.D., McGregor, R.G., Blowes, D.W., Benner, S.G., and Mountjoy, K. (2002). A permeable reactive barrier for treatment of heavy metals. Groundwater *40*, 59–66.

Mann, V. (2012). Laboratory Studies of Biobarrier Technology in Fractured Rock. PhD Thesis, Queen's University, Kingston, ON, Canada.

Meng, F., Su, G., Hu, Y., Lu, H., Huang, L.-N., and Chen, G.-H. (2014). Improving nitrogen removal in an ANAMMOX reactor using a permeable reactive biobarrier. Water Res. *58*, 82–91.

Michalsen, M.M., Peacock, A.D., Smithgal, A.N., White, D.C., Spain, A.M., Sanchez-Rosario, Y., Krumholz, L.R., Kelly, S.D., Kemner, K.M., and McKinley, J. (2009). Treatment of nitric acid-, U (VI)-, and Tc (VII)-contaminated groundwater in intermediate-scale physical models of an *in situ* biobarrier. Environ. Sci. Technol. *43*, 1952–1961.

Miller, K.D., Johnson, P.C., and Bruce, C.L. (2001). Full-scale in-situ biobarrier demonstration for containment and treatment of MTBE. Remediation J. *12*, 25–36.

Moon, J.-W., Goltz, M.N., Ahn, K.-H., and Park, J.-W. (2003). Dissolved organic matter effects on the performance of a barrier to polycyclic aromatic hydrocarbon transport by groundwater. J. Contaminant Hydrol. *60*, 307–326.

Morris, K.A. (2009). Applications of mulch biowalls—three case studies. Remediation J. *20*, 59–68.

Muñoz, R., Köllner, C., and Guieysse, B. (2009). Biofilm photobioreactors for the treatment of industrial wastewaters. J. Hazard. Mater. *161*, 29–34.

National Research Council (1994). Alternatives for Ground Water Cleanup (National Academies Press, Washington, DC, USA).

Nielsen, P.H., and Jahn, A. (1999). Extraction of EPS. In Microbial Extracellular Polymeric Substances, Wingender, J., Neu, T.R., and Flemming, H.-C., eds. (Springer-Verlag, Berlin, Germany), pp. 49–72.

O'Hannesin, S.F., and Gillham, R.W. (1998). Long-term performance of an in situ 'iron wall' for remediation of VOCs. Groundwater *36,* 164–170.

Obiri-Nyarko, F., Grajales-Mesa, S.J., and Malina, G. (2014). An overview of permeable reactive barriers for *in situ* sustainable groundwater remediation. Chemosphere *111,* 243–259.

Ortega-Calvo, J.-J., and Saiz-Jimenez, C. (1998). Effect of humic fractions and clay on biodegradation of phenanthrene by a Pseudomonas fluorescens strain isolated from soil. Appl. Environ. Microbiol. *64,* 3123–3126.

Öztürk, Z., Tansel, B., Katsenovich, Y., Sukop, M., and Laha, S. (2012). Highly organic natural media as permeable reactive barriers: TCE partitioning and anaerobic degradation profile in eucalyptus mulch and compost. Chemosphere *89,* 665–671.

Pan, X., Liu, J., and Zhang, D. (2010). Binding of phenanthrene to extracellular polymeric substances (EPS) from aerobic activated sludge: A fluorescence study. Colloid. Surface. B *80,* 103–106.

Pandey, G., and Jain, R.K. (2002). Bacterial chemotaxis toward environmental pollutants: role in bioremediation. Appl. Environ. Microbiol. *68,* 5789–5795.

Parales, R.E., Ditty, J.L., and Harwood, C.S. (2000). Toluene-degrading bacteria are chemotactic towards the environmental pollutants benzene, toluene, and trichloroethylene. Appl. Environ. Microbiol. *66,* 4098–4104.

Park, Y.-D., Park, C.-S., and Park, J.-W. (2010). Interaction between iron reducing bacteria and nano-scale zero valent iron. J. Environ. Eng. Manage. *20,* 233–238.

Pesce, S., Morin, S., Lissalde, S., Montuelle, B., and Mazzella, N. (2011). Combining polar organic chemical integrative samplers (POCIS) with toxicity testing to evaluate pesticide mixture effects on natural phototrophic biofilms. Environ. Pollut. *159,* 735–741.

Phillips, D., Nooten, T.V., Bastiaens, L., Russell, M., Dickson, K., Plant, S., Ahad, J., Newton, T., Elliot, T., and Kalin, R. (2010). Ten year performance evaluation of a field-scale zero-valent iron permeable reactive barrier installed to remediate trichloroethene contaminated groundwater. Environ. Sci. Technol. *44,* 3861–3869.

Pintelon, T.R., Picioreanu, C., van Loosdrecht, M., and Johns, M.L. (2012). The effect of biofilm permeability on bio-clogging of porous media. Biotechnol. Bioeng. *109,* 1031–1042.

Powell, R.M., Puls, R.W., Blowes, D., Gillham, R., and Schultz, D. (1998). Permeable reactive barrier technologies for contaminant remediation. NASA, United States Environmnetal Protection Agency.

Puglisi, E., Cappa, F., Fragoulis, G., Trevisan, M., and Del Re, A.A. (2007). Bioavailability and degradation of phenanthrene in compost amended soils. Chemosphere *67,* 548–556.

Puls, R.W., Blowes, D.W., and Gillham, R.W. (1999). Long-term performance monitoring for a permeable reactive barrier at the US Coast Guard Support Center, Elizabeth City, North Carolina. J. Hazard. Mater. *68,* 109–124.

Quintelas, C., Sousa, E., Silva, F., Neto, S., and Tavares, T. (2006). Competitive biosorption of ortho-cresol, phenol, chlorophenol and chromium (VI) from aqueous solution by a bacterial biofilm supported on granular activated carbon. Process Biochem. *41,* 2087–2091.

Richardson, J.P., and Nicklow, J.W. (2002). In situ permeable reactive barriers for groundwater contamination. Soil Sediment Contamination *11,* 241–268.

Rodrigues, A.C., Wuertz, S., Brito, A.G., and Melo, L.F. (2005). Fluorene and phenanthrene uptake by Pseudomonas putida ATCC 17514: kinetics and physiological aspects. Biotechnol. Bioeng. *90,* 281–289.

Salanitro, J., Spinnler, G., Maner, P., Tharpe, D., Pickle, D., Wisniewski, H., Johnson, P., and Bruce, C. (2001). In situ bioremediation of MTBE using biobarriers of single or mixed cultures. Sixth International In Situ and On Site Bioremediation Symposium (San Diego, CA, USA).

Sandoval-Carrasco, C.A., Ahuatzi-Chacón, D., Galíndez-Mayer, J., Ruiz-Ordaz, N., Juárez-Ramírez, C., and Martínez-Jerónimo, F. (2013). Biodegradation of a mixture of the herbicides ametryn, and 2, 4-dichlorophenoxyacetic acid (2, 4-D) in a compartmentalized biofilm reactor. Bioresource Technol. *145,* 33–36.

Saponaro, S., Negri, M., Sezenna, E., Bonomo, L., and Sorlini, C. (2009). Groundwater remediation by an *in situ* biobarrier: a bench scale feasibility test for methyl tert-butyl ether and other gasoline compounds. J. Hazard. Mater. *167,* 545–552.

Scherer, M.M., Richter, S., Valentine, R.L., and Alvarez, P.J. (2000). Chemistry and microbiology of permeable reactive barriers for *in situ* groundwater clean up. Crit. Rev. Microbiol. *26,* 221–264.

Schipper, L., and Vojvodić-Vuković, M. (1998). Nitrate removal from groundwater using a denitrification wall amended with sawdust: field trial. J. Environ. Qual. *27,* 664–668.

Schorer, M., and Eisele, M. (1997). Accumulation of inorganic and organic pollutants by biofilms in the aquatic environment. Water Air Soil Pollut. *99,* 651–659.

Schwarzenbach, R.P., Gschwend, P.M., and Imboden, D.M. (2005). Environmental Organic Chemistry (John Wiley & Sons, Somerset, NJ, USA).

Seo, Y., and Bishop, P.L. (2007). Influence of nonionic surfactant on attached biofilm formation and phenanthrene bioavailability during simulated surfactant enhanced bioremediation. Environ. Sci. Technol. *41,* 7107–7113.

Seo, Y., and Bishop, P.L. (2008). The monitoring of biofilm formation in a mulch biowall barrier and its effect on performance. Chemosphere *70,* 480–488.

Seo, Y., Lee, W.-H., Sorial, G., and Bishop, P.L. (2009). The application of a mulch biofilm barrier for surfactant enhanced polycyclic aromatic hydrocarbon bioremediation. Environ. Pollut. *157,* 95–101.

Shen, H., and Wilson, J.T. (2007). Trichloroethylene removal from groundwater in flow-through columns simulating a permeable reactive barrier constructed with plant mulch. Environ. Sci. Technol. *41,* 4077–4083.

Singh, R., Paul, D., and Jain, R.K. (2006). Biofilms: implications in bioremediation. Trends Microbiol. *14,* 389–397.

Singleton, I. (1994). Microbial metabolism of xenobiotics: fundamental and applied research. J. Chem. Technol. Biotechnol. *59,* 9–23.

Stelmack, P.L., Gray, M.R., and Pickard, M.A. (1999). Bacterial adhesion to soil contaminants in the presence of surfactants. Appl. Environ. Microbiol. *65*, 163–168.

Taylor, S.W., and Jaffé, P.R. (1990). Biofilm growth and the related changes in the physical properties of a porous medium: 3. Dispersivity and model verification. Water Resources Res. *26*, 2171–2180.

Teerakun, M., Reungsang, A., Lin, C.-J., and Liao, C.-H. (2011). Coupling of zero valent iron and biobarriers for remediation of trichloroethylene in groundwater. J. Environ. Sci. *23*, 560–567.

Thiruvenkatachari, R., Vigneswaran, S., and Naidu, R. (2008). Permeable reactive barrier for groundwater remediation. J. Ind. Eng. Chem. *14*, 145–156.

Tiehm, A. (1994). Degradation of polycyclic aromatic hydrocarbons in the presence of synthetic surfactants. Appl. Environ. Microbiol. *60*, 258–263.

Van Nooten, T., Springael, D., and Bastiaens, L. (2008). Positive impact of microorganisms on the performance of laboratory-scale permeable reactive iron barriers. Environ. Sci. Technol. *42*, 1680–1686.

Velimirovic, M., Chen, H., Simons, Q., and Bastiaens, L. (2012). Reactivity recovery of guar gum coupled mZVI by means of enzymatic breakdown and rinsing. J. Contaminant Hydrol. *142*, 1–10.

Vesela, L., Nemecek, J., Siglova, M., and Kubal, M. (2006). The biofiltration permeable reactive barrier: practical experience from Synthesia. Int. Biodeterioration Biodegradation *58*, 224–230.

Viamajala, S., Peyton, B.M., Gerlach, R., Sivaswamy, V., Apel, W.A., and Petersen, J.N. (2008). Permeable reactive biobarriers for *in situ* Cr (VI) reduction: bench scale tests using Cellulomonas sp. strain ES6. Biotechnol. Bioeng. *101*, 1150–1162.

Vogan, J., Focht, R., Clark, D., and Graham, S. (1999). Performance evaluation of a permeable reactive barrier for remediation of dissolved chlorinated solvents in groundwater. J. Hazard. Mater. *68*, 97–108.

Volkering, F., Breure, A., and Rulkens, W. (1997). Microbiological aspects of surfactant use for biological soil remediation. Biodegradation *8*, 401–417.

Wang, S.-M., and Tseng, S.-K. (2009). Dechlorination of trichloroethylene by immobilized autotrophic hydrogen-bacteria and zero-valent iron. J. Biosci. Bioeng. *107*, 287–292.

Wang, W., Wang, W., Zhang, X., and Wang, D. (2002). Adsorption of p-chlorophenol by biofilm components. Water Res. *36*, 551–560.

Wang, Z., Hessler, C.M., Xue, Z., and Seo, Y. (2012). The role of extracellular polymeric substances on the sorption of natural organic matter. Water Res. *46*, 1052–1060.

Waybrant, K., Blowes, D., and Ptacek, C. (1998). Selection of reactive mixtures for use in permeable reactive walls for treatment of mine drainage. Environ. Sci. Technol. *32*, 1972–1979.

Weathers, L.J., Parkin, G.F., and Alvarez, P.J. (1997). Utilization of cathodic hydrogen as electron donor for chloroform cometabolism by a mixed, methanogenic culture. Environ. Sci. Technol. *31*, 880–885.

Wei, X., Fang, L., Cai, P., Huang, Q., Chen, H., Liang, W., and Rong, X. (2011). Influence of extracellular polymeric substances (EPS) on Cd adsorption by bacteria. Environ. Pollut. *159*, 1369–1374.

Wei, Z., and Seo, Y. (2010). Trichloroethylene (TCE) adsorption using sustainable organic mulch. Journal of hazardous materials *181*, 147–153.

Whang, L.-M., Liu, P.-W.G., Ma, C.-C., and Cheng, S.-S. (2008). Application of biosurfactants, rhamnolipid, and surfactin, for enhanced biodegradation of diesel-contaminated water and soil. J. Hazard. Mater. *151*, 155–163.

Wick, L.Y., Pasche, N., Bernasconi, S.M., Pelz, O., and Harms, H. (2003). Characterization of multiple-substrate utilization by anthracene-degrading Mycobacterium frederiksbergense LB501T. Appl. Environ. Microbiol. *69*, 6133–6142.

Wilkin, R.T., Puls, R.W., and Sewell, G.W. (2003). Long-term performance of permeable reactive barriers using zero-valent iron: Geochemical and microbiological effects. Groundwater *41*, 493–503.

Wingender, J., Neu, T.R., and Flemming, H.-C. (1999). Microbial Extracellular Polymeric Substances: Characterization, Structure, and Function (Springer-Verlag, Berlin, Germany).

Wolfaardt, G., Lawrence, J., Robarts, R., and Caldwell, D. (1995). Bioaccumulation of the herbicide diclofop in extracellular polymers and its utilization by a biofilm community during starvation. Appl. Environ. Microbiol. *61*, 152–158.

Wrenn, B. (2004). Enhanced reductive dechlorination through biological interaction with zero-valent iron. Federal Remediation Technologies Roundtable Meeting, Arlington, Virginia, USA.

Writer, J.H., Ryan, J.N., and Barber, L.B. (2011). Role of biofilms in sorptive removal of steroidal hormones and 4-nonylphenol compounds from streams. Environ. Sci. Technol. *45*, 7275–7283.

Wuertz, S., Bishop, P.L., and Wilderer, P.A. (2003). Biofilms in Wastewater Treatment: An Interdisciplinary Approach (IWA Publishing, London, UK).

Xue, Z., and Seo, Y. (2013). Impact of chlorine disinfection on redistribution of cell clusters from biofilms. Environ. Sci. Technol. *47*, 1365–1372.

Xue, Z., Sendamangalam, V.R., Gruden, C.L., and Seo, Y. (2012). Multiple roles of extracellular polymeric substances on resistance of biofilm and detached clusters. Environ. Science Technol. *46*, 13212–13219.

Xue, Z., Hessler, C.M., Panmanee, W., Hassett, D.J., and Seo, Y. (2013). inactivation mechanism is affected by capsular extracellular polymeric substances reactivity with chlorine and monochloramine. FEMS Microbiol. Ecol. *83*, 101–111.

Xue, Z., Lee, W.H., Coburn, K.M., and Seo, Y. (2014). Selective reactivity of monochloramine with extracellular matrix components affects the disinfection of biofilm and detached clusters. Environ. Sci. Technol. *48*, 3832–3839.

Yerushalmi, L., Manuel, M.F., and Guiot, S.R. (1999). Biodegradation of gasoline and BTEX in a microaerophilic biobarrier. Biodegradation *10*, 341–352.

# Comparison of the Degradation Activity of Biofilm-associated versus Planktonic Cells 12

Masaaki Morikawa and Kenji Washio

## Abstract

Biofilm formation is a common microbial strategy to construct and maintain a favourable niche in stressful natural environments. Consequently, the application of biofilms in bioproduction and bioremediation processes should be a natural and rational choice (Morikawa, 2006). In this chapter, we describe two examples that demonstrate the usefulness of biofilm-associated cells in bioremediation, and provide basic mechanisms that should advance environmental biotechnologies. First, *Pseudoalteromonas* strains were shown to be dominant biofilm forming bacteria in fish-farming fields and some produced extracellular proteases in a biofilm-dependent manner. The results suggest that *Pseudoalteromonas* bacteria living in the biofilm community, contribute in part to remove excess proteinaceous material from the sediment sludge and maintain good water quality in fish farms. Second, biofilm-associated cells of *Pseudomonas stutzeri* T102 did not degrade naphthalene during initial hours of incubation but later they degraded it even faster than planktonic cells. This delayed but elevated degradation activity of the biofilm-associated cells could be attributed to 'super-activated cells' as they detach from biofilms. T102 biofilm-associated cells were capable of surviving for 10 weeks in natural petroleum-contaminated soils; by then T102 planktonic cells were mostly extinct. Naphthalene degradation activity in the soils that had been inoculated with T102 biofilms was indeed higher than that observed in soils inoculated with T102 planktonic cells. It would appear that biofilms may provide a constant stream of super-active contaminant degrading bacteria which could may be used in bioremediation applications.

## Efficacy of biofilm formation by protease producing bacteria and their role in maintaining water quality in marine fish farms

Biofilms are structured microbial communities formed on the surface of solid materials or interfaces (Watnick and Kolter, 2000). It has been pointed out that environmental microorganisms exist predominantly as biofilms and gain high tolerance to physical, chemical, and biological stresses (Gorbushina and Broughton, 2009). Decreases in cell growth rate and metabolic activities are one of the key attributes observed in biofilm-associated bacteria. Biofilm-associated cells exhibit specific gene expression profiles, often controlled by complex quorum sensing systems, or dormancy (Gilbert *et al.*, 1990; Donlan and Costerton, 2002; Rani *et al.*, 2007; Dickschat, 2010). There are reports that biofilm-associated cells are more dormant and inactive than planktonic cells (Werner *et al.*, 2004; Lewis, 2007). Our previous understanding was that the biofilm-associated cells should not actively produce large quantities of extracellular enzymes, except for those responsible for biofilm dispersal (Kaplan, 2010).

Fish farming is considered an economically feasible and perhaps environmentally friendly approach to save natural marine resources and there is considerable interest in exploring smart technologies to enhance the productivity and minimize the environmental impacts of this industry (van Rijn, 1996). One of the bottlenecks for successful fish farming is raising the survival rates of juvenile fish. Fish fry are susceptible to various diseases and sensitive to environmental changes. For this reason, management of their improved viability is important to increase the final yield of fish production.

However, deteriorations in water quality, caused by leftover food and fish faeces are a ubiquitous problem in fish farming. High concentrations of organic matter from these substrates cause the depletion of oxygen by the proliferation of aerobic microorganisms, emission of sulfide and methane by anaerobic bacteria, and may cause outbreaks of pathogenic species (Verschuere *et al.*, 2000). In this regard, quick removal of excess nutrients from fish farm sediments is very important to maintain the health of juvenile fish.

In marine environments, biofilms are ubiquitous on all aquatic surfaces and the microbial communities within them may exhibit a different microbial community structure as compared to organisms present within the water column (Patel *et al.*, 2003). There are empirical reports that survival rates of fish fry, such as *Epinephelus bruneus* and *Takifugu rubripes*, are increased when powders of foraminiferous limestone are submerged into the fish farm or tank (personal communication; Minami-izu sea-farming centre, Shizuoka, Japan and Momoshima sea-farming centre, Hiroshima, Japan). The exact mode of these treatments is unclear; however, we often observed that the sediments of fish farms were covered with thick biofilms. We hypothesized that the biofilm-associated microorganisms contribute to purify the water and examined this possibility.

## *Pseudoalteromonas* bacteria produce extracellular protease in a biofilm-dependent manner

Natural biofilm samples were collected from the sediments of two fish-farming centres in Japan and further developed under laboratory conditions. It was revealed that *Pseudoalteromonas* strains such as SB-A1, -A2, -B1, -B2, -E1, -G1, and -G2 were predominant in the sediment biofilms (Fig. 12.1; Iijima *et al.*, 2009). We expected that some of these biofilm-associated bacteria would have high proteolytic activities that would be useful for the swift degradation of proteinaceous substances present in bottom sediments. It has been reported that *Pseudoalteromonas* are generally found in association with marine eukaryotes and display anti-bacterial, bacteriolytic, agarolytic and algicidal activities (Holmström and Kjelleberg, 1999). Of particular interest, it was also shown that *Pseudoalteromonas undina* strain VKM-124 had static activities against pathogenic *Vibrio* sp. and potential applications of this strain for use in aquaculture systems have been considered (Maeda *et al.*, 1997; Tubiash *et al.*, 1970).

To determine whether isolated *Pseudoalteromonas* strains produce extracellular proteases, culture supernatants were spotted onto peptonized agar medium containing 1% casein and assayed for protease production by observing halo formation around the colonies (Fig. 12.2). Samples were prepared from bacterial cultures grown under shaking (representative of planktonic growth) and standing (preferred by biofilm formation) conditions. Caseinolytic (casein hydrolysing) halos were observed around the spots of SB-A1, -B1, -E1, and -G1, demonstrating that these strains were active producers of extracellular proteases. Interestingly, their protease productivity was higher in standing than in shaking culture conditions. In contrast to

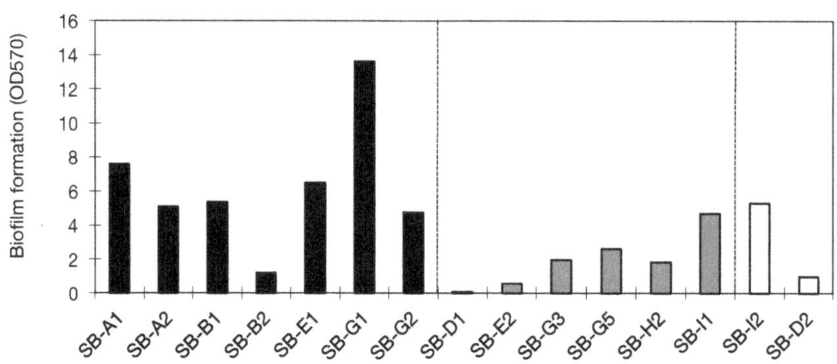


**Figure 12.1** Biofilm formation by bacteria isolated from fish farm sediments. Biofilm formation was calculated by the crystal violet staining method. Symbols are black bar, *Pseudoalteromonas* sp., Grey bar, *Vibrio* sp., and White bar, *Halomonas* sp.

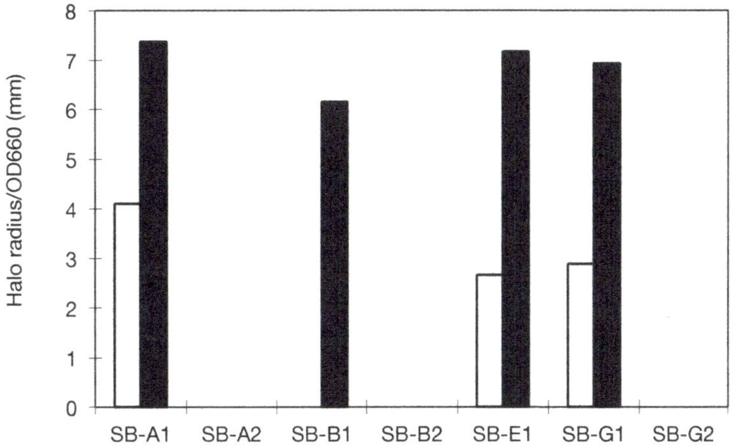


**Figure 12.2** Specific protease activities in the culture supernatants (5.0 ml) of *Pseudoalteromonas* strains grown at shaking or standing condition (30°C, 2 days). The radius of caseinolytic halos were measured and each value divided by the OD600nm value of the culture. Symbols are white bar, shaking culture, and black bar, standing culture.

moderate inductions of protease activities during standing cultures of SB-A1, -E1, and -G1 (ca. 2-fold inductions), SB-B1 strictly and significantly produced proteases only when incubated under standing culture, and not shaking culture conditions. For further analysis of extracellular proteases produced by SB-B1, we used zymograph analysis. As shown in Fig. 12.3, protein samples prepared from the culture supernatants of SB-B1 showed gelatinolytic activity. The intensity of the band was much lower for cells grown under shaking conditions; whereas, it was visible when SB-B1 were grown as standing cells. These results suggest that biofilm formation in SB-B1 appears to trigger the production of a protease with molecular mass of 38 kDa.

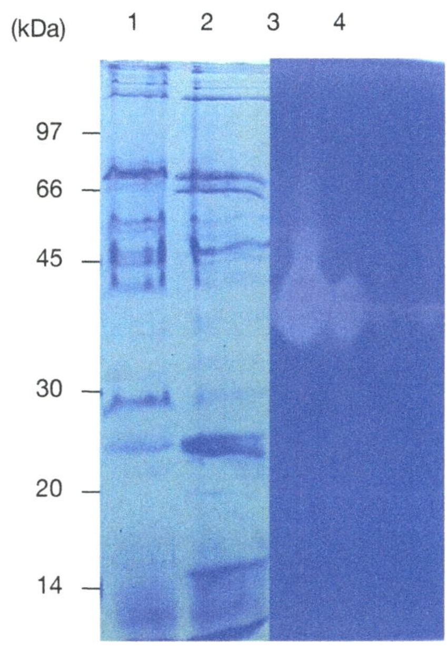


**Figure 12.3** Protein samples prepared from culture supernatants of strain SB-B1 grown under shaking (lanes 1 and 3) or standing (lanes 2 and 4) conditions (30°C, 2 days) were separated on a 12.5% SDS-PAGE gel (left panel). A gelatinolytic band in a zymograph gel indicates protease activity (38 kDa, right panel).

## Identification of biofilm-dependent metalloprotease I from *Pseudoalteromonas* sp. SB-B1

We obtained a genome fragment encoding a metalloprotease from SB-B1, designated as BmpI (biofilm metalloprotease I, 38 kDa). The open reading frame of *bmpI* gene was 2184 bp with a GTG start codon and encoded a protein of 727 amino acids. Several extracellular proteases were previously documented in *Pseudoalteromonas* strains (Xiong *et al.*, 2007), where biochemical properties of the SB-B1 protease were well matched with a monomeric type of metalloprotease, such as the protease CP1 of *Pseudoalteromonas* sp. CP76 (38 kDa) (Sánchez-Porro *et al.*, 2003), EmpI of *Pseudoalteromonas* sp. A28 (38 kDa) (Lee *et al.*, 2002) and MCP-02 of *Pseudoalteromonas* sp. SM9913 (36 kDa) (Chen *et al.*, 2003). The amino acid sequence of BmpI indeed shared high similarity with those of the protease MCP-02 of *Pseudoalteromonas* sp. SM9913 (96% identity), EmpI of *Pseudoalteromonas* sp. A28 (80% identity), vibriolysin of *Vibrio parahaemolyticus* AQ3810 (51% identity), and LasB elastase of *Pseudomonas aeruginosa* PAO1 (50% identity). LasB esterase has been reported to be responsible for formation of the 3D structure of biofilms (Tielen *et al.*, 2010).

Standing cultures of SB-B1 cells exhibited the same exponential growth rate but reached a final cell density which was less than that of shaking cultured cells (Fig. 12.4A). When an aliquot of filter-purified culture supernatant was spotted on a casein-containing agar plate, caseinolytic clear zones were observed at later stage of standing cultivation (48 h). However, it was not detectable with shaking cultivation, even at culture densities higher than obtained under the standing condition (48 h). It is suggested that growth stages as well as other unknown factors control protease productions in SB-B1. To examine how expression of *bmpI* is regulated in SB-B1 cells, total RNAs were extracted at the different growth stages and subjected to RT-PCR analysis using *bmpl* specific primers (Fig. 12.4B). The *bmpI* mRNA was abundant at the exponential phase of shaking condition and declined into the stationary phase. During standing cultivation, transcript levels of *bmpI* increased after the onset into

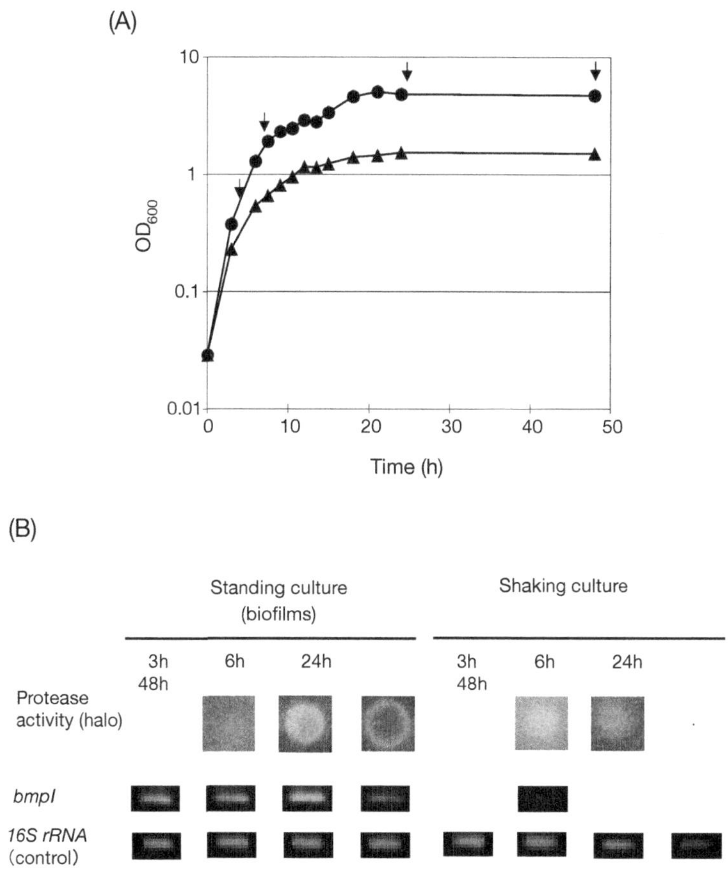


**Figure 12.4** Growth phase and protease production by *Pseudoalteromonas* sp. SB-B1. (A) Growth curves of strain SB-B1 grown at shaking or standing conditions. Cultivated cells were dispersed by vortex mixing and the optical densities (OD600) in a peptonized medium determined throughout the growth cycle at 30°C. The experiment was repeated three times and average values are shown. Symbols are circle, shaking culture; triangle, standing culture. Arrows indicate the time point of sampling for RT-PCR analysis. (B) Protease production in filter-purified culture supernatants and RT-PCR analysis of the bmpl gene at the different stages of SB-B1 growth cultivated under shaking or standing conditions. The PCR products in a 1.5% agarose gel were visualized by ethidium bromide staining. The samples of 16S rRNA gene fragments are included to confirm equal sample loading onto each lane.

the stationary phase and were maximal after 2 days of incubation. The continuous appearance of *bmpI* mRNA seemed to link with higher protease activities observed in the standing culture supernatants. This suggests that the protease activity stimulated by standing cultivation is, at least in part, attributed to the biofilm-dependent expression of the *bmpI* gene.

SB-B1 is phylogenetically related to *Pseudoalteromonas* sp. SM9913, which is a cold-adapted strain isolated from deep-sea sediment. SM9913 produces two extracellular proteases, named MCP-01 and MCP-02 (Chen *et al.*, 2003). MCP-01 is a novel type of serine protease in the subtilase subfamily (Chen *et al.*, 2007). MCP-02 is a mesophilic metalloprotease that is homologous to BmpI. At the optimum temperature (12°C), SM9913 produced much more MCP-01 than MCP-02. Purified MCP-01 showed psychrophilic properties and was considered to be adapted to the cold environment, whereas the role of MCP-02 remained unknown. As previously described, the production of BmpI was induced by growth conditions suitable for biofilm formation. SM9913 and SB-B1 are commonly isolated from sea sediments, where both strains may reside as biofilms. This suggests that BmpI-like proteases may act on the community formation of multi-species biofilms, determining their structure, maintenance and dispersal. Although regulatory processes leading to BmpI production have not been elucidated, SB-B1 must recruit BmpI during biofilm formation to enable the biofilm cells to utilize the surrounding protein efficiently. This, in turn may contribute in part to the rapid removal of excess proteinaceous matter in the sediment sludge of fish farms. Practical applications of biofilm technology in aquaculture systems have been discussed in terms of their control of deleterious organisms or to serve as biofilters to absorb and biodegrade excess nutrients (Qian *et al.*, 2007).

## Efficacy of biofilm formation by naphthalene-degrading bacteria for bioremediation

Among the strategies to clean up pollutants it is widely recognized that biological treatments, so called bioremediation technologies, have the advantage over chemical and physical treatments in terms of their compatibility with global cycles, thus minimizing environmental impacts (Alexander, 1999). Active on-site bioremediation includes technologies that activate indigenous microbial populations, *biostimulation*, or introduce specific competent foreign bacteria to the contaminated site, *bioaugmentation*. Although bioaugmentation is expected to be a quicker and more effective technology than biostimulation, low fitness and poor colonization of the introduced bacteria to the contaminated sites often make efficacy of bioaugmentation poor (Bouchez *et al.*, 2000; El-Fantroussi and Agathos, 2005). The activity and viability of the introduced bacteria often decreases at the contaminated sites when compared with those observed under laboratory conditions, probably due to diverse environmental stresses (van Veen *et al.*, 1997; Thompson *et al.*, 2005). These include predation by protozoa, competition with other bacteria, unfavourable pH and temperature conditions, and poor availability of nutrients and oxygen. Catabolite repression by other organic compounds also decreases the expression of degradation genes. However, bioaugmentation can become more effective if both microbial ecology and population sizes are taken into account (Dejonghe *et al.*, 2001; Silva *et al.*, 2004; El-Fantroussi and Agathos, 2005). One of the options for effective clean-up methods of contaminated sites is the use of carrier materials so as to maintain sufficient activity of inoculants during prolonged periods. There are several reports demonstrating that immobilized or encapsulated cells effectively degrade pollutants at contaminated sites (Moslemy *et al.*, 2002; Mrozik and Piotrowska-Seget, 2010). However, the cellular and molecular mechanisms underlying these technologies are largely unknown.

There are several reports that demonstrate the efficacy of biofilm formation by useful bacteria and their use in bioremediation and biotransformation technologies (Qureshi *et al.*, 2005; Singh *et al.*, 2006; Yamaga *et al.*, 2010). However, to our best knowledge there was no report that directly compares degradation activities and persistence in contaminated soils inoculated with single species biofilms or planktonic cells (Shimada *et al.*, 2012). When compared with artificially immobilized and encapsulated cells, the advantage of biofilms is that a high density and tolerance of the constituent cell is naturally achieved. Moreover, biofilms can be introduced to contaminated sites free of supports that might cause additional environmental pollution. The secondary purpose of the study was to examine the potential of model biofilms towards developing bioaugmentation technology.

*Pseudomonas stutzeri* are ubiquitous and useful environmental bacteria exhibiting high degradation abilities for harmful chemical compounds including biocides, halogenated alcohols, and hydrocarbons (Lalucat *et al.*, 2006). Among the tested laboratory strains, *P. stutzeri* T102, which was isolated from the sludge of an oil reservoir tank in Okinawa (Hirano *et al.*, 2004), formed biofilms on the surface of various materials including polypropyrene, polystyrene, polyethylene terephthalate, acrylic, and glass bottles. *P. stutzeri* T102 is capable of degrading PAHs, such as naphthalene and dibenzothiophene. When using a batch culture system for biofilm formation, the amount of T102 biofilms continued increasing even after 120 h (Fig. 12.5A; Shimada *et al.*, 2012). Thus, we chose T102 as a model bacterial strain for further experiments. Gross *et al.* (2007) tested 69 bacterial strains for biofilm forming capacity and showed that many of the best biofilm formers belonged to *Pseudomonas*. Our experimental result is consistent with their result. Since T102 is a facultative aerobic strain, it formed ring-shape biofilms near the air–liquid interface on the inner surface of glass bottles (Fig. 12.5B). However, they formed rather uniform biofilms in small plastic tubes, such as 1.5 ml polypropyrene tubes and 96-well polystyrene plates probably due to the small depth of the liquid phase, better aeration condition, reduced hydrophobicity and zeta potential of the tube materials (Fig. 12.5C). It may be worth to note that a major outer membrane protein, OmpA exhibited opposite effects on the biofilm formation in glass

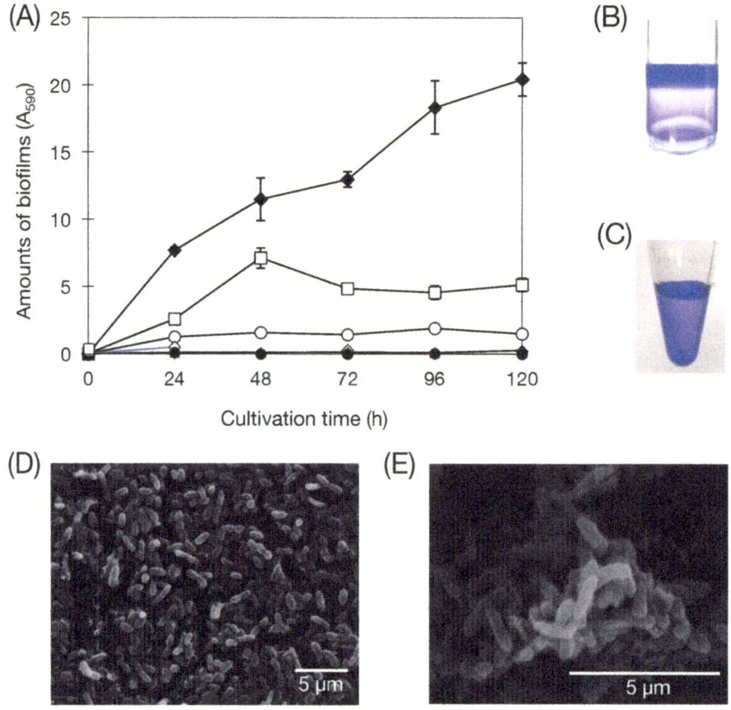


**Figure 12.5** Biofilm formation by naphthalene degrading *Pseudomonas stutzeri* T102 (Shimada *et al.*, 2012). (A) Biofilms were formed in 1.5 ml polypropyrene tubes containing 0.5 ml Y medium. Y medium contained (g/l) 2.0 g Bacto tryptone (Difco), 1.0 yeast extract and 2.0 NaCl, 0.25 $KH_2PO_4$, 0.25 $K_2HPO_4$, 0.1 Ca $(CH_3COO)_2{\cdot}H_2O$, and 0.1 Mg $(CH_3COO)_2{\cdot}4H_2O$. Data points are the averages of triplicate assays and error bars represent standard deviations. Symbols are as follows: Closed diamond, *P. stutzeri* T102 (30°C); open square, *Gordonia* sp. C3 (30°C); open circle, *Oleomonas sagaranensis* HD1 (30°C); open diamond, *Rhodococcus* sp. TMP2 (20°C); closed circle, *Arthrobacter* sp. CAB1 (30°C). (B) Crystal violet staining of *P. stutzeri* T102 biofilms in a 20 ml glass bottle; C. in an 1.5 ml polypropyrene test tube. D. and E. Scanning electron micrographs. All biofilms were formed at 30°C for 48 h.

bottles and plastic tubes where surface hydrophobicity is different (Ma and Wood, 2009). This suggests that each environmental bacterium has different surface condition requirements for optimal biofilm formation. Scanning electron microscopy demonstrated that biofilms of T102 (Fig. 12.5C) are also not flat but highly structured (Figs 12.5D and 5E).

## Naphthalene degradation by T102 biofilms and planktonic cells

There are reports that biofilm-associated cells are more dormant and inactive than planktonic cells (Werner *et al.*, 2004; Lewis, 2007; Rani *et al.*, 2007). This feature of biofilms partly explains their high resistance to environmental stresses (Gilbert *et al.*, 1990). It was thus a concern that T102 biofilms might exhibit less naphthalene degradation activity than planktonic cells. In order to examine this possibility, the activities of T102 biofilm and planktonic cells were compared in a pure liquid culture system (Fig. 12.6). Biofilms of T102 were formed in glass bottles containing Y-medium. After cultivation at 30°C for 24 h without shaking, free planktonic cells were carefully removed from the bottles and the biofilm formed on the inside walls (ca $2 \times 10^7$ CFUs) was rinsed once with sterile water. The biofilms were soaked in mineral medium (BM-medium) containing 20 ppm naphthalene and the bottles were tightly sealed with butyl rubber stoppers and incubated at 30°C without shaking. Planktonic cells were separately grown in a shaking flask. Mid-exponential phase cells were harvested by centrifugation and washed and suspended with a small amount of medium. BM-medium containing the planktonic cells at the same CFU value (ca $2 \times 10^7$ CFUs) as the biofilm samples were prepared in glass bottles with 20 ppm naphthalene added. The bottles were sealed and kept at 30°C as described previously. All the samples were prepared in triplicate for each sampling time and a negative control included that contained no cells. Sample bottles were taken every 4 h until 16 h and then every 8 h until 48 h. Remaining amounts of naphthalene in the samples were analysed by gas chromatography.

Experimental results revealed that T102 biofilms did not degrade naphthalene during the first 4 h but after that time they degraded naphthalene faster than

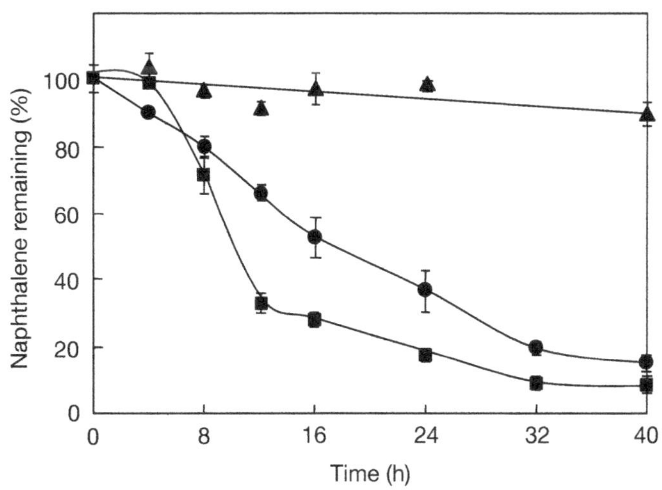


**Figure 12.6** Comparison of naphthalene degradation activity of T102 biofilm and planktonic cell cultures. Data points are the average of triplicate assays; error bars represent standard deviations. BM-medium is a minimal medium used for naphthalene degradation experiments of *P. stutzeri* T102, that contained (g/l) 1 $(NH_4)_2SO_4$, 24.0 $Na_2HPO_4 \cdot 12H_2O$, 9.1 $KH_2PO_4$, 0.2 $MgCl_2 \cdot 6H_2O$, and 0.1% (v/v) vitamin mixture and trace elements before inoculation. Vitamin mixture contained (mg/l) 0.6 biotin, 0.6 folic acid, 59 inositol, 59 *p*-amino benzoic acid, 59 riboflavin, 120 calcium pantothenate, 120 niacin, 120 pyridoxine–HCl and 120 thiamine-HCl. Trace elements contained (mg/l) 1 $CoCl_2 \cdot 6H_2O$, 3 $CuSO_4 \cdot 5H_2O$, 7 $NaMoO_4 \cdot 2H_2O$, 15 $H_3BO_3$, 17 $MnSO_4 \cdot H_2O$, 29 $ZnSO_4 \cdot 7H_2O$, and 2000 EDTA. pH of the media was adjusted to 7.2 by 2 M NaOH before autoclaving. Degradation of naphthalene was started from 20 mg/l (20 ppm) in BM-medium. Initial CFUs for biofilm and planktonic cell cultures were 21,470,000 and 23,360,000, respectively. Symbols are as follows: Triangle, no cells; square, T102 biofilm cell culture; circle, T102 planktonic cell culture.

planktonic cells. This interesting observation was reproducible in several independent experiments. T102 biofilms degraded 14 ppm or 70% of the initial naphthalene (20 ppm) in 12 h and exhibited a maximum degradation rate of 1.7 ppm/h between 4 and 12 h. On the other hand, T102 planktonic cells started to degrade naphthalene immediately after incubation. The degradation rate was almost constant, about 0.5 ppm/h for 16 h, and it took 28 h for the degradation of 14 ppm of naphthalene. About 1.6 ppm (8%) of the initial naphthalene was absorbed to the butyl rubber septum of the bottle cap and remained un-degradable. This means that naphthalene was almost completely eliminated from the culture of T102 biofilms after 32 h.

We hypothesized that initial delay for degradation by biofilms may be the time that it takes the naphthalene to penetrate through the extracellular matrix of the biofilm so that the genes responsible for degradation of naphthalene might be induced.

## Comparison of expression levels of nahA in T102 biofilms and planktonic cells

It is known that *nahAa*, *nahAb*, *nahAc* and *nahAd* form an operon and encode a ferredoxin, ferredoxin oxidoreductase, and naphthalene dioxygenase large and small subunits, respectively (Ensley *et al.*, 1982; Morikawa, 2010). All these genes are essential for the aerobic degradation of naphthalene and related compounds. Thus we analysed the expression level of *nahAc*, which encodes the large subunit of naphthalene dioxygenase, in T102 biofilms and planktonic cells (Fig. 12.7). Total RNA was extracted from T102 cells and cDNA was synthesized from DNase-treated RNA as previously described (Takei *et al.*, 2008). RT-PCR amplification was performed using a primer set targeting *nahAc* in T102. The PCR products were separated on 1.5% agarose gels and stained by ethidium bromide. Contrary to our hypothesis, *nahAc* in T102 biofilms expressed *nahAc* at constant levels from the start of incubation through to 40 h. Constant gene expression was also confirmed for *nahAa*, *nahAb*, and *nahAd* (results not shown). These results suggest that the initial lag and subsequent high naphthalene degradation periods exhibited by T102 biofilms are not due to changing gene expression levels of *nahAabcd*. The *nahAabcd* operon in *Pseudomonas putida* has been reported to be inducible under regulation of *nahR* (Park *et al.*, 2002). In this experiment expression of *nahA* genes was observed at 0 h, before naphthalene was added to the medium. This suggests that *nahR* does not exist or does not function to regulate naphthalene degradation in *P. stutzeri* T102, which has been isolated from the bottom sludge of a petroleum reservoir tank where naphthalene is always abundant.

When we carefully observed the culture, we noticed that a part of T102 detached and grew planktonically in the culture bottle containing T102 biofilms. Thus, we carefully separated these detached cells from biofilms, and examined the expression level of *nahAc*. It was found that the expression level of *nahAc* in the detached cells was clearly and significantly higher than that in biofilms from 12 to 20 h (Fig. 12.8). This result allowed us to conclude that degradation of naphthalene was largely due to the activity of detached cells rather than biofilms. The naphthalene degradation activities of T102 biofilms, detached, and planktonic cells, were determined as 0.02, 1.1 and 0.3 pg $CFU^{-1}$ $h^{-1}$, respectively. The degradation activity of the biofilms was much lower than planktonic cells probably due to depletion of oxygen in biofilms where cells were densely packed and due to poor naphthalene penetration through the matrix that encases biofilm-associated cells.

## Fitness of T102 biofilms and planktonic cells in petroleum-contaminated soils

The observation that T102 biofilms were capable of producing highly active detached cells prompted us to test their performance in soils contaminated with natural petroleum. Petroleum contaminated soils were taken from Ishikari petroleum field (Hokkaido, Japan) and used for following experiments without sterilization. For biofilm *samples*, T102 biofilms were formed in advance at 30°C for 24 h in screw cap polypropyrene tubes containing 0.4 ml medium. The liquid culture containing planktonic cells was removed from the tubes leaving biofilms on the inside wall. Contaminated soils, 0.5 g, and the same amount of filtered cell free spent medium from planktonic cell cultures were added to each tube. Then, two different conditions were set for biofilm samples, one was the 'intact biofilm samples' with no treatment, the other was 'dispersed biofilm samples' which were vortexed

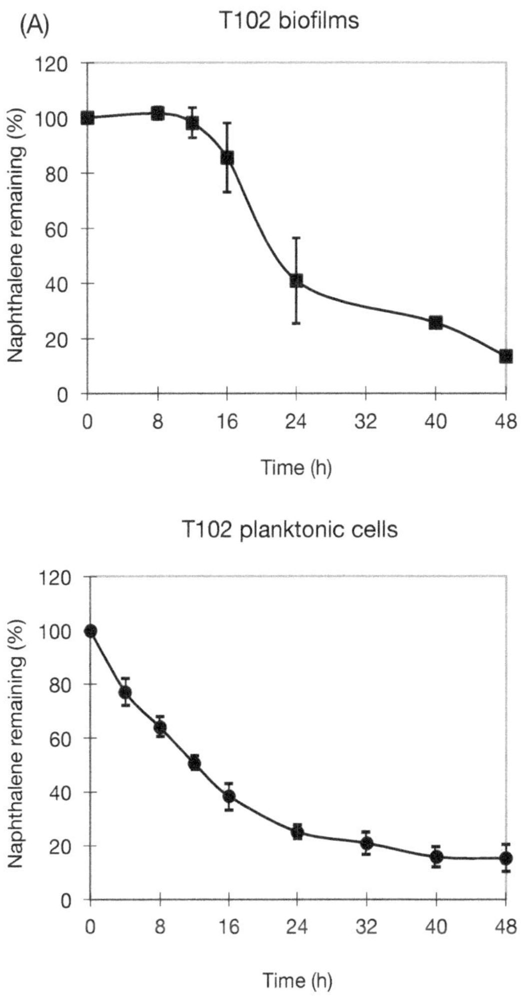


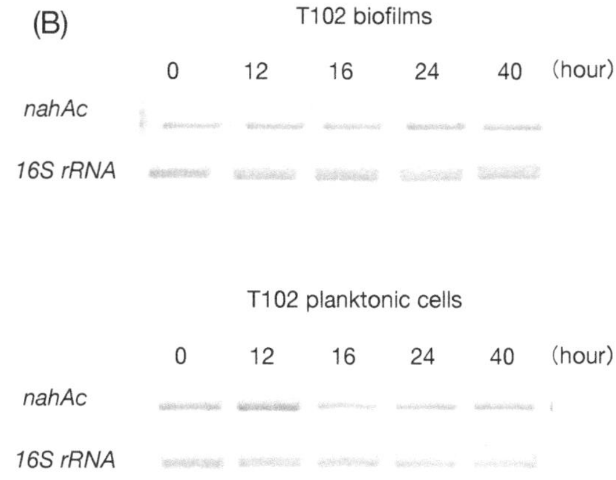


**Figure 12.7** (A) Naphthalene degradation activity of T102 biofilm (upper) and planktonic (lower) cell cultures, respectively. Degradation of naphthalene was started from 20 mg/l (20 ppm) in BM-medium. Initial CFUs were 18,850,000 (biofilm culture) and 22,120,000 (planktonic cell culture). Data points are the average of triplicate assays; error bars represent standard deviations. (B) RT-PCR analyses of *nahAc* and 16S rRNA gene expression in the T102 biofilms (upper) and planktonic cells (lower), respectively.

for 2 s to detach and disperse the biofilm pieces into the soils. For planktonic cell samples, 0.5 g of the soils in 2-ml tubes were supplemented with mid-log phase planktonic cell cultures containing similar CFUs to above biofilm samples. CFUs in the samples were adjusted to previously determined CFUs $OD_{600}$ values. A soil sample with no inoculation of T102 cells was also prepared as a negative control. The soil samples inoculated with 'intact biofilms', 'dispersed biofilms', planktonic cells, or no inoculation were incubated at 30°C using the naphthalene vapour exposure method (Park and Crowley, 2006). Naphthalene vapour was supplied by placing a small quantity of naphthalene crystals in the middle of Petri dishes which also contained sample tubes that were loosely capped and placed on wet paper to avoid desiccation. The Petri dishes containing sample tubes and naphthalene crystals were sealed with parafilm, which is permeable to oxygen.

To estimate the fitness of *P. stutzeri* T102 in the soils, we used the community DNA fingerprinting method denaturant gradient gel electrophoresis (DGGE) to assess variation in bacterial community composition among samples. Although this method is semi-quantitative due to different efficiencies of DNA extraction and PCR depending on the bacterial strains, it is useful to analyse population changes of single strains over the time. Moreover, distribution of the DNA bands in DGGE gels provides a rough overview of the bacterial community dynamics upon introduction of T102 biofilm and planktonic cells to the soils. Total DNA was prepared from the culture and PCR amplified 16S rRNA gene fragments were separated on an 8% polyacrylamide gel with a denaturing gradient

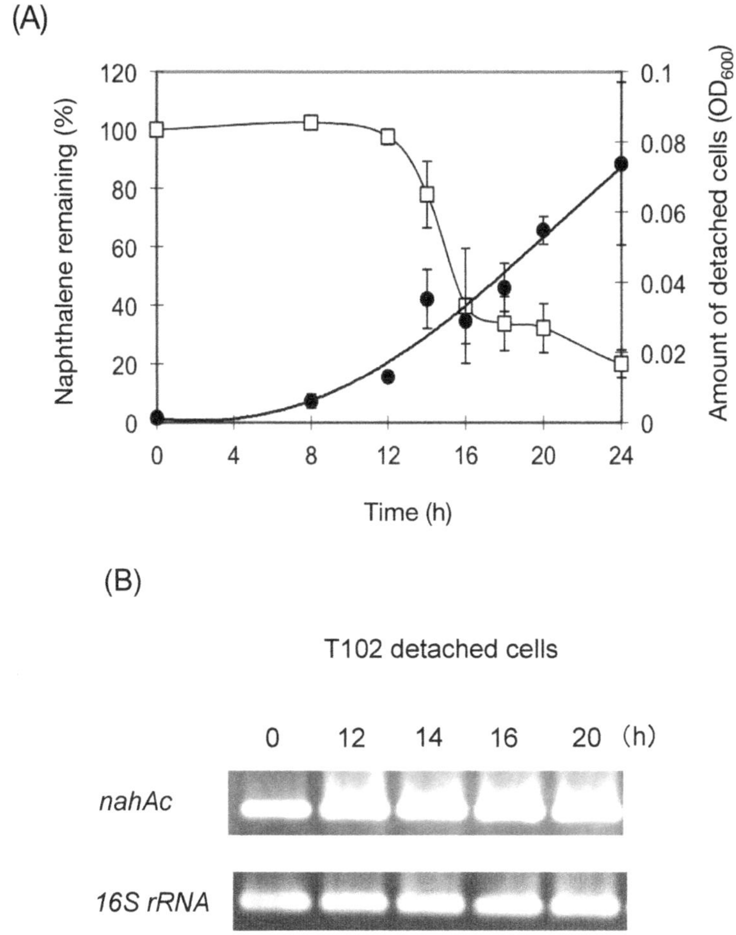


**Figure 12.8** (A) Naphthalene degradation activity (open square) and the amount of detached cells (closed circle) in a T102 biofilms flask. Degradation of naphthalene was started from 20 mg/l (20 ppm) in BM-medium. Data points are the average of triplicate assays and error bars represent standard deviations. (B) RT-PCR analyses of nahAc and 16S rRNA gene expression in the detached cells. Only cells sampled at 0 h were taken from biofilms.

of urea and formamide ranging from 20% to 50%. The density of the DNA bands corresponding to *P. stutzeri* T102 16S rRNA genes were estimated using imaging software.

We observed that the 'dispersed biofilm' sample maintained a rather dense T102 DNA band for 10 weeks and kept over 40% of its original density level for 8 weeks (Fig. 12.9A). The density of T102 DNA bands of 'planktonic cells' and 'intact biofilms' samples decreased more rapidly than that of 'dispersed biofilms', and they almost disappeared (ca 2% remained) after 10 weeks. The decrease in the amount of T102 DNA bands was more significant within the first 3 weeks than later for all experimental conditions. The amount of T102 DNA in the 'intact biofilms' sample continuously decreased over the period. This may be because the nutrients and oxygen were more rapidly depleted around the sessile 'intact biofilms' than 'planktonic cells' and 'dispersed biofilms'. The duration for which each sample kept over 20% of the initial density of T102 DNA was 10, 5, and 2 weeks for 'dispersed biofilms', 'intact biofilms', and 'planktonic cells' inoculates, respectively. These results indicate that 'dispersed biofilms' are more tolerant and stable than 'intact biofilms' and 'planktonic cells' in the petroleum contaminated soils. A band of increasing intensity, indicated by an asterisk, in Fig. 12.9B may be an indigenous naphthalene degrader since it also appeared in

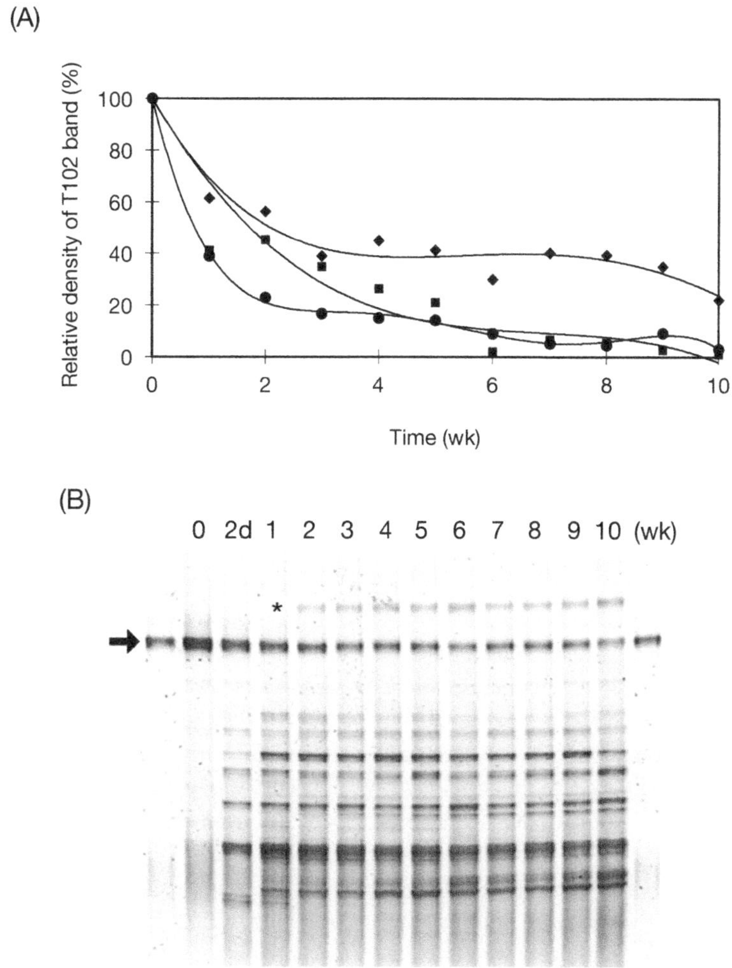


**Figure 12.9** (A) Comparison of the fitness of T102 cells inoculated into contaminated soils. The relative density of 16S rRNA gene fragments relating to T102 in each DGGE gel is plotted in the graph. Symbols are as follows: Diamond, dispersed T102 biofilms; square, intact T102 biofilms; circle, planktonic cells. (B) DGGE gel data for the dispersed biofilms experiment from 0 to 10 weeks. The arrow indicates the position for the 16S rRNA gene fragments associated with T102.

the sample without T102 inoculation (data not shown).

The above experimental results are consistent with previous knowledge that biofilm-associated cells are generally more tolerant to environmental stresses than their planktonic counterparts. The environmental robustness of biofilm-associated cells could be attributed to both specific gene expression and physicochemical toughness provided to the densely packed cells by extracellular matrices (Chang *et al.*, 2007; Bossé *et al.*, 2010; Halan *et al.*, 2011). Next, we examined the naphthalene degradation activities of contaminated soils containing either T102 biofilm or planktonic cells.

## Naphthalene degradation activity of soils containing T102 biofilms and planktonic cells

As shown above, the liquid culture system inoculated with T102 biofilms exhibited naphthalene degradation activity comparable to or even higher than with T102 planktonic cells (Fig. 12.6). PCR-DGGE analysis suggested that populations of T102 remained higher in soils inoculated with 'dispersed biofilms' rather than the 'planktonic cells' during incubation for 10 weeks in contaminated soils. Thus, we expected that the soil sample containing 'dispersed biofilms' should exhibit higher degradation activities than 'planktonic cells' over the period. But it was difficult to measure naphthalene degradation directly in the contaminated soil samples because the soils contained large amounts of various hydrocarbons and the signal/noise ratio in GC analysis was quite low. We decided to measure naphthalene degradation activity of each soil sample in BM-medium containing an additional 100 ppm naphthalene (Fig. 12.10). The entire soil sample, 0.5 g, in the 1.5 ml polypropyrene tube was transferred into a 50 ml screw-capped glass bottle containing 20 ml of BM-medium and 100 ppm naphthalene in methanol solution. Bottles were tightly closed and incubated at 30°C for 36 h. Extraction of the remaining naphthalene and GC/FID analysis were performed as previously described.

It was demonstrated that 'dispersed biofilms' and 'planktonic cells' initially degraded 48 and 52 ppm naphthalene in 36 h, respectively. Their degradation activities gradually decreased as the incubation time increased. When their activities were compared after 9 weeks of incubation, 'dispersed biofilms' still degraded 19 ppm naphthalene while 'planktonic cells' and no inoculation samples degraded only < 0.1 ppm. Petroleum-contaminated soils that were taken from Ishikari oil field contained various hydrocarbon compounds such as straight-, branched- and cyclo-alkanes, and PAHs including naphthalene. Thus, it was not surprising that the soils contained significant amounts of naphthalene-degrading bacteria. The degradation of naphthalene with no inoculation could be attributed to these indigenous bacteria. Densitometry of the T102 bands of 'dispersed biofilms' and 'planktonic cells' in DGGE gel revealed that 35 and 9% of the initial band intensities remained after 9 weeks, respectively. These results suggest that the specific naphthalene degradation activity of 'dispersed biofilms' including detached cells was much higher than 'planktonic cells' in the petroleum contaminated soils after 9 weeks.

## Conclusion

Biofilms confer microbial cells with high resistance to environmental stresses and facilitate their survival in complex microenvironments that help generate diverse cellular heterogeneity and activities. In the first example, *Pseudoalteromonas* bacteria formed significant biofilms on the sediments under sea fish farms. Four out of the six strains produced more proteases in biofilms than in planktonic cultures, suggesting their contribution to rapid degradation of proteinaceous sediments to avoid deterioration of water quality. Moreover, BmpI protease from strain SB-B1 was shown to be produced in a strict biofilm-dependent manner. In the second example, the naphthalene degradation rate of *P. stutzeri* T102 biofilms was initially low but later became higher than that of T102 planktonic cells. The rapid degradation activity of biofilm cultures

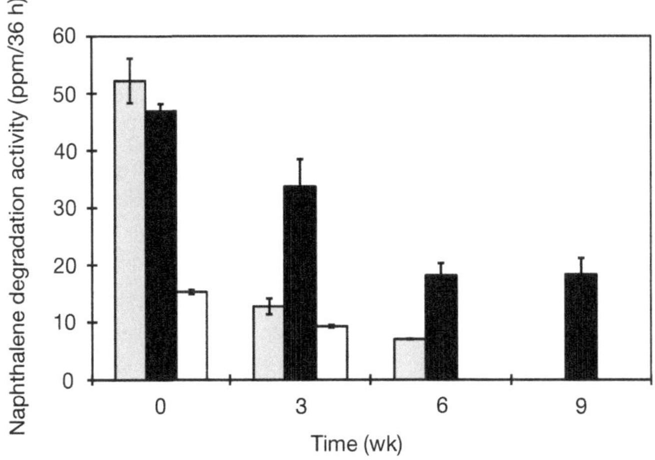


**Figure 12.10** Naphthalene degradation activity of soils with T102 'planktonic cells' (shaded bar), 'dispersed biofilms' (closed bar), and no T102 cells (open bar). Data points are the average of triplicate assays and error bars represent standard deviations.

could be attributed to producing detached cells. It was shown that biofilms 'act as aircraft carriers' that launch superactive fighter cells which may contribute to the continued degradation of harmful environmental contaminants. T102 cells were also shown to be more durable and active in petroleum contaminated soils when they are introduced as 'dispersed biofilms' rather than as 'planktonic cell' suspensions. These experimental results suggest that inoculation of contaminated sites with pollutant degrading biofilms may enhance bioaugmentation as a durable and effective bioremediation technology.

## References

Alexander, M. (1999). Biodegradation and Bioremediation, 2nd edn (Academic Press, London, UK).

Bossé, J.T., Sinha, S., Li, M.S., O'Dwyer, C.A., Nash, J.H., Rycroft, A.N., Kroll, J.S., and Langford, P.R. (2010). Regulation of pga operon expression and biofilm formation in *Actinobacillus pleuropneumoniae* by sigmaE and H-NS. J. Bacteriol. *192*, 2414–2423.

Bouchez, T., Patureau, D., Dabert, P., Juretschko, S., Doré, J., Delgenès, P., Moletta, R., and Wagner, M. (2000). Ecological study of a bioaugmentation failure. Environ. Microbiol. *2*, 179–190.

Chang, W.S., van-de-Mortel, M., Nielsen, L., Nino-de-Guzman, G., Li, X., and Halverson, L.J. (2007). Alginate production by *Pseudomonas putida* creates a hydrated microenvironment and contributes to biofilm architecture and stress tolerance under water-limiting conditions. J. Bacteriol. *189*, 8290–8299.

Chen, X.L., Zhang, Y.Z., Gao, P.J., and Luan, X.W. (2003). Two different proteases produced by a deep-sea psychrotrophic bacterial strain, *Pseudoalteromonas* sp. SM9913. Mar. Biol. *143*, 989–993.

Chen, X.L., Xie, B.B., Lu, J.T., He, H.L., and Zhang, Y. (2007). A novel type of subtilase from the psychrotolerant bacterium *Pseudoalteromonas* sp. SM9913: catalytic and structural properties of deseasin MCP-01. Microbiology *153*, 2116–2125.

Dejonghe, W., Boon, N., Seghers, D., Top, E.M., and Verstraete, W. (2001). Bioaugmentation of soils by increasing microbial richness: missing links. Environ. Microbiol. *3*, 649–657.

Dickschat, J.S. (2010). Quorum sensing and bacterial biofilms. Nat. Prod. Rep. *27*, 343–369.

Donlan, R.M., and Costerton, J.W. (2002). Biofilms: survival mechanisms of clinically relevant microorganisms. Clin. Microbiol. Rev. *15*, 167–193.

El-Fantroussi, S., and Agathos, S.N. (2005). Is bioaugmentation a feasible strategy for pollutant removal and site remediation? Curr. Opin. Microbiol. *8*, 268–275.

Ensley, B.D., Gibson, D.T., and Laborde, A.L. (1982). Oxidation of naphthalene by a multicomponent enzyme system from *Pseudomonas* sp. strain NCIB 9816. J. Bacteriol. *149*, 948–954.

Gilbert, P., Collier, P.J., and Brown, M.R. (1990). Influence of growth rate on susceptibility to antimicrobial agents: biofilms, cell cycle, dormancy, and stringent response. Antimicrob. Agents Chemother. *34*, 1865–1868.

Gorbushina, A.A., and Broughton, W.J. (2009). Microbiology of the atmosphere–rock interface: how biological interactions and physical stresses modulate a sophisticated microbial ecosystem. Annu. Rev. Microbiol. *63*, 431–450.

Gross, R., Hauer, B., Otto, K., and Schmid, A. (2007). Microbial biofilms: new catalysts for maximizing productivity of long-term biotransformations. Biotechnol. Bioeng. *98*, 1123–1134.

Halan, B., Schmid, A., and Buehler, K. (2011). Real-time solvent tolerance analysis of *Pseudomonas* sp. strain VLB120dC catalytic biofilms. Appl. Environ. Microbiol. *77*, 1563–1571.

Hirano, S., Kitauchi, F., Haruki, M., Imanaka, T., Morikawa, M., and Kanaya, S. (2004). Isolation and characterization of *Xanthobacter polyaromaticivorans* sp. nov. 127W that degrades polycyclic and heterocyclic aromatic compounds under extremely low oxygen conditions. Biosci. Biotechnol. Biochem. *68*, 557–564.

Holmström, C., and Kjelleberg, S. (1999). Marine *Pseudoalteromonas* species are associated with higher organisms and produce biologically active extracellular agents. FEMS Microbiol. Ecol. *30*, 285–293.

Iijima, S., Washio, K., Okahara, R., and Morikawa, M. (2009). Biofilm formation and proteolytic activities of *Pseudoalteromonas* bacteria that were isolated from fish farm sediments. Microbial. Biotechnol. *2*, 361–369.

Kanamori, T., Rashid, N., Morikawa, M., Atomi, H., and Imanaka, T. (2002). *Oleomonas sagaranensis* gen. nov., sp. nov., represents a novel genus in the alpha-proteobacteria. FEMS Lett., *217*, 255–261.

Kaplan, J.B. (2010). Biofilm dispersal: mechanisms, clinical implications, and potential therapeutic uses. J. Dent. Res. *89*, 205–218.

Kato, T., Haruki, M., Imanaka, T., Morikawa, M., and Kanaya, S. (2001). Isolation and characterization of psychrotrophic bacteria from oil reservoir water and oil sands. Appl. Microbiol. Biotechnol. *55*, 794–800.

Kunihiro, N., Haruki, M., Takano, K., Morikawa, M., and Kanaya, S. (2005). Isolation and characterization of *Rhodococcus* sp. strains TMP2 and T12 that degrade 2,6,10,14-tetramethylpentadecane (pristane) at moderately low temperatures. J. Biotechnol. *115*, 129–136.

Lalucat, J., Bennasar, A., Bosch, R., García-Valdés, E., and Palleroni, N.J. (2006). Biology of *Pseudomonas stutzeri*. Microbiol. Mol. Biol. Rev. *70*, 510–547.

Lee, S.O., Kato, J., Takiguchi, N., Kuroda, A., Ikeda, T., Mitsutani, A., and Ohtake, H. (2000). Involvement of an extracellular protease in algicidal activity of the marine bacterium Pseudoalteromonas sp. strain A28. Appl. Environ. Microbiol. *66*, 4334–4339.

Lewis, K. (2007). Persister cells, dormancy and infectious disease. Nat. Rev. Microbiol. *5*, 48–56.

Ma, Q and Wood, T.K. (2009). OmpA influences *Escherichia coli* biofilm formation by repressing cellulose production through the CpxRA two-component system. Environ. Microbiol. *11*, 2735–2746.

Maeda, M., Nogami, K., Kanematsu, M., and Hirayama, K. (1997). The concept of biological control methods in aquaculture. Hydrobiologia *358*, 285–290.

Morikawa, M. (2006). Beneficial biofilm formation by industrial bacteria *Bacillus subtilis* and related species. J. Biosci. Bioeng. *101*, 1–8.

Morikawa, M. (2010). Dioxygen activation responsible for oxidation of aliphatic and aromatic hydrocarbon compounds: current state and variants. Appl. Microbiol. Biotechnol. *87*, 1596–1603.

Morikawa, M., and Imanaka, T. (1993). Isolation of a new mixotrophic bacterium which can fix $CO_2$ and assimilate aliphatic and aromatic hydrocarbons anaerobically. J. Ferment. Bioeng. *76*, 280–283.

Moslemy, P., Neufeld, R.J., and Giot, S.R. (2002). Biodegradation of gasoline by gellan gum-encapsulated bacterial cells. Biosci. Bioeng. *80*, 175–184.

Mrozik, A., and Piotrowska-Seget, Z. (2010). Bioaugmentation as a stratergy for cleaning up soils contaminated with aromatic compounds. Microbiol. Res. *182*, 2675–2679.

O'Toole, G., Kaplan, H.B., and Kolter, R. (2000). Biofilm formation as microbial development. Annu. Rev. Microbiol. *54*, 49–79.

O'Toole, G.A., Pratt, L.A., Watnick, P.I., Newman, D.K., Weaver, V.B., and Kolter, R. (1999). Genetic approaches to study of biofilms. Methods Enzymol. *310*, 91–109.

Park, W., Jeon, C.O., and Madsen, E.L. (2002). Interaction of NahR, a LysR-type transcriptional regulator, with the alpha subunit of RNA polymerase in the naphthalene degrading bacterium, *Pseudomonas putida* NCIB 9816-4. FEMS Microbiol. Lett. *213*, 159–165.

Park, J.W., and Crowley, D.E. (2006). Dynamic changes in *nahAc* gene copy numbers during degradation of naphthalene in PAH-contaminated soils. Appl. Microbiol. Biotechnol. *72*, 1322–1329.

Patel, P., Callow, M.E., Joint, I., and Callow, J.A. (2003). Specificity in the settlement – modifying response of bacterial biofilms towards zoospores of the marine alga Enteromorpha. Environ. Microbiol. *5*, 338–349.

Qian, P.Y., Lau, S.C., Dahms, H.U., Dobretsov, S., and Harder, T. (2007). Marine biofilms as mediators of colonization by marine macroorganisms: implications for antifouling and aquaculture. Mar. Biotechnol. *9*, 399–410.

Rani, S.A., Pitts, B., Beyenal, H., Veluchamy, R.A., Lewandowski, Z., Davison, W.M., Buckingham-Meyer, and K., and Stewart, P.S. (2007). Spatial patterns of DNA replication, protein synthesis, and oxygen concentration within bacterial biofilms reveal diverse physiological states. J. Bacteriol. *189*, 4223–4233.

van Rijn, J. (1996). The potential for integrated biological treatment systems in recirculating fish culture – a review. Aquaculture *139*, 181–201.

Sánchez-Porro, C., Mellado, E., Bertoldo, C., Antranikian, G., and Ventosa, A. (2003). Screening and characterization of the protease CP1 produced by the moderately halophilic bacterium *Pseudoalteromonas* sp. strain CP76. Extremophiles *7*, 221–228.

Shemesh, M., Kolter, R., and Losick, R., 2010. The biocide chlorine dioxide stimulates biofilm formation in *Bacillus subtilis* by activation of the histidine kinase KinC. J. Bacteriol. *192*, 6352–6356.

Silva, E., Fialho, A.M., Sá-Correia, I., Burns, R.G., and Shaw, L.J. (2004). Combined bioaugmentation and biostimulation to cleanup soil contaminated with high concentrations of atrazine. Environ. Sci. Technol. *38*, 632–637.

Singh, R., Paul, D., and Jain, R.K. (2006). Biofilms: implications in bioremediation. Trends Microbiol. *14*, 389–397.

Takei, D., Washio, K., and Morikawa, M. (2008). Identification of alkane hydroxylase genes in Rhodococcus sp. strain TMP2 that degrades a branched alkane. Biotechnol. Lett. *30*, 1447–1452.

Thompson, I.P., van-der-Gast, C.J., Ciric, L., and Singer, A.C. (2005). Bioaugmentation for bioremediation: the challenge of strain selection. Environ. Microbiol. *7*, 909–915.

Tielen, P., Rosenau, F., Wilhelm, S., Jaeger, K.E., Flemming, H.C., and Wingender, J. (2010). Extracellular enzymes affect biofilm formation of mucoid *Pseudomonas aeruginosa*. Microbiology *156*, 2239–2252.

Tubiash, H.S., Colwell, R.R., and Sakazaki, R. (1970). Marine vibrios associated with bacillary necrosis, a disease of larval and juvenile bivalve mollusks. J. Bacteriol. *103*, 271–272.

van Veen, J.A., van Overbeek, L.S., and van Elsas, J.D. (1997). Fate and activity of microorganisms introduced into soil. Microbiol. Mol. Biol. Rev. *61*, 121–135.

Verschuere, L., Rombaut, G., Sorgeloos, P., and Verstraete, W. (2000). Probiotic bacteria as biological control agents in aquaculture. Microbiol. Mol. Biol. *64*, 655–671.

Watnick, P., and Kolter, R. (2000). Biofilms, city of microbes. J. Bacteriol. *13*, 20–26.

Werner, E., Roe, F., Bugnicourt, A., Franklin, M.J., Heydorn, A., Molin, S., Pitts, B., and Stewart, P.S. (2004). Stratified growth in *Pseudomonas aeruginosa* biofilms. Appl. Environ. Microbiol. *70*, 6188–6196.

Xiong, H., Song, L., Xu, Y., Tsoi, M.Y., Dobretsov, S., and Qian, P.Y. (2007). Characterization of proteolytic bacteria from the Aleutian deep-sea and their proteases. J. Ind. Microbiol. Biotechnol. *34*, 63–71.

Yamaga, F., Washio, K., and Morikawa, M. (2010). Sustainable biodegradation of phenol by *Acinetobacter calcoaceticus* P23 isolated from the rhizosphere of duckweed *Lemna aoukikusa*. Environ. Sci. Technol. *44*, 6470–6474.

# Using Microbial Biofilms to Enhance the Phytoremediation of Contaminants in Soil and Water – A Trial for Sustainable Phenol Degradation by Duckweed-colonizing Biofilms

# 13A

Masaaki Morikawa, Fumiko Yamaga, Kazuya Suzuki, Koki Kurashina, Kyoko Miwa and Kenji Washio

## Abstract

Here, we demonstrate how conventional passive phytoremediation technologies can be transformed into active bioremediation technologies using growth-promoting, plant associated bacteria. Phenol-degrading bacteria were isolated from the surface of duckweed, *Lemna aoukikusa* (currently *Lemna aequinoctialis*), using an enrichment culture method. One of the isolates, *Acinetobacter calcoaceticus* P23, exhibited an excellent ability to degrade phenol and formed robust biofilms under laboratory conditions. P23 rapidly colonized the surface of sterilized duckweed and enhanced its growth. A long-term test using P23 colonized *L. aoukikusa* showed that the continuous removal of phenol could be attributed to a beneficial symbiotic interaction between duckweed and its surface bacterium. The rational reconstruction of multispecies microbial communities, including taxa with pollutant-degrading and plant growth-promoting activities, holds much promise for the development of innovative phytoremediation technologies.

## Introduction

Phytoremediation is a defined as the use of green plants to remove pollutants from the environment or to render them harmless (Salt *et al.*, 1998). Phytoremediation can be categorized by phytoextraction, phytodegradation, rhizofiltration, phytostabilization, and phytovolatilization depending on the mechanism of pollutant removal or transformation. The phytodegradation of organic pollutants is mostly thought to be caused by the activity of microorganisms in the rhizosphere, an area of enriched microbial activity close to the surface of plant roots. This form of phytodegradation is sometimes termed rhizoremediation (Kuiper *et al.*, 2004). Plant roots support the growth and metabolic activities of a wide variety of microorganisms that may also have a profound effect on the growth and/or health of plants. The advantage of phytodegradation is that bacteria in the rhizosphere benefit from the nutrients provided by plant root exudates and oxygen exchange in the root system, enhancing the rate and extent of pollutant degradation (Chaudhry *et al.*, 2005). Phytodegradation in soil has been reported for a variety of organic pollutants, such as chlorinated ethenes, polycyclic aromatic hydrocarbons, and polychlorinated biphenyls (Kuiper *et al.*, 2004).

While phytoremediation has typically focused on the remediation of contaminated soil and sediment, contamination of environmental water is a serious issue for conservation of the biosphere. Pollutants in water are more diffusive and at lower concentrations than those in soils. Thus, the immobilization and stimulation of contaminant degrading bacteria remains critical for the complete restoration of contaminated aquatic environments (Beun *et al.*, 1999). The microbial rhizosphere is ubiquitous in natural aquatic environments, as it is in the soil

(Rai, 2008). Previous studies have therefore sought to accelerate degradation of organic chemicals using the activity of microbial communities present in the rhizosphere of aquatic plants such as the giant duckweed *Spirodela polyrhiza* (Toyama *et al.*, 2009; Toyama *et al.*, 2006). Here, we focus on small duckweed, *Lemna aoukikusa*, and its surface bacteria associations, whereas in part B of this chapter, Toyama *et al.* describe the successful degradation of various phenolic endocrine-disrupting compounds by reed *Phragmites australis*–rhizosphere bacteria associations (Fig. 13A.1).

Duckweeds, typified by the family *Lemnaceae*, are small free-floating aquatic angiosperm plants without distinct stems and leaves. The whole plant body is reduced to form a flat small leaf-like structure called a frond. This unique morphological feature of duckweeds provides them with substantial rhizosphere and phyllosphere communities that may interact directly with the plant and surrounding aquatic environment. It is noteworthy that chlorophylls are contained not only in the frond but also in the root of duckweed. It is then possible that oxygen produced by photosynthetic activity is directly supplied from the roots to rhizobacteria, thereby elevating the activities of rate-limiting mono- and dioxygenase enzymes which are essential for the breakdown of many organic pollutants (Morikawa, 2010).

Phenolic compounds are common chemicals that are widely used in many manufacturing processes and are a major pollutant of industrial wastewaters (Watanabe *et al.*, 1996). These compounds are toxic to particular aquatic species (Brown *et al.*, 1967) and are also harmful to human health when ingested or inhaled even at very low concentrations (Pedersen *et al.*, 2002). Therefore, it is valuable to develop strategies for the efficient and continuous biodegradation of phenols with minimum impact to the environment. Here, we present an example of developing a beneficial interaction between duckweed and its surface bacteria which are shown to be useful for the continuous and ecologically safe degradation of phenol.

## Isolation of phenol-degrading bacteria from the surface of the duckweed, *Lemna aoukikusa*

Mature duckweed plants (*Lemna aoukikusa*) were collected from a small pond in the botanical garden of Hokkaido University (Sapporo, Japan). The duckweed plants were gently washed with sterile water and cultivated in a flask containing Hoagland medium (Arnon and Hoagland, 1939) supplemented with 200 mg/l phenol at 25°C with a day/night photoperiod of 16 h/8 h at 8500 lux. After 3 weeks of incubation, ten fronds of the duckweed were washed with sterilized water and then sonicated to disperse the surface-attached bacteria. The bacterial suspension was serially diluted with sterilized water and spread onto basal salt medium (BM) (Kunihiro *et al.*, 2005) agar plates supplemented with phenol as the sole carbon source. Colonies that appeared on the plates were picked and subcultured several times to obtain pure isolates of phenol degrading bacteria. Finally, five strains exhibited marked abilities to degrade phenol (Fig. 13A.2).

(A) (B) (C)

**Figure 13A.1** Aquatic plants described in this chapter. (A) *Lemna* sp., (B) *Spirodela polyrhiza*, and (C) *Phragmites australis.* Bars indicate the size of 1 cm (A, B) and 1 m (C). Photos were taken by Drs. T. Toyama and K. Mori (University of Yamanashi).

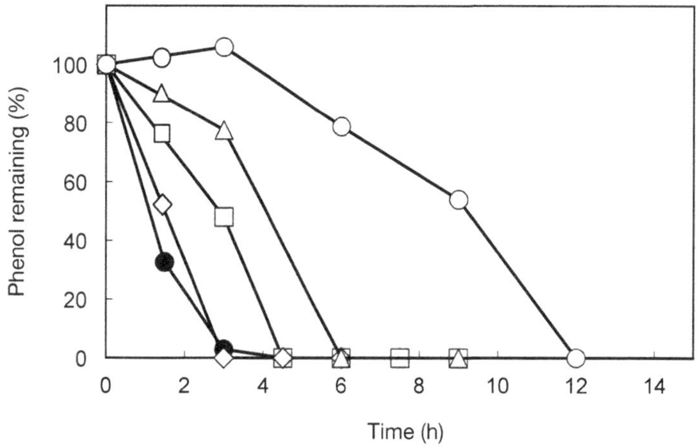


**Figure 13A.2** Degradation of phenol by bacteria isolated from the surface of the duckweed *L. aoukiuksa*. Degradation tests of 20 mg $l^{-1}$ phenol in Luria–Bertani (LB) medium were performed at 25°C with shaking at 150 rpm. Phenol concentrations were quantified over time by high-performance liquid chromatography (HPLC) analysis using a Cosmosil 5$C_{18}$-AR II column and a mobile phase containing 63% (v/v) water and 37% (v/v) acetonitrile. The detector wavelength was set at 254 nm. Each symbol shows the phenol degradation attributed to a single strain. Symbols are: P02, square; P11c, diamond; P22, triangle; P23, closed-circle; P24a, open-circle.

Partial 16S rRNA gene fragments were sequenced for the five strains which revealed that these strains belong to *Pseudomonas* (P02), *Rhodococcus* (P11c), and *Acinetobacter* (P22, P23, and P24a). Further analysis revealed that P22, P23, and P24a clustered tightly with *A. rhizosphaerae*, *A. calcoaceticus*, and *A. tandoii*, respectively.

The property of adherence is a critical trait of surface associated bacteria, including in the rhizosphere (Ramey *et al.*, 2004). Biofilm formation, which is an indicator of adherence to solid surfaces, was first examined in each strain using a crystal violet assay (Fig. 13A.3; Iijima *et al.*, 2009). Each strain was grown in a 1.5-ml polypropylene tube without shaking for 24 h in Hoagland medium, basal medium, or Luria–Bertani (LB) medium. The three *Acinetobacter* strains formed robust biofilms when grown in the LB medium, whereas *Pseudomonas* sp. P02 and *Rhodococcus* sp. P11c formed marginal biofilms under all of the tested conditions. These results are in accordance with the classification of the genus *Acinetobacter*, species of which are non-motile and grow attached to a variety of solid surfaces in nature (Juni, 1978). It has been reported that a toluene-degrading *Acinetobacter* sp. Tol5 has a highly hydrophobic cell surface and preferably adheres to a solid phase (Ishii *et al.*, 2004).

Because P23 exhibited the best phenol degradation and good biofilm formation abilities among the tested strains, we hypothesized that P23 could be a potential candidate for the effective biodegradation of phenol. Thus, we used this strain for further experiments. When we examined the ability of this strain to utilize a range of carbon sources, almost all of the tested compounds, including glucose, maltose, cellobiose, sucrose, trehalose, inositol, 5-ketogluconic acid, pyruvate, arabinose, mannitol, arabitol, sorbitol, galacturonic acid, benzene, toluene, failed to support the growth of P23, except for phenol and ethanol. Because phenolic compounds are major exudates of plant roots (Chaudhry *et al.*, 2005), we hypothesized that the surface of duckweed should provide favourable growth conditions for P23.

## Preparation of duckweed surface dominated by *Acinetobacter* P23, the *L. aoukikusa*/P23 system

To eliminate or reduce the abundance of indigenous microorganisms that coexist with duckweeds, *L. aoukikusa* fronds were either (1) submerged in 0.5% (v/v) NaClO/0.05% (v/v) Triton X-100 and gently stirred for 30 min or (2) treated with a

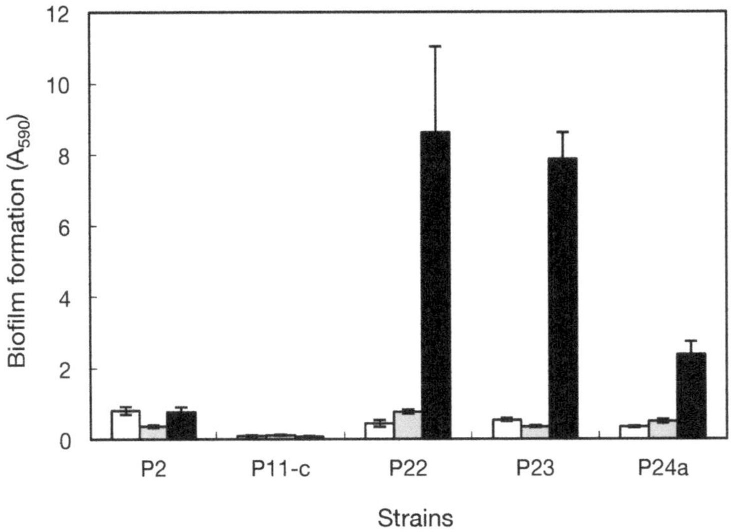


**Figure 13A.3** Biofilm formation by phenol-degrading bacteria. Bacterial strains were grown at 25°C in (white) Hoagland medium, (grey) basal medium (BM), or (black) LB medium. Biofilm formation was calculated using a crystal violet staining method. Phenol (20 mg $l^{-1}$) was added to Hoagland and BM medium as a carbon source. The error bars represent the mean ± SD ($n = 3$).

sonicator for 5 seconds, six times. These plants were then washed twice for 10 min with sterilized water and transferred into Hoagland medium. P23 cells from the late exponential growth phase, cultured in LB media supplemented with 2 mg/l phenol, were recovered by centrifugation at 10,000 × g for 10 min and suspended in Hoagland medium. This cell suspension (1 ml) and 30 surface-cleansed fronds were transferred into a 100-ml flask with 50 ml of Hoagland medium supplemented with 20 mg/l phenol (final $OD_{600} = 0.3$) and incubated under the day/night photoperiod for 72 h at 25°C. During incubation of duckweed plants with P23 in Hoagland medium, most of the P23 cells adhered to the duckweed, and culture turbidities ($OD_{600}$) significantly declined from 0.3 at 0 h to 0.02 at 72 h (Fig. 13A.4A).

When the *L. aoukiuksa*/P23 cultures were stained with the LIVE/DEAD BacLight kit and observed by fluorescence microscopy, many green fluorescent spots, indicating living bacteria, were observed on the surface of the roots at 72 h, whereas there was less green fluorescence on the *L. aoukiuksa* plants without P23 (Fig. 13A.4B – green fluorescence is visible as white in this black and white image). The bacterial cells were dispersed from the *L. aoukiuksa*/P23 plants by sonication and spread on LB agar plates so that we could confirm that P23 was the dominant strain of adhered bacteria (confirmed by 16S rRNA gene sequence identity; data not shown). Thus, the small colonies and biofilms formed on the root of duckweeds should be that of P23. We further measured the number of attached bacterial cells separately for the leaf-like part of fronds and the roots. It was revealed that most P23 (about 90%) are colonized on the leaf-like frond, or $3.0 \times 10^6$ CFUs/frond rather than the roots which had $3.2 \times 10^5$ CFUs/root. This probably reflects that the surface area of the frond is much larger than that of roots.

## Sustainable biodegradation of phenol by *L. aoukikusa*/P23

To determine whether the reinforcement of the specific single species P23 on the surface-cleansed *L. aoukiuksa* plants, namely *L. aoukiuksa*/P23 system, could be useful for effective biodegradation, several treatment systems of *L. aoukiuksa* with P23 were tested for long-term phenol degradation ability (Fig. 13A.5). These were natural *L. aoukiuksa* (intact as they are), surface-cleansed *L. aoukiuksa*, P23, and *L. aoukiuksa* inoculated with P23. Ten fronds of the *L. aoukikusa*/P23 prepared as described above were transferred into a 100-ml flask containing 50 ml of Hoagland medium supplemented with 40 mg/

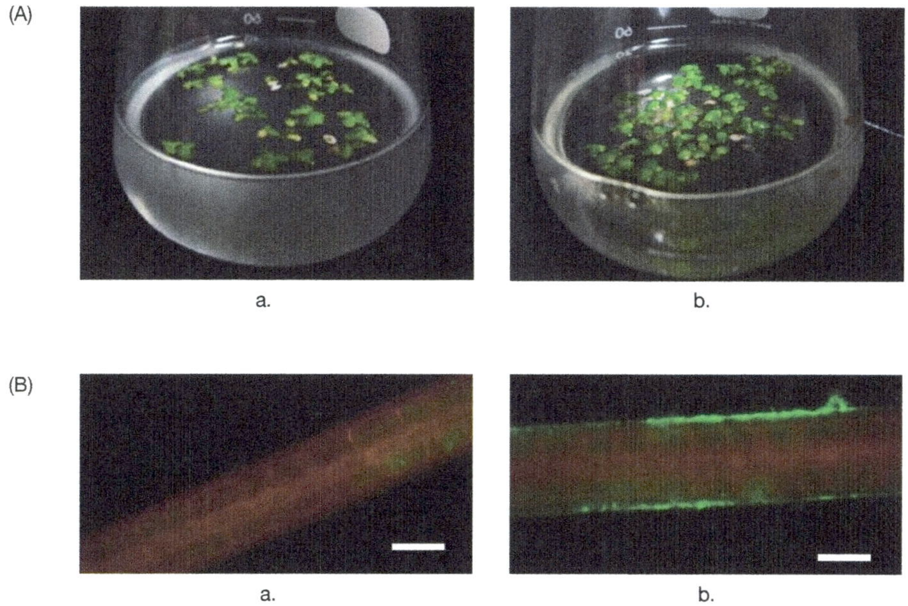


**Figure 13A.4** Rapid attachment and colonization of *A. calcoaceticus* P23 on the surface of *L. aoukiuksa*. (A) Changes in culture turbidity during the mixed cultivation of *L. aoukiuksa* and P23. At 0 h (a), the $OD_{600}$ was 0.3; at 72 h (b), the $OD_{600}$ decreased to 0.02. (B) LIVE/DEAD staining showed more live, biofilm associated cells were present on plant surfaces after 72 h when incubated with P23 (b-right) than when P23 was not included (a-left). The presence of live bacterial cells is shown by the lighter colours in this fluorescence microscopy image. Bacterial cells attached to duckweed roots were observed using a fluorescence microscope (BZ-9000; Keyence, Osaka, Japan). Scale bars, 100 µm.

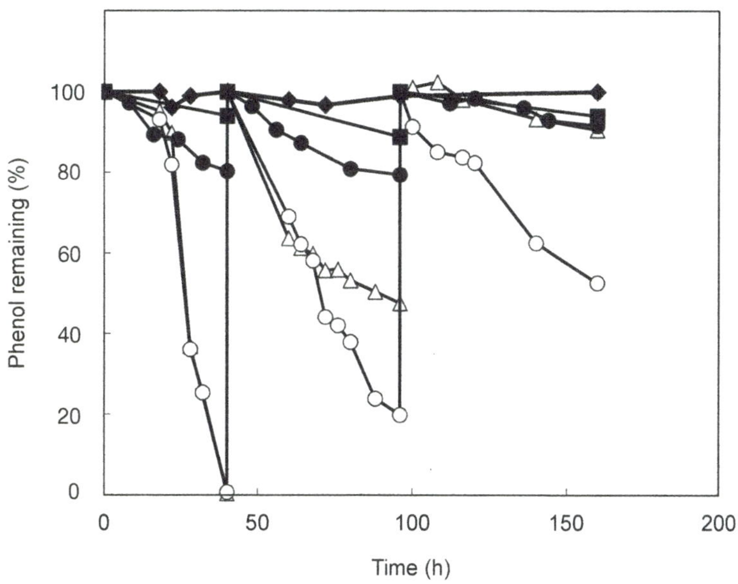


**Figure 13A.5** Long-term performance test of phenol degradation by *Lemna* and P23. The elimination of phenol was examined for the different *Lemna* and P23 combination systems. Symbols are (open-circle) Lemna/P23; (closed-circle) P23 only; (open-triangle) natural Lemna; (closed-square) sterilized Lemna; (closed-diamond) control with medium only. P23 grown without *Lemna* was tested using ten times more cells than was observed to be attached to *Lemna*. Control was set for the experiment without *Lemna* and P23. Each data point is the average of triplicate experiments.

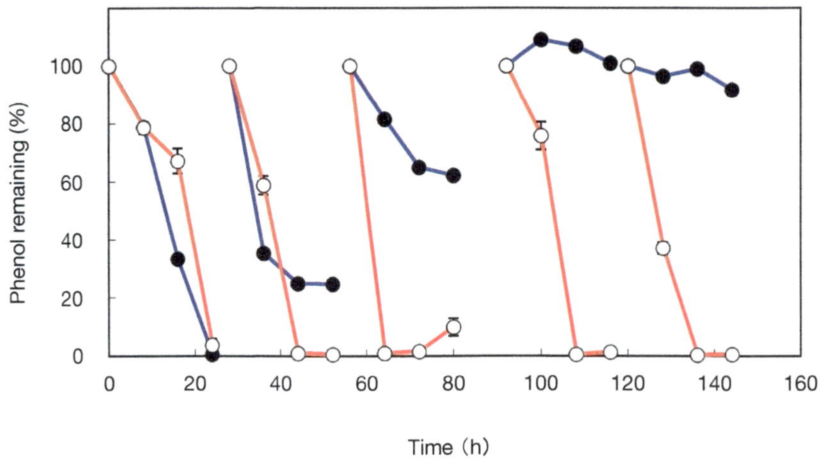


**Figure 13A.6** Long-term performance test of phenol degradation by *L. aoukiuksa*/P23 either (closed-circle) in a flask condition or (open-circle) in series of flask transfers. Degradation was started at 40 mg $l^{-1}$ phenol.

ml phenol and further cultivated under the photoperiod at 25°C. For the degradation test by P23 alone, $0.5 \times 10^{10}$ cells, or ten times of the number of P23 cells that adhered to 10 fronds as counted by an assay of colony-forming units at 72 h, were applied. Culture medium (0.25 ml) was collected at different times and subjected to HPLC analysis to determine phenol concentrations. After 40 h and 88 h, the culture volume and the concentration of phenol was re-adjusted to 50 ml and 40 mg/l, respectively.

In the first period of 40 h, the intact *L. aoukiuksa* and the *L. aoukiuksa*/P23 systems completely eliminated 40 mg/l phenol in the culture, whereas neither the surface-sterilized duckweed nor P23 alone effectively eliminated phenol. These observations are consistent with a previous report for giant duckweed *Spirodela polyrhiza* that phenol degradation activity of sterilized duckweed plant is negligible and that indigenous rhizobacteria of natural duckweed are responsible for degrading phenol (Toyama *et al.*, 2006). In the second period (40–96 h), there was roughly 80% and 50% elimination of phenol by *L. aoukiuksa*/P23 and the natural duckweed, respectively. P23 alone showed less degradation ability, accounting for the elimination of ca. 20% of the phenol. At the end of the third period (160 h), *L. aoukiuksa*/P23 maintained a 50% phenol elimination level, whereas the other systems eliminated less than 10% of the phenol. The long-term performance test (96–160 h) under nutrient-starvation conditions resulted in a severe reduction of phenol degradation activity for all combinations except for the *L. aoukiuksa*/P23 system. These results also suggest that indigenous plant surface bacteria capable of degrading phenol were rapidly attenuated, probably because of competition in the complex microbial community. The exact mechanisms are unclear; however, these experimental results indicate that the durable and continuous removal of phenol by the *L. aoukiuksa*/P23 system may be attributed to the beneficial interaction between P23 and the duckweed. Decrease in the phenol degradation activity of the *L. aoukiuksa*/P23 system during the periods was due to decreased numbers of P23 cells, from $5.1 \times 10^5$ CFUs (day 0) to $1.6 \times 10^4$ CFUs (day 7). This happened probably because metabolic wastes were accumulated and/or minerals for plant growth were depleted in the flasks.

In order to address this point, *L. aoukiuksa*/P23 were transferred after every 24 h to a new Hoagland medium containing 40 mg $ml^{-1}$ phenol. It was clearly shown that the *L. aoukiuksa*/P23 system maintained high phenol degradation activity for one week, over five transfer cycles (Fig. 13A.6). The number of P23 cells that attached to *L. aoukiuksa* plants was increased by roughly ten fold, from $5.7 \times 10^4$ CFUs (day 0) to $5.3 \times 10^5$ CFUs (day 7). In contrast, degradation activity of *L. aoukikusa*/P23 system in a flask without transfer was dramatically decreased after 3 days, three cycles.

## Growth promotion of duckweed by P23 colonization

During the long-term performance test of the *L. aoukiuksa*/P23 system in a flask (Fig. 13A.5), we noticed that the number of fronds markedly increased as compared with the system containing surface-cleansed *L. aoukiuksa* plants without P23 (Fig. 13A.7). The number of *L. aoukiuksa*/P23 fronds increased by 3.7-fold at 160 h, whereas the surface-cleansed *L. aoukiuksa* plants increased by only ca. 2.2-fold. This result clearly indicates that P23 promoted *L. aoukiuksa* frond growth, especially multiplication. It should be noted that the growth promoting activity of P23 for duckweeds was also observed in the phenol free Hoagland medium. This result eliminates a possibility that the growth promotion is caused by the removal of toxic phenol by P23. *A. calcoaceticus* SE370, that was isolated from soil produces plant growth-promoting hormones (gibberellins) and also carries phosphate-solubilizing activity (Kang *et al.*, 2009). However, our preliminary result indicates that the plant growth-promoting factor of P23 is not gibberellins (data not shown). Moreover, phosphate-solubilizing activity of the associated bacteria may be not the case in the liquid culture systems that we adopted. Further studies are needed to clarify the environmental factors and cellular responses of P23 that provide beneficial functions for the duckweed. To our best knowledge, P23 is the first growth-promoting bacterium identified from the duckweed plants in the genus *Lemna*.

## Future perspective

In this study, a beneficial symbiotic interaction was demonstrated between a phenol-degrading/plant growth-promoting *A. calcoaceticus* P23 and the duckweed *Lemna aoukikusa*. Although the application of this system to natural and field-scale polluted environments or wastewater treatment systems is still challenging, the results suggest the potential use of aquatic plant-associated bacteria to achieve efficient and sustainable phenol removal. The enhancement of plant growth by such bacteria should enable us to increase the total efficacy of phytoremediation technologies. Another advantage of using bacteria which colonize floating plants, including duckweeds, is the ease with which these bacteria and plants may be recovered after site remediation. In fact, the turbidity of *L. aoukikusa*/P23 culture remained at an $OD_{600}$ < 0.01 during the long-term performance test over 160 h. This feature should therefore be useful for reducing risks of microbial community disruption caused by the introduction of foreign bacteria. When we used completely surface sterilized *L. aoukikusa*, the number of P23 cells attached to the plant was significantly decreased. This observation suggests that unknown microorganisms remained on the

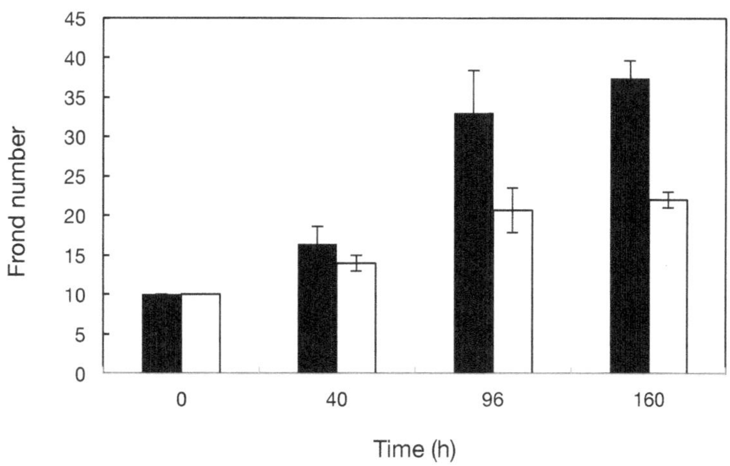


**Figure 13A.7** The effects of *A. calcoaceticus* P23 on the growth of *L. aoukiuksa*. Frond number was compared between (black) surface-cleansed *L. aoukikusa* carrying P23 (*L. aoukikusa*/P23); and (white bar) surface-cleansed *L. aoukikusa* only, during growth. Significant growth-promotion was observed for *L. aoukiusa*/P23. Error bars are 1 × standard deviation ($n$ = 3).

surface-washed plants that helped the attachment of P23 to the plants. The next challenge maybe to design beneficial multi-species microbial community biofilms on plant surfaces, further advancing opportunities for sustainable water purification technologies.

## References


Arnon, D.I., and Hoagland, D.R. (1939). A comparison of water culture and soil as media for crop production. Science *89*, 512–514.

Beun, J.J., Hendriks, A., van Loosdrecht, M.C.M., Morgenroth, E., Wilderer, P.A., and Heijnen, J.J. (1999). Aerobic granulation in a sequencing batch reactor. Water Res. 33, 2283–2290.

Brown, V.M., Jordan, D.H.M., and Tiller, B.A. (1967). The effect of temperature on the acute toxicity of phenol to rainbow trout in hard water. Water Res. *1*, 587–594.

Chaudhry, Q., Blom-Zandstra, M., Gupta, S., and Joner, E.J. (2005). Utilising the synergy between plants and rhizosphere microorganisms to enhance breakdown of organic pollutants in the environment. Environ. Sci. Pollut. Res. Int. *12*, 34–48.

Iijima, S., Washio, K., Okahara, R., and Morikawa, M. (2009). Biofilm formation and proteolytic activities of *Pseudoalteromonas* bacteria that were isolated from fish farm sediments. Microbial. Biotech. 2, 361–369.

Ishii, S., Koki, J., Unno, H., and Hori, K. (2004). Two morphological types of cell appendages on a strongly adhesive bacterium, *Acinetobacter* sp. strain Tol 5. Appl. Environ. Microbiol. 70, 5026–5029.

Juni, E. (1978). Genetics and physiology of *Acinetobacter*. Annu. Rev. Microbiol. 32, 349–371.

Kang, S.M., Joo, G.J., Hamayun, M., Na, C.I., Shin, D.H., Kim, H.Y., Hong, J.K., and Lee, I.J. (2009). Gibberellin production and phosphate solubilization by newly isolated strain of *Acinetobacter calcoaceticus* and its effect on plant growth. Biotechnol. Lett. *31*, 277–281.

Kuiper, I., Lagendijk, E.L., Bloemberg, G.V., and Lugtenberg, B.J. (2004). Rhizoremediation: a beneficial plant–microbe interaction. Mol. Plant. Microbe. Interact. *17*, 6–15.

Kunihiro, N., Haruki, M., Takano, K., Morikawa, M., and Kanaya, S. (2005). Isolation and characterization of *Rhodococcus* sp. strains TMP2 and T12 that degrade 2,6,10,14-tetramethylpentadecane (pristane) at moderately low temperatures. J. Biotechnol. *115*, 129–136.

Larkin, M.A., Blackshields, G., Brown, N.P., Chenna, R., McGettigan, P.A., McWilliam, H., Valentin, F., Wallace, I.M., Wilm, A., Lopez, R., *et al.* (2007). Clustal W and Clustal X version 2.0. Bioinformatics 23, 2947–2948.

Morikawa, M. (2010). Dioxygen activation responsible for oxidation of aliphatic and aromatic hydrocarbon compounds: current state and variants. Appl. Microbiol. Biotechnol. *87*, 1596–1603.

Pedersen, G., Brynskov, J., and Saermark, T. (2002). Phenol toxicity and conjugation in human colonic epithelial cells. Scand. J. Gastroenterol. 37, 74–79.

Rai, P.K. (2008). Heavy metal pollution in aquatic ecosystems and its phytoremediation using wetland plants: an ecosustainable approach. Int. J. Phytoremed. *10*, 131–158.

Ramey, B.E., Koutsoudis, M., von Bodman, S.B., and Fuqua, C. (2004). Biofilm formation in plant–microbe associations. Curr. Opin. Microbiol. 7, 602–609.

Salt, D.E., Smith, R.D., and Raskin, I. (1998). Phytoremediation. Annu. Rev. Plant Physiol. Plant Mol. Biol. *49*, 643–668.

Toyama, T., Yu, N., Kumada, H., Sei, K., Ike, M., and Fujita, M. (2006). Accelerated aromatic compounds degradation in aquatic environment by use of interaction between *Spirodela polyrrhiza* and bacteria in its rhizosphere. J. Biosci. Bioeng. *101*, 346–353.

Toyama, T., Sei, K., Yu, N., Kumada, H., Inoue, D., Hoang, H., Soda, S., Chang, Y.C., Kikuchi, S., Fujita, M., and Ike, M. (2009). Enrichment of bacteria possessing catechol dioxygenase genes in the rhizosphere of *Spirodela polyrrhiza*: a mechanism of accelerated biodegradation of phenol. Water Res. *43*, 3765–3776.

Watanabe, S., Hino, S., and Takahashi, N. (1996). Responses of activated sludge to an increase in phenol loading. J. Ferment. Bioeng. *82*, 522–524.

# Using Microbial Biofilms to Enhance the Phytoremediation of Contaminants in Soil and Water – The Sustainable Biodegradation of Phenolic Endocrine-disrupting Chemicals by Bacteria in the Rhizosphere of *Phragmites australis*

13B

Tadashi Toyama and Kazuhiro Mori


## Abstract

Rhizoremediation, the degradation and removal of pollutants by exploiting the elevated microbial activity in the rhizosphere of plants, has gained much attention in recent years. In this second part of this chapter, we discuss the use of two rhizobacteria (*Sphingobium fuliginis* TIK1 and *Sphingobium* sp. IT4) colonizing the root of the common reed *Phragmites australis*, and the use of this association for the sustainable treatment of water polluted with phenolic endocrine-disrupting chemicals (EDCs). Strains TIK1 and IT4, isolated from *Phragmites* rhizospheres, degrade various phenolic EDCs (4-alkylphenols and bisphenols). The root exudates of *Phragmites* appeared to facilitate the degradation activity of these, using the exudates as a source of carbon and energy for cell growth. Additionally, the roots provide a large surface area for bacterial colonization. *Phragmites* plants inoculated with these bacteria (i.e. both *Phragmites*–strain TIK1 and *Phragmites*–strain IT4 associations) were able to simultaneously and repeatedly remove various phenolic EDCs from polluted water. We conclude that the use of hydroponic systems using *Phragmites*–TIK1 and *Phragmites*–IT4 associations would be an effective strategy for the sustainable treatment of polluted waters contaminated by mixtures of phenolic EDCs.


## Introduction

4-alkylphenols, including 4-nonylphenol (4-NP), 4-*tert*-octylphenol (4-*tert*-OP), and 4-*tert*-butylphenol (4-*tert*-BP), have been used in the production of nonylphenol polyethoxylate surfactants, phenolic resins, and polycarbonate resins. 4-alkylphenols are persistent, toxic compounds classified as endocrine-disrupting chemicals (EDCs); they are capable of interfering with the hormonal systems of numerous organisms (Ying *et al.*, 2002; Soares *et al.*, 2008). Bisphenol A (BPA: 2,2-bis[4-hydroxyphenyl] propane) is a high-production-volume chemical used to make polycarbonates and epoxy resins. Other bisphenol analogues (bisphenols), such as bisphenol B (BPB: 2,2-bis[4-hydroxyphenyl] butane), bisphenol E (BPE: bis[4-hydroxyphenyl] ethane), bisphenol F (BPF: bis[4-hydroxylphenyl] methane), bisphenol P (BPP: 2,2-bis[4-hydroxy-3-methylphenyl]propane), and bisphenol S (BPS: bis[4-hydroxyphenyl]sulfone), are also used in the generation of resins and plastics (Chen *et al.*, 2002). These bisphenols are well-known EDCs with estrogenic activity (Chen *et al.*, 2002; Crain *et al.*, 2007). The widespread and extensive use of these phenolic EDCs has led to their release into aquatic environments. Phenolic EDCs are often found in effluents of sewage and wastewater treatment plants and in surface waters and sediments (Heemken *et al.*, 2001; Ko *et al.*, 2007; Jonkers *et al.*, 2009, 2010). It is therefore necessary to establish technologies for decontaminating waters polluted with phenolic EDCs. Although phenolic EDCs can be degraded by physicochemical methods such as ozonation and advanced oxidation processes (Gültekin and Ince 2007; Nakada *et al.*, 2007), these treatments are generally costly and consume much energy. Biological processes are a potential cost-effective option for the treatment of waters polluted with phenolic EDCs and on-site purification of aquatic environments polluted by phenolic EDCs.

Rhizoremediation, the degradation and removal of pollutants by exploiting the elevated microbial activity in the rhizosphere of plants, is a cost-effective and environmentally friendly remediation technology (Shaw and Burns 2001; Chaudhry *et al.*, 2005). The rhizosphere was first defined by Lorenz Hiltner as the zone influenced by the root system (Hiltner 1904). The fact that the microbial biomass and activity in the rhizosphere are greater than that in the non-rhizosphere zone led to the coining of the term 'the rhizosphere effect'. Plant roots supply numerous compounds (e.g. sugars, amino acids, organic acids, vitamins, mucilage, siderophores, aromatic compounds, and enzymes) that have biological activity, such as enzyme induction, or can be used as sources of carbon, nitrogen, or phosphorus by microbes in the rhizosphere (Shaw and Burns 2001). Plant roots also provide a large surface area that microbes can colonize, and they transport oxygen to aerobic microbes in the rhizosphere (Shaw and Burns 2001). The rhizosphere effect may therefore be responsible for the highly elevated rates of metabolic degradation of organic compounds often observed in the rhizosphere.

In the last decade, accelerated biodegradation of recalcitrant organic pollutants such as petroleum hydrocarbons (Radwan *et al.*, 1995), pesticides (Hsu and Bartha 1979), chlorinated solvents (Walton and Anderson 1990), and polycyclic aromatic hydrocarbons (Binet *et al.*, 2000) in the rhizosphere of terrestrial plant–soil systems has been revealed. Recently, the rhizospheres of an emergent aquatic plant, *Phragmites australis* (common reed), and the floating aquatic plants *Spirodela polyrhiza* (giant duckweed) and *Lemna aequinoctialis* (lesser duckweed) have also been shown to be associated with the acceleration of biodegradation in both water and sediment contaminated with single aromatic compounds (Toyama *et al.*, 2006; Yamaga *et al.*, 2010) including BPA (Toyama *et al.*, 2009) and 4-NP (Toyama *et al.*, 2011). The rhizosphere of both terrestrial and aquatic systems is therefore a 'hotspot' for bacterial growth and aerobic biodegradation. However, rhizoremediation processes are often limited by the presence of persistent, toxic pollutants. The introduction of specific pollutant-degrading bacteria to the rhizosphere has great potential to enhance the efficacy of rhizoremediation (Shaw and Burns, 2001; Chaudhry *et al.*, 2005; Yamaga *et al.*, 2010), and bacteria that efficiently colonize roots and degrade pollutants markedly enhance the process. Introduction of such bacteria into the rhizosphere can make possible the design of ideal plant–bacteria associations for target pollutant degradation.

Here, we focus on and explain the treatment of waters polluted with phenolic EDCs by use of specially designed plant–bacteria associations (*Phragmites australis* in association with its rhizobacteria *Sphingobium fuliginis* TIK1 or *Sphingobium* sp. IT4). We describe how the use of hydroponic systems containing *Phragmites*–strain TIK1 and *Phragmites*–strain IT4 associations can be an effective strategy for the sustainable treatment of waters polluted by various phenolic EDCs (Toyama *et al.*, 2010, 2011, 2013).

## Biodegradation of 4-*tert*-BP and branched 4-NP in the rhizosphere of *Phragmites* and isolation of 4-*tert*-BP-degrading *Sphingobium fuliginis* TIK1 and 4-NP-degrading *Sphingobium* sp. IT4 from the rhizosphere

We have observed the accelerated removal of 4-*tert*-BP (Toyama *et al.*, 2010) and branched 4-NP (Toyama *et al.*, 2011) in a *Phragmites* rhizosphere sediment system collected from a natural pond, whereas both 4-*tert*-BP and 4-NP persisted in non-rhizosphere sediment. In the rhizosphere sediment associated with a *Phragmites* plant, 91% of 4-*tert*-BP initially added to the sediment (initial concentration, 25 mg 4-*tert*-BP/kg dry sediment) was removed over 42 days, and 53% of 4-NP added to the sediment (initial concentration, 25 mg 4-NP/kg dry sediment) was removed in the same period. In addition, we isolated 4-*tert*-BP-degrading *Sphingobium fuliginis* TIK1 (Toyama *et al.*, 2010) and 4-NP-degrading *Sphingobium* sp. IT4 (Toyama *et al.*, 2011) from the *Phragmites* rhizosphere. In contrast, we did not observe substantial degradation of 4-*tert*-BP and 4-NP in the non-rhizosphere sediment of the same pond, and our efforts to enrich and isolate 4-*tert*-BP-degrading bacteria and 4-NP-degrading bacteria from the non-rhizosphere sediment were unsuccessful. We concluded that 4-*tert*-BP-degrading bacteria and 4-NP-degrading bacteria were present only in the rhizosphere sediment. The successful enrichment and isolation of 4-*tert*-BP-degrading bacteria and 4-NP-degrading bacteria from the rhizosphere might be attributable to 'the rhizosphere effects' of *Phragmites*.

## Abilities of strains TIK1 and IT4 to degrade phenolic EDCs, and growth of strains TIK1 and IT4 with organic compounds exuded by *Phragmites* roots

Strain TIK1 can utilize 4-*tert*-BP as a sole carbon source via phenolic ring hydroxylation and a *meta*-cleavage pathway (Toyama *et al.*, 2010). Strain IT4 can utilize branched 4-NP or 4-*tert*-OP as a sole carbon source via an *ipso*-substitution mechanism (Toyama *et al.*, 2011). The abilities of strains TIK1 and IT4 to degrade 4-alkylphenols and bisphenols are summarized in Table 13B.1. Strain TIK1 can degrade various 4-alkylphenols and bisphenols via phenolic ring hydroxylation followed by a *meta*-cleavage pathway. Strain IT4 can also degrade 4-alkylphenols and bisphenols by an *ipso*-substitution mechanism. The two strains have wide scope for degradation of phenolic EDCs and are potentially useful for the treatment of waters polluted with multiple species of phenolic EDCs. We examined the effects of organic compounds released from *Phragmites* roots on cell growth and the degradation activity of strains TIK1 and IT4. We prepared *Phragmites* root extracts from crushed roots of 2-month-old, sterile *Phragmites* plants (Toyama *et al.*, 2013). Cells of each strain were incubated in 100 ml of basal salt medium containing the sterilized root extract (140 mg/l of dissolved organic carbon) at 28°C and 120 rpm. The two strains were able to grow with *Phragmites* root extract as a sole source of carbon. The yield coefficients of cell mass were 0.74 mg dry-weight/mg carbon of root extract for strain TIK1 and 0.61 mg dry-weight/mg carbon of root extract for strain IT4 (Toyama *et al.*, 2013). For comparison, when strain TIK1 used 4-*tert*-BP or glucose as a sole carbon source, the yield coefficients of cell growth for 4-*tert*-BP and glucose were 0.91 mg dry-weight/mg carbon of 4-*tert*-BP and 1.05 mg dry-weight/mg carbon of glucose, respectively. When strain IT4 used 4-*tert*-OP or glucose as a sole carbon source, the yield coefficients of cell growth for 4-*tert*-OP and glucose were 0.18 mg dry-weight/mg carbon of 4-*tert*-OP and 0.87 mg dry-weight/mg carbon of glucose, respectively. In addition, strains TIK1 and IT4 grown with root organic compounds were highly capable of degrading various phenolic EDCs (Table 13B.1). Therefore, organic compounds from *Phragmites* roots function as carbon and energy sources for cell growth and enhance the degradation activity of the two strains, the result being elevated cell numbers and activities of the bacteria.

## Construction of *Phragmites*–TIK1 or *Phragmites*–IT4 associations and degradation of phenolic EDCs by the *Phragmites*–TIK1 or *Phragmites*–IT4 associations

We tried to prepare *Phragmites* associations with strain TIK1 or IT4 (*Phragmites*–TIK1 and *Phragmites*–IT4 associations) using the following simple method (Toyama *et al.*, 2013). Cells of each strain were suspended at $OD_{600} = 0.3$ (about 0.15 mg/dry-weight ml) in 200 ml of sterilized Hoagland solution in a 500-ml flask containing a 2-month-old plant (about 0.05 g dry root biomass and 0.14 g dry above-ground biomass). The flask containing the plant and cells of each strain was incubated statically for 1 h. The cell suspension was then removed from the flask, and the roots were gently washed three times with sterilized Hoagland solution to remove suspended cells. The *Phragmites* plant was then presumptively termed either *Phragmites*–TIK1 or *Phragmites*–IT4 associations dependent on the strain it was inoculated with. To determine whether the two strains had successfully colonized the roots and were able to express degradation activity on the root surfaces of the previously sterile *Phragmites*, we conducted a phenolic EDC degradation assay (five cycles, each with a 24-h reaction time) with the *Phragmites*–TIK1 or *Phragmites*–IT4

**Table 13B.1** *Sphingobium fuliginis* strain TIK1 and *Sphingobium* sp. strain IT4 display exceptionally high abilities to degrade phenolic EDCs (Toyama *et al.*, 2010, 2011, 2013)

| Phenolic EDC | Strain TIK1, transformation ratio (%)[a] | Strain IT4, transformation ratio (%)[a] |
|---|---|---|
| 4-*n*-Butylphenol | 100.0 | 70.2 |
| 4-*sec*-Butylphenol | 100.0 | 95.5 |
| 4-*tert*-Butylphenol | 100.0 | 100.0 |
| 4-*n*-Pentylphenol | 100.0 | 68.0 |
| 4-*tert*-Pentylphenol | 100.0 | 98.9 |
| 4-*n*-Hexylphenol | 100.0 | 82.2 |
| 4-*n*-Heptylphenol | 100.0 | 96.1 |
| 4-*n*-Octylphenol | 100.0 | 91.6 |
| 4-*tert*-Octylphenol | 98.0 | 100.0 |
| 4-*n*-Nonylphenol | 100.0 | 91.9 |
| Branched 4-nonylphenol | 64.0 | 92.6 |
| Bisphenol A | 100.0 | 100.0 |
| Bisphenol B | 99.0 | 100.0 |
| Bisphenol E | 100.0 | 100.0 |
| Bisphenol F | 100.0 | 100.0 |
| Bisphenol P | 78.2 | 91.1 |
| Bisphenol S | 100.0 | 100.0 |

[a]Transformation ratio (%) refers to the per cent reduction in the concentration of the phenolic EDC in cultures after a 24-h incubation relative to the concentration in a sterile control. Concentrations were obtained from HPLC chromatograms.

association. One *Phragmites*–TIK1 or *Phragmites*–IT4 association was placed in a 500-ml flask with 200 ml of Hoagland solution containing phenolic EDCs (5 mg/l BPA, 5 mg/l BPS, 5 mg/l 4-*tert*-BP, 2.5 mg/l 4-*tert*-OP, and 2.5 mg/l branched 4-NP). The flask was then statically incubated in a chamber (28 ± 1°C, 8000 lux, 16-h light and 8-h dark cycle). After 24 h, the EDCs–Hoagland solution was removed completely from the flask and replaced with 200 ml of fresh EDCs–Hoagland solution. This 24-h cycle was repeated a total of five times in triplicate.

After preparation of a *Phragmites*–TIK1 or *Phragmites*–IT4 association, each strain easily colonized the roots, strain TIK1 at a density of $9.08 \pm 2.32 \times 10^{10}$ CFU/g dry root and IT4 at a density of $3.13 \pm 0.83 \times 10^{10}$ CFU/g dry root. At the end of five cycles, the populations on the roots remained at $5.37 \pm 0.83 \times 10^{11}$ CFU/g dry root for TIK1 and $3.24 \pm 0.24 \times 10^{11}$ CFU/g dry root for strain IT4 (Toyama *et al.*, 2013). When the roots of a *Phragmites*–TIK1 or *Phragmites*–IT4 association were stained using a LIVE/DEAD BacLight assay (Molecular Probes-Invitrogen, San Diego, CA) and observed by fluorescence microscopy, many green fluorescent spots, indicative of small colonies or biofilms of living bacteria, were observed on the surface of each root after five cycles (Fig. 13B.1).

The changes in EDC concentrations during the degradation experiments are shown in Fig. 13B.2. The *Phragmites*–TIK1 and *Phragmites*–IT4 associations both consistently removed all EDCs from the EDC-dosed Hoagland solutions within the 24-h reaction time throughout all five cycles. In the control experiments with sterile *Phragmites*, the concentrations of BPA, BPS, and 4-*tert*-BP did not decline at all; 4-*tert*-OP declined by about 20%, and 4-NP declined by about 30% within the 24-h reaction time throughout all five cycles (Fig. 13B.2). Strains TIK1 and IT4 therefore appear capable of sustainably colonizing sterilized *Phragmites* roots and degrading phenolic EDCs for at least 120 h, presumably with support from organic compounds exuded from the *Phragmites* roots.

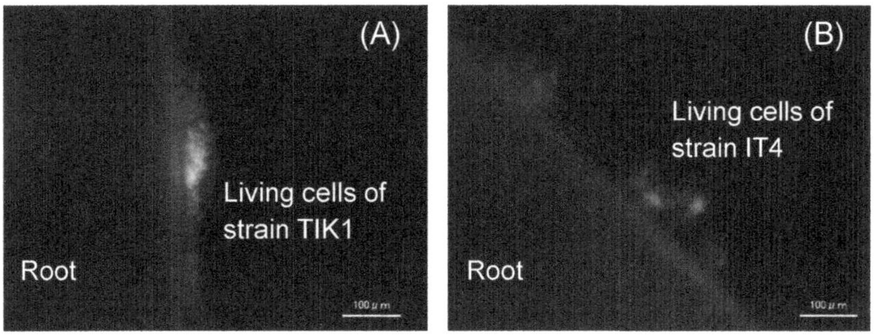


**Figure 13B.1** Staining for live bacteria and observation by fluorescence microscopy of *Phragmites*–TIK1 or *Phragmites*–IT4 roots. (A) Main root of *Phragmites*–TIK1 association. (B) Main root of *Phragmites*–IT4 association. The high intensity of green fluorescence, visible as bright white in this image, indicates living bacteria (Toyama *et al.*, 2013).

## Sustainable removal of phenolic EDCs from polluted effluent by use of a sequencing batch reactor (SBR) system containing *Phragmites*–TIK1 or *Phragmites*–IT4 associations

We conducted a laboratory-scale sequencing batch reactor (SBR) experiment using the *Phragmites* associations to treat 500-ml samples of effluent polluted with EDCs (10 cycles, each with a 12-h reaction time) to evaluate the potential for using *Phragmites*–TIK1 and *Phragmites*–IT4 associations to treat waters polluted with phenolic EDCs (Toyama *et al.*, 2013). Effluent polluted with EDCs was prepared by dissolving 5 mg/l BPA, 5 mg/l BPS, 5 mg/l 4-*tert*-BP, 2.5 mg/l 4-*tert*-OP, and 2.5 mg/l branched 4-NP into a secondary-treated effluent sample of sewage treatment plant. The *Phragmites*–TIK1 and *Phragmites*–IT4 associations were prepared by inoculating roots of 3-month-old, non-sterile *Phragmites* (about 2.5 g dry root biomass and 10 g dry above-ground biomass) with cells of either strain TIK1 or strain IT4 at $OD_{600} = 0.3$. EDC-polluted effluent (500 ml) was placed in the bottle with one *Phragmites*–TIK1 or one *Phragmites*–IT4 association. The bottle was statically incubated in a chamber (28±1°C, 8000 lux, 16-h light and 8-h dark cycle). After 12 h, the EDC-polluted effluent was completely removed from the flask and replaced with 500 ml of fresh, EDC-polluted effluent. This 12-h cycle was repeated 10 times in triplicate.

Uninoculated *Phragmites* removed small amounts of BPA (1.9–10.3%), BPS (0.5–3.8%), 4-*tert*-BP (1.7–4.1%), 4-*tert*-OP (22.0–26.8%), and 4-NP (31.2–33.6%) from the effluent (Table 13B.2). Partial removal of these EDCs from polluted effluent in the presence of uninoculated *Phragmites* might have resulted from adsorption

**Table 13B.2** Removal of phenolic EDCs by *Phragmites*–TIK1, *Phragmites*–IT4, and uninoculated *Phragmites* in sequencing batch reactors. Values shown are the reduction in contaminant concentrations at the end of a reaction cycle as a percentage of the concentration at the beginning of the cycle. The initial concentration was 5.0 mg/l for BPA, BPS, and 4-*tert*-BP and 2.5 mg/l for 4-*tert*-OP and branched 4-NP (Toyama *et al.*, 2013)

| | *Phragmites*–TIK1 | | | *Phragmites*–IT4 | | | Uninoculated *Phragmites* | | |
|---|---|---|---|---|---|---|---|---|---|
| | Cycle number | | | Cycle number | | | Cycle number | | |
| Contaminant | 1 | 5 | 10 | 1 | 5 | 10 | 1 | 5 | 10 |
| BPA | 100 | 100 | 100 | 100 | 100 | 100 | 1.9 | 2.6 | 9.0 |
| BPS | 100 | 100 | 100 | 100 | 100 | 100 | 1.7 | 3.5 | 1.4 |
| 4-*tert*-BP | 100 | 100 | 100 | 100 | 100 | 100 | 2.1 | 2.5 | 4.1 |
| 4-*tert*-OP | 100 | 100 | 100 | 100 | 100 | 100 | 23.4 | 26.8 | 25.2 |
| Branched 4-NP | 100 | 100 | 100 | 100 | 100 | 100 | 31.8 | 33.6 | 33.2 |

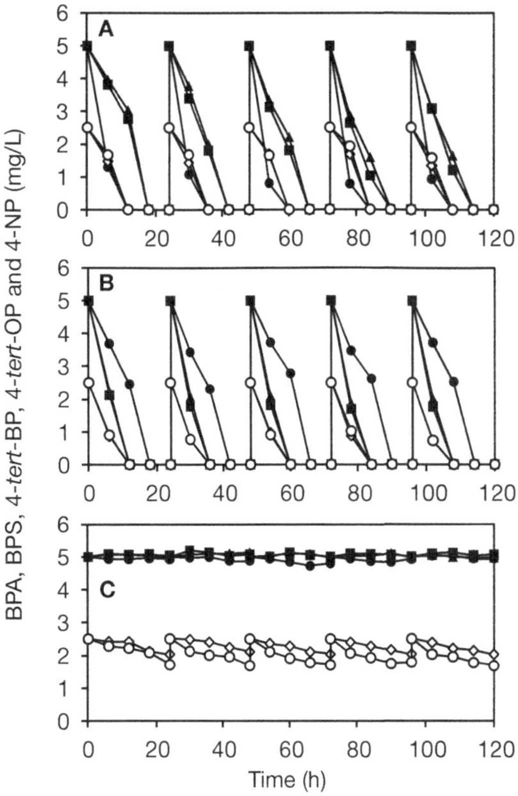


**Figure 13B.2** Changes in concentrations of the phenolic EDCs BPA (■), BPS (▲), 4-*tert*-BP (•), 4-*tert*-OP (◊), and branched 4-NP (○) in Hoagland solution. The changes reflect degradation in the presence of (A) *Phragmites*–TIK1 association, (B) *Phragmites*–IT4 association, and (C) sterile and uninoculated *Phragmites*. Values are means from triplicate experiments. The Hoagland solution containing the EDCs was replaced with fresh solution every 24 h (Toyama *et al.*, 2013).

onto the roots and biodegradation by indigenous bacteria in the secondary effluent. In contrast, both the *Phragmites*–TIK1 and *Phragmites*–IT4 associations repeatedly achieved complete removal of all phenolic EDCs from the dosed water throughout all 10 cycles (Table 13B.2). TIK1 and IT4 were able to thoroughly colonize the surface of non-sterile *Phragmites* roots. Both *Phragmites*–TIK1 and *Phragmites*–IT4 associations could therefore be used for long-term, simultaneous removal of phenolic EDCs from polluted effluent. We conclude that the use of hydroponic systems using *Phragmites*–TIK1 and *Phragmites*–IT4 associations would be a successful strategy for the sustainable treatment of polluted water containing various phenolic EDCs.

## References


Binet, P., Portal, J.M., and Leyval, C. (2000). Dissipation of 3-6-ring polycyclic aromatic hydrocarbons in the rhizosphere of ryegrass. Soil Biol. Biochem. *32*, 2011–2017.

Chaudhry, Q., Blom-Zandstra, M., Gupta, S., and Joner, E.J. (2005). Utilizing the synergy between plants and rhizosphere microorganisms to enhance breakdown of organic pollutants in the environment. Environ. Sci. Pollut. Res. *12*, 34–48.

Chen, M.T., Ike, M., and Fujita, M. (2002). Acute toxicity, mutagenicity, and estrogenicity of bisphenol-A and other bisphenols. Environ. Toxicol. *17*, 80–86.

Crain, D.A., Eriksen, M., Iguchi, T., Jobling, S., Laufer, H., LeBlanc, G.A., and Guillette, Jr., L.J. (2007). An ecological assessment of bisphenol-A: Evidence from comparative biology. Reprod. Toxicol. *24*, 225–239.

Gültekin, I., and Ince, N.H. (2007). Synthetic endocrine disruptors in the environment and water remediation by advanced oxidation processes. J. Environ. Manage. *85*, 816–832.

Heemken, O.P., Reincke, H., Stachel, B., and Theobald, N. (2001). The occurrence of xenoestrogens in the Elbe River and the North Sea. Chemosphere *45*, 245–259.

Hiltner, L. (1904). Uber neue erfahrungen und problem auf dem gebiet der bodenbakteriolgie und unter besonderes berucksichtigung der grundugungen und brauche. Arb. Dtsch. Landwirt. Ges. Berl *98*, 59–78.

Hsu, T.S., and Bartha, R. (1979). Accelerated mineralization of two organophosphate insecticides in the rhizosphere. Appl. Environ. Microbiol. *37*, 36–41.

Jonkers, N., Kohler, H.-P.E., Dammshäuser, A., and Giger, W. (2009). Mass flows of endocrine disruptors in the Glatt River during varying weather conditions. Environ. Pollut. *157*, 714–723.

Jonkers, N., Sousa, A., Galante-Oliveira, S., Barroso, C.M., Kohler, H.-P.E., and Giger, W. (2010). Occurrence and source of selected phenolic endocrine disruptors in Ria de Aveiro, Portugal. Environ. Sci. Pollut. Res. *17*, 834–843.

Ko, E.-J., Kim, K.-W., Kang, S.-Y., Kim, S.-D., Bang, S.-B., Hamm, S.-Y., and Kim, D.-W. (2007). Monitoring of environmental phenolic endocrine disrupting compounds in treatment effluents and river waters, Korea. Talanta *73*, 674–683.

Nakada, N., Shinohara, H., Murata, A., Kiri, K., Managaki, S., Sato, N., and Takada, H. (2007). Removal of selected pharmaceuticals and personal care products (PPCPs) and endocrine-disrupting chemicals (EDCs) during sand filtration and ozonation at municipal sewage treatment plant. Water Res. *41*, 4373–4382.

Radwan, S., Sorkhoh, N., and el-Nemr, I. (1995). Oil biodegradation around roots. Nature, *376*, 302.

Shaw, L.J., and Burns, R.C. (2001). Biodegradation of organic pollutants in the rhizosphere. Adv. Appl. Microbiol. *53*, 1–60.

Soares, A., Guieysse, B., Jefferson, B., Cartmell, E., and Lester, J.N. (2008). Nonylphenol in the environment: A critical review on occurrence, fate, toxicity and treatment in wastewaters. Environ. Int. *34*, 1033–1049.

Toyama, T., Yu, N., Kumada, H., Sei, K., Ike, M., and Fujita, M. (2006). Accelerated aromatic compounds degradation in aquatic environment by use of interaction between *Spirodela polyrhiza* and bacteria in its rhizosphere. J. Biosci. Bioeng. *101*, 346–353.

Toyama, T., Sato, Y., Inoue, D., Sei, K., Chang, Y.-C., Kikuchi, S., and Ike, M. (2009). Biodegradation of bisphenol A and bisphenol F in the rhizosphere sediment of *Phragmites australis*. J. Biosci. Bioeng. *108*, 147–150.

Toyama, T., Momotani, N., Ogata, Y., Miyamori, Y., Inoue, D., Sei, K., Mori, K., Kikuchi, S., and Ike, M. (2010). Isolation and characterization of 4-*tert*-butylphenol-utilizing *Sphingobium fuliginis* strains from *Phragmites australis* rhizosphere sediment. Appl. Environ. Microbiol. *76*, 6733–6740.

Toyama, T., Murashita, M., Kobayashi, K., Kikuchi, S., Kanunari, S., Tanaka, Y., Ike, M., and Mori, K. (2011). Acceleration of nonylphenol and 4-*tert*-octylphenol degradation in sediment by *Phragmites australis* and associated rhizosphere bacteria. Environ. Sci. Technol. *45*, 6524–6530.

Toyama, T., Ojima, T., Tanaka, Y., Mori, K., and Morikawa, M. (2013). Sustainable biodegradation of phenolic endocrine-disrupting chemicals by *Phragmites australis*-rhizosphere bacteria association. Water Sci. Technol. *68*, 522–529.

Walton, B.T., and Anderson, T.A. (1990). Microbial degradation of trichloroethylene in the rhizosphere: potential application to biological remediation of waste sites. Appl. Environ. Microbiol. *56*, 1012–1016.

Yamaga, F., Washio, K., and Morikawa, M. (2010). Sustainable biodegradation of phenol by *Acinetobacter calcoaceticus* P23 isolated from the rhizosphere of duckweed *Lemna aoukikusa*. Environ. Sci. Technol. *44*, 6470–6474.

Ying, G.-G., Williams, B., and Kookana, R. (2002). Environmental fate of alkylphenols and alkylphenol ethoxylates – a review. Environ. Int. *28*, 215–226.

# Index

Lightning Source UK Ltd.
Milton Keynes UK
UKOW07f0253140416

272218UK00004B/7/P

9 781910 190296